Landscape Series

Volume 24

Aims and Scope

Springer's innovative Landscape Series is committed to publishing high-quality manuscripts that approach the concept of landscape from a broad range of perspectives. Encouraging contributions on theory development, as well as more applied studies, the series attracts outstanding research from the natural and social sciences, and from the humanities and the arts. It also provides a leading forum for publications from interdisciplinary and transdisciplinary teams.

Drawing on, and synthesising, this integrative approach the Springer Landscape Series aims to add new and innovative insights into the multidimensional nature of landscapes. Landscapes provide homes and livelihoods to diverse peoples; they house historic—and prehistoric—artefacts; and they comprise complex physical, chemical and biological systems. They are also shaped and governed by human societies who base their existence on the use of the natural resources; people enjoy the aesthetic qualities and recreational facilities of landscapes, and people design new landscapes.

As interested in identifying best practice as it is in progressing landscape theory, the Landscape Series particularly welcomes problem-solving approaches and contributions to landscape management and planning. The ultimate goal is to facilitate both the application of landscape research to practice, and the feedback from practice into research.

More information about this series at http://www.springer.com/series/6211

Christina von Haaren • Andrew A. Lovett
Christian Albert
Editors

Landscape Planning with Ecosystem Services

Theories and Methods for Application in Europe

Springer

Editors
Christina von Haaren
Institute of Environmental Planning
Leibniz University Hannover
Hannover, Germany

Christian Albert
Institute of Environmental Planning
Leibniz University Hannover
Hannover, Germany

Andrew A. Lovett
School of Environmental Sciences
University of East Anglia
Norwich, UK

ISSN 1572-7742 ISSN 1875-1210 (electronic)
Landscape Series
ISBN 978-94-024-1679-4 ISBN 978-94-024-1681-7 (eBook)
https://doi.org/10.1007/978-94-024-1681-7

This Springer imprint is published by the registered company Springer Nature B.V.
The registered company address is: Van Godewijckstraat 30, 3311 GX Dordrecht, The Netherlands

To the landscape planners of tomorrow

Preface

In an era where the Internet dominates knowledge acquisition, it seems obvious to ask the question "why a book about landscape planning methodologies?" We decided to write and edit this book for two main reasons:

1. Practitioners and students of landscape planning take in huge amounts of unconnected information. Only with difficulty can students put papers about separate aspects into a context that enables them to use methods critically and in a targeted manner.
2. The question "how to evaluate the landscape?" is often answered by landscape planners and others involved in environmental planning or impact studies across European countries in different ways. However, the purposes and tasks are often very similar, as indeed are many of the relevant methods in a European context.

Thus, the overall motive for this book is to provide orientation in the information jungle. We also feel that many of the pressing environmental challenges can be best addressed by combining the practical orientation of landscape planning with concepts and approaches from the burgeoning literature on ecosystem services and natural capital. In this book, we discuss how these two fields can be integrated and review hands-on methods which, in principle, are applicable in all European countries. A feature of this book is that an emphasis is placed on combing evaluations based on legal norms with those based on public preferences, including economic approaches. Furthermore, over 45 authors from different disciplines have adopted a common framework for discussing their methodologies. This ensures a consistency of material for the reader which, in turn, assists in combining different elements in practical applications.

It has taken a long time to complete this book, and many people have supported our vision with advice, energy, creativity, and sheer hard work. We would particularly like to thank all of the chapter authors and reviewers. Advice from the editors of the Springer Landscape Series and publishing staff (particularly Nel van der Werf) is also much appreciated.

Financial support from several research grants has assisted with our work on a number of chapters in the book. In particular, Andrew A. Lovett would like to acknowledge the support from the UK Economic and Social Research Council (award ES/L011859/1 for the Business and Local Government Data Research

Centre) and the UK Natural Environment Research Council (award NE/M019713/21 for the ADVENT project). Christian Albert has been supported by a grant from the German Federal Ministry of Education and Research (BMBF) for the PlanSmart research group (funding code: 01UU1601A).

We are also very grateful to the team of people in Hannover and Norwich who have helped with tasks such as proofreading, redrawing diagrams, checking references, formatting, and the multiple other tasks that are involved in preparing the final version of a manuscript. Our sincere thanks go to Martha Graf, Judith McAlister-Hermann, Zhiyuan Peng, Sascha Vandrey, Anna-Lena Vollheyde, Louise von Falkenhayn, Eick von Ruschkowski, and Trudie Dockerty. Special thanks are due to Ingrid Albert for acting as coordinator of this team and making sure the plates kept spinning.

An international collaboration of this type has involved considerable travel and visits to our respective universities. We would therefore like to thank the Hotel in Herrenhausen for providing a "home from home" for Andrew A. Lovett during visits to Hannover and to Gilla and Lena Sünnenberg for hosting Christina von Haaren during stays in Norwich.

We hope that readers of this book will gain as much insight from reading it as we have benefited from planning and writing it.

Hannover, Germany
Norwich, UK
Hannover, Germany
September 2018

Christina von Haaren
Andrew A. Lovett
Christian Albert

Contents

Contributors

Christian Albert Institute of Environmental Planning, Leibniz University Hannover, Hannover, Germany

Rachelle Alterman Technion – Israel Institute of Technology, Samuel Neaman Institute for National Policy Research, Haifa, Israel

Peter Stubkjær Andersen Department of Geosciences and Natural Resource Management, Landscape Architecture and Planning, University of Copenhagen, Frederiksberg, Denmark

Jan Barkmann Fachbereich Gesellschaftswissenschaften, University of Applied Sciences Darmstadt, Darmstadt, Germany

Thiemen Boll Dezernat für Natur- und Landschaftsschutz, Fischerei, Bezirksregierung Münster, Münster, Germany

Irene Bouwma Alterra Wageningen UR, Wageningen, The Netherlands

Birte Bredemeier Institute of Environmental Planning, Leibniz University Hannover, Hannover, Germany

Jan Bug Landesamt für Bergbau, Energie und Geologie, Hannover, Germany

Institute for Physical Geography and Landscape Ecology, Leibniz University Hannover, Hannover, Germany

Miguel A. Cebrián-Piqueras Institute of Environmental Planning, Leibniz University Hannover, Hannover, Germany

Richard J. Cooper School of Environmental Sciences, University of East Anglia, Norwich, UK

Carolin Galler Niedersächsisches Ministerium für Umwelt, Energie und Klimaschutz, Hannover, Germany

Adrienne Grêt-Regamey Swiss Federal Institute of Technology (ETH) Zurich, Institute for Spatial and Landscape Development, Zurich, Switzerland

Roy Haines-Young Fabis Consulting Ltd, Nottingham, UK

Peter Haus Grün Stadt Zürich, Zürich, Switzerland

Julian Helfenstein Department of Environmental Systems Science, Swiss Federal Institute of Technology (ETH) Zurich, Zürich, Switzerland

Johannes Hermes Institute of Environmental Planning, Leibniz University Hannover, Hannover, Germany

Sylvia Herrmann Institute of Environmental Planning, Leibniz University Hannover, Hannover, Germany

Kevin M. Hiscock School of Environmental Sciences, University of East Anglia, Norwich, UK

Daniela Kempa Institute of Environmental Planning, Leibniz University Hannover, Hannover, Germany

Felix Kienast Swiss Federal Institute for Forest, Snow and Landscape Research WSL, Birmensdorf, Switzerland

Hermann Klug Interfaculty Department of Geoinformatics – Z_GIS, University of Salzburg, Salzburg, Austria

Eva Konkoly-Gyuró Institute of Forest Resource Management and Rural Development, University of Sopron, Sopron, Hungary

Sebastian Krätzig Institute of Environmental Planning, Leibniz University Hannover, Hannover, Germany

Christoffer Lange-Kabitz Institute of Environmental Planning, Leibniz University Hannover, Hannover, Germany

Andrew A. Lovett School of Environmental Sciences, University of East Anglia, Norwich, UK

Rodrigo Caracciolo Martins Institute of Environmental Planning, Leibniz University Hannover, Hannover, Germany

Paul Opdam Land Use Planning Group and Alterra Wageningen UR Nature and Society Group, Wageningen University, Wageningen, The Netherlands

Claudia Palmas Institute of Environmental Planning, Leibniz University Hannover, Hannover, Germany

Marion Potschin Fabis Consulting Ltd, Nottingham, UK

Eeva Primmer Environmental Policy Centre, Finnish Environment Institute (SYKE), Helsinki, Finland

Steffen Reichel Interfaculty Department of Geoinformatics – Z_GIS, University of Salzburg, Salzburg, Austria

Sónia Carvalho Ribeiro Universidade Federal de Minas Gerais, Belo Horizonte, Brazil

Michael Rode Institute of Environmental Planning, Leibniz University Hannover, Hannover, Germany

Stefan Rüter Leibniz University Hannover, Institute of Environmental Planning, Hannover, Germany

Christian Schleyer University of Innsbruck, Institute of Geography, Innsbruck, Austria

Olaf Schroth Hochschule Weihenstephan-Triesdorf, Freising, Germany

Catharina Schulp Department of Earth Sciences, Faculty of Earth and Life Science, Environmental Geography Group, VU University Amsterdam, Amsterdam, The Netherlands

Vivek Shandas Portland State University, Portland, OR, USA

Hiroyuki Shimizu Nagoya University, Nagoya, Japan

Gisela Sünnenberg School of Environmental Sciences, University of East Anglia, Norwich, UK

Amy Thomas Centre for Ecology & Hydrology, Environment Centre Wales, Bangor, Gwynedd, UK

Jenni A. Turner School of Environmental Sciences, University of East Anglia, Norwich, UK

Christina von Haaren Institute of Environmental Planning, Leibniz University Hannover, Hannover, Germany

Bartlett Warren-Kretzschmar Department of Landscape Architecture and Environmental Planning, Utah State University, Logan, UT, USA

Raphael Weber Institute of Environmental Planning, Leibniz University Hannover, Hannover, Germany

Klara J. Winkler Natural Resource Sciences, McGill University, Ste-Anne-de-Bellevue, Canada

Abbreviations

ES	Ecosystem Services
AEM	Agri-Environmental Measures
AEP	Agri-Environmental Payments
ARIES	Artificial Intelligence for Ecosystem Services
ASLA	American Society of Landscape Architects
AWC	Available Water Capacity
BLM	Bureau of Land Management (Oregon)
CaBA	Catchment-Based Approach
CAP	Common Agricultural Policy
CBA	Cost-Benefit Analysis
CBD	Convention on Biological Diversity
CEA	Cost-Effectiveness Analysis
CEC	Cation Exchange Capacity
CES	Cultural Ecosystem Services
CICES	Common International Classification of Ecosystem Services
CITES	Convention on International Trade in Endangered Species of wild fauna and flora
CORINE	Coordination of Information on the Environment
CPRE	Campaign to Protect Rural England
CSA	Critical Source Area
CSF	Catchment Sensitive Farming
CWS	County Wildlife Site
DEFRA	Department for Environment, Food and Rural Affairs
DEM	Digital Elevation Model
DLM	Digital Landscape Model
DPSIR	Driving forces, Pressures, State, Impacts and Responses (framework)
DTC	Demonstration Test Catchment
DTM	Digital Terrain Model
ECRR	European Centre for River Restoration
EEA	European Environment Agency
EFG	European Federation of Geologists
EFU	Exclusive Farm/or Forest Use zone
EIA	Environmental Impact Assessment

ELC	European Landscape Convention
ELCAI	European Landscape Character Assessment Initiative
ES	Ecosystem Service(s)
ESEE	Economic, Social, Environmental, and Energy assessment framework (Oregon)
EQO	Environmental Quality Objective
EU	European Union
EUCC	Coastal and Marine Union
EUNIS	European Nature Information System
EVRI	Environmental Valuation Reference Inventory
FARMSCOPER	FARM Scale Optimisation of Pollutant Emissions Reduction
FIO	Faecal Indicator Organism
FLPMA	Federal Land Policy and Management Act (USA)
GAP	Geodiversity Action Plan
GAEC	Good Agricultural and Environmental Condition
GCR	Geological Conservation Review
GDP	Gross Domestic Product
GFP	Good Farming Practice
GHG	Greenhouse Gas
GIS	Geographical Information System
GPP	Gross Primary Productivity
HDP	Habitat Development Potential
GPS	Global Positioning System
IA	Impact Assessments
ICAAM	Instituto de Ciências Agrarias e Mediterrânicas
INSPIRE	Infrastructure for Spatial Information in the European Community
InVEST	Integrated Valuation of Ecosystem Services and Tradeoffs
IPBES	Intergovernmental Science-Policy Platform on Biodiversity and Ecosystem Services
IPCC	Intergovernmental Panel on Climate Change
IRENA	Indicator Reporting on the Integration of Environmental Concerns into Agricultural Policy
ISO	International Organization for Standardization
IUCN	International Union for Conservation of Nature
IUGS	International Union of Geological Science
JRC	European Commission Joint Research Centre
LAC	Landscape Aesthetics Capacity
LCA	Landscape Character Assessment
LCDC	Land Conservation and Development Commission (Oregon)
LDU	Landscape Description Units
LGAP	Local Geodiversity Action Plan
LiDAR	Light Detection and Ranging
LIM	Landscape Information Model
LP	Landscape Planning

LUCC	Land Use and Cover Change
LULUCF	Land Use, Land-Use Change, and Forestry
MA (also MEA)	Millennium Ecosystem Assessment
N	Nitrogen
NCA	National Character Area
NCP	Nature's Contribution to People
NEPA	National Environmental Policy Act (USA)
NGO	Non-governmental Organization
NOAA	National Oceanic and Atmospheric Administration
NVZ	Nitrate Vulnerable Zone
NWRM	Natural Water Retention Measures
OECD	Organisation for Economic Co-operation and Development
OSM	OpenStreetMap
P	Phosphorous
PES	Payments for Ecosystem Services
PNAP	Política Nacional Arquitetura e Paisagem
PPP	Polluter Pays Principle
PTF	Pedotransfer Function
PV	Photovoltaic
RBZ	Riparian Buffer Zone
RDP	Rural Development Programme
RIGS	Regionally Important Geological and Geomorphological Sites
RMP	Resource Management Plan
SDG	Sustainable Development Goal(s)
SEA	Strategic Environmental Assessment
SMR	Statutory Management Requirements
SQR	Soil Quality Rating
SRP	Soluble Reactive Phosphorus
SSSI	Site of Special Scientific Interest
STW	Sewage Treatment Works
SUDs	Sustainable Urban Drainage Systems
SWAT	Soil & Water Assessment Tool
TEEB	The Economics of Ecosystems and Biodiversity
TEEB-DE	Naturkapital Deutschland – The Economics of Ecosystems and Biodiversity
TEV	Total Economic Value
UAA	Utilized Agricultural Area
UK-NEA	UK National Ecosystem Assessment
UN	United Nations
UNCED	United Nations Conference on Environment and Development
UNEP	UN Environment Programme
UNESCO	United National Educational, Scientific and Cultural Organization
UNFCCC	United Nations Framework Convention on Climate Change
US EPA	United States Environmental Protection Agency

USLE	Universal Soil Loss Equation
USGS	United States Geological Survey
VGI	Volunteered Geographic Information
WFD	Water Framework Directive
WMO	World Meteorological Organization
WTA	Willingness to Accept
WTO	World Trade Organization
WTP	Willingness to Pay

Glossary

Anthropogenic impacts Impacts resulting from human activities (TEEB, online).

Assessment "The analysis and review of information for the purpose of helping someone in a position of responsibility to evaluate possible actions or think about a problem. Assessment means assembling, summarising, organising, interpreting, and possibly reconciling pieces of existing knowledge and communicating them so that they are relevant and helpful to an intelligent but inexpert decision-maker (Parson 1995)" (cited according to Maes et al. (2013)). Assessment includes inventory and evaluation (see below).

Assets Economic resources (TEEB 2010).

Benefits Positive change in well-being from the fulfilment of needs and wants (TEEB 2010).

Biodiversity "The variability among living organisms from all sources, including inter alia terrestrial, marine, and other aquatic ecosystems and the ecological complexes of which they are part, this includes diversity within species, between species, and of ecosystems (cf. Article 2 of the Convention on Biological Diversity, 1992)" (cited according to Maes et al. (2013)).

Biotope An ecological area that supports a particular range of biological communities (TEEB, online).

Criterion A standard on which a judgment or decision may be based (Merriam-Webster, online).

Cultural ecosystem services The nonmaterial benefits people obtain from ecosystems through spiritual enrichment, cognitive development, reflection, recreation, and aesthetic experience, including knowledge systems, social relations, and aesthetic values (MA 2005).

Delivered ecosystem services Delivered ecosystem services represent the totality of ecosystem contributions that may provide benefits to humans today or in the future (but need not necessarily be used today). Delivered ecosystem services were previously termed "offered ecosystem services" in von Haaren et al. (2014). Some publications refer to delivered ES as ecosystem services potentials, but we opt to refer to them as delivered ES to convey that they are actually delivered (although not used at present).

Direct use value (of ecosystems) The benefits derived from the services provided by an ecosystem that are used directly by an economic agent. These include consumptive uses (e.g., harvesting goods) and nonconsumptive uses (e.g., enjoyment of scenic beauty). Agents are often physically present in an ecosystem to receive direct use value (MA 2005).

Driver Any natural or human-induced factor that directly or indirectly causes a change in an ecosystem (MA 2005).

Economic valuation The process of expressing a value for a particular good or service in a certain context (e.g., of decision-making) in monetary terms (TEEB 2010).

Ecosystem A dynamic complex of plant, animal, and microorganism communities and their nonliving environment interacting as a functional unit (MA 2005). For practical purposes, it is important to define the spatial dimensions of concern (quoted from Maes et al. 2013).

Ecosystem function Subset of the interactions between biophysical structures, biodiversity, and ecosystem processes that underpin the capacity of an ecosystem to provide ecosystem services (adapted from TEEB 2010). The term "ecosystem functions" can be used in both a descriptive and a normative sense (cf. Jax 2010, Spangenberg et al. 2014). In this book, we use ecosystem functions in its descriptive sense as a subset of ecosystem processes, elements, etc. with meaning for ES (de Groot et al. 2010). A normative interpretation of ecosystem functions would be similar to "delivered ES." We acknowledge that the normative use of the term "functions" carries the risk of confusion with the descriptive meaning (von Haaren et al. 2014).

Ecosystem process Any change or reaction which occurs within ecosystems, physical, chemical, or biological. Ecosystem processes include decomposition, production, nutrient cycling, and fluxes of nutrients and energy (MA 2005).

Ecosystem services The direct and indirect contributions of ecosystems to human well-being (TEEB 2010). We acknowledge that the MA (2005) referred to ecosystem services as "the benefits that people obtain from ecosystems," but we find it important to make a distinction between ecosystem services and the benefits they provide (cf. definition of "benefit"). The Intergovernmental Science-Policy Platform on

Biodiversity and Ecosystem Services (IPBES) recently proposed a new definition of ecosystem services as nature's contributions to people (NCP) (Díaz et al. 2015; Pascual et al. 2017). NCP considers all "positive contributions or benefits, and occasionally negative contributions, losses or detriments, that people obtain from nature." NCP thus builds on prior conceptualizations of ecosystem services but stronger emphasizes other knowledge systems and world views (Pascual et al. 2017).

Ecosystem state The physical, chemical, and biological condition of an ecosystem at a particular point in time (Maes et al. 2013).

Ecosystem status A classification of ecosystem state among several well-defined categories. It is usually measured against time and compared to an agreed target in EU environmental directives (e.g., Habitats Directive, Water Framework Directive, Marine Strategy Framework Directive) (Maes et al. 2013).

Evaluation To assign a value by comparing the state with a standard or benchmark. This may include determining the level of significance or importance (cf. chapter 2.1).

Existence value The value that individuals place on knowing that a resource exists, even if they never use that resource (also sometimes known as conservation value or passive use value) (MA 2005).

Externality A consequence of an action that affects someone other than the agent undertaking that action and for which the agent is neither compensated nor penalized through the markets. Externalities can be positive or negative (MA 2005).

Governance (of ecosystems) The process of regulating human behavior in accordance with shared ecosystem objectives. The term includes both governmental and nongovernmental mechanisms (TEEB, online).

Habitat The physical location or type of environment in which an organism or biological population lives or occurs. Terrestrial or aquatic areas distinguished by geographic, abiotic, and biotic features, whether entirely natural or seminatural (Maes et al. 2013).

Human well-being A context- and situation-dependent state, comprising basic material for a good life, freedom and choice, health and bodily well-being, good social relations, security, peace of mind, and spiritual experience (MA 2005).

Indicator Observed value representative of a phenomenon to study. In general, indicators quantify information by aggregating different and multiple data. Oftentimes, an indicator is a more indirect, or proxy, measure of property of interest. The indicator can help in measuring the status of a criterion. The resulting information is therefore synthesized (Maes et al. 2013, altered).

Indirect use value The benefits derived from the goods and services provided by an ecosystem that are used indirectly by an economic agent. For example, an agent at some distance from an ecosystem may derive benefits from drinking water that has been purified as it passed through the ecosystem (MA 2005).

Intrinsic value The value of someone or something in and for itself, irrespective of its utility for someone else (MA 2005).

Inventory An itemized list of current asset (Merriam-Webster, online).

Landscape Landscape means an area, as perceived by people, whose character is the result of the action and interaction of natural and/or human factors (European Landscape Convention (ELC), Council of Europe 2000, Art. 1). In Chap. 1, we add further remarks on our more specific interpretation.

Landscape planning Landscape planning is understood as a "strong forward-looking action to enhance, restore or create landscapes" (European Landscape Convention (ELC), Council of Europe 2000, Art. 1).

Mapping Mapping can be interpreted in two ways: (i) as a spatial representation of information and (ii) as a categorization of information and relationships.

Market failure The inability of a market to capture the correct values of ecosystem services (MA 2005).

Measure (or measurement) This term is used in several different ways. It can mean (i) an action intended to achieve a particular objective, (ii) a standard or unit of measurement, or (iii) the act of measuring (Merriam-Webster, online).

Mitigation (or restoration) cost The cost of mitigating the effects of the loss of ecosystem services or the cost of getting those services restored (TEEB, online).

Monetary valuation see *Economic valuation.*

Natural capital An economic metaphor for the limited stocks of physical and biological resources found on earth (MA 2005). Natural capital provides the basis for delivering ecosystem services. It is a subset of ecosystem components and processes that have a particular meaning for delivery of ecosystem services (e.g., specific soils) (altered from TEEB, online).

Nonuse value Benefits which do not arise from direct or indirect use. Examples of nonuse value are benefits which people gain from knowing that something exists (e.g., existence value).

Precautionary principle The management concept stating that in cases "where there are threats of serious or irreversible damage, lack of full scientific certainty shall not be used as a reason for postponing cost-effective measures to prevent environmental degradation," as defined in the Rio Declaration (MA 2005). "The precautionary principle is detailed in Article 191 of the Treaty on the Functioning of the European Union. It aims at ensuring a higher level of environmental protection through preventative decision-taking in the case of risk" (European Commission 2000).

Provisioning services The products obtained from ecosystems, including genetic resources, food and fiber, and fresh water (MA 2005).

Regulating services The benefits obtained from the regulation of ecosystem processes, including the regulation of climate, water, and some human diseases (MA 2005).

Responses Human actions, including policies, strategies, and interventions, to address specific issues, needs, opportunities, or problems. In the context of ecosystem management, responses may be of legal, technical, institutional, economic, and behavioral nature and may operate at various spatial and time scales (MA 2005).

Scale The measurable dimensions of phenomena or observations. Expressed in physical units, such as meters, years, population size, or quantities moved or exchanged. In observation, scale determines the relative fineness and coarseness of different detail and the selectivity among patterns these data may form (MA 2005).

Sensitivity Likelihood of change in state in response to pressures and responses.

Stakeholder A person, group, or organization that has a stake in the outcome of a particular activity (TEEB, online).

Supporting services Ecosystem services that are necessary for the maintenance of all other ecosystem services. Some examples include biomass production, production of atmospheric oxygen, soil formation and retention, nutrient cycling, water cycling, and provisioning of habitat (MA 2005).

Sustainability A characteristic or state whereby the needs of the present and local population can be met without compromising the ability of future generations or populations in other locations to meet their needs (MA 2005).

Sustainable flow (of ecosystem services) The availability of ecosystem services to yield a continuous benefit to present generations while maintaining its potential to meet the needs and aspirations of future generations (MA 2005).

Threshold A point or level at which new properties emerge in an ecological, economic, or other system, invalidating predictions based on mathematical relationships that apply at lower levels. For example, species diversity of a landscape may decline steadily with increasing habitat degradation to a certain point and then fall sharply after a critical threshold of degradation is reached. Human behavior, especially at group levels, sometimes exhibits threshold effects. Thresholds at which irreversible changes occur are especially of concern to decision-makers (MA 2005).

Total economic value The value obtained from the various constituents of utilitarian value, including direct use value, indirect use value, option value, quasi-option value, and existence value (TEEB, online).

Trade-offs of ecosystem services The way in which one ecosystem service relates to or responds to a change in another ecosystem service. These trade-offs can be either positive or negative (altered from TEEB, online).

Valuation The process of expressing a value for a particular good or service in a certain context (e.g., of decision-making) usually in terms of something that can be counted, often money, but also through methods and measures from other disciplines (sociology, ecology, and so on) (MA 2005).

Value Relative worth, utility, or importance, something (such as a principle or quality) intrinsically valuable or desirable (Merriam-Webster, online).

Vulnerability Exposure to contingencies and stress and the difficulty in coping with them (TEEB, online).

References

Council of Europe. (2000). *European landscape convention*. Florence: Council of Europe.

de Groot, R. S., Alkemade, R., Braat, L., et al. (2010). Challenges in integrating the concept of ecosystem services and values in landscape planning, management and decision making. *Ecological Complexity, 7*, 260–272.

Díaz, S., Demissew, S., Carabias, J., et al. (2015). The IPBES conceptual framework – Connecting nature and people. *Current Opinion in Environment Sustainability, 14*, 1–16.

European Commission. (2010). *Communication from the commission on the precautionary principle*. http://eur-lex.europa.eu/legal-content/EN/TXT/?uri=LEGISSUM%3Al32042. Accessed 17 Jan 2018.

Jax, K. (2010). *Ecosystem functioning*. Cambridge: Cambridge University Press.

MA. (2005). *Ecosystems and human wellbeing: Current state and trends* (Vol. 1). Washington, DC: Island Press.

Maes, J., Teller, A., Erhard, M., et al. (2013). *Mapping and Assessment of Ecosystems and their Services. An analytical framework for ecosystem assessments under action 5 of the EU biodiversity strategy to 2020*. Publications office of the European Union, Luxembourg

Merriam-Webster. (online). https://www.merriam-webster.com. Accessed 17 Jan 2018.

Parson, E. A. (1995). Integrated assessment and environmental policy making, in Pursuit of usefulness. *Energy Policy, 23*(4/5), 463–476.

Pascual, U., Balvanera, P., Díaz, S., et al. (2017). Valuing nature's contributions to people: The IPBES approach. *Current Opinion in Environment Sustainability, 26*(27), 7–16.

Spangenberg, J. H., von Haaren, C., & Settele, J. (2014). The ecosystem service cascade: Further developing the metaphor. Integrating societal processes to accommodate social processes and planning, and the case of bioenergy. *Ecological Economics, 104*, 22–32.

TEEB. (2010). *The economics of ecosystems and biodiversity: Ecological and economic foundation*. Cambridge: Earthscan.

TEEB. (online). http://www.teebweb.org/resources/glossary-of-terms. Accessed 17 Jan 2018.

von Haaren, C., Albert, C., Barkmann, J., et al. (2014). From explanation to application: Introducing a practice-oriented ecosystem services evaluation (PRESET) model adapted to the context of landscape planning and management. *Landscape Ecology, 29*, 1335–1346.

Part I
Landscape Planning with Ecosystem Services

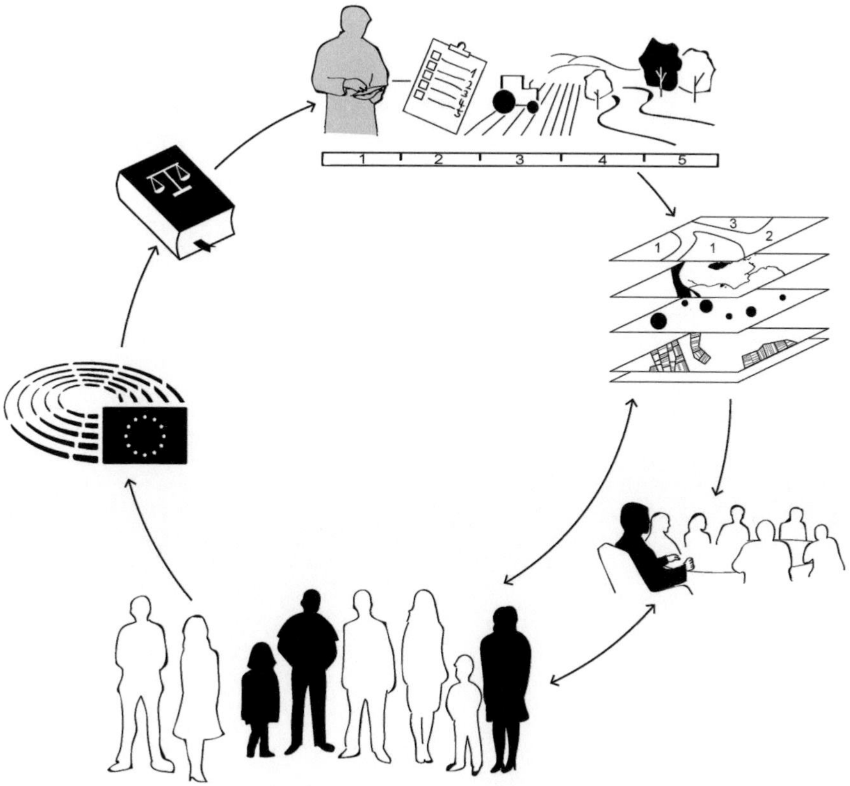

This part of the book is about the context and theory of landscape planning. It provides understanding about: (i) the nature and purpose of landscape planning in societal decision making, (ii) the foundational principles for methods to assess ecosystem services and design response options, (iii) the different values that underpin the evaluation of biodiversity and ecosystem services, and (iv) the process of planning including the role of public participation.

Landscape Planning and Ecosystem Services: The Sum is More than the Parts

1

Christina von Haaren, Andrew A. Lovett, and Christian Albert

Abstract

Landscapes provide a broad range of ecosystem services that are crucial for many aspects of human well-being. However, this provision is increasingly under threat from a variety of economic, social and environmental changes. Many of these are manifested in unsustainable land uses. Integrative and proactive environmental planning is needed to address these challenges and can be achieved by combining the conceptual strengths of the ecosystem services approach with the practical and implementation-orientated focus of landscape planning.

Keywords

Sustainability · Environmental impacts · Integrated planning · Economic valuation · Spatial scale

1.1 The Need for More Integrated Environmental Planning

Human well-being depends in many ways on maintaining the stock of natural resources which deliver the ecosystem services from which humans benefit, such as productivity of soils, flood water retention or beautiful landscapes. However, the continued flow of these services is increasingly threatened by unsustainable land uses. This is becoming particularly evident on regional and local scales. Many land

C. von Haaren (✉) · C. Albert
Institute of Environmental Planning, Leibniz University Hannover, Hannover, Germany
e-mail: haaren@umwelt.uni-hannover.de; albert@umwelt.uni-hannover.de

A. A. Lovett
School of Environmental Sciences, University of East Anglia, Norwich, UK
e-mail: a.lovett@uea.ac.uk

C. von Haaren et al. (eds.), *Landscape Planning with Ecosystem Services*, Landscape Series 24, https://doi.org/10.1007/978-94-024-1681-7_1

uses compete within the same area and can produce harmful environmental impacts. Particular threats exist to those public environmental goods whose values are not well-represented in economic markets or whose deterioration will only affect future generations. As market forces alone are not sufficient, effective means for local and regional planning are needed in order to safeguard scarce natural resources, coordinate land uses and create sustainable landscape structures.

European law already includes a set of instruments to protect different environmental goods and services. Many of these are reflected in the planning framework, important examples being the Strategic Environmental Assessment (SEA) and Environmental Impact Assessment (EIA) Directives. In addition, several proactive planning approaches are implemented across the EU, such as the Water Framework, Flood Protection and Habitats Directives. However, these are quite sectoral in nature and do not fully exploit the synergies that could be achieved under a more integrated and multifunctional landscape perspective. Furthermore, existing methods for environmental assessments are often not especially appropriate for practical application. Elaborate models used in science may have data requirements that cannot be met in some regions and the results often have a degree of detail that is too complex for implementation-oriented measures. Up to now, an EU Directive regarding proactive overall environmental planning is still lacking.

1.2 Landscape Planning can Help Fill the Gap

As of January 2018 thirty eight countries had ratified the European Landscape Convention (ELC) (Council of Europe 2018), thus committing themselves to implementing landscape planning. *Landscape* is defined by the ELC (Chapter I article 1f) as an "area, as perceived" (and, we would like to add, 'as understood') "by people". The character of the landscape is 'the result of the action and interaction of natural and/or human factors'. This definition highlights the human influence on landscape; also that a landscape is socially constructed, a selection of the 'real' world shaped by human capacities to perceive, measure and understand. In addition, the idea of landscape has a scale connotation since it does not refer to small areas like habitats (which may be landscape components), although there is no precise agreement on how large a landscape should be. There are landscapes in which human impact is minimal (natural landscapes) and those that are predominately shaped by humans (cultural landscapes). The term landscape stems from medieval times where it meant a territory, area or region (Burckhardt 1995; Tress and Tress 2001). It has been used in common language, particularly in English, as referring to pleasant surroundings. In science, landscape was first used in the eighteenth century by Humboldt, who defined it as the "total character of a region of the earth" (Neef and Neef 1977).

Landscape planning is understood here in line with the definition in the European Landscape Convention as 'strong forward-looking action to enhance, restore and create landscapes'. Acknowledging that understandings of landscape planning vary between countries, a broad definition is adopted here in order to cover different legal and cultural landscape planning frameworks. An inventory of landscape planning tasks would include the production of place-based environmental information, reconciliation of competing land uses, protection, redevelopment, management and monitoring of natural and cultural assets and the development of strategic thinking about land use and management (Sell and Zube 1986; Leitão and Ahern 2002; Ogrin 2010). Furthermore, landscape planning should not only improve the citizen's and politician's understanding of the consequences of planned actions, but also contribute to setting priorities for policy implementation (see BenDor et al. 2017 for US land use planning). This understanding of landscape planning encompasses environmental planning and partly overlaps with what is understood in some countries by 'land use planning'.

Planning is interpreted in this book as both the result and the activity of making a plan and preparing its realisation. In our understanding, a plan is no longer a static piece of paper, but a database of geographical information, attributes and criteria adaptable to new conditions and reflecting uncertainties. The process of planning includes 'using cultural and scientific knowledge' (ASLA 2018) and the translation of scientifically generated results into implementable measures in a manner that bridges the gap between science and politics.

At present, landscape planning has not been introduced in all European states in the way that the ELC suggests and the approaches adopted are quite diverse (Kozová and Finka 2010). There are some European countries where landscape planning has been established for decades (e.g. Germany, the Netherlands and Switzerland). In these cases landscape planning is used as both an integrated source of information (e.g. for reactive instruments such as EIA) and to provide strategic guidance for landscape development. Other states have started to integrate the relevant content into environmentally-oriented spatial planning or supplemented spatial planning with strategic environmental impact assessments (Wende et al. 2011). In England, initiatives such as the Catchment-Based Approach (CaBA) (https://www.catchmentbasedapproach.org) and the recent 25 Year Environment Plan (Defra 2018) have introduced more integrative place-based thinking compared to a previous sectoral emphasis.

While the ratification and implementation of the ELC has initiated more landscape planning and methodological exchange in European countries (Kovács et al. 2013), there are other kinds of environmental planning which also have the potential to offer the same integrative and spatially-explicit perspective as landscape planning. In this book, therefore, landscape planning is also used as a shorthand term for all kinds of environmental planning dealing with holistic frameworks for multiple environmental resources and services.

1.3 Ecosystem Services: The Communicative Turn in Environmental Protection

Alongside a growing awareness of landscape planning, recent years have seen more interest in the concept of ecosystem services at national and European scales. The purpose has been to better communicate the link between nature and human wellbeing, especially to highlight the importance of this to policy and decision makers (Daily et al. 2009; Albert et al. 2014; Mascarenhas et al. 2014). In many cases, the introduction of the ecosystem services concept has been accompanied by a stronger emphasis on economic reasoning.

In contrast to the ancient origins of landscape, the term ecosystem stems from the much younger science of ecology (Tansley 1935). In general terms an ecosystem can be described as consisting of living organisms and the non-living components of their environment at any scale, in which there are continuous fluxes of matter and energy in an interactive open system (see Willis 1997; Smith and Smith 2012). In this book we include human influences as part of ecosystems, although this is a matter of dispute in the scientific community. In principle, an ecosystem has no defined scale or spatial delineation since these depend on the research question under investigation. The connection between ecosystem and landscape is underpinned by an early remark of Whittaker in relation to the classification of natural communities, (1962: 125), who observed, that "the ecosystem conception suggests a multifactorial or landscape approach to classification" (after Willis 1997). In comparison to landscape and ecosystem, the term environment (also often used in this book) has a wider definition since it includes the whole world surrounding humans, including the societal context with which they interact.

The term 'ecosystem services' is used ambiguously in the literature. Divergent definitions exist with overlapping and sometimes conflicting meanings. Differences in definitions refer to the terms used, the concepts applied to these terms, the ecosystem services classification systems considered, and how actual ecosystem services are defined (von Haaren and Albert 2011; Albert et al. 2016).

Despite this ambiguity, the definitions applied in three major international assessments provide a good overview and orientation as these are most often referred to, and applied, in planning applications. The Millennium Ecosystem Assessment (MA 2005), the first global assessment of the state of ecosystems and biodiversity, defined ecosystem services as "the benefits people obtain from ecosystems". A few years later, the international study on The Economics of Ecosystems and Biodiversity (TEEB 2010) provided a refined definition of ecosystem services as the "direct and indirect contributions of ecosystems to human well-being". By doing so, TEEB emphasized the role of ecosystem services for human well-being and disentangled the concept of ecosystem services from the benefits they provide. Most recently, the Intergovernmental Platform on Biodiversity and Ecosystem Services (IPBES) adopted a new definition of ecosystem services as nature's contributions to people (NCP) (Diaz et al. 2015; Pascual et al. 2017). NCP considers all "positive contributions or benefits, and occasionally negative contributions, losses or detriments that people obtain from nature". As such, NCP relates to the ecosystem services term, but includes stronger acknowledgment of the diversity of worldviews, knowledge systems and values (Pascual et al. 2017; Diaz et al. 2018). The recent introduction of the NCP term has sparked substantial scientific discussion (e.g. Braat 2018; Maes et al. 2018; Peterson et al. 2018) and it remains to be seen what role this new term will play in future research and application.

Action 5 of the EU Biodiversity Strategy to 2020 calls upon member states to map and assess the state of ecosystems and their services in their national territories. These accounting and reporting systems are intended to include the economic value of services which presents further challenges in terms of the way in which multiple natural and human capital assets combine to support flows of services and associated benefits (e.g. hydroelectric power requires stream flow and the application of human expertise to construct the necessary generation technology, Fisher et al. 2009). Properties such as biodiversity and geodiversity are particularly difficult in this respect because on one hand they are part of the basic underpinning natural capital of ecosystems yet also contribute to particular ES, especially those related to natural and cultural heritage. There is consequently considerable potential for double counting and this is partly why up to now ES assessment has been predominantly at the national scale (e.g. UK-NEA or TEEB-DE) and not adapted to regional and local needs. For instance, many important economic values are spatially specific, and indeed this is what is required for local and regional decision making, yet the derivation of such values involves further technical complexities (Bateman et al. 2013). In general, there are challenges in translating natural capital and ecosystem service ideas into practice and this has created a situation in which some planning and management practitioners are reluctant to use the concepts (Albert et al. 2014) and several initiatives have sought to address the problems (e.g. see the Natural Capital Committee (2017) workbook).

1.4 Combining the Strengths of Landscape Planning and Ecosystem Services

Obviously, there is complementarity between landscape planning and the ecosystem services concept. Linking landscape planning and ecosystem services creates a two-way benefit: landscape planning is strong in producing area-specific results, which can be incorporated into implementation mechanisms such as legally-binding land use planning, protected area designations or targeted agri-environmental schemes. The ecosystem services concept does not yet provide a fully developed system of assessment methodologies which are applicable in practice on regional or local scale, nor are there established means of implementation. However, a strength of the ecosystem services concept lies in making the connection between the status of natural assets and human well-being more explicit, as well as the use of economic valuation which can resonate with a range of public and private sector decision makers. Economic analysis also has a capacity to cast a wider perspective on environmental problems and help reveal the influence of driving forces on pressures and the state of ecosystems. Furthermore the economic perspective, more than analysis in landscape planning and related physio-geographical approaches, focusses on individual preferences and benefits, which can help validate the acceptability of environmental planning goals. Thus linking landscape planning and the ecosystem services concept can be regarded as prototypical for the concept of usable science, which is guided by the needs of decision making (Ford et al. 2013).

This two-way benefit is also reflected in terms of methodologies. A full ecosystem services assessment should not rely primarily on current, perhaps volatile, preferences and monetary values, which as yet cannot fully capture the non-use values of ecosystems. If these long-term or non-use values are to be adequately included, the methodologies from environmental planning need to be incorporated into a toolbox for ecosystem services assessment. In such circumstances, a large methodological overlap exists between landscape planning and the ecosystem services approach. All in all, for an ecosystem services-informed landscape planning, a consistent compendium of methodologies would be of great added value. The potential of merging the approaches and the mission to contribute practicable and consistent methodologies for a wide range of applications in landscape planning, as well as in other environmental assessments, has motivated the authors to write this book. Individual articles scattered amongst the journal literature do not provide sufficient orientation and cannot do justice to this goal.

References

Albert, C., Hauck, J., Buhr, N., et al. (2014). What ecosystem services information do users want? Investigating interests and requirements among landscape and regional planners in Germany. *Landscape Ecology, 29*, 1301–1313. https://doi.org/10.1007/s10980-014-9990-5.

Albert, C., Galler, C., Hermes, J., et al. (2016). Applying ecosystem services indicators in landscape planning and management: The ES-in-planning framework. *Ecological Indicators, 61*, 100–113.

ASLA. (2018). *Glossary*. Landscape architecture terms. https://www.asla.org/nonmembers/publicrelations/glossary.htm. Accessed 30 Mar 2018.

Bateman, I. J., Harwood, A. R., Mace, G. M., et al. (2013). Bringing ecosystem services into economic decision-making: Land use in the United Kingdom. *Science, 341*(6141), 45–50.

BenDor, T. K., Spurlock, D., Woodruff, S. C., et al. (2017). A research agenda for ecosystem services in American environmental and land use planning. *Cities, 60*, 260–271.

Braat, L. C. (2018). Five reasons why the science publication "Assessing nature's contributions to people" (Díaz et al 2018) would not have been accepted in ecosystem services. *Ecosystem Services, 30*, A1–A2.

Burckhardt, L. (1995). Landschaft ist transitorisch. Zur Dynamik der Kulturlandschaft. *Laufener Seminarberichte, 4*(95), 31–36.

Council of Europe. (2018). *European landscape convention*. https://www.coe.int/en/web/landscape/home. Accessed 17 Jan 2018.

Daily, G. C., Polasky, S., Goldstein, J., et al. (2009). Ecosystem services in decision making: Time to deliver. *Frontiers in Ecology and the Environment, 7*(1), 21–28.

Defra. (2018). *A green future: Our 25 year plan to improve the environment* (Policy paper). Available via www.gov.uk. https://www.gov.uk/government/uploads/system/uploads/attachment_data/file/693158/25-year-environment-plan.pdf. Accessed 30 Mar 2018.

Díaz, S., Demissew, S., Carabias, J., et al. (2015). The IPBES conceptual framework — Connecting nature and people. *Current Opinion in Environment Sustainability, 14*, 1–16.

Díaz, S., Pascual, U., Stenseke, M., et al. (2018). Assessing nature's contributions to people. *Science, 359*(6373), 270–272.

Fisher, B., Turner, R. K., & Morling, P. (2009). Defining and classifying ecosystem services for decision making. *Ecological Economics, 68*(3), 643–653.

Ford, J. D., Knight, M., & Pearce, T. (2013). Assessing the 'usability' of climate change research for decision-making: A case study of the Canadian international polar year. *Global Environmental Change, 23*(5), 1317–1326.

Kovács, K. F., Sallay, A., Jombach, S., et al. (2013, April). Landscape in the spatial planning system of European countries. In *Conference paper*. https://www.researchgate.net/publication/243460260_Landscape_in_the_spatial_planning_system_of_European_countries. Accessed Jan 2018.

Kozová, M., & Finka, M. (2010). Landscape development planning and management systems in selected European countries. *The Problems of Landscape Ecology, 28*(28), 101–110.

Leitão, A. B., & Ahern, J. (2002). Applying landscape ecological concepts and metrics in sustainable landscape planning. *Landscape and Urban Planning, 59*(2), 65–93.

MA. (2005). *Millennium ecosystem assessment ecosystems and human well-being: synthesis*. Washington, DC: Island Press.

Maes, J., Burkhard, B., & Geneletti, D. (2018). Ecosystem services are inclusive and deliver multiple values. A comment on the concept of nature's contributions to people. *One Ecosystem, 3*, e24720.

Mascarenhas, A., Ramos, T. B., Haase, D., et al. (2014). Integration of ecosystem services in spatial planning: A survey on regional planners' views. *Landscape Ecology, 29*(8), 1287–1300.

Natural Capital Committee. (2017). *How to do it: A natural capital workbook*. https://assets.publishing.service.gov.uk/government/uploads/system/uploads/attachment_data/file/608852/ncc-natural-capital-workbook.pdf. Accessed 5 May 2018.

Neef, E., & Neef, V. (Eds.). (1977). *Brockhaus Handbuch "Sozialistische Landeskultur": Umweltgestaltung, Umweltschutz; mit einem ABC*. Leipzig: VEB Brockhaus-Verlag.

Ogrin, D. (2010). Quo vadis, topos europaeus? *Landscape, 21*(2010), 63–69.

Pascual, U., Balvanera, P., Díaz, S., et al. (2017). Valuing nature's contributions to people: The IPBES approach. *Current Opinion in Environment Sustainability, 26*, 7–16.

Peterson, G. D., Harmackova, Z. V., Meacham, M., et al. (2018). Welcoming different perspectives in IPBES: "Nature's contributions to people" and "Ecosystem services". *Ecology and Society, 23*(1), 39.

Sell, J. L., & Zube, E. H. (1986). Perception of and response to environmental change. *Journal of Architectural and Planning Research, 3*(1986), 33–54.

Smith, T. M., & Smith, R. L. (2012). *Elements of ecology* (8th ed.). Boston: Benjamin Cummings.

Tansley, A. G. (1935). The use and abuse of vegetational concepts and terms. *Ecology, 16*, 284–307.

TEEB. (2010). *The economics of ecosystems and biodiversity: Mainstreaming the economics of nature: A synthesis of the approach, conclusions and recommendations of TEEB*. www.teebweb.org. Accessed 28 June 2018.

Tress, B., & Tress, G. (2001). Capitalising on multiplicity: A transdisciplinary systems approach to landscape research. *Landscape and Urban Planning, 57*(3–4), 143–157.

von Haaren, C., & Albert, C. (2011). Integrating ecosystem services and environmental planning: Limitations and synergies. *International Journal of Biodiversity Science, Ecosystem Services & Management, 7*(3), 150–167. https://doi.org/10.1080/21513732.2011.616534.

Wende, W., Wojtkiewicz, W., Marschall, I., et al. (2011). Putting the plan into practice: Implementation of proposals for measures of local landscape plans. *Landscape Research, 37*(4), 483–500.

Whittaker, R. H. (1962). Classification of natural communities. *The Botanical Review, 28*(11), 1–239.

Willis, A. J. (1997). The ecosystem: An evolving concept viewed historically. *Functional Ecology, 11*, 268–271.

Objectives and Structure of the Book 2

Christina von Haaren, Andrew A. Lovett,
and Christian Albert

Abstract

The objective of this book is to provide an introduction and overview of relevant concepts, methods and techniques for landscape planning with ecosystem services in Europe. It presents a new, ecosystem services-informed, approach to landscape planning that constitutes both a framework and toolbox for students and practitioners to address the environmental and landscape challenges of the twenty-first century. The book is structured into six parts which broadly follow the well-known Driving forces, Pressures, State, Impacts and Responses (DPSIR) framework for describing human-environment relationships. Part I introduces key theories, concepts, and methodological foundations for landscape planning and ecosystem services. Drivers and pressures instigating landscape change in Europe are discussed in Part II. Part III outlines methods for assessing states and impacts of various components of biodiversity and ecosystem services. Approaches for deriving response measures are the focus on Part IV. Part V addresses communication in landscape planning, while Part VI provides international perspectives and an outlook on the future prospects of landscape planning. This chapter also outlines the types of questions addressed in the book and discusses an example of how it might be used in landscape planning practice.

Keywords

Landscape planning · Ecosystem services · Integrated planning · Role of landscape planning · DPSIR

C. von Haaren (✉) · C. Albert
Institute of Environmental Planning, Leibniz University Hannover, Hannover, Germany
e-mail: haaren@umwelt.uni-hannover.de; albert@umwelt.uni-hannover.de

A. A. Lovett
School of Environmental Sciences, University of East Anglia, Norwich, UK
e-mail: a.lovett@uea.ac.uk

C. von Haaren et al. (eds.), *Landscape Planning with Ecosystem Services*,
Landscape Series 24, https://doi.org/10.1007/978-94-024-1681-7_2

2.1 Objectives of the Book

Human well-being is heavily dependent on maintaining the stock of natural resources which deliver the services from which human's benefit. However, these resources and flows of services are increasingly threatened by changes in land use. Market forces alone are not sufficient to address these problems, so effective local and regional planning is required in order to safeguard scarce natural resources, coordinate land uses and create sustainable landscape structures.

This book argues that a solution to such challenges in Europe can be found by merging the landscape planning tradition with ecosystem services concepts. Landscape planning has strengths in recognition of public benefits and implementation mechanisms, while the ecosystem services approach makes the connection between the status of natural assets and human well-being more explicit. It can also provide an economic perspective, focused on individual preferences and benefits, which helps to validate the acceptability of environmental planning goals. Thus linking landscape planning and ecosystem services provides a two-way benefit, creating a usable science to meet the needs of local and regional decision making.

The main objective of this book is therefore to provide an introduction and overview of relevant concepts, methods and techniques for landscape planning with ecosystem services in Europe. It presents a new, ecosystem services-informed, approach to landscape planning that constitutes both a framework and toolbox for students and practitioners to address the environmental and landscape challenges of the twenty-first century.

The choice of landscape planning concepts and methods presented in this book reflects their applicability at local to regional levels of decision-making and the types of data that are usually available at these scales. Furthermore, we focus on the need to deliver spatially-explicit assessments or proposals as decision-support, and to incorporate existing objectives and values as derived from democratically legitimized laws or expressed in participatory processes. While the concepts and methods have been selected with landscape planning in mind, they may also be applicable in other sectors including, for example, spatial or conservation planning, implementation of the EU Water Framework Directive, environmental impact assessments and initiatives regarding mitigation banking or impact compensation.

2.2 Book Structure

The structure of the book as illustrated in Fig. 2.1 helps students and practitioners to easily find the methods designed for tackling landscape planning challenges and needed for practical planning that discuss a broad topic such as methods for assessing and evaluating ecosystem services and biodiversity, for designing response objectives and measures, or for facilitating public participation.

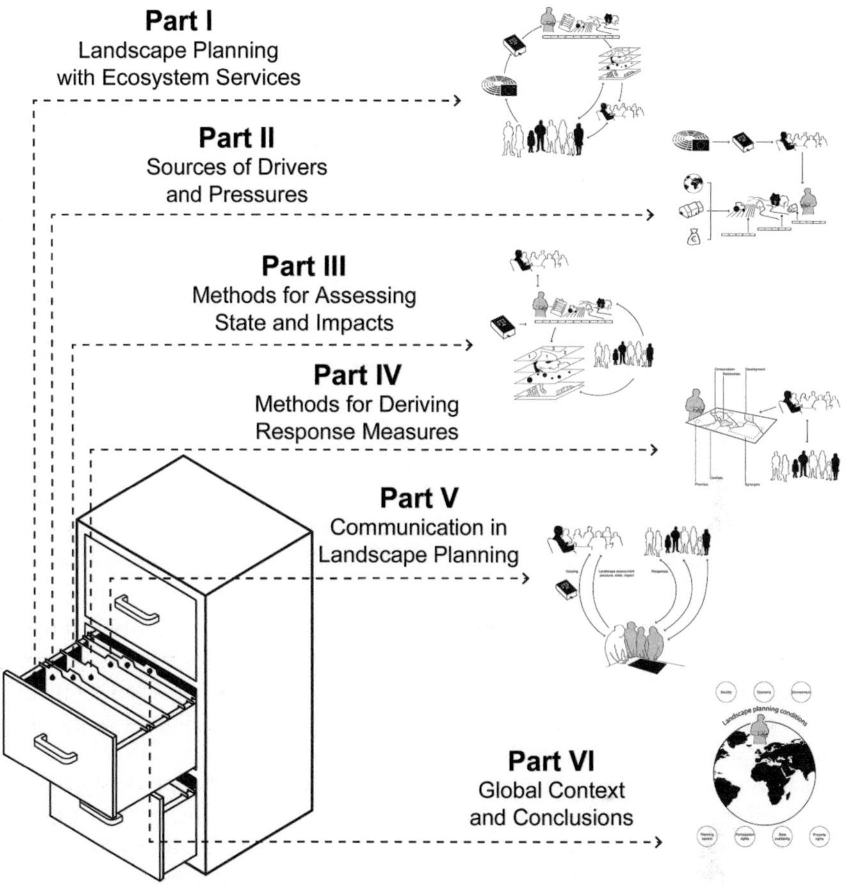

Fig. 2.1 The organisation of parts in the book. This book is analogous to a filing cabinet, providing orientation (i.e. context and theory) as well as hands-on technical guidance (methods and procedures) to practitioners and students on how to prepare landscape plans. Each part of the book is represented by a particular file in the cabinet, consisting of a set of chapters

The formal structure of the book consists of six parts which broadly follow the well-known Driving forces, Pressures, State, Impacts and Responses (DPSIR) framework (Smeets and Weterings 1999) for describing human-environment relationships. Part I introduces the topics of landscape planning and ecosystem services and the objective of this book (von Haaren et al., Chaps. 1 and 2) and sets out key objectives, theories and methodological considerations (von Haaren et al., Chap. 3). The sets of values, including legal, economic and social factors, that underpin landscape planning are discussed (von Haaren et al., Chap. 4), while Lovett and Sünnenberg (Chap. 5) review available spatial data sources and Kempa and Lovett (Chap. 6) provide an introduction to the use of Geographic Information Systems (GIS) technology.

Part II explores the drivers and pressures shaping landscape change in Europe. It begins with an overview of key pressures (Kienast et al., Chap. 7), provides insights into EU policies as key drivers of ecosystem services protection or enhancement (Schleyer et al., Chap. 8), and introduces methods for the practical assessment of pressures in landscape planning exercises (Albert et al., Chap. 9).

Methods for assessing current states or impacts for different components of biodiversity and ecosystem services are reviewed in Part III. Bug et al. (Chap. 10) focus on food, materials and energy. Cooper and Hiscock (Chap. 11) discuss catchment water resources. Palmas et al. (Chap. 12) contribute a chapter on methods for assessing renewable energy production capacities and goods. Regional climate regulation capacities are introduced by Klug and Reichel (Chap. 13). Greenhouse gas storage and sequestration is addressed by Thomas and Schulp (Chap. 14). Ribeiro and colleagues (Chap. 15) review available methods for assessing aesthetic landscape capacities. Geodiversity is discussed by Turner (Chap. 16), while von Haaren et al. (Chap. 17) examine methods for assessing habitat development potentials. Rüter and Opdam (Chap. 18) consider methods for assessing habitat capacities. The issue of multifunctionality evaluation is portrayed by Andersen et al. (Chap. 19). Finally, the issues and methods associated with economic valuation are examined by Lovett (Chap. 20).

Part IV of the book concerns approaches for deriving response measures. General considerations and guidelines for developing such measures are outlined by Albert et al. (Chap. 21). Subsequently, Cebrián-Piqueras (Chap. 22) outlines measures for protecting soil-related ecosystem services, and mitigation measures for water pollution and flooding are reviewed by Cooper et al. (Chap. 23). The various measures available for safeguarding and enhancing landscape aesthetics are described in Albert et al. (Chap. 24). The final chapters in this part of the book address measures for biodiversity (Lange-Kabitz et al., Chap. 25), for multifunctional landscape development (Galler et al., Chap. 26), and for using Leitbilder and scenarios (Albert et al., Chap. 27).

Issues of communication and participation are central to many aspects of landscape planning. Part V addresses these topics through a contribution by Krätzig et al. (Chap. 28) on techniques for participatory approaches and a chapter by Warren-Kretzschmar and von Haaren (Chap. 29) regarding the role of design.

The final chapters in Part VI provide some wider context. Shandas et al. (Chap. 30) introduce perspectives on landscape planning from outside of Europe. To conclude, Albert et al. (Chap. 31) provide a synthesis and outline some potential future prospects for landscape planning.

Taken together, the chapters in the book aim to answer a series of fundamental questions that arise in landscape planning. Additional supporting material can be found on a dedicated website at https://www.umwelt.uni-hannover.de/LPwithES. Tables 2.1 and 2.2 provide a summary of the questions addressed in each part of the book.

Table 2.1 Key questions addressed in Parts I–III of the book

Book part	Questions addressed
Part I: Landscape planning with ecosystem services	Overall question How can landscape planning with ecosystem services be understood and conceptualized? Subsidiary questions What is landscape planning? What are ecosystem services and how can planning incorporate them? How is the assessment process organized? What are the values that underpin evaluation of options? What types of spatial data are available and how can a GIS be used to process them?
Part II: Sources of drivers and pressures	Overall question What are the key driving forces and pressures for change that landscape planning should address? Subsidiary questions What are the main pressures on landscapes in Europe? How have EU policies and standards shaped ecosystem services provision? How can pressures be assessed in landscape planning?
Part III: Methods for assessing state and impacts	Overall question Which methods can be used to assess current states or impacts for different components of biodiversity and ecosystem services Subsidiary questions How can the status of provisioning, regulating and cultural services be assessed? What is geodiversity and how can it be evaluated? What are the methods for evaluating habitat development potentials and capacities? How can multifunctionality be assessed? What are the methods and challenges associated with the economic valuation of ecosystem services?

2.3 How Might This Book be Used?

We envisage that the main audience for this book will be students of landscape and environmental planning, together with practitioners looking for an appropriate method to solve a practical problem. To provide a perspective on the organization and content of this book for these prospective readers we therefore offer the following (fictitious) example concerning a young landscape planner, Lena, who has recently started work in a consultancy office and has been tasked with helping to develop a landscape plan for a nearby municipality. When she picks up this book, in what respects can it help and where can relevant insights be found?

What is the nature of the task? Lena looks at the European Landscape Convention and national legislation and finds that the landscape has important cultural, ecological, environmental and social dimensions, as well as constituting a resource underpinning economic activities and public interests. Thus, landscape planning is a means to protect, manage and improve landscapes for these beneficial purposes.

Table 2.2 Key questions addressed in Parts IV–VI of the book

Book part	Questions addressed
Part IV: Deriving response measures	Overall question Which response options and measures are available to safeguard and enhance biodiversity and ecosystem services delivery? Subsidiary questions What general principles need to be considered in developing response measures? Which measures are appropriate for provisioning, regulatory and cultural services? What are the options for supporting and enhancing biodiversity? How can multifunctional landscape development be planned and assessed? What methods are available for scoping landscape futures?
Part V: Communication in landscape planning	Overall question How can landscape planning outcomes be appropriately communicated? Subsidiary questions What methods and tools are available to facilitate public participation in landscape planning processes? How can we best integrate design elements and approaches into landscape planning?
Part VI: Global context and conclusions	Overall question How does landscape planning in Europe compare with other developed economies and what are the prospects for the future? Subsidiary questions How transferrable are the European experience and the approaches discussed in this book to other developed economies? What might landscape planning in Europe be like in 2030?

This sounds good, but is very general. For more information, she picks up this book and looks at the conceptual framework in Chap. 3 that gives a more detailed description of the purpose and role of landscape planning. However, the public nature of landscape planning raises several additional questions: what are the public interests, are they already defined or do I need to identify them in each case? Furthermore, what role do all the individual interests and preferences of the people in my municipality play? Answers to these types of questions are discussed in Chap. 4. The framework in Chaps. 3 and 4 will also help Lena to understand that a municipal landscape plan is not a stand-alone solution. There may be plans or policies on regional or even national level, which address transboundary problems that cannot be tackled only by her municipality. She also appreciates that environmental impacts, which harm nobody except the person who causes them, are of no interest for landscape planning. If somebody wants to trim the bushes in their garden into the shape of Mickey Mouse, so what? Nobody else will be bothered and we do not judge by our own taste. Only if the garden owners will harm the public interest, for instance by extracting too much ground water for watering their lawn and plants, will the provisions of the landscape plan start to become relevant. Thus, this book does not support the design of individually-used green space. However, the methodologies discussed may complement design activities or vice versa (see Chap. 29). If Lena is asked to undertake a design task for an individual or a small group of people she should better ask her landscape design colleague next door or consult the book by van den Brink et al. (2016) to better understand what landscape architecture is about.

Now the real work can begin! Lena needs to organize the landscape planning process and find data to support the necessary assessments. She has learnt in Chap. 3 about the different steps and feedback loops in planning and that public participation has to start very early. But how to do it? Chap. 28 will help her to identify the most appropriate approach. What about the data? Will the data that she finds on the internet about soil, geology, water resources, flora, fauna, habitats and climate be suitable for the assessment tasks? Can citizens contribute to the evidence base? Which technologies should she use for processing the data and how could using GIS help her? She turns to Chaps. 5 and 6 and is relieved to see that even in European countries with traditionally limited national data on environmental issues there are now pan-European databases, which she can use as a starting point and process in her GIS.

At the initial meeting with the municipality Lena had also learnt about the goals, which the local policy makers and council staff wanted the landscape plan to address. These included answers to such questions as: What are the most urgent environmental problems to solve? How can we increase tourism and become a more attractive place for residents? How might the municipality best achieve a transition to renewable energy sources by 2050? Where is the local landscape especially beautiful, valued by local people, or providing valuable ecosystem services that should be protected or enhanced? Can Lena respond to these challenges and identify methods that will enable her to assess the current delivery of ecosystem services, as well as assessing the impact of existing or proposed changes in the landscape? She is pleased to find in the central part of this book (Chaps. 10, 11, 12, 13, 14, 15, 16, 17, 18, 19 and 20) a variety of approaches and simple techniques that can be employed, despite some imperfect databases, to undertake an uncertain, but probably sufficient, assessment to support local decisions about land use change. Furthermore, she also finds information about more sophisticated methods and models that could be used when the required input data are available or if very sensitive problems have to be solved with a high degree of confidence in the results.

Now she can identify the areas that have a high priority for protection, for instance against urban development, perhaps because they deliver highly valuable provisioning services, are favoured and heavily used for recreational activities, or are multifunctional and serve a variety of purposes. In some cases, it will also be possible to quantify the financial impact of a proposed change in land use compared to the status quo or other possible alternatives. After many meetings with citizens, land managers, other planners and local policy makers Lena goes on to create a landscape and environmental information system for the municipality, which not only contains maps of the current state and present pressures on the landscape, but is also enriched with local knowledge, accepted in terms of the reliability of the findings and, perhaps, in terms of the conclusions to be drawn.

Proposing measures and initiatives which can contribute to management or respond to problems is now much more straightforward. Again, she turns to Chap. 28, but also to Chap. 29, in order to learn how she can include the public in a process to derive measures that have community support and where it might be helpful to involve her colleague from landscape design. She can also draw upon a large list of possible response measures depending on the diagnosis regarding the state of the landscape (Chaps. 21, 22, 23, 24, and 25). Furthermore, she is able to identify situations in which multifunctional measures could be more efficient in terms of using the scarce financial and land resources (Chap. 26). Possible future development paths for the municipality will be shown in scenario form (Chap. 27) in order to support decision-making regarding preferred pathways. Lena will also need to talk to local audiences and colleagues from other professional fields in order to determine the best instruments to achieve the different objectives and measures. This step in landscape planning is crucial for gaining political support and providing the necessary financial or legal means for implementation. In vain, she looks for that final support in this book! It is not covered, not because it is unimportant, but because so much can depend on the local political or economic context and the authors thought they should focus this book on methodologies. Lena will therefore need to apply her own local knowledge and propose implementation options in her final presentation of the plan to the municipality. She is also proud to report, that some citizens – enthused by their involvement in the participation process – have already started a small, self-organized, project to restore a pond in their neighbourhood to a more natural state. In addition, the different agencies responsible for agriculture, nature conservation and water protection have started to work together on multifunctional agri-environmental measures in order to spend the available budget from EU funds more effectively. The council accepts the landscape plan and agrees that many of the proposals should be integrated into the local development plan. In addition, the regional planning board wants to adopt some ideas more widely and also utilise the example of the excellent GIS database, which Lena created according to European data documentation standards (Chap. 6).

Lena is now curious whether the approaches she has employed could work in other parts of the world. The examples in Chap. 30 demonstrate that there are many opportunities for landscape planning and her skills and knowledge could be widely employed. However, a typology of different planning systems demonstrates that the legal and political framing conditions for planning and environmental assessment are rather varied and consequently it would be a matter of adapting her expertise to the local context.

References

Smeets, E., & Weterings, R. (1999). *Environmental indicators: Typology and overview* (Technical report No 25. EEA). Copenhagen.

van den Brink, A., Bruns, D., Tobi, H., et al. (Eds.). (2016). *Research in landscape architecture – Methods and methodologies*. London: Routledge.

Theories and Methods for Ecosystem Services Assessment in Landscape Planning

3

Christina von Haaren, Andrew A. Lovett, and Christian Albert

Abstract

This chapter introduces the key theoretical and methodological concepts for landscape planning in Europe. A short portrait of landscape planning and its contribution to supporting sustainable landscape development provides insights into the capabilities of an integrative environmental planning tool that cuts across different sectors and levels of decision-making. The chapter then presents landscape planning procedures following the so-called DPSIR framework – Driving forces, Pressures, the State of the landscape, Impacts, and potential Response options. A subsequent discussion outlines how the concept of ecosystem services can be adapted to best integrate with the practice-oriented focus of landscape planning. Finally, the chapter provides some guidance on methodological aspects of landscape planning for ecosystem services, acknowledging the multiple types of values, scale issues, and the need for comparability of results, communication of uncertainties and transparency in the derivation of responses.

Keywords

Landscape planning · Ecosystem services · Methodology · DPSIR

The original version of this chapter was revised: The Fig. 3.3 has been updated now. The correction to this chapter is available at https://doi.org/10.1007/978-94-024-1681-7_32

C. von Haaren (✉) · C. Albert
Institute of Environmental Planning, Leibniz University Hannover, Hannover, Germany
e-mail: haaren@umwelt.uni-hannover.de; albert@umwelt.uni-hannover.de

A. A. Lovett
School of Environmental Sciences, University of East Anglia, Norwich, UK
e-mail: a.lovett@uea.ac.uk

C. von Haaren et al. (eds.), *Landscape Planning with Ecosystem Services*, Landscape Series 24, https://doi.org/10.1007/978-94-024-1681-7_3

3.1 Introduction

Biodiversity and ecosystem services are under pressure in Europe, as is particularly obvious at regional and local levels. Different land uses and conservation needs compete and there is a need to mitigate conflicts and to coordinate and optimize land use patterns in a sustainable way (cf. Chap. 7). Landscape planning can contribute to minimizing conflicts and delivering solutions if it is based on sound ecological data, a legitimized evaluation of the ES in the landscape and takes into account the preferences and knowledge of the local population. To be effective, landscape planning proposals need to be supported and implemented by decision makers, stakeholders and the public.

The choice of theories, concepts and methods to be applied in landscape planning is thus driven by consideration of the conditions for local and regional implementation of actions for the conservation and sustainable use of biodiversity and ecosystem services (for overview on planning theories see Hillier and Healey 2010; Allmendinger 2017; van den Brink et al. 2017). Landscape planning methods therefore need to provide results in a transparent and comparable way, and they need to provide assessments, valuation and proposals that integrate across the diverse and fragmented implementation contexts as reflected by various sectors and levels of decision-making (Leitão and Ahern 2002; Selman 2006; Albert et al. 2016a, b; BenDor et al. 2017). The aim of this chapter is to introduce key theoretical and methodological concepts with relevance for landscape planning in Europe. The chapter thus provides the theoretical background and describes the application context of all procedures and methods presented in the subsequent sections of this book.

3.2 Landscape Planning in a Nutshell

The definition of landscape planning applied in this book follows the European Landscape Convention (Council of Europe 2000: art. 1), characterizing it as "strong forward-looking action to enhance, restore and create landscapes" (see Chap. 1). Given this broad understanding, landscape planning arguably provides a proactive approach for bridging the fragmented efforts relating to the conservation and sustainable use of biodiversity and ecosystem services across different sectors and levels of decision making (cf. Selman 2010). In most European countries, there is a form of planning system comprising spatial, urban development and conservation planning activities that oversees, for example, the implementation of the Water Framework Directive or the Habitats Directive. Landscape planning can contribute to this process by either supplying a multifunctional, environmental perspective or by using the information available to provide an integrated multifunctional concept of landscape development. For this purpose, landscape planning must generate *a comprehensive, spatially-explicit information base* that supports the precautionary consideration and integration of biodiversity and ecosystem services into land use decision processes and fosters efficient implementation. The potential users of information generated by landscape planning are policy makers, stakeholders and the public. For application and implementation, landscape planning needs to provide place-based outcomes in the form of maps – particularly on local and regional scales (cf. Ogrin 1994; Gruehn and Kenneweg 1998; Reinke 2002; Nassauer and Opdam 2008).

Landscape planning plays an important role in combining *proactive* and *reactive* instruments with the overall objective of mainstreaming the consideration of biodiversity and ecosystem services in all spatially relevant decisions by public authorities or private project investors (Fig. 3.1). *Proactive* planning supports the implementation of conservation efforts by area protection and maintenance e.g. by adoption of agri-environmental measures (AEM) as well as restoration of impaired landscapes. Furthermore, it supplies an information base with data, evaluations and objectives relating to ecosystem services, which can support *reactive* instruments. *Reactive* planning is triggered by programme activities or projects and seeks to adapt resulting land use changes to the principles of environmentally-friendly development e.g. through the screening process in a Strategic Environmental Assessment (SEA) or Environmental Impact Assessment (EIA).

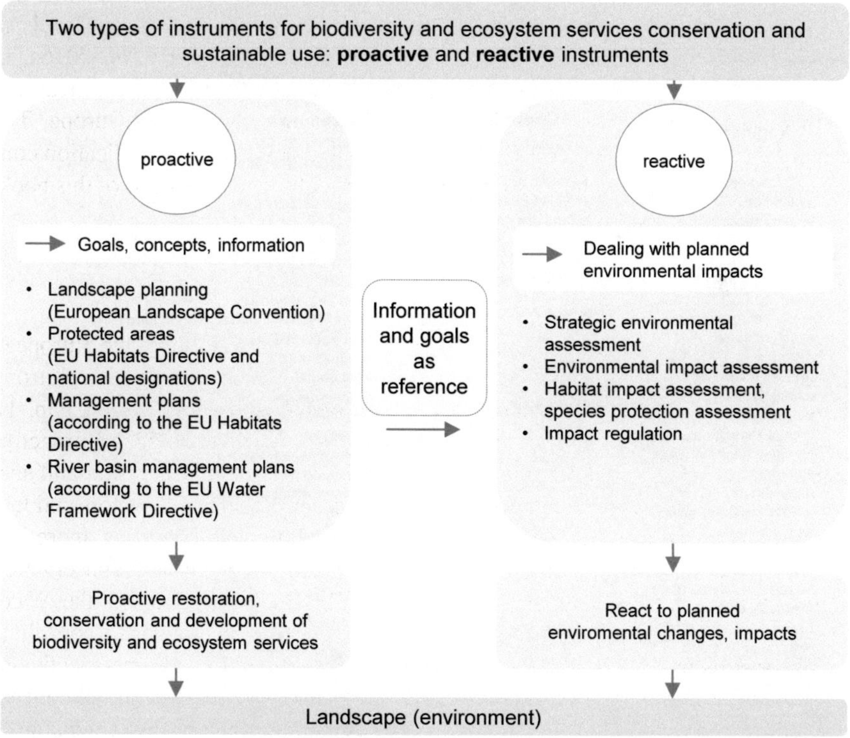

Fig. 3.1 Two types of instrument for considering ecosystem services in spatial decisions. Landscape planning is important for proactively pursuing environmental goals and as an information and evaluation basis for instruments which respond to planned interventions such as environmental impact assessments and offset mechanisms

Both types of instrument can use the methods presented in this book concerning the assessment of biodiversity and ecosystem services and for deriving appropriate response measures. The multifunctional scope of landscape planning is especially broad with regard to taking into account all ecosystem services that are relevant as public resources and in striving for multifunctional measures where efficient (Termorshuizen et al. 2007; Galler et al. 2015). Landscape planning includes: (i) identifying synergies and conflicts between different ES as well as with land uses; (ii) proposing needs for change and possible solutions; (iii) and considering the preferences and needs of those impacted by decisions. Thus, landscape planning supports political and regulatory decisions, public participation and social learning as well as the valorisation of ES in commercial markets (Fig. 3.2). Cooperation of the different sector administrations is fostered by identifying synergetic interests and multifunctional measures (Chap. 19), which is important in terms of efficiently spending public money.

As a consequence of landscape planning's orientation towards decision support, the *spatial extent and delineation of the planning areas* is identical to the areas of jurisdiction on the different administrative levels (Albert et al. 2017). This implies

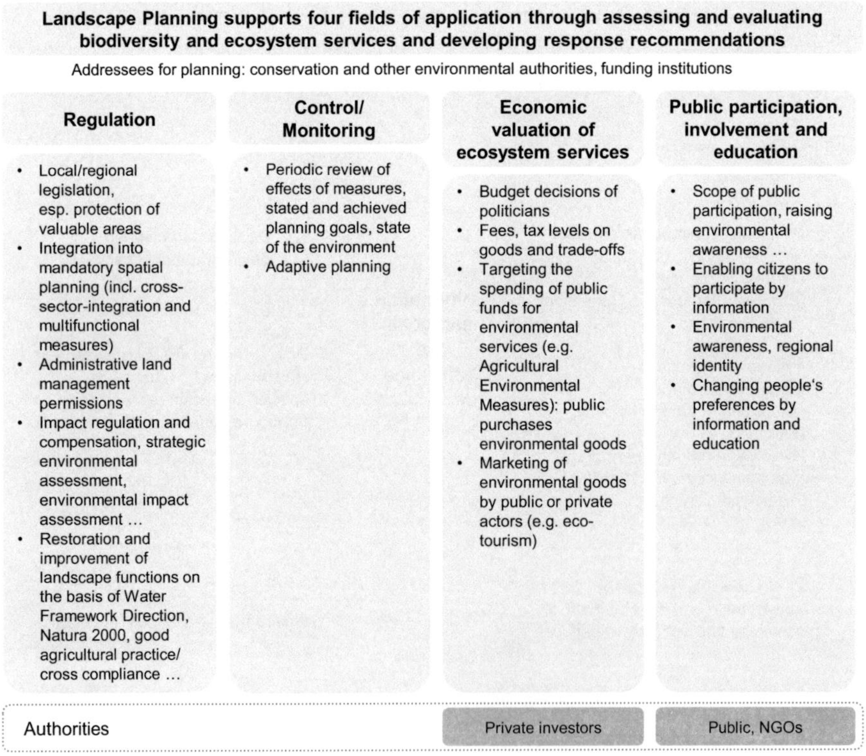

Fig. 3.2 Practical applications of landscape planning. The environmental information system, the objectives and management measures can be used by different stakeholders and in diverse contexts

that those aspects which are relevant (e.g. for a whole river catchment) should be addressed on a political decision level high enough to regulate upstream as well as downstream effects and actors (Fig. 3.3). Additionally, scarce natural resources (e.g. water provision, rareness of species) have to be assessed and considered on every decision level in order to prevent the destruction of resources by the tyranny of the small decisions (Odum 1982).

Planning at regional and local levels should consider and adhere to the framework conditions and objectives passed down from higher political (and planning) levels. Examples of such supra-local objectives are habitats or species protected in the European network of Natura 2000 sites, or the objectives laid down in plans to implement the Water Framework Directive. These supra-local objectives may not be open for local discussions or amendment, which is especially important to note during participation processes. Regional and local landscape planners should, in turn, highlight the issues for which they are responsible and be accountable for the implications of their decisions. Typical examples of such issues are spatial frameworks for urban development and zoning, regionally endangered species, local recreation amenities, and measures to operationalise higher level objectives (cf. Albert et al. 2017).

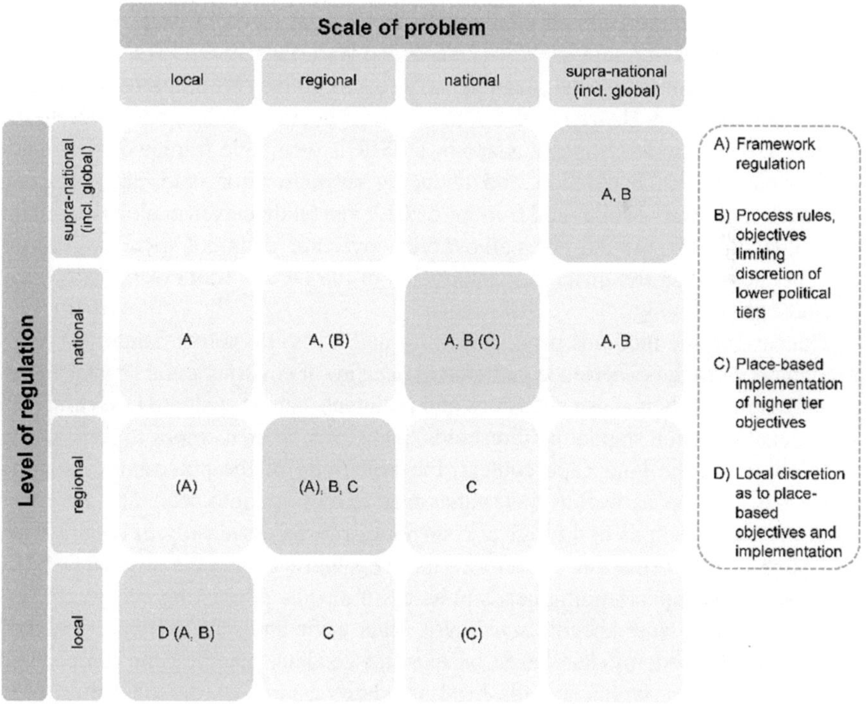

Fig. 3.3 Defining the decision space of landscape planning. Tasks on different planning tiers are determined by the scale of the problem and associated responsibilities. Projects with cross-boundary impacts or trans-boundary ecosystems (such as river catchments) need to be considered at higher planning tiers with authority that covers the whole relevant area. (von Haaren 2016: 171, amended)

A consequence of adopting a proactive approach and of matching the territories of political administrations is that landscape planning has blanket coverage and includes all types of landscapes whether obviously at risk or not, a feature which is specifically highlighted by the European Landscape Convention. This broad scope enables landscape planning to function as an environmental 'health check' for municipalities and regions.

3.3 DPSIR: A Framework for Assessment and Identification of Responses in Landscape Planning

The methods adopted in a particular landscape planning exercise should be selected and designed according to the purpose, possible responses and resources for implementation. A suitable framework which reflects this implementation-driven approach for determining the content of a plan is the widely used Driving forces, Pressures, State, Impacts and Responses (DPSIR) model (originally proposed by Smeets and Weterings (1999) in a report to the EEA, published 1997) (Fig. 3.4).

DPSIR represents a framework for studying casual relationships between socio-economic activities and the environment (Tscherning et al. 2012). Environmental indicators are required for all elements of this causal chain in order to meet the information needs of policy makers (Smeets and Weterings 1999). A range of different frameworks for landscape planning exist (e.g. Steinitz 1993; Steiner 2000; Kato and Ahern 2008; von Haaren et al. 2008) but all relate, more or less obviously, to the general DPSIR model. Slightly adapted, DPSIR is a suitable framework for landscape analysis, ES evaluation and deducing responses for landscape planning (Schößer et al. 2010; Müller and Burkhard 2012; van Oudenhoven et al. 2012; Albert et al. 2016a, b). Figure 3.5 gives an overview over the methodological approaches used for describing the different components of the DPSIR framework, as applied in landscape planning.

Concerning the methods used to identify and assess *pressures,* landscape planners can refer to the experience gathered in decades of environmental impact analyses. Pressures such as noise emissions and pollutants can be evaluated as a first step using legal emission standards (thresholds). However, when it comes to considering their impact in the landscape context, the sensitivity of the potentially impaired ecosystem services as well as their value need to be taken into account. Less regulated pressures such as hydrological changes can be assessed only in combination with such *state* information. Therefore – more explicitly than in the original DPSIR-concept – landscape planning needs to assess the *value* of existing ecosystem services and the pressure-specific *sensitivity*. State value and sensitivity are analysed by (indicator-based) models based on existing geodata, mapping the terrain, and evaluation models, which include legal standards as well as default values (e.g. federal/regional averages). Due to this approach and differing slightly from the original DPSIR-model, in landscape planning *impact* is conceptualized as part of *state,* which may include impairments from past activities. These are identified by the presence of pressures and a landscape state which contradicts societal objectives

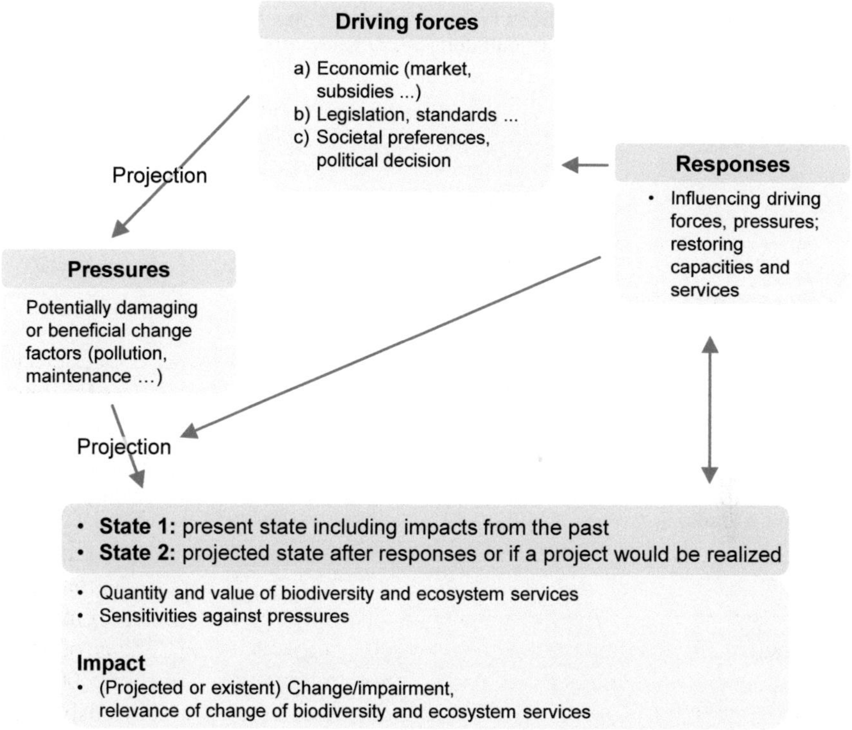

Fig. 3.4 Concept for modelling and assessing the state of ES, the need for action and possible responses in landscape planning. (Based on EEA 2011, adapted for landscape planning, cf. Albert et al. 2016a, b)

and thresholds for ecosystem services conservation. Reversible harmful impacts can be handled as triggers for rehabilitation. For example, the rapid eutrophication of a lake is highlighted by an abundance of algae which results in a low rating of the state of the lake. This should trigger the search for potential polluters (pressures) und suitable responses. Standardized impact analyses also offer the possibility to change the input data relating to pressure and thus generate state-scenarios about the impacts of different land use options.

The DPSIR concept involves deducing *responses* or implementation measures from knowledge about D, P, S, I and to use these insights as starting points to improve the delivery of ecosystem services. Possible responses can be found in Part IV of this book. For example, such recommendations may include changing local taxes (drivers), reducing commuting or private car use (pressures) or building amphibian tunnels which limit animal loss (state and impact). Methods include drawing from an information base about measures and their effect on preserving, maintaining, rehabilitating or developing biodiversity and ecosystem services. Assessing the effect of multifunctional measures and their optimized allocation is an important aspect of generating space-efficient and cost-saving planning

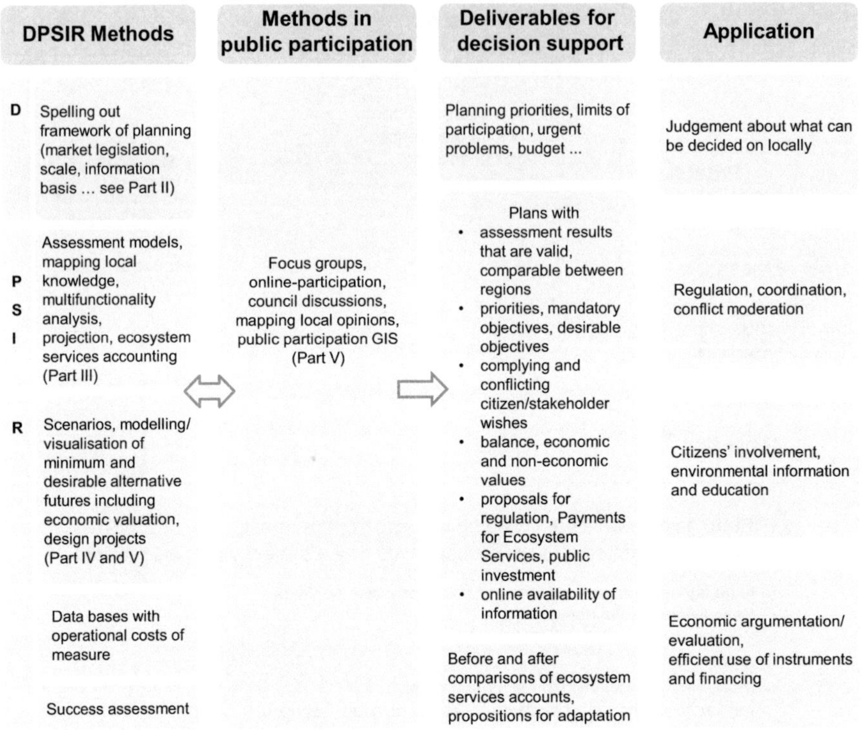

Fig. 3.5 Methods and the resulting deliverables for decision support and implementation. Parts II–V refer to sections of this book

solutions. Prioritizing objectives and measures draws on both the evaluation of the ecosystem services and the urgency of action due to projected impacts. Finally, the evaluation of the success of response measures may be reflected in a change of state (from condition 1 to condition 2, etc.).

Participation of stakeholders and the public should be part of the entire planning process and across all steps of the DPSIR model. Suitable participation methods must promote the elicitation and integration of local knowledge as well as active involvement within the assessment and planning process. Methods for facilitating participation include face-to-face events (e.g. town hall meetings) and online consultation through tools such as interactive maps and citizen mapping. Different techniques of visualizing scenarios or alternative futures support communication and a common understanding of the planning proposals (e.g. Albert et al. 2012; Steinitz 2012). In addition, desired alternative response options can be combined with design approaches and thus may be part of bottom-up participation (von Haaren et al. 2014b). More detail on participation techniques can be found in Part V of this book.

3.4 The Process of Landscape Planning and the Role of Participation

Landscape planners not only produce the content of a landscape plan but also organize and facilitate participation and decision processes. In principle, the different planning processes involved can be structured along the components of the DPSIR framework, accompanied by many feed-back loops and systematic public participation throughout the entire process (Fig. 3.6).

3.4.1 Scoping

The first phase of proactive environmental planning is characterised by a scoping process. City or regional officials, stakeholders and planners come together to identify urgent problems in the area, goals for future development of the region and the possible contribution of landscape planning, as well as drivers from higher policy levels. Such drivers cannot be changed in local landscape planning but may be addressed in strategy building for implementation or for defining the limits of

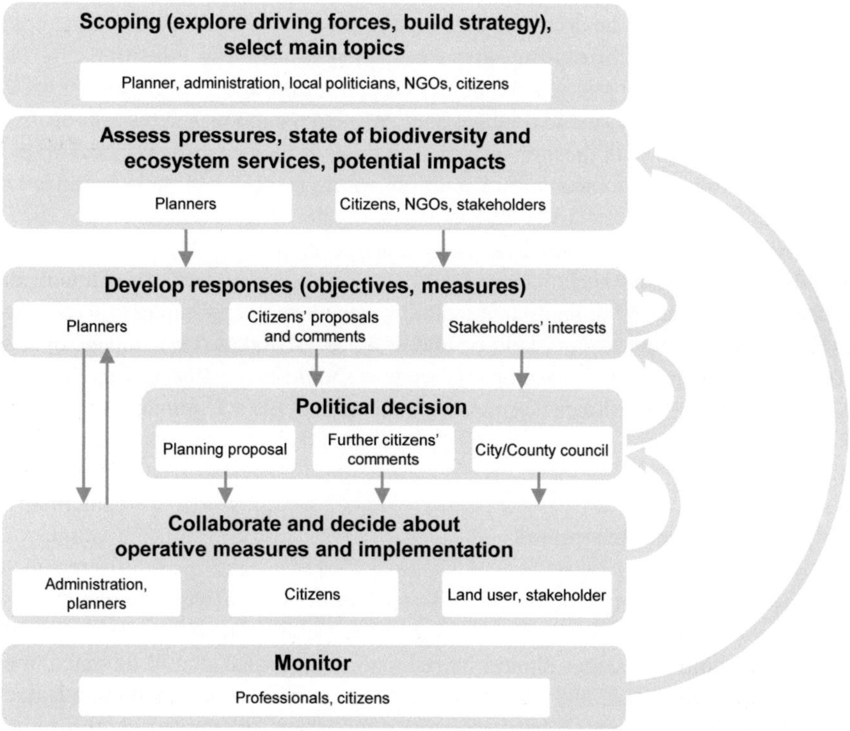

Fig. 3.6 The landscape planning process includes many feedback loops. The planner has to organize and facilitate this process. The whole process is accompanied by public participation

participation. Drivers include governmental regulations and standards which are both a basis of assessment and a driving force for changing pressures, for example if law enforcement is activated. Other drivers are market forces, such as product prices, which will influence the actions of land users. Financial incentives e.g. through EU programmes or purchaser preferences also fall in this category (see Part II of this book). In addition, the national planning system will be relevant. It defines the content of landscape and other land use planning and whether an integrated approach, covering all ecosystem services instead of only biodiversity or landscape aesthetics, can be pursued.

In this early phase of the planning process, some implementation activities should be started to motivate citizens to participate and to maintain the impetus during the planning process. The best way to do this is to initiate small projects which will show quick results. The restoration of a creek to a more natural state is an example which gives landowners and citizens the opportunity to discuss and decide about locations and design.

3.4.2 Assessment

The next phase of the planning process is the inventory and evaluation of the state and prospects for the landscape, biodiversity, and ecosystem services. This comprises an assessment of existing and foreseen pressures and their impacts. The inclusion of the public and stakeholders is crucial for the acceptability of the whole plan. The objective is to avoid doubts about correctness and bias in the approach, to acquaint the public with the new information base, to include as much local knowledge as possible, and to account for multiple values (as now also acknowledged in major assessments, cf. TEEB 2010; Maes et al. 2012; Pascual et al. 2017). Landowners and farmers can be a particularly sensitive stakeholder group. It is mandatory that those stakeholders get very area-specific information and the opportunity to comment on the landscape planning inventory, for example, as to the designation of their land as grassland or arable fields. Mistakes can result in legal or financial consequences, for example if an area should be legally protected, or if a cross-check with the direct payment system of the EU's Common Agricultural Policy is performed.

3.4.3 Develop Responses

The inventory and evaluation of the landscape and ES are the basis for the response measures proposed to decision makers and the public. These measures indicate where and which pressures should be reduced, which sites should be maintained and possibly protected, and which impacted areas should be rehabilitated. Each response should also have a level of priority for action. Prioritising responses and arguing about the basic needs for protection or rehabilitation, must draw on a sound inventory and evaluation of the present and projected states of the ES. The response objectives and measures should be framed and presented according to the needs of different interested parties and possible means of implementation. For example, spatial planners will adopt propositions better if the ES objectives have been translated into the planning categories of the regional plan. Citizens are likely to welcome 3D visualisations portraying the visual consequences of a neighbourhood development or renewable energy developments in the landscape; the nature conservation authority needs information about habitat and species rareness combined with a proposal for protection priorities or recommendations on where to allocate incentives for landscape maintenance. Again, participation is crucial in this phase.

3.4.4 Implementation

Implementation can be initiated by a political decision of the regional or municipal council. An operational plan will include timelines, financing and priorities. Authorities can use landscape planning as basis for quick decision making about activities and projects with possible impact on the environment. For farmers, landscape planning outcomes can provide a basis for locating agri-environmental measures on their farm. Private sector developers or investors may draw on mitigation measures proposed in the landscape plan and demonstrate the success of their investment in nature to the public. Finally, environmental agencies can update the digital data base to include recent changes. This process can be regarded as ongoing *adaptive planning*, in which measures are altered according to landscape changes, unforeseen conditions or the outcome of evaluations.

A similar approach to that sketched out here is the framework proposed by Steinitz (1990) that structures landscape planning along key questions to be answered in each phase of the work. This framework also follows the steps of inventory, evaluation, prognosis and determination of advice. It is influenced by a design approach and particularly emphasises feedback loops which are necessary to refine the study question, choose appropriate methods, and finally implement the study. The process flow may go back to any previous phase if evidence in the current phase indicates the need for corrections or modifications. Feedback from stakeholders and officials plays a particularly important role in this iterative process. Such enhanced flexibility is particularly important if the planning process must be performed quickly and with a limited supporting evidence base.

3.5 Incorporating Ecosystem Services Concepts Into Landscape Planning

As outlined in Chap. 1, the concept of 'ecosystem services' is defined in various ways in the literature. This book draws upon many of the existing definitions and concepts, but adapts them to the specific requirements of landscape planning implementation (de Groot et al. 2010; von Haaren et al. 2014a; Spangenberg et al. 2014; Albert et al. 2016a, b).

Landscape planning concentrates on the elements and processes of an ecosystem which are relevant for human needs. Thus, ecosystem services in landscape planning represent a selection of the properties of the real world, driven by our abilities to understand and survey, and by our preferences and needs. This approach is different from basic ecological science, which strives to understand the processes and structure of ecosystems. Given this specific perspective, the understanding of ecosystem services applied in this book (Fig. 3.7) includes both the currently delivered but unused provisions by nature (final ES in UK NEA 2011) as well as ecosystem services which are actually utilized (termed goods by UK NEA 2011). The delivered ecosystem services represent the totality of ecosystem contributions that may provide benefits to humans today or in the future, but need not necessarily be used today. In other studies, these types of service are referred to as capacities or functions (Haines-Young and Potschin 2010; Potschin and Haines-Young 2016; cf. TEEB 2010). However, these terms seem to be more difficult to communicate to politicians as they are more abstract and refer to a 'potential' rather than an existing and already valuable resource. The provision of delivered services is dependent on appropriate underlying ecosystem elements (hereafter termed *natural capital*), including processes and structures as well as geo-and biodiversity. The utilized ecosystem services are those that are actually turned into goods or directly consumed by humans. This transformation often requires human input (UK NEA 2011), with examples being fertilizer, energy, pesticide, labour, infrastructure or knowledge (cf. Burkhard et al. 2014). The resulting benefits are impacts on actual human well-being, individual or collective, stemming from the direct or indirect contributions of delivered and/or utilized ES. Examples for the different categories included in Fig. 3.7 are as follows:

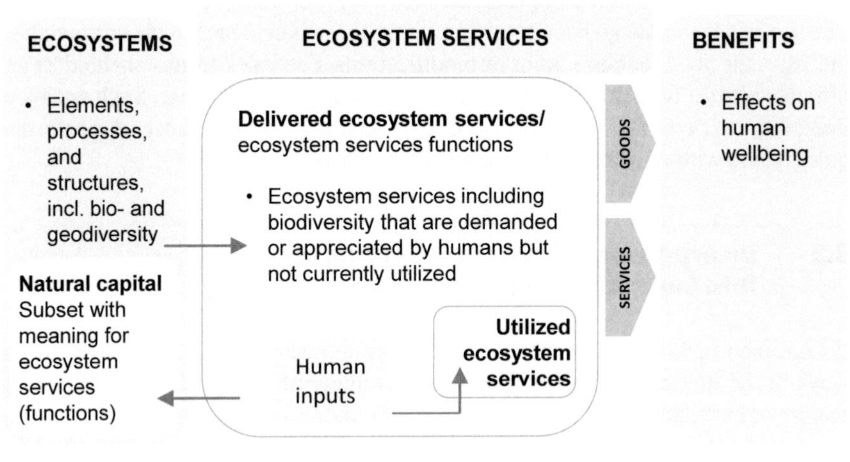

Fig. 3.7 Proposed ecosystem service concepts and terminology for landscape planning

- *Ecosystem Elements and Processes (also termed natural capital here, including ecosystem assets)* – primary production, water cycling, nutrient cycling, soil formation, weathering, ecological interaction, evolutionary processes
- *Delivered Ecosystem Services* – production capacity for food, renewable energies, pollination, water retention, clean water supply, GHG sequestration
- *Utilized Ecosystem Services* – food, drinking water, energy supply, flood control, air pollution mitigation, climate regulation, recreation amenities
- *Benefits* – health, good nutritional status, security, education, enjoyment, happiness

Protection of delivered ecosystem services is governed primarily through objectives and standards as described in legislation (representing shared societal values) and then interpreted and made more specific by planners (Fig. 3.8). This legal basis is essential for applications in planning and decision-processes to ensure the legitimacy of objectives classified as mandatory, their transparency and a fair balancing of public and private/individual interests. In contrast, utilized ecosystem services tend to be assessed from an individual perspective and are represented by other economic measures (e.g. crop yields or sale values) or preferences which can be captured through socio-economic valuation methods. These different forms of evaluation are further discussed in Chap. 4.

Analysis of both delivered and utilized ecosystem services allows for presentation of different and complementary perspectives to inform planning and decision-making processes, enabling consideration of the public, legal perspective alongside the economic and individual perspective. Evaluation is therefore based on a range

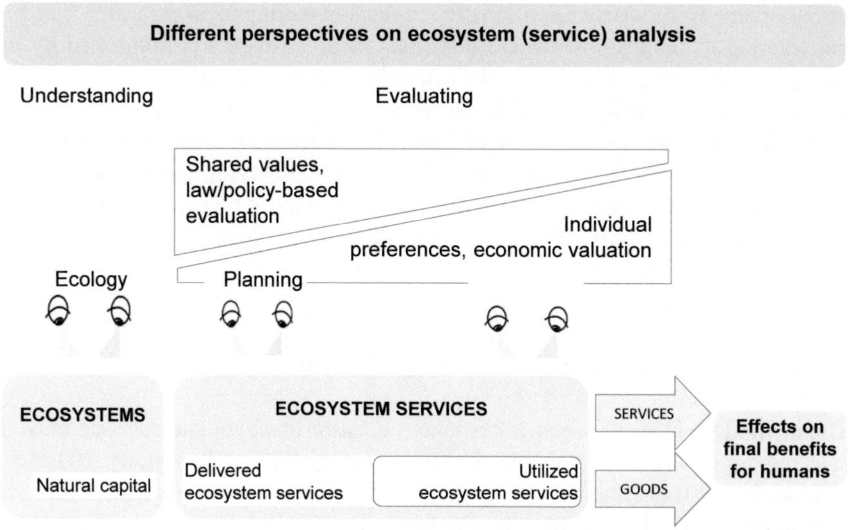

Fig. 3.8 Different values underpinning the assessment of ecosystem services. (cf. von Haaren et al. 2014a)

of values. This helps to reduce the risk of the economic valuation (and thus of commodification of nature) becoming the priority, which is feared by many scientists and practitioners (Albert et al. 2014; Schröter et al. 2014). Valuable delivered ecosystem services should be protected even if the benefits only accrue to future generations. However, including both the individual (economic) perspective is valuable for the participation process, as well as for deciding when market instruments are the right choice for policy responses.

3.6 Methodological Issues in Landscape Planning

Based on our experience there are a number of issues that need attention in almost any landscape planning exercise. These include transparency in the methods adopted and the normative judgments made, ensuring comparability of assessment results, considering the applicability of methods at different spatial scales, communicating uncertainty in findings and justifying choices of response measures (cf. von Haaren and Albert 2011; Selman 2006; von Haaren et al. 2008). These requirements are elaborated on below and can be considered as the checklist for landscape planning exercises.

3.6.1 Distinguishing Scientific and Normative Components

Planning and decision support methods almost always consist of both scientific and normative components. These two components need to be distinguished from each other in order to give policy makers and citizens the opportunity to understand and discuss them, particularly the normative aspects of setting local priorities. The initial framing of both problems and questions to be answered is influenced by the normative basis of a society, as is the selection of ecosystem aspects to be mapped and assessed. The methods used for inventory compilation are invariably scientific, while the evaluation of outcomes and choice of responses is driven by normative standards. Actual implementation is mainly driven by scientific and practical knowledge. This mixture should be reflected in planning practice by clearly separating the inventory and evaluation phases, and by making any subjective planning decisions transparent within the methodological workflow.

3.6.2 Selecting and Implementing Methods

The methods to be used, whether bespoke (i.e. tailor-made) or standardised must be selected according to the intended application of the results (Merry 2011: 89; Fukuda-Parr 2014) (Fig. 3.9). In reality, a combination of both standardized and tailor-made methods will often be the best solution. For example, evaluating the visual quality of a landscape by first using a nationwide calibrated/normalized scale provides citizens with information about the value of different areas compared to

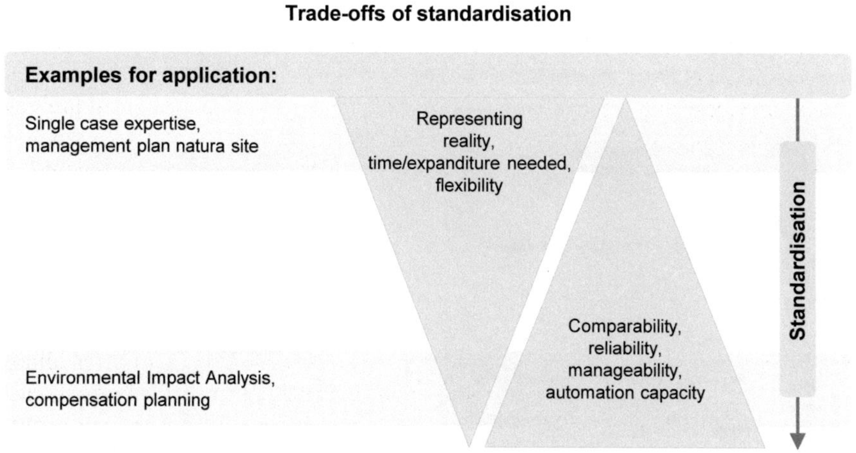

Fig. 3.9 Trade-offs between bespoke and standardised methods

the national mean. Such information could be relevant to assessing the potential to attract nature-based tourism. If national level data is then amended with more local information, e.g. from citizen surveys of preferred places or the application of an adapted, local preference scale, it can provide a valuable contribution to place-specific recreation planning.

Whether a method is bespoke or standardised it is important that the workflow of steps is thoroughly documented. Fig. 3.10 shows an example of the type of approach that should be followed. In this case the objective is sustainable use of groundwater and the first stage is to create an inventory of the existing state across a region. Since the groundwater recharge rate cannot be directly measured it needs to be modelled based on soil type, slope and precipitation. The results in terms of estimated recharge rate are then compared to standards in order to evaluate differences in state and determine priorities for action.

Bespoke evaluation methods, often including the elicitation of local preferences, allow for flexibility, adapting planning to local needs and including specific local parameters and indicators. In general, one advantage of a tailor-made approach over a standardised method is the higher accuracy of the results, especially with respect to quantification of ecosystem services. In addition, tailor-made methods allow for eliciting individual values or interests of local citizens and facilitate engagement in the participation process.

In contrast, standardised methods rely on consistent evaluation factors and their application follows a strict, pre-defined procedure and data format. One advantage of standardised methods is the comparability of the results across different regions and users. However, there is no 'one fits all' solution and the trade-off between flexibility and standardisation (Adams et al. 2016: 143) must be considered. Wherever possible, landscape planners should prefer standardised over bespoke methods because they allow for inter-area comparisons. Planners usually need to prioritise some areas in comparison with others – be that on regional, national or global scale.

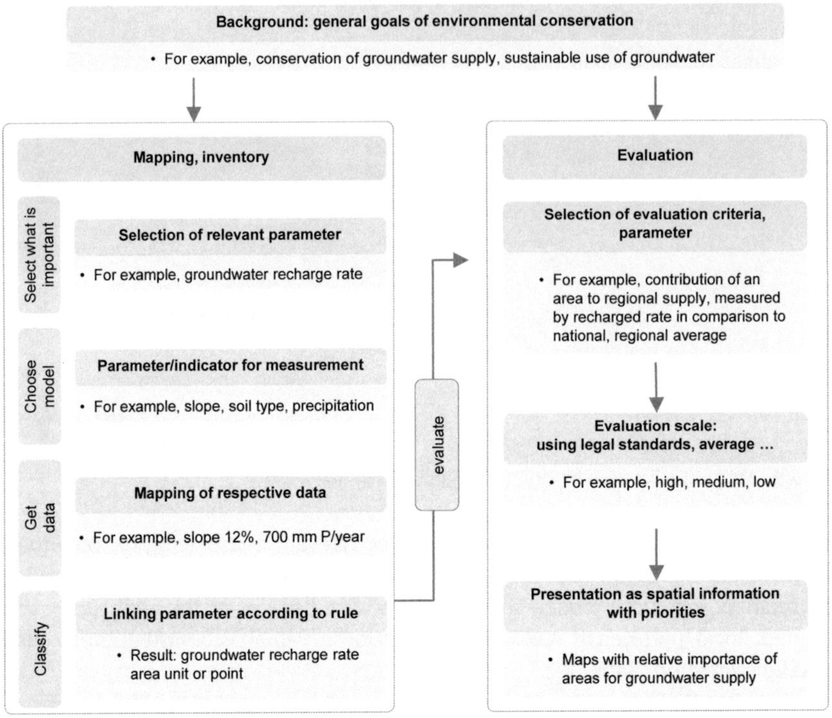

Fig. 3.10 Workflow example to assess the state of ecosystem services

In such decisions about priorities, standardised methods allow for comparison and any lower level of accuracy in the assessment of individual sites is often acceptable. In other words, achieving high quantitative accuracy in ecosystem services assessment is less relevant in landscape planning if prioritising areas or actions is the main purpose and all results have a similar level of accuracy. Similarly, if payments are connected to the quantitative outcomes, using less exact results may be inconsequential so long as every individual or organisation is treated equally in resulting implementation processes. The planner should also recognize if some implementation options require exact quantitative assessment outcomes. Examples would be whether there is an exceedance of a pollution threshold, or how much a polluter should pay if their land use related greenhouse gas (GHG) emissions are included in a greenhouse gas trading system. In these cases, calculations must be as exact as possible to treat polluters equally.

Standardisation and a detailed description of the methods also helps ensure that the results under the same condition will be repeatable and independent of who carries out the method. The results will not necessarily be objective in a strict sense (like the laws of physics), but they can be considered *neutral* as they (ideally) are independent of the specific preferences, biases or abilities of the person applying the method. Even evaluation standards and criteria for issues which are usually considered subjective, like landscape aesthetics, can be neutral in this sense. Nevertheless, in cases of forecasting, even if several individuals come to the same result, this may be fatally inaccurate. Therefore, in choosing the methods, it is important to strive for as much validity as the application purpose requires and communicate any uncertainties associated with the results.

3.6.3 Appreciating the Properties of Assessment Scales

Transforming the results of the inventory or evaluation to an assessment scale may require summation using quantitative or qualitative measurements. By measuring the properties of ecosystems, we summarise the vast complexity of nature and landscapes into classes, transforming them into statements that are meaningful for scientists, the public, or decision makers. Depending on the nature of the properties which we want to measure, and depending on the purpose intended, we can use four types of scales: nominal, ordinal, interval and ratio (Stevens 1946; Chrisman 1998) (Fig. 3.11).

Nominal scales are used when the categories of an inventory are of equal importance (without any order or hierarchy) or consist of only two classes (e.g. protected or not protected).

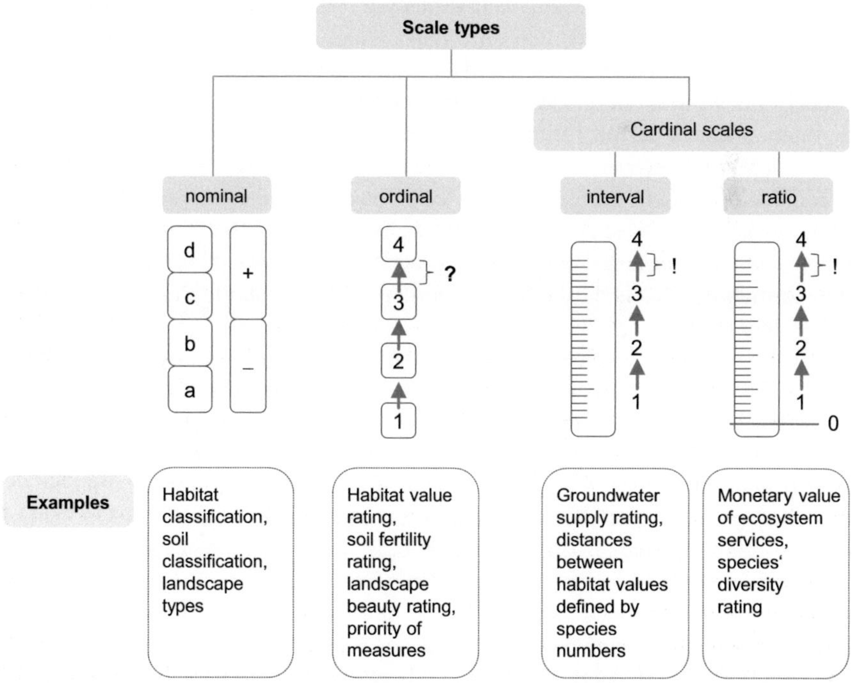

Fig. 3.11 Types of scales for ecosystem services assessments. (According to Stevens 1946; Chrisman 1998)

The *ordinal scale* implies an order or ranking amongst the different classes. Landscape planning frequently uses ordinal scales in evaluation, for example to assign values to habitat types from 'very rare' to 'very common'. However, it is important to note that on an ordinal scale the intervals (i.e. amounts of difference) between the classes are undefined. We cannot tell whether a habitat classed as 'very rare' is twice as important as one rated as 'rare'.

Interval and ratio scales are very similar and can be subsumed under the term *cardinal scale*. Both are characterised by the interval between points on a scale being known. The two types only differ in that the ratio scale has an absolute (natural) zero point. A classic interval scale would be for example the measurement of temperature in degrees Centigrade (°C). Here, the differences between degrees are defined and equal, but the zero point is arbitrary (i.e. 0 °C is a temperature and does not represent no heat). In contrast, a characteristic such as species richness can be measured on a ratio scale (no species at all being the zero point). Cardinal scales are used in planning if definite quantities are needed as an assessment outcome and the measurement system permits quantification (*compare Porter* 1994). An example is the calculation of the phosphorous loss from a sub-catchment into a river to assess its contribution to the total pollutant load of the water body.

The scale types allow different reclassifications and calculations to be performed. On nominal scales, it is possible to reclassify individual categories by summarising or grouping them into new classes (e.g. subsume habitat types into habitat groups). Ordinal scales do not allow any reclassification operations beyond the ones already possible for nominal scales. On an ordinal scale, each level stands for a relative quality or priority. Planners using ordinal scales therefore need to take care that the order of the scale is not disturbed by regrouping.

Interval scales permit linear transformations such as addition, subtraction and multiplication. However, the absence of a true zero means that ration calculations (i.e. division) are not meaningful (i.e. 20 °C is not twice as hot as 10 °C). A ratio scale does allow for such proportional transformations (see Chrisman 1998 for more about permissible statistics). In practice, awareness of these scale-specific transformation rules is particularly important when accounting or monetisation is the desired outcome.

The case of habitat value demonstrates how the barrier between ordinal and cardinal scales may be overcome in certain cases. Habitat value is often represented on an ordinal scale (e.g. low, medium, high). However, there may be a need to derive a numeric habitat value for a particular area. Examples include the calculation of the total habitat value for a farm with the aim of comparing it to other farms, or to a modelled prediction of the same area after improvement measures. One simple option, assuming spatial data are available, is to calculate the proportion of area occupied by different categories (e.g. the percentage of the farm rated as high habitat value). A more refined approach would be to match the ordinal categories with quantitative data (such as numbers of species present e.g. Bredemeier et al. 2015) and then use these to define the distances between the points on the scale. Another possibility is to set a standard where the ordinal levels are distributed evenly across the cardinal scale. If adopted, this approach must be agreed by the agencies or (political) authorities with responsibility for the application of the method and for the area where it will be used.

3.6.4 Considerations Regarding Spatial Scale

The realm in which a particular landscape planning process is carried out should be taken into account in two ways: in terms of the resolution of the available spatial data and the detail of the assessments required. In general, a method should only be applied at the spatial level for which it was originally developed. If a particular method is applied across a much larger region, the required quality of data may be difficult to find. Conversely, applications of a method in a smaller area than originally intended may generate results that are too generalised for robust planning purposes. However, as noted already, the best achievable detail is not always necessary. In some cases, the planning process may be overloaded with the amount of content or precision and, wherever possible, the amount of details should be adapted to the decision level. Also, as described earlier (in Sect. 3.2), many evaluation standards or objectives are adopted from higher decision levels. This is especially important if ecosystems such as rivers or national habitat networks cross the jurisdiction boundaries of planning authorities. In such cases, it can be difficult for local decision makers to judge the wider implications of their actions and there is again a case for adopting standardised evaluation methods to support consistency amongst relevant agencies.

3.6.5 Assessing and Communicating Uncertainties

The data bases used in landscape planning often have some limitations and this has implications for the confidence that can be placed in assessment results (Grêt-Regamey et al. 2013; Neuendorf et al. 2018). Nevertheless, decisions about actions and future developments will need to be made despite possible gaps in information. Using imperfect data for analyses in landscape planning is invariably better than taking no action at all. However, it does mean that it is important to assess and communicate the levels of uncertainty in inventories, evaluations and projections.

Considering possible future conditions introduces further uncertainty. Such assessments can be undertaken in landscape planning in several different ways. These include:

- *Predictions* over short time spans are generally based on sound scientific knowledge of what will happen and have relatively high certainty. The probability of a particular event occurring (e.g. a flood of particular magnitude) can often be calculated.
- *Deductive forecasting* is based on well-established and verified hypotheses. Again, the probability of a particular outcome can be calculated.

- *Projections* and *scenarios* are least certain and based on assumptions regarding plausible development pathways. The probability of events cannot be calculated. The methods applied in this type of approach are trend-extrapolation (using data from the past to estimate the future), analogy-projection (results from other cases are transferred to new situations) and expert interviews (e.g. opinions on how the future will unfold). Scenarios may also be based on goals for the future.

Uncertainties in assessments and projections can be calculated in various ways (see Chap. 6, Neuendorf et al. 2018). Many different forms of media have been used to communicate uncertainties including text, images, and dynamic visualisations (Appleton et al. 2004; Janssen et al. 2005). Another common way to express uncertainty is the use of scenarios to illustrate different plausible trajectories (Schenk and Lensink 2007). In addition, uncertainties can be considered by monitoring and adaptation of objectives as an on-going process accompanying implementation (see also Steinitz 2012: 119).

3.6.6 Deriving Response Options in a Transparent Manner

Transparency in aggregating and interpreting evaluation results should be the leading principle in the phase of deciding upon response options. As illustrated in Fig. 3.12 several 'rules' can be used to interpret the state and impact information, helping to achieve clarity in the derivation of priorities. An initial step is to evaluate the state conditions and, as discussed already, this will draw upon both scientific expertise and normative judgements. Once such evaluations and ratings have been made the key next step is to distinguish between those phenomena where mandatory objectives apply (e.g. those set by EU objectives or national laws) and those where improvements are desirable (or even not required at all). Where mandatory obligations exist the priority is usually to maintain and protect very valuable assets or to restore impaired systems, since otherwise there may be consequences such as fines or other enforcement actions. With desirable objectives there is more discretion about whether and how they are achieved, though a common approach would be to protect areas of high value before contemplating restoration or development initiatives.

With discretionary objectives it is particularly important to undertake participatory activities and engage creativity to generate measures that are in accordance with people's needs, which create local and regional identity and can be communicated by collecting design ideas (see the change models of Steinitz 2012; von Haaren et al. 2014b).

Planners need to find appropriate ways to communicate the results of assessment exercises to decision makers, stakeholders, and the public. In this context, an ongoing discussion concerns the role of ecosystem services in communicating

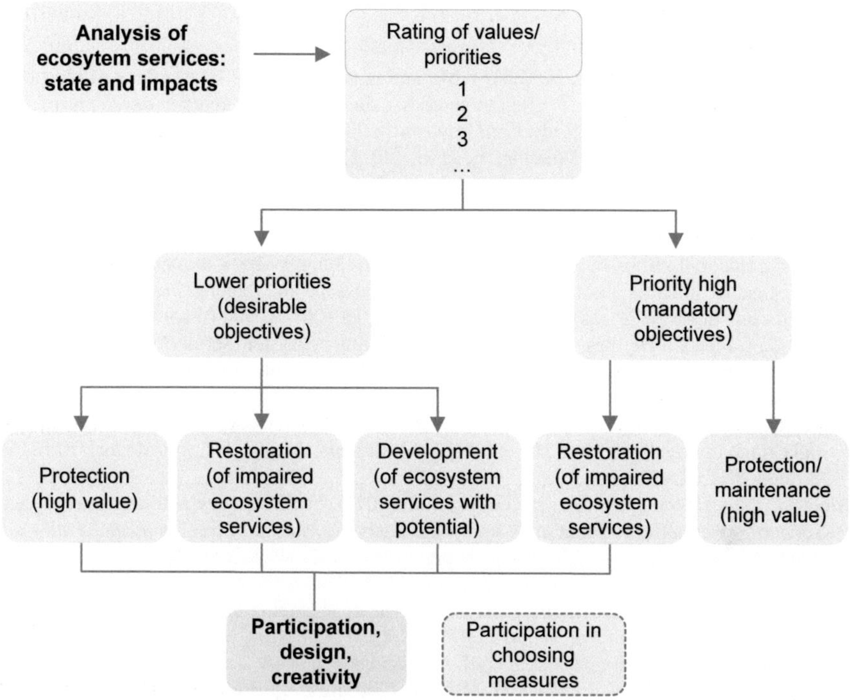

Fig. 3.12 General 'rules' for deducing response options

landscape planning outcomes. For example, it could be that politicians and citizens would better understand or accept particular measures or objectives if aggregated performance measures of ecosystem services delivery and use were provided. Examples of such situations would include the comparison of different variants of a new road corridor and the accounting of the total environmental performance of a region. Politicians usually ask for summary arguments that are easy to use. Performing such summations leads into a dilemma, as it usually implies 'comparing apples and pears'. The answer depends on the complexity of the original results and the purpose for which the information will be used. In particular the consequences an oversimplified result could generate must be carefully considered. Methodologically, multicriteria aggregation is a scaling problem because properties which are classified on different value scales have to be unified on one common scale. One example is the presentation of phenomena as monetary values and this is not without its challenges (see Chaps. 4 and 20). Another potentially problematic situation is the transformation of ordinal assessments to cardinal scales (see Sect. 3.6.3).

In order to express the scientific reservation which often accompany the aggregation process, it may help to present comparisons both using one or more overall scales and additional text, referring to individual ecosystem services, for an aggregate evaluation of state and changes in ecosystem services (compare Bateman et al. 2013).

References

Adams, B., Bissio, R., & Judd, K. (2016). Measuring accountability: The politcs of indicators. In B. Adams, R. Bissio, C. Y. Ling, et al. (Eds.), *Spotlight on sustainable development 2016* (pp. 141–147). Beirut: Suva Reflection Group on the 2030 Agenda for Sustainable Development.

Albert, C., Zimmermann, T., Knieling, J., et al. (2012). Social learning can benefit decision-making in landscape planning: Gartow case study on climate change adaptation, Elbe valley biosphere reserve. *Landscape and Urban Planning, 105*, 347–360. https://doi.org/10.1016/j.landurbplan.2011.12.024.

Albert, C., Hauck, J., Buhr, N., et al. (2014). What ecosystem services information do users want? Investigating interests and requirements among landscape and regional planners in Germany. *Landscape Ecology, 29*, 1301–1313. https://doi.org/10.1007/s10980-014-9990-5.

Albert, C., Galler, C., Hermes, J., et al. (2016a). Applying ecosystem services indicators in landscape planning and management: The ES-in-Planning framework. *Ecological Indicators, 61*(1), 100–113. https://doi.org/10.1016/j.ecolind.2015.03.029.

Albert, C., Bonn, A., Burkhard, B., et al. (2016b). Towards a national set of ecosystem service indicators: Insights from Germany. *Ecological Indicators, 16*, 38–48. https://doi.org/10.1016/j.ecolind.2015.08.050.

Albert, C., von Haaren, C., Othengrafen, F., et al. (2017). Scaling policy conflicts in ecosystem services governance: A framework for spatial analysis. *Journal of Environmental Policy and Planning, 2017*(19), 574–592. https://doi.org/10.1080/1523908X.2015.1075194.

Allmendinger, P. (2017). *Planning theory*. London: Palgrave.

Appleton, K., Lovett, A., Dockerty, T., et al. (2004). Representing uncertainty in visualisations of future landscapes. In *Proceedings of ISPRS XXXV Congress*, www.isprs.org/proceedings/XXXV/congress/comm4/papers/385.pdf. Accessed 8 Aug 2018.

Bateman, I. J., Harwood, A. R., Mace, G. M., et al. (2013). Bringing ecosystem services into economic decision-making: Land use in the United Kingdom. *Science, 341*(6141), 45–50.

BenDor, T. K., Spurlock, D., Woodruff, S. C., et al. (2017). A research agenda for ecosystem services in American environmental and land use planning. *Cities, 60*, 260–271. https://doi.org/10.1016/j.cities.2016.09.006.

Bredemeier, B., von Haaren, C., Rüter, S., et al. (2015). Evaluating the nature conservation value of field habitats: A model approach for targeting agri-environmental measures and projecting their effects. *Ecological Modelling, 295*, 113–122.

Burkhard, B., Kandziora, M., Hou, Y., et al. (2014). Ecosystem service potentials, flows and demand – Concepts for spatial localisation, indication and quantification. *Landscape Online, 34*, 1–32.

Chrisman, N. R. (1998). Rethinking levels of measurement for cartography. *Cartography and Geographic Information Systems, 25*, 231–242.

Council of Europe. (2000). *European landscape convention*. Florence: Council of Europe.

de Groot, R. S., Alkemade, R., Braat, L., et al. (2010). Challenges in integrating the concept of ecosystem services and values in landscape planning, management, and decision-making. *Ecological Complexity, 7*(3), 260–272.

Fukuda-Parr, S. (2014). Global goals as a policy tool. Intended and unintended consequences. *Journal of Human Development and Capabilities, 15*(2–3), 118–131.

Galler, C., von Haaren, C., & Albert, C. (2015). Optimizing environmental measures for landscape multifunctionality: Effectiveness, efficiency and recommendations for agri-environmental programs. *Journal of Environmental Management, 151*, 243–237.

Grêt-Regamey, A., Brunner, S. H., Altwegg, J., et al. (2013). Facing uncertainty in ecosystem services-based resource management. *Journal of Environmental Management, 127*, 145–154. https://doi.org/10.1016/j.jenvman.2012.07.028.

Gruehn, D., & Kenneweg, H. (1998). *Berücksichtigung der Belange von Naturschutz und Landschaftspflege in der Flächennutzungsplanung. Ergebnisse aus dem F+E-Vorhaben 808 06 011 des Bundesamtes für Naturschutz*. Münster: Landwirtschaftsverlag.

Haines-Young, R., & Potschin, M. (2010). The links between biodiversity, ecosystem services and human well-being. In D. G. Raffaelli & C. L. J. Frid (Eds.), *Ecosystem ecology – A new synthesis* (pp. 110–139). Cambridge: Cambridge University Press.

Hillier, J., & Healey, P. (2010). *The Ashgate research companion to planning theory: Conceptual challenges for spatial planning*. New York: Routledge.

Janssen, P. H. M., Petersen, A. C., van der Sluijs, J. P., et al. (2005). A guidance for assessing and communicating uncertainties. *Water Science and Technology, 52*(6), 125–131.

Kato, S., & Ahern, J. (2008). "Learning by doing": Adaptive planning as a strategy to address uncertainty in planning. *Journal of Environmental Planning and Management, 51*, 543–559.

Leitão, A. B., & Ahern, J. (2002). Applying landscape ecological concepts and metrics in sustainable landscape planning. *Landscape and Urban Planning, 59*(2), 65–93.

Maes, J., Egoh, B., Willemen, L., et al. (2012). Mapping ecosystem services for policy support and decision making in the European Union. *Ecosystem Services, 1*, 31–39.

Merry, S. E. (2011). Measuring the world indicators, Human rights, and Global governance. *Current Anthropology, 52*, 83–95.

Müller, F., & Burkhard, B. (2012). The indicator side of ecosystem services. *Ecosystem Services, 1*, 26–30.

Nassauer, J. I., & Opdam, P. (2008). Design in science: Extending the landscape ecology paradigm. *Landscape Ecology, 23*, 633–644.

Neuendorf, F., von Haaren, C., & Albert, C. (2018). Assessing and coping with uncertainties in landscape planning: An overview. *Landscape Ecology, 33*(6), 861–878.

Odum, W. (1982). Environmental degradation and the tyranny of small decisions. *Bioscience, 32*(9), 728–729.

Ogrin, D. (1994). Landscape architecture and its articulation into landscape planning and landscape design. *Landscape and Urban Planning, 30*, 131–137.

Pascual, U., Balvanera, P., Díaz, S., et al. (2017). Valuing nature's contributions to people: The IPBES approach. *Current Opinion in Environment Sustainability, 26–27*, 7–16. https://doi.org/10.1016/j.cosust.2016.12.006.

Porter, T. M. (1994). Making things quantitative. *Science in Context, 7*(3), 389–407.

Potschin, M., & Haines-Young, R. (2016). Defining and measuring ecosystem services. In M. Potschin, R. Haines-Young, R. Fish, et al. (Eds.), *Routledge handbook of ecosystem services* (pp. 25–44). London/New York: Routledge.

Reinke, M. (2002). *Qualität der kommunalen Landschaftsplanung und ihre Berücksichtigung in der Flächennutzungsplanung im Freistaat Sachsen*. Berlin: Logos Verlag.

Schenk, N. J., & Lensink, S. M. (2007). Communicating uncertainty in the IPCC's greenhouse gas emissions scenarios. *Climatic Change, 82*(3–4), 293–308.

Schößer, B., Helming, K., & Wiggering, H. (2010). Assessing land use change impacts – A comparison of the SENSOR land use function approach with other frameworks. *Journal of Land Use Science, 5*, 159–178. https://doi.org/10.1080/1747423X.2010.485727.

Schröter, M., van der Zanden, E. H., van Oudenhoven, A. P. E., et al. (2014). Ecosystem services as a contested concept: A synthesis of critique and counter-arguments. *Conservation Letters, 7*(6), 514–523.

Selman, P. H. (2006). *Planning at the landscape scale*. Oxon: Routledge.

Selman, P. H. (2010). Centenary paper: Landscape planning – Preservation, conservation and sustainable development. *The Town Planning Review, 81*, 381–406.

Smeets, E., & Weterings, R. (1999) *Environmental indicators: Typology and overview* (Technical report no 25). Copenhagen: EEA.

Spangenberg, J. H., von Haaren, C., & Settele, J. (2014). The ecosystem service cascade: Further developing the metaphor. Integrating societal processes to accommodate social processes and planning, and the case of bioenergy. *Ecological Economics, 104*, 22–32. https://doi.org/10.1016/j.ecolecon.2014.04.025.

Steiner, F. (2000). *The living landscape: An ecological approach to landscape planning* (2nd ed.). New York: McGraw-Hill.

Steinitz, C. (1990). A framework for theory applicable to the education of landscape architects (and other environmental design professionals). *Landscape Journal, 9*, 136–143.

Steinitz, C. (1993). A framework for theory and practice in landscape planning. *GIS Europe, 2*, 42–45.

Steinitz, C. (2012). *A framework for geodesign: Changing geography by design*. Redlands: ESRI Press.

Stevens, S. (1946). On the theory of scales of measurement. *Science, 103*(2684), 677–680. https://doi.org/10.1126/science.103.2684.677.

TEEB. (2010). The economics of ecosystems and biodiversity: Mainstreaming the economics of nature: A synthesis of the approach, conclusions and recommendations of TEEB. www.teebweb.org

Termorshuizen, J. W., Opdam, P., & van den Brink, A. (2007). Incorporating ecological sustainability into landscape planning. *Landscape and Urban Planning, 79*, 374–384.

Tscherning, K., Helming, K., Krippner, B., et al. (2012). Does research applying the DPSIR framework support decision making? *Land Use Policy, 29*, 102–110.

UK NEA. (2011). *The UK national ecosystem assessment: Synthesis of the key findings*. Cambridge: UNEP-WCMC.

van den Brink, A., Bruns, D., Tobi, H., et al. (2017). *Research in landscape architecture: Methods and methodology*. New York: Routledge.

van Oudenhoven, A. P. E., Petz, K., Alkemade, R., et al. (2012). Framework for systematic indicator selection to assess effects of land management on ecosystem services. *Ecological Indicators, 21*, 110–122.

von Haaren, C. (2016). Wie viel und welche Natur braucht der Mensch im Anthropozän? In W. Haber, M. Held, & M. Vogt (Eds.), *Die Welt im Anthropozän. Erkundungen im Spannungsfeld zwischen Ökologie und Humanität* (pp. 165–177). München: oekom verlag.

von Haaren, C., & Albert, C. (2011). Integrating ecosystem services and environmental planning: Limitations and synergies. *International Journal of Biodiversity Science, Ecosystem Services & Management, 7*(3), 150–167.

von Haaren, C., Galler, C., & Ott, S. (2008). *Landscape planning – The basis of sustainable landscape development*. Bonn: Bundesamt für Naturschutz (BfN).

von Haaren, C., Albert, C., Barkmann, J., et al. (2014a). From explanation to application: Introducing a practice-oriented ecosystem services evaluation (PRESET) model adapted to the context of landscape planning and management. *Landscape Ecology, 29*, 1335–1346.

von Haaren, C., Warren-Kretzschmar, B., Milos, C., et al. (2014b). Opportunities for design approaches in landscape planning. *Landscape and Urban Planning, 130*, 159–170.

The Basis of Evaluation: Legal, Economic and Social Values

4

Christina von Haaren and Andrew A. Lovett

Abstract

A range of different societal values need to be considered when evaluating ecosystem services and their changes in relation to both landscape planning and assessing specific planning proposals. These are typically expressed through legislative or preference-based frameworks, the latter involving economic or social methodologies. This chapter describes the various types of societal values and the core principles of the different legislative and preference-based frameworks, their strengths and weaknesses, and how they can be combined in the assessment of proposals.

Keywords

Environmental values · Instrumental values · Intrinsic values · Legislative evaluation · Preference-based evaluation

4.1 Introduction: Basic Types of Values

This chapter discusses the basic motives, values and guidelines through which the evaluation of changes in ecosystem services (ES) associated with different planning proposals can be undertaken. It is essential for the legitimisation and communication of planning decisions that the value basis of ES evaluation is made transparent to different stakeholders (Grêt-Regamey et al. 2008; de Groot et al. 2010; Albert

C. von Haaren (✉)
Institute of Environmental Planning, Leibniz University Hannover, Hannover, Germany
e-mail: haaren@umwelt.uni-hannover.de

A. A. Lovett
School of Environmental Sciences, University of East Anglia, Norwich, UK
e-mail: a.lovett@uea.ac.uk

C. von Haaren et al. (eds.), *Landscape Planning with Ecosystem Services*, Landscape Series 24, https://doi.org/10.1007/978-94-024-1681-7_4

et al. 2014). Without an evaluation of the quantity and quality of ES provision we cannot answer fundamental planning questions such as: Is the amount and quality of ES provided in a landscape sufficient, as measured by our objectives at local and national scales? Where are the ES provided, what is their quality, and how can we set priorities for their management? For example, defining a scale to evaluate water bodies against specific criteria and a target status to achieve, enables planners to designate water bodies as of different quality and prioritize measures to improve the situation.

The values which underpin any evaluation, and which guide the objectives proposed in environmental planning, can be categorised in different ways. There are those which emphasize the benefits which nature provides for humans (*instrumental* or *utilitarian* values) and those which focus on the moral imperatives for nature conservation regardless of use (*intrinsic* values). These two perspectives ('nature for people' versus 'nature for itself') are sometimes regarded as conflicting, though they need not be (Goulder and Kennedy 2011; Mace 2014; Pearson 2016). *Instrumental* values can be subdivided into *use* and *non-use* values (Pearce and Turner 1990; Pascual et al. 2010). *Use* values can be direct or indirect, while *non-use* values stem from knowing that some feature of an environment (e.g. a species or habitat) will continue to exist. *Option* value is a form of use value that relates to the importance that people give to the future availability of ES for personal benefit and so occupies something of an intermediate position between use and non-use. Intrinsic and non-use values have some similarities in that both value nature irrespective of human use, but non-use is based on human needs for well-being and therefore potentially open to quantification (and monetization), unlike intrinsic values in the strict sense (Pearson 2016).

The above discussion highlights that there is an underlying continuum of values that needs to be taken into account in any evaluation (cf. Pascual et al. 2017). Two basic approaches are used to implement such values in planning processes: (i) *legislative evaluation* which is founded in (democratically) legitimized standards about the desirable state of ES and (ii) *economic or social evaluation* based primarily on the current preferences of people (e.g. as expressed in market prices). Both approaches reflect aspects of human well-being and a combination is often required in landscape planning to provide a convincing basis for decision making. In addition, both may be relevant for rating a particular ecosystem service and may well provide different perspectives on the current or desired extent of an individual ES.

Legislative Evaluation In this approach the standards used for rating the quantity and quality of ES stem from environmental goals and standards set out in legislation, political agendas and professional guidelines. Societal ethics and general definitions of human wellbeing should drive the political processes leading to such legislation. These law-based evaluations tend to focus on non-use values or use values associated with collective utilisation, whose preservation may not be reflected in individual interests or the operation of market mechanisms. Particularly important examples of the latter are *externalities*, where the impacts (positive or negative) of an investment decision or action are not fully taken into account (Fisher et al. 2015).

The legislative guidelines used in planning are commonly derived from international agreements (including the European treaty and associated legislation), national laws and state law as well as regional and local policies with decreasing spatial applicability. In addition, sublegal policies can be used as "soft" guidelines with less binding character. The most basic forms are: (i) state and pressure-related guidelines referring to the maintenance of natural capital and biodiversity (ii) process-related guidelines such as the precautionary principle, participation requirements and the principle of proportionality. Specific standards established in laws and official documents can be used for rating and prioritising ES in practical landscape evaluation and for prioritising objectives. They may be binding thresholds (e.g. for GHG emissions) defining the required path for sustainable development but are often sufficiently broad to leave scope for public preferences and discussion to shape the specific outcome.

Economic or Social Evaluation This approach can take a variety of forms ranging from monetary valuation to deliberative processes utilising other metrics (e.g. species diversity or hazard reduction) (Gómez-Baggethun et al. 2016). Ultimately, all of these are based on actual human preferences (though expressed in different ways) and reflect the importance of the benefits derived from the ES concerned. Such methods of evaluation also tend to be more robust where the use (direct or indirect) of goods or services is involved. However, the appropriateness of different preference-based techniques is likely to vary according to factors such as the type of benefit and the degree of controversy associated with how it is assessed (Kenter 2016). This issue is discussed further in Chap. 20.

As noted earlier, a combination of these two perspectives is often required in landscape planning. However, this does not mean that they should be somehow merged into a single evaluation scale. Most commonly the legislative element will serve to set the limits of any influence from an economic or social-based evaluation. For instance, there may be mandatory requirements (e.g. for habitat protection) or a degree of legal restriction (e.g. in a floodplain) that essentially determine the use of a parcel of land. On the other hand, where such legal, political and expert-based standards are absent then economic and social considerations are likely to drive planning decisions. In some respects the most complex (and contested) circumstances are where the legal framework sets certain objectives (e.g. standards to be attained) but there is considerable discretion about how these are achieved. In such circumstances it becomes particularly important to consider the economic trade-offs between objectives and the distributional consequences (who gains or might require compensation) associated with any decision (Sikor et al. 2016). Since the legal and preference-base evaluations represent different perspectives on ES, both approaches need to be considered in many decisions for implementation-oriented landscape planning. Using a policy appraisal *balance sheet* (Turner 2016), can help to structure the evaluation process in such cases and help set priorities in spatial planning. This ensures that the basis of any ultimate decision is as transparent as possible.

The following sections expand on the above introduction, particularly in terms of core principles and the settings in which different types of evaluations are likely to be most applicable.

4.2 Overview of the Two Basic Approaches

Planners regularly find themselves in situations in which they have to present and discuss their evaluations of different planning proposals. Be it in a session with political representatives, regional or local administrators or stakeholders, planners are expected to be able to give convincing, transparent explanations for the priorities assigned to particular services.

The results of the evaluations have to be credible and replicable. They are rarely based on scientific analysis alone but are also rooted in different societal and individual preferences for the protection of nature. Often the question in environmental planning is "Which values should guide us?" (de Genaro 2012: xi). Being clear about the motives and fundamental values behind an evaluation is crucial for connecting to people's shared and individual convictions and to create common ground for the basic goals of environmental protection. Additionally, informing people about the sources from which evaluation results stem creates credibility and provides the information necessary for defining the space for participation and local decision making.

Fig. 4.1 provides an overview of different underlying values and how they link through to frameworks for evaluation. The values relate to the components of Total Economic Value (TEV) (see Pascual et al. 2010; UK National Ecosystem Assessment 2011), ranging from direct use values, through intermediate option values, into aspects of non-use such as bequest and existence values. The model is potentially general enough to be reconciled with nature conservation evaluation ethics (e.g. Ott 2000) if it is broadly interpreted – responding to the proposition of Jax et al. (2013) to use different values and valuation languages.

An important feature of Fig. 4.1 is the contrast in the ability of the two main frameworks to reflect different types of values. Preference-based methods, particularly economic ones, tend to be more powerful and robust where use values are involved, especially if the goods or services concerned are traded in markets (Fisher et al. 2015). Applications to non-use values are quite possible, but outcomes tend to be more variable according to factors such as the techniques employed and the degree of controversy associated with the proposed policy or management change. In the latter case a more deliberative approach, drawing upon other social sciences and possibly using a non-monetary metric, may be more effective (Kenter 2016). Legislative frameworks provide a means of implementing collective societal preferences (e.g. regarding forms of non-use values) as well as beliefs which are shared by many people and represent the convictions of most politicians. Thus they can also incorporate the assumed interests of future generations (Agenda 21) as well as the interests and rights of minorities. This makes them a suitable basis for evaluating the quality and quantity of delivered ES without considering the actual uses and

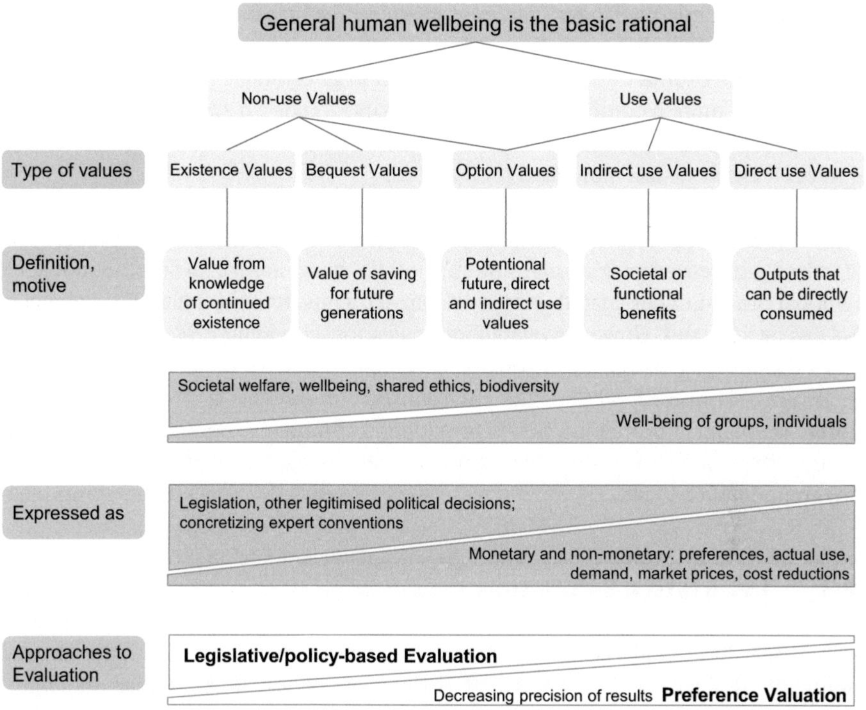

Fig. 4.1 Types of values and the capacities of legislative and preference-based evaluation frameworks to represent them in assessments of ES provision. (Sources: von Haaren 1988; Hampicke 1992; Engel et al. 2008; Pascual et al. 2010; UK National Ecosystem Assessment 2011)

interests which may be local or volatile. Intrinsic (sometimes referred to as biocentric or physiocentric) values which consider nature as having a right to exist irrespective of function can be part of societal morality (e.g. animal welfare) or individual beliefs. However, if not formally legitimised (e.g. through legislation), these are difficult to incorporate in public planning because the necessary prioritisation of places and services is not possible on the basis of everything being equally valuable (von Haaren 1988; Goulder and Kennedy 2011).

Both the legislative and the social-economic approach relate to important aspects of human well-being and both should be used in landscape planning to convincingly support decision making and to respond to the requirements of sustainable development. These approaches may be applied to rate delivery of the same ecosystem service but each will offer a different perspective. The legislative approach is needed even if an economic assessment generates very convincing evidence (e.g. a change in provisioning services such as production of food or drinking water). This is because legal standards are an important means of maintaining productive capacity for many types of provisioning services (e.g. Forest Stewardship Council 2016). It should be noted that the robustness of the results may differ with regard to the type of ES. For example, it is difficult to evaluate biodiversity and some cultural services by economic methods. However there is an extensive literature on non-market valuation methods (e.g. for recreation, Sen et al. 2014) enabling other ES to be taken into account. A further consideration is related to governance since a legally-based evaluation is more a component of a (necessarily) top down approach to protecting collective values, while economic/social valuation fits better into a bottom-up approach emphasising individual preferences.

The strength of using the legislative framework is that methodologically this can extend to evaluating ES on nominal, ordinal or cardinal scales (e.g. deviation from standards, observed maximum/minimum of ES provision or average national or regional provision per spatial unit; see Chap. 3). Thus both the state and changes in ES can be assessed. However, evaluation scales for individual ES (such as biodiversity or water provision) are diverse and cannot be readily compared or used for generating an overall value of the environment. In contrast, a strength of an economic (monetary) evaluation lies in providing a common denominator for summarizing different ES and thus the means of calculating an overall assessment of environmental changes.

4.3 Legislative Evaluation

The value base underlying legislative evaluations derives from legal norms, politically-defined official goals and their specification in particular standards. Predominantly legal and political value definitions are strong at covering common welfare, which may not be sufficiently represented through the (sum of) actual preferences of the living population or through market mechanisms. Legislative evaluation thus usually focusses on non-use-values (i.e. existence, bequest, option values) or on use-values for collectively utilised resources, whose preservation is not guaranteed by individual interests (or their summation) and the market. Aspects of intrinsic value (e.g. animal welfare, conservation of biodiversity) can also be legitimized by legislation if they can be understood and in principle accepted by a sufficient proportion of the population (Hampicke 1992).

From an economic perspective, in liberal societies individuals are fundamentally entitled to make an unconstrained use of the resources they have at hand. This fundamental freedom, in principle, includes the use of environmental resources for consumption, production, and exchange with other economic agents. With respect to exchange transactions, economic theory as well as legal principles (contract law) are based on the idea that competent actors engage in voluntary, uncoerced transactions. However, many production and consumption choices can result in negative environmental effects upon other individuals or future generations. When such external effects occur, unconstrained production and consumption choices impinge on the right of other actors to make unconstrained used of their resources, e.g. to 'quietly enjoy' their respective resources. The principles of voluntary exchange are violated, and markets cannot be expected to result in a fair attribution of the advantages and disadvantages of the individual production and consumption activities.

On some occasions clear cause-effect relationships exist between the actions of one individual and the harm to a good in the legal possession of another individual or one that future generations might need. For a subset of these cases, the Common Law systems in the Anglo-Saxon tradition offer the institution of tort law. Importantly, the tort of nuisance can be invoked if an actor neglects reasonable duties of care, and foreseeable harm is done to neighbours or proximate others. Under strict liability, negligence is not required, and the burden of proof for the cause-effect-relation is relaxed. Similar legal principles can also be found in the continental Civil Law system.

The EU Environmental Liability Directive is a case in which the EU mandates member countries to enact an equivalent of statutory tort law that addresses the liability for environmental damages resulting mainly from industrial operations. Similarly, the EU product liability rules constitute tort law. For cases beyond tort, several states have defined environmental crimes to be punished under criminal law. For example, water pollution is a crime punishable with up to 5 years in prison under German Umweltkriminalitätsgesetz (UKG).

Criminal, tort and liability law are reactive, however, and focused on some sort of individually attributable agency. They do not mandate that private and/or public land use decisions are made to minimize negative impacts on biodiversity or ES. However, EU and national legislation also include precautionary standards or objectives, which can be used as a yardstick for evaluation and determining responses in landscape planning. Evaluation methods, such as those used in spatial or landscape planning, provide frameworks for private and public sector land use decision making. Decisions about housing development, road construction, groundwater extraction or deforestation are made through legal procedures, as are those about the protection of sites. The basis of such decisions is evaluation of the environmental properties of localities and projected impacts following the proposed changes.

4.3.1 Legitimization of Legislative Values

How Do Legislative Values Evolve?

In democratic countries – such as the EU member states – the process of establishing legalised values is formalised. As polls about opinions and preferences of the population may be very volatile, in a representative democracy the elected politician is obliged to follow his/her conscience. Politicians can contribute to the definition of general goals and standards about the desired state of ES in the interest of their constituents and wider society. In this process, politicians can consider moral principles, ethical goals regarding human wellbeing (for instance as laid down in a constitution or international conventions such as that on human rights, or the Sustainable Development Goals, United Nations 2015), interests of minorities or future generations, goals set by higher political levels as well as economic considerations. The information needed for making such decisions concerning ES provision particularly concerns the value, scarcity and endangerment of natural resources at the relevant level of political decision making. Even if reality may leave much to be desired in democratic decision-making processes, there seems to be no obviously better procedure. Bottom-up participation processes can be a part of this decision making but lack democratic legitimacy to completely replace a legitimised elected institution. Such standard setting also may include guidelines about which restrictions landowners must accept as part of their mandatory societal obligations and which ones they should be compensated for by society. General rules about the importance of property rights are crucial for the implementation of planning objectives. Place-based ES can be evaluated on a legislative basis according to how much they deviate from defined targets and this then sets the direction required to achieve sustainable development.

Such processes of defining values and standards can take place on every level of the political system. In a hierarchical framework, the lower levels must take into account the objectives and standards established by those higher in the system. Individual interests or those of specific groups are acknowledged and considered in such legal procedures through the participation process but usually not by a systematic inventory of individual and collective interests. The latter is the strength of the economic valuation approach.

Degrees of Legitimacy

In assessing the landscape and deducing responses, the planner should appreciate the degree of legitimacy held by the values used in evaluation. The degree of legitimacy has consequences for the extent to which the objectives and measures in the landscape plan are binding. High legitimacy of the underlying values and standards combined with robust assessment results generally means there is limited scope for changing response measures e.g. by public participation.

Usually the degree of legitimacy increases according to the specificity with which the objective is expressed in legislation (legitimization increases starting from general statement, qualitative description, through to quantitative standards). Furthermore, the extent to which the objective is binding is also relevant (political decision, law, reinforced by penalties) (Fig. 4.2).

A well-defined, quantitative legal standard is most specific and binding and there is little room for discussion e.g. whether the emission of contaminants beyond the threshold value should be stopped or not. However, only some of the values embodied in legislation are presented as quantitative standards. If precise standards are missing in the official documents, science and policy experts may translate general values into sub-statutory guidelines which can be used for the practical assessment of ES. If no such guidelines exist, the individual planner may flesh out a general legal principle or term into a scale for local evaluation, but in such cases legitimacy will be lower (compare Merry 2011; Fukuda-Parr 2014) In addition, legitimacy has a geographic dimension. While some goals are globally applicable – such as those of the Convention on Biodiversity – others have only national validity. Usually, the

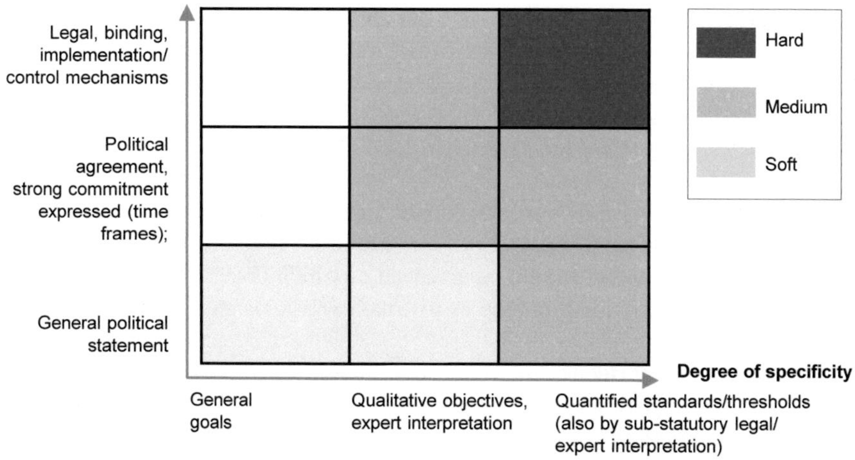

Fig. 4.2 A classification of values used in landscape evaluation. Hard: high acceptability, applicable to everybody, applied to core evaluation; Medium: typically based on widely accepted political statements, e.g. signed by many international parties and made specific by expert interpretation; Soft: based on politically legitimate statements and thus supra-individual legitimacy, but operationalisation is left to individual cases

goals or standards set by a high political level are not only applicable across this jurisdiction, but are also binding for the lower political levels, which may formulate their objectives in the framework of the national goals. However, often the national goals are more general and need to be operationalised on the regional to local level depending on the scale of the problem to be addressed.

Weak legitimacy needs to be made transparent for political decision makers as well as during public participation activities. The process of interpreting values and, in particular, generating standards and objectives needs to follow as transparent a procedure as possible. If the planner has to directly deduce ES evaluation from general principles without the help of intermediate legitimised standards, then the vagueness of this process and any normative judgments should to be indicated. Generally, a norm is defined as "a standard of appropriate behaviour for actors with a given identity" (Finnemore and Sikking 1998: 891). The derivation of the rules for our use of nature should still be traceable back to the general values laid down in legislation.

When the goals proposed in landscape planning go beyond legal minimum standards, it is particularly important that the non-binding character of the proposed objectives is stressed and that the underlying motives and reasons for these goals are transparent and open to public discussion. Planners and decision makers should thus be aware of their own beliefs and intentions and they should know what legitimate alternatives there are. Without such clarity a participation process at local or regional level is likely to encounter difficulties.

4.3.2 Guidelines and Standards for the Evaluation of Proposals

A wide variety of guidelines, criteria and standards have been set by legislation on European or national scales which can be used for the evaluation of planning propositions. Basically, there are two types:

1. *State and pressure guidelines* (and derived standards) refer directly to the state of the environment (see Chap. 3 for discussion of these terms). They define *what* quality of environment should be achieved or which objects are valuable. State guidelines also provide direction as to which pressures and impacts on ES are (not) acceptable.
2. *Process guidelines* define *how* objectives about the desired state of the environment or the restriction of land uses should be generated. In landscape planning we are interested particularly in the process leading to area-specific objectives and measures.

The most important guidelines for ES management are presented below. The relevant environmental standards are presented with descriptions of the methods for the assessment of status and impact (see Part III) or proposed response measures (see Part IV).

4.3.2.1 State and Pressure Guidelines and Standards

State and pressure guidelines and standards may be found in international declarations as well as in national and subnational legislation and expert standards. Usually the more general international principles are made more specific in national legislation. The overall guidelines which may guide evaluation as well as setting objectives for landscape planning are:

- Maintenance of natural capital in the context of sustainable development (Agenda 21; United Nations 2015), and
- Maintenance of bio- and geo-diversity, namely the diversity of habitats and biocenosis, of species and the gene pool within species, and of landscapes and ecosystems (CBD 1993).

(a) ***Maintenance of Natural Capital***

Sustainable development is widely defined as that which "meets the needs of the present without compromising the ability of future generations to meet their own needs" (United Nations 1987). Interpreted strictly this represents strong sustainability in the sense that the natural capital must not decrease and cannot be substituted by social or economic capital (Cato 2009) (Fig. 4.3). However, there are arguments for allowing substitution by other capital provided that over time the aggregate level of natural capital does not decline (Helm 2015).

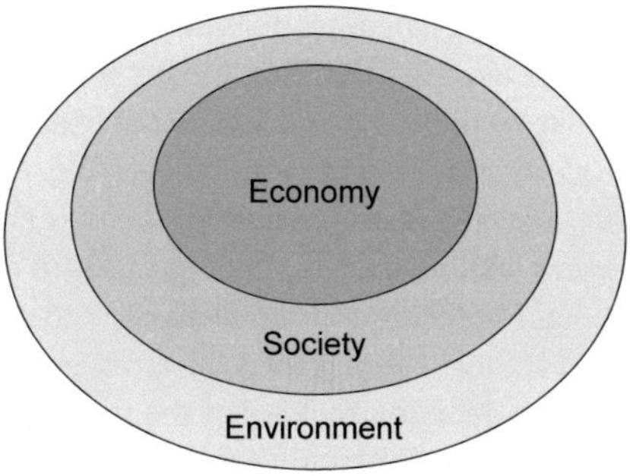

Fig. 4.3 Strong sustainability presumes that natural capital is not substitutable by human capital. Both economy and society are constrained by environmental limits. (Adapted from Cato 2009)

The basic guidelines for maintaining natural capital have been translated into the following rules for practical decision-making (cf. Daly 1999; Goodland and Daly 1995).

1. The rate of use of renewable raw materials must not be greater than their rate of replenishment. This basic rule includes not only materials such as wood but also biodiversity, if this is considered as a raw material.
2. The rate of use of non-renewable raw materials must not be greater than their rate of substitution by physically and functionally equivalent renewable raw materials or by increased productivity of both renewable and non-renewable resources.
3. The rate of deposition of materials must not be greater than their rate of assimilation. Materials introduced into the environment should thus be evaluated in terms of absorption capacity, with all functions taken into consideration.
4. Biological diversity must be maintained. This includes the protection and development of ecosystems, habitats and species diversity and genetic diversity in the landscape.
5. The time frame of human activity impacting on the environment must be kept in balance with the time frame of natural processes relevant to the ability of the environment to respond.
6. The intensity of use should be adapted to local conditions (functions, sensitivity of ecosystem processes) in order to be ecologically responsible and sustainable, thus resulting in a 'differentiated' intensity of use (Haber 1972; Delzeit et al. 2018).
7. Human health should not be endangered or exposed to unwarranted risk.
8. Ideally, all people should have an adequate right to use the world's resources. For example, every person has the right to a standard of living adequate for the health and well-being of himself and of his family according to the Universal Declaration of Human Rights (1948) and the Resolution on the Human Right to Water and Sanitation (United Nations General Assembly, resolution A/RES/64/292) recognizes the right to safe and clean drinking water and sanitation as a human right.

The above guidelines can be directly applied in practical planning and ES evaluation if nothing more concrete is at hand. For instance, the rate of soil erosion can be compared against the rate of soil (re)generation in a specific place. Thus the rules can be used for defining the local meaning of sustainable development.

Nevertheless, there are many examples of these guidelines being disregarded in both in human history and today (e.g. see Diamond 2011). Huge amounts of materials, for instance embedded water included in goods, are transported all over the world (Allen 2011). This represents a redistribution of resources without regard for replacing those non-renewables with comparable amounts of renewable substances (cf. Haber 1994: 169 ff.). Consequently, landscape planning has to constantly remind decision makers of the principle: 'think globally, act locally'.

In practical planning it also needs to be recognised that much international legislation (e.g. European Directives) is often not explicit regarding the ES to be conserved or the particular qualities of an ecosystem. However, usually the lawmaker implies that a certain quality of the environment should be conserved. For instance, in the EIA Directive "significant, derogatory environmental effects" are mentioned. Without describing a specific desirable environmental state, the identification of a possible change as "derogatory" is problematic. Careful interpretation of other relevant environmental laws is therefore needed to provide a basis for implementation.

(b) ***Maintenance of Biodiversity and Geodiversity***

Biodiversity has already been mentioned above as part of natural capital. More specifically, the Convention on Biological Diversity (CBD 1993) can be considered as a general normative basis that is accepted worldwide. The CBD is underpinned by many international treaties, which contain very particular standards in terms of defining protected species or habitats (e.g. Bern Convention, CITES, the Ramsar and Alpine Conventions, EU Birds and Habitat Directives). Within Europe, the EU Habitat and Birds Directives are particularly relevant to LP. On a state level the protected species as well as classifications of certain types of biotopes as endangered ('red listed') may be used as a standard for biodiversity (habitat and species) assessment. In landscape planning, these goals and standards are then applied in order to rate and prioritise a specific place.

Nevertheless, the CBD and European guidelines still have to be interpreted for practical purposes at regional and local scales. Their general aim is to safeguard, or achieve, the greatest possible biological diversity and range of landscapes, in order to foster a wide variety of processes and functions, including aesthetic value. Abiotic ecosystem elements are also important in this objective for 'biodiversity' which, therefore, should be referred to as geo- and biodiversity (see Chap. 16). Diversity, in itself, is not of value but should be assessed in the context of its ability to shape an ecosystem and in the context of the relevant spatial scale. The aim cannot be to increase the number of species in a landscape without regard to the combination of species typical for a site. The global demand for biodiversity does not imply that the greatest possible diversity of species should be attained even at the scale of habitat or patch. The natural community of a biotope may be relatively homogenous or species-poor. However, from a landscape-scale perspective, or in the greater biogeographic area, it may be rare and thus contribute to landscape or global biodiversity. A reed bed, for example, can make a landscape more valuable and diverse if this type of habitat is rare, despite the fact that it contains only a few species per area unit. A useful framework for describing this relationship between species diversity, spatial scale and ecological patterns, is the classification of diversity into seven categories. These categories range from the point diversity of a small, or microhabitat, generally of 10–100 square meters, to the epsilon diversity of a broad region of differing landscapes, generally of 1,000,000–100,000,000 hectares (Whittaker 1977; Jennings 1996). However, for practical environmental conservation and thus landscape planning, the level on which political decisions are made as to whether something should be protected is even more important as a scale of reference. Determining whether a species or a geological formation is rare, endangered and protected globally, in Europe, nationally, regionally or locally, is important to establish its value in terms of the normative concept of biodiversity. The purpose is to maintain global, regional and local geodiversity and biodiversity in characteristic constellations and with regard both to the diversity of regions, landscapes and ecosystems as well as to the number of species above the local level (i.e. gamma diversity, Whittaker 1977). This concept includes a demand for temporal diversity provided by ecosystems in different developmental stages and the resulting diverse processes in ecosystems (Beierkuhnlein 2003).

4.3.2.2 Process Guidelines

Three basic guidelines, which define the *how* of evaluation and decision support in landscape planning can be derived from international conventions, treaties or European Directives:

1. The precautionary principle (European Treaty of Amsterdam), requires and permits states to take decisions for safeguarding the environment as if the impact on the environment is uncertain. It prescribes that preventive action should be taken and priority should be given to rectifying damage at source.
2. Democracy and participation in decision-making are required by several pieces of international legislation (e.g. Agenda 21, EU SEA and EIA Directives, Aarhus Convention, European Landscape Convention).
3. Proportionality, a general criterion of fairness and justice, which demands that the costs or restrictions imposed by a decision are weighted against the benefits, features in several national constitutions and also in legislation such as the EU Water Framework Directive.

(a) ***The Precautionary Principle***

Acting on the *precautionary principle* in local or regional planning situations, means that risks should be minimised. In consequence, alterations to ecosystems and their services should be judged according to their reversibility. Irreversible damage to the productivity and functionality of nature and landscapes should be avoided; for example, the destruction of biotopes which take a very long time to develop. The reversibility of an impact thus can be used as a criterion for impact assessment. If the guidelines for nature conservation must be overruled, strategic impact assessment (EU Directive 2001/42/EC) and national rules, such as the German impact regulation (§ 18 of the German Nature Conservation Act), require that society provides, as a minimum, a substitute for the lost functions. The precautionary principle also has implications for uncertainty, assigning the burden of proof on the polluter that a pressure will not harm the environment.

(b) ***Role of Participation in Evaluating ES and Implementing Results***

The requirement for participation strives to assure a democratic process and inclusion of as many people as possible in decisions affecting environmental developments; it should be possible to make use of public knowledge, experiences and desires for their landscape (cf. Agenda 21; Diaz et al. 2015). Furthermore, mutual exchange of information and the process of collaborating on an issue provide an opportunity to increase the public's sense of responsibility towards nature and to better adapt goals to local contingencies. The principle of participation also takes into account the fact that there is no 'correct' and complete existing theory about the degree and manner to which nature can be used. Limits to the use of nature are marked only by cornerstones (e.g. thresholds for pollution, protected species) and a general framework, which form an avenue for sustainable development (Rogall 2004). These mark the area in which participation, stakeholder interests and local political processes interact on decisions about environmental objectives and measures. Such a process requires that nature conservation authorities and planners do the following:

- ensure that the normative aspects of nature conservation goals are clear to themselves and to others, and that the planning is transparent;
- include alternative goals;
- clarify the limits of local decision-making competence in processes occurring at the local level and identify which areas/measures are subject to local discretion.

This approach has consequences for the form and structure of practical planning objectives and measures in landscape plans. For instance, the minimum environmental standards and targets which are not negotiable in local or regional participation processes have to be clearly distinguished from possible measures or objectives, which might be desirable, but open to changes through public dialogue.

Goals have to be set for all planning levels. In practice, this means for example, that all goals with effects for which the local political level can be held responsible should be decided on at that level (principle of subsidiarity defined in Article 5 of the treaty establishing the European Community). Ultimately, in a representative democracy, those with political responsibility make decisions about the goals of public landscape planning. However, local citizens should be included as far as possible to incorporate their insights. On their own property however, people can make decisions according to their individual interests while remaining within the limits of the restrictions set by public demands. At the local level, the limits to the decisions of public bodies are found where interests concerning the good of the community impinge on those of other levels (region, country, state, etc.). At present, these limits (for example, limits to groundwater extraction beyond replenishment from an aquifer which crosses municipal boundaries) are determined chiefly by EU and national legislation.

(c) ***The Principle of Proportionality***

In every decision process, especially if public capital or restrictions on land owners are involved, the *proportionality* of investments or actions versus the expected benefit has to be taken into account. In household or administrative law this demand is usually established as an overarching procedural guideline, which refers to every public decision. In such an evaluation, decision makers have to consider all necessary and relevant information in order to judge whether, for instance, environmental benefits will justify investing in protective measures or, on the other hand, if the added social-economic value of a development justifies an environmental loss. Even one of the strongest acts of environmental legislation, the European Habitat Directive, has the principle of proportionality incorporated into it. Proportionality helps to define exceptions to the strong protection of certain species and habitats if the planned project serves superior societal goals and if the impact on nature can be compensated.

In order to clarify whether a landscape/spatial planning measure complies with the guideline of proportionality it should be demonstrated that there is a legitimate aim for the measure; that it is suitable to achieve the aim; there is no less onerous way of achieving the aim, and that the measure is reasonable given the competing interests of different groups involved (Craig and de Burca 2011).

4.4 Economic or Social Valuation

As discussed earlier methods of economic or social valuation are based on human preferences. They tend to be particularly effective in terms of decisions on ES provision where *use values* are important. This is also why economic valuations are often seen as a powerful means of communicating environmental consideration to decision makers and stakeholders (Albert et al. 2016, 2017). The dominant approach to date has been to express preferences in terms of monetary values, but this tends to provide a more robust and effective basis for decision-making in some circumstances than others. The discussion below therefore begins by considering the arguments for monetary valuation followed by an assessment of situations where economic-based methods are likely to be more or less appropriate. This leads on to an overview of approaches drawing upon other social sciences which are still concerned with preferences but adopt more deliberative or interpretative methodologies.

Arguments for monetary valuation often focus on the variety of units in which the provision of ES can be measured and the complexity of trade-offs where a proposed development may increase some but reduce others. There is consequently an advantage in making the trade-offs between different services more comparable through a common unit of measurement (Fisher et al. 2015). Money has particular strengths in the latter respect since it is a pure unit of exchange and one that is highly familiar to both political decision-makers and society at large. Adopting such a unit also helps to place the value of spending on welfare generating ES on a comparable basis to other investment options (Badura et al. 2016). Furthermore, it simplifies the comparison of the distribution of costs and benefits between different stakeholders. It has been widely acknowledged that a failure to include the economic value of ES within decision-making has been an important factor in their worldwide decline (e.g. Millennium Ecosystem Assessment 2005; TEEB 2010) and the use of valuation has also been highlighted as important in raising the profile of ES and natural capital for the profitability of private sector businesses (van Beukering et al. 2015).

If monetary valuation of ES possesses advantages then it also needs to be recognised that such exercises involve a number of complexities. In the first instance, it is often the interaction of multiple ES with manmade capital that provides particular benefits to human societies (Bateman et al. 2011a). For the purposes of economic valuation it is therefore important not to double count various 'intermediate' or 'supporting' services and instead focus on those 'final' services which produce benefits that can increase human welfare (Fisher et al. 2009).

A second issue is that valuation methodologies tend to be more robust in assessing marginal changes in ES provision (i.e. an increase or decrease) than the overall state of ES or the total value of an ecosystem or region (Rolls and Sunderland 2014; Badura et al. 2016). There are studies which have sought to estimate the total value of global or national ecosystems (e.g. Costanza et al. 2014) and these can help raise awareness of environmental qualities or degradation. However it is rare that a policy proposal involves the complete loss (or creation) of an ecosystem and at a global scale there is an argument that calculating the total value of ES is not meaningful i.e. it must be infinite since human life could not exist without the natural world (Bockstael et al. 2000).

The third consideration is the type of good or service involved. If it is *rival* (use diminishes availability for everyone else) and *excludable* (access can be prevented) then it is more likely to be exchanged in some form of market (Fisher et al. 2015). Where ES are traded in this way (e.g. many types of provisioning service) then it is fundamentally easier to convert a proposed change in provision to a monetary value. Of course, many ES are public rather than private goods and not subject to market exchange. Nevertheless, if there is some form of direct use or *revealed preference* linked to human behaviour (e.g. visits to open-access recreation sites) then there are techniques of non-market valuation that can be used to derive economic values for the benefits associated with ES provision. Other, *stated preference* methods can be used to explore hypothetical choices through public surveys (Badura et al. 2016).

These have the advantage that they can explore values related to non-use (e.g. existence values). However, there is also considerable debate about incorporating the needs of future generations who cannot be asked for their preferences and also the general robustness of such methods. Sometimes survey respondents are asked about situations where they are not especially well-informed and the design of the question format can have a considerable influence on the results obtained. Chap. 20 discusses these economic valuation methods in more detail.

Conducting economic valuation studies can involve considerable expense and time. In many practical planning situations there may not be the resources available to complete specific surveys or valuation methodologies at particular sites. Consequently an increasingly common approach is to undertake some form of 'benefit transfer' where ES values estimated in another study are applied to the site(s) of interest. This is reasonably straightforward where uniform rates can be employed (e.g. the annual GHG sequestration of a hectare of broadleaf woodland) but much more difficult where valuations are context-dependent (e.g. the recreation value of a woodland is likely to vary according to the availability of substitute destinations). In the latter situations transferring a function which estimates a relationship between the site characteristics (including those of the surrounding area) and the value of ES provision may be more effective (e.g. Bateman et al. 2011b; Sen et al. 2014). However, there is still a need for care in applying such functions because the area (and population) over which benefits are aggregated can have a considerable influence on the results obtained.

Alternatives to economic approaches have become more prominent with the increasing recognition that for some types of ES provision there can be a pluralism of values, each of which in principle may be correct and yet conflict with each other (Gómez-Baggethun et al. 2016). These situations are particularly likely where there is a lack of common understanding, or indeed controversy, surrounding the aims or impacts of a proposed policy or measure (Kenter 2016). They also tend to occur with some aspects of non-use values, especially certain cultural ES which tend to be quite context-specific.

In such circumstances, methods based on individually-elicited preferences can generate highly variable results and there may be advantages in a workshop-based discussion or deliberative exercise to obtain more agreement about how issues are framed and outcomes evaluated. These methods can be used to derive monetary valuations but if the ES under consideration is one where a commodity metaphor is less obvious then it may be fruitful for group dynamics to adopt a non-monetary approach (e.g. a multi-criteria assessment framework) to provide a broader perspective on the decision involved. This also tends to be the case if the evaluation is on a local scale and there are stakeholders with strongly opposing interests so that wider issues of ethics, justice and trade-offs need to be considered (Jax 2016; Turner 2016). Further discussion regarding the use of deliberative and non-monetary valuation methods is provided by Kenter (2016).

4.5 Integrating Legislative and Preference-Based Evaluation Frameworks in Landscape Planning

As discussed earlier, legislative-based frameworks tend to be top-down in form and typically represent a societal expression of collective use or non-use values. Economic or social evaluations are more bottom-up and represent individual preferences in the form of monetary values or other metrics. Frequently, the legislative element will serve to set the limits of any influence from an economic or social-based evaluation. However, the balance of power between the two elements can vary appreciably from place-to-place, reflecting international differences between countries in the power of the state and also within countries in terms of the relative influence of federal, state and local legislation (see Chap. 30).

It is quite possible for legislative and individual preference based frameworks to present contrasting views on the desired extent of particular ES provision in a locality. If this is the case then the outcome is likely to depend on the degree of discretion that exists at the relevant level of planning governance. More commonly, however, the legal framework sets certain restrictions regarding the zoning of land uses along with standards to be attained elsewhere. Economic and social preferences then provide the means for evaluating how best these can be achieved. However, given the increasingly complex and contested nature of such decisions it is becoming less likely that a single method of preference evaluation will be sufficient and that more diverse and flexible forms of decision support will be required to achieve satisfactory outcomes. Turner (2016) outlines one such 'balance sheet' approach with three main elements – a cost-benefit analysis informed by monetary valuations as a starting point, supplemented by consideration of equity and distributional consequences followed by group-based deliberative processes to address trade-off and compensation issues. The need for, and relative importance, of these three elements is likely to vary from case to case, with the economic analysis tending to be more important in setting national or regional priorities and the consideration of ethical, trade-off and possible compensation issues being key for locally acceptable solutions. There is consequently no single, simple rule for integrating perspectives, but there is invariably a need to consider a range of values through several different evaluation frameworks. In making such comparisons approaches such as the 'balance sheet' can help clarify the information requirements and structure the stages of decision-making.

References

Albert, C., Aronson, J., Fürst, C., et al. (2014). Integrating ecosystem services in landscape planning: Requirements, approaches, and impacts. *Landscape Ecology, 29*(8), 1277–1285.

Albert, C., Galler, C., Hermes, J., et al. (2016). Applying ecosystem services indicators in landscape planning and management: The ES-in-Planning framework. *Ecological Indicators, 61*(1), 100–113. https://doi.org/10.1016/j.ecolind.2015.03.029.

Albert, C., Schröter-Schlaack, C., Bernd Hansjürgens, B., et al. (2017). An economic perspective on land use decisions in agricultural landscapes: Insights from the TEEB Germany study. *Ecosystem Services, 25*, 69–78.

Allen, J. A. (2011). *Virtual water: Tackling the threat to our planet's most precious resource*. London: Taurus.

Badura, T., Bateman, I. J., Agarwala, M., et al. (2016). Valuing preferences for ecosystem-related goods and services. In M. Potschin, R. Haines-Young, R. Fish, et al. (Eds.), *Routledge handbook of ecosystem services* (pp. 228–242). Oxford: Routledge.

Bateman, I. J., Mace, G. M., Fezzi, C., et al. (2011a). Economic analysis for ecosystem service assessments. *Environmental and Resource Economics, 48*, 177–218.

Bateman, I. J., Brouwer, R., Ferrini, S., et al. (2011b). Making benefit transfers work: Deriving and testing principles for value transfers for similar and dissimilar sites using a case study of the non-market benefits of water quality improvements across Europe. *Environmental and Resource Economics, 50*, 365–387.

Beierkuhnlein, C. (2003). Der Begriff der Biodiversität. *Nova Acta Leopoldina, 87*, 51–71.

Bockstael, E. N., Freeman, A. M., Kopp, R. J., et al. (2000). On measuring economic values for nature. *Environmental Science & Technology, 34*, 1384–1389.

Cato, M. (2009). *Green economics*. London: Earthscan.

CBD. (1993). *Convention on biological diversity*. https://www.cbd.int/convention/text/. Accessed 16 Dec 2016.

Costanza, R., de Groot, R., Sutton, P., et al. (2014). Changes in the global value of ecosystem services. *Global Environmental Change, 26*, 152–158.

Craig, P., & de Burca, G. (2011). *EU law* (5th ed.). Oxford: Oxford University Press.

Daly, H. E. (1999). *Wirtschaft jenseits von Wachstum: die Volkswirtschaftslehre nachhaltiger Entwicklung*. Salzburg: Anton Pustet.

de Genaro, I. (Ed.). (2012). *Value: Sources and readings of a key concept of the globalized world*. Leiden: Brill.

de Groot, R. S., Alkemade, R., Braat, L., et al. (2010). Challenges in integrating the concept of ecosystem services and values in landscape planning, management, and decision-making. *Ecological Complexity, 7*(3), 260–272.

Delzeit, R., Lewandowski, I., Arslan, A., et al. (2018). *How the sustainable intensification of agriculture can contribute to the sustainable development Goals* (Working paper 18/1). Stuttgart: German Committee Future Earth.

Diamond, J. (2011). *Collapse: How societies choose or fail to survive* (Updated ed.). London: Penguin.

Diaz, S., Demissew, S., Carabias, J., et al. (2015). The IPBES conceptual framework—Connecting nature and people. *Current Opinion in Environment Sustainability, 14*, 1–16.

Engel, S., Pagiola, S., & Wunder, S. (2008). Designing payments for environmental services in theory and practice: An overview of the issues. *Ecological Economics, 62*, 663–674.

Finnemore, M., & Sikkink, K. (1998). International norm dynamics and political change. *International Organization, 52*, 887–917.

Fisher, B., Turner, R. K., & Morling, P. (2009). Defining and classifying ecosystem services for decision making. *Ecological Economics, 68*, 643–653.

Fisher, B., Naidoo, R., & Ricketts, J. (2015). *A field guide to economics for conservationists*. Greenwood: Roberts and Company.

Forest Stewardship Council. (2016). *FSC certification*. https://ic.fsc.org/en/certification. Accessed 22 Aug 2016.

Fukuda-Parr, S. (2014). Global goals as a policy tool. Intended and unintended consequences. *Journal of Human Development and Capabilities, 15*(2–3), 118–131.

Gómez-Baggethun, E., Barton, D. N., Berry, P., et al. (2016). Concepts and methods in ecosystem services valuation. In M. Potschin, R. Haines-Young, R. Fish, et al. (Eds.), *Routledge handbook of ecosystem services* (pp. 99–111). Oxford: Routledge.

Goodland, R., & Daly, H. (1995). Universal environmental sustainability and the principle of integrity. In L. Westra & J. Lemons (Eds.), *Perspectives on ecological integrity* (pp. 102–124). Dordrecht: Kluwer.

Goulder, L. H., & Kennedy, D. (2011). Interpreting and estimating the value of ecosystem services. In P. Kareiva, H. Tallis, T. H. Ricketts, et al. (Eds.), *Natural capital: Theory & practice of mapping ecosystem services* (pp. 15–33). Oxford: Oxford University Press.

Grêt-Regamey, A., Walz, A., & Bebi, P. (2008). Valuing ecosystem services for sustainable landscape planning in mountain regions. *Mountain Research and Development, 28*(2), 156–165.

Haber, W. (1972). Grundzüge einer ökologischen Theorie der Landnutzungsplanung. *Innere Kolonisation, 24*, 294–298.

Haber, W. (1994). Nachhaltige Entwicklung-aus ökologischer Sicht. *Zeitschrift für angewandte Umweltforschung, 7*(1), 9–13.

Hampicke, U. (1992). *Ökologische Ökonomie*. Opladen: Westdeutscher Verlag.

Helm, D. (2015). *Natural capital*. Oxford: Oxford University Press.

Jax, K. (2016). Ecosystem services and ethics. In M. Potschin, R. Haines-Young, R. Fish, et al. (Eds.), *Routledge handbook of ecosystem services* (pp. 301–303). Oxford: Routledge.

Jax, K., Barton, D., Chan, K., et al. (2013). Ecosystem services and ethics. *Ecological Economics, 93*, 260–268.

Jennings, M. D. (1996). Some scales for describing biodiversity. *GAP Bulletin*, p. 5. http://pubs.usgs.gov/gap/05/report.pdf. Accessed 23 Aug 2016.

Kenter, J. O. (2016). Deliberative and non-monetary valuation. In M. Potschin, R. Haines-Young, R. Fish, et al. (Eds.), *Routledge handbook of ecosystem services* (pp. 271–288). Oxford: Routledge.

Mace, G. M. (2014). Whose conservation? *Science, 345*, 1558–1560.

Merry, S. E. (2011). Measuring the world indicators, human rights, and global governance. *Current Anthropology, 52*, S83–S95.

Millennium Ecosystem Assessment. (2005). *Ecosystems and human well-being*. Washington, DC: Island Press.

Ott, K. (2000). Umweltethik – Einige vorläufige Positionsbestimmungen. In K. Ott & M. Gorke (Eds.), *Spektrum der Umweltethik* (pp. 13–39). Marburg: Metropolis-Verlag.

Pascual, U., Muradian, R., Brander, L., et al. (2010). The economics of valuing ecosystem services and biodiversity. In P. Kumar (Ed.), *The economics of ecosystems and biodiversity: Ecological and economic foundations* (pp. 183–256). London: Earthscan.

Pascual, U., Balvanera, P., Díaz, S., et al. (2017). Valuing nature's contributions to people: The IPBES approach. *Current Opinion in Environment Sustainability, 26–27*, 7–16. https://doi.org/10.1016/j.cosust.2016.12.006.

Pearce, D., & Turner, R. K. (1990). *Economics of natural resources and the environment*. London: Harvester Wheatsheaf.

Pearson, R. G. (2016). Reasons to conserve nature. *Trends in Ecology & Evolution, 31*, 366–371.

Rogall, H. (2004). *Ökonomie der Nachhaltigkeit – Handlungsfelder für Politik und Wirtschaft*. Berlin: Springer.

Rolls, S., & Sunderland, T. (2014). *Microeconomic evidence for the benefits of investment in the environment 2 (MEBIE2)* (Natural England research reports, number 057). Bristol: Natural England.

Sen, A., Harwood, A., Bateman, I. J., et al. (2014). Economic assessment of the recreational value of ecosystems: Methodological development and national and local application. *Environmental and Resource Economics, 57*, 233–249.

Sikor, T., Martin, A., Fisher, J., et al. (2016). Ecosystem services and justice. In M. Potschin, R. Haines-Young, R. Fish, & R. K. Turner (Eds.), *Routledge handbook of ecosystem services* (pp. 299–301). Oxford: Routledge.

TEEB. (2010). *The economics of ecosystems and biodiversity: Ecological and economic foundations*. London: Earthscan.

Turner, R. K. (2016). The 'balance sheet' approach with adaptive management for ecosystem services. In M. Potschin, R. Haines-Young, R. Fish, et al. (Eds.), *Routledge handbook of ecosystem services* (pp. 289–298). Oxford: Routledge.

UK National Ecosystem Assessment. (2011). *The UK national ecosystem assessment: Synthesis of the key findings*. Cambridge: UNEP-WCMC.

United Nations. (1987). *Our common future, world commission on environment and development.* Oxford: Oxford University Press.

United Nations. (2015). *Transforming our world – The 2030 agenda for sustainable development.* https://sustainabledevelopment.un.org/topics/sustainabledevelopmentgoals. Accessed 16 Dec 2016.

van Beukering, P. J. H., Brouwer, R., & Koetse, M. J. (2015). Economic values of ecosystem services. In J. Bouma & P. J. H. van Beukering (Eds.), *Ecosystem services: From concept to practice* (pp. 89–107). Cambridge: Cambridge University Press.

von Haaren, C. (1988). Beitrag zu einer normativen Grundlage für praktische Zielentscheidungen im Arten- und Biotopschutz. *Landschaft + Stadt, 20*, 97–106.

Whittaker, R. H. (1977). Species diversity in land communities. *Evolutionary Biology, 10*, 1–67.

Data Sources for Assessments

5

Andrew A. Lovett and Gisela Sünnenberg

Abstract
A combination of technical and societal factors has resulted in major changes in the extent and availability of spatial data in the past two decades. This chapter discusses the evolution of spatial data infrastructures, emphasising the increasing availability of freely-accessible data via geoportals. An overview is provided of sources likely to be of particular value for those involved in landscape planning and the assessment of ecosystem services. This includes information on terrain, geology, soils, land cover, biodiversity, protected areas, population density and a variety of socio-economic variables. Issues associated with data integration and uncertainty are also considered, particularly the need for caution when using sources derived at a variety of spatial scales or where systems are highly dynamic.

Keywords
Spatial data · Open data · Geoportals · Uncertainty

5.1 Introduction

As emphasised in Chap. 3 the availability of a spatially-explicit information base is central to the integration of ecosystem services (ES) assessments into landscape planning. This is primarily because location is such a useful dimension through which to compare or combine different variables. In addition, the value of the benefits provided by the ES in an area is often dependent on their spatial context, including the availability of substitutes, connectivity to other resources and

A. A. Lovett (✉) · G. Sünnenberg
School of Environmental Sciences, University of East Anglia, Norwich, UK
e-mail: a.lovett@uea.ac.uk; g.sunnenberg@uea.ac.uk

C. von Haaren et al. (eds.), *Landscape Planning with Ecosystem Services*, Landscape Series 24, https://doi.org/10.1007/978-94-024-1681-7_5

proximity of human populations (e.g. Bateman et al. 2013). Thankfully, the availability of spatial data and software tools (such as GIS) to support such assessments and planning activities is now extensive, with progressive improvements in both the quality of information and the ease with which it can be obtained. Since many relevant databases are global in coverage this situation applies across Europe, though there are inevitably some differences both between and within countries.

This chapter focuses on spatial data resources that cover much, if not all, of Europe. There may well be national, regional or local databases which are superior to those discussed, but the objective is to provide pointers to resources that could be used to support analysis and implementation anywhere in Europe. It is also important to note that the URLs given were correct as of August 2017, but could change in the future. For this reason, copies of the tables included below will be maintained on the website mentioned in Chap. 2 and additional resources added as they become available. Since it is also likely that many applications will involve integration of data from a range of sources, the final part of the chapter presents a checklist of issues to consider in order to ensure that the information used is as *fit for purpose* as possible.

5.2 Factors Influencing the Extent and Availability of Spatial Data

Changes in the provision of digital spatial data over recent decades have been driven by a combination of technical and societal factors. In the technical sphere three key developments have been:

- The widespread availability of civilian signals from Global Navigation Satellite Systems (GNSS) such as GPS, GLONASS and GALILEO to provide <5 m positional accuracy on a variety of receivers (including smartphones). This has transformed the ability to undertake mapping exercises, including the capacity for individuals and communities to create their own databases (Milner 2016).
- Improvements in aircraft and satellite-based sensors to enhance mapping of topography, land cover and other properties of the earth's surface. Freely accessible global topographic databases have advanced from an approximately 1 km grid cell resolution in the late 1990s to 30 m today while land cover databases now exist as 300 m (global coverage) and 100 m (European) products (Lillesand et al. 2015).
- Enhancements in digital infrastructure to support: (i) the dissemination and downloading of spatial data via geoportals, (ii) the creation of user-generated content (e.g. http://www.openstreetmap.org), and (iii) the provision of data layers through web mapping services (e.g. http://discomap.eea.europa.eu/).

On a societal level, the main change has been for much more government information to be placed in the public domain (i.e. freely available and not restricted by copyright or contractual constraints on use) as part of national spatial data infrastructures (NSDI) (Longley et al. 2015, p 426–432). The extent of such *Open Data* developments has varied among countries, but has been generally stimulated by recognition of the increasing importance of location-based information and the potential economic benefits from easier integration and sharing via internet-based technologies. Within the EU further direction has been provided by the INSPIRE Directive (2007/2/EC, Infrastructure for Spatial Information in the European Community) which entered into force in May 2007. This outlines principles for a Spatial Data Infrastructure to support European Community environment policies and sets a number of obligations for member states to support data exchange, sharing and re-use. Data charges and licensing are permitted by the Directive, but it also requires that certain categories of key information "… are made available under conditions which do not unduly restrict their extensive use" (INSPIRE Directive, Paragraph 6, see European Commission 2007). Progress towards the implementation of INSPIRE has varied between countries (see European Data Portal 2016), but in examples such as France, Germany, Spain and the UK it has resulted in both mapping agencies and government departments releasing many holdings or products as Open Data (e.g. https://www.geoportail.gouv.fr/, https://www.govdata.de/, http://sig.magrama.es/geoportal/ or https://data.gov.uk/).

The combination of these developments has also meant that there is now much more diversity in terms of sources of spatial data. This is true both in terms of providers (e.g. public sector, private businesses or crowdsourced information) and how content is accessed. While many data sets are still downloaded and accessed locally, the ability of widely-used GIS software such as ArcGIS (http://www.esri.com/) or QGIS (http://qgis.org/en/site/) to support online mapping services means that applications are often now based on distributed databases with suppliers spread across a number of locations. This flexibility in data retrieval has obvious advantages, but also means that it is important to ensure that the sources used are compatible (e.g. through appropriate coordinate transformations) and *fit for purpose* (in terms of resolution and attribute definitions). These issues are discussed further in Sect. 5.4.

5.3 Overview of Spatial Data Sources

There are now many different *geoportals* which provide access to digital spatial data. In some instances, these are quite broad in content, while others focus on a particular theme. Most sites support a search function through either a catalogue or map viewer and data can be downloaded either immediately or following user registration. In all the examples listed below the data are available free of charge and generally under some form of public or open data license allowing a wide range of uses provided due acknowledgement is given to the source.

The sites listed in Table 5.1 are general-purpose and good places to begin when a spatial database needs to be constructed from scratch. Most have been created by international organisations (e.g. the UN Environment Programme – UNEP) or government agencies (e.g. the US Geological Survey – USGS, European Environment Agency). The USGS Earth Explorer is particularly good as an easy-to-use interface to an enormous range of global imagery resources covering several decades.

Table 5.1 General purpose global or European data portals

Name	Content and Internet address
Global coverage	
UNEP Environmental Data Explorer	Data used by UNEP and partners for a range of environment assessments – http://geodata.grid.unep.ch/
NOAA National Centers for Environmental Information (NCEI)	US government site providing access to a range of geophysical information, including on natural hazards – https://www.ngdc.noaa.gov/
USGS Earth Resources Observation and Science (EROS) Data Centre	US government site providing access to many types of elevation products and satellite imagery – http://eros.usgs.gov/find-data
USGS EarthExplorer	Data viewer providing access to many types of satellite imagery and elevation products – https://earthexplorer.usgs.gov/
OpenStreetMap	User-created and freely downloadable vector mapping, content varies spatially, but can be very detailed – http://www.openstreetmap.org
European coverage	
European Data Portal	Collection of metadata on public sector information available on portals across European countries – https://www.europeandataportal.eu/
European Environment Agency	Data catalogue for European Environment Agency holdings – https://www.eea.europa.eu/data-and-maps
European Data Centres	Links to data centres on issues such as air pollution, biodiversity, climate change, land use and water quality – https://www.eea.europa.eu/data-and-maps/european-data-centres/european-data-centres

Note: Compiled from information available in August 2017

OpenStreetMap (OSM) is a crowdsourced world map. Beginning in 2004 it now has over three million registered users and is one of the most prominent examples of volunteered geographic information (VGI) (Sui et al. 2013). The database is a good source on many types of land use, built infrastructure and transport networks and though the level of content varies from place-to-place there are many examples (especially in Europe) where it is more detailed or up-to-date than government or commercial mapping. Extracts of the data can be downloaded from the OSM website, but there are also websites that provide subsets of the data for particular countries or regions such as http://download.geofabrik.de/.

Information on topography is a key input for landscape planning and is something that has been revolutionised by data from satellite sensors (e.g. synthetic aperture radar) or airborne LiDAR (light detection and ranging) (Maune 2007; Lillesand et al. 2015). Since 2000 the combination of the Space Shuttle Radar Topography Mission (SRTM) and ASTER GDEM project have created ~30 m resolution global elevation databases which have now been combined to generate a specific EU-DEM (see details in Table 5.2). All of these sources are surface models (i.e. including reflections from buildings and vegetation) rather than just the bare-earth terrain and are known to have some problems with spikes and voids, but there have been a number of initiatives to correct such issues (e.g. http://www.cgiar-csi.org/data/

Table 5.2 Sources of terrain, geology, soils and climate data

Name	Content and Internet address
Global coverage	
NASA SRTM	~30 m global DEM from space shuttle radar topography mission in 2000 – http://www2.jpl.nasa.gov/srtm/
ASTER GDEM	~30 m global DEM from stereo processing of ASTER imagery from the Terra satellite – https://asterweb.jpl.nasa.gov/gdem.asp
WorldClim	Baseline ~1 km resolution bioclimatic variables based on averages for 1970–2000 – http://worldclim.org/version2
IPCC Data Distribution Centre	Access to a wide variety of observed and modelled future climate data – http://www.ipcc-data.org/index.html
European coverage	
EU-DEM	Hybrid ~30 m product based on SRTM and ASTER GDEM data fused by a weighted averaging approach as part of the EU Copernicus programme – https://www.eea.europa.eu/data-and-maps/data/eu-dem
European Geological Data Infrastructure	Access to European and national datasets from the geological survey organisations of Europe – http://www.europe-geology.eu/
European Soil Database	Database with many soil properties in vector and raster (1 km resolution) formats – http://esdac.jrc.ec.europa.eu/resource-type/european-soil-database-soil-properties
European Soil Threats	Databases on soil erosion, organic carbon decline, compaction, salinization, soil biodiversity decline etc. – http://esdac.jrc.ec.europa.eu/resource-type/soil-threats-data

Note: Compiled from information available in August 2017

srtm-90m-digital-elevation-database-v4-1) so that their quality is increasingly reliable. There is also now a commercial elevation product at 12 m cell resolution (http://www.intelligence-airbusds.com/worlddem/), which indicates the standard that can be expected for public domain information in the future.

LiDAR can generate topographic data at much finer resolutions (e.g. <1 m compared to 30 m for SRTM) which is particularly useful for representing built infrastructure or where water management (e.g. flooding) is a key concern. In many parts of Europe the LiDAR data created by government agencies is now in the public domain and can be obtained from national level portals (e.g. https://data.gov.uk/dataset/lidar-composite-dtm-2m1), though at present coverage tends to be better in urban or coastal areas than rural hinterlands.

Details of baseline climate parameters and future projections are relatively straightforward to obtain at a global level due to the work of initiatives such as the Intergovernmental Panel on Climate Change (IPCC). For geology the data provision is more fragmented with coarse 1:5 million scale maps available at a European level and more detailed information accessible from individual national surveys (e.g. http://www.igme.es/default.asp or http://www.bgs.ac.uk/opengeoscience/). Assessments of soils have been greatly enhanced by initiatives from the European Commission Joint Research Centre (JRC) so that there is a 1 km resolution database of soil properties, as well as specific layers on issues such as erosion or organic content.

Table 5.3 Sources of land cover, biodiversity and protected area data

Name	Content and Internet address
Global coverage	
USGS Land Cover Institute	Extensive catalogue of land cover data sets – https://landcover.usgs.gov/landcoverdata.php
Global Land Cover Facility	Access to a range of products derived from satellite imagery – http://www.landcover.org/data/
ESA GlobCover	~300 m resolution data for 2005–6 and 2009 – http://due.esrin.esa.int/page_globcover.php
ESA Climate Change Initiative Land Cover Project	~300 m resolution annual global land cover time series from 1992 to 2015 – http://maps.elie.ucl.ac.be/CCI/viewer/
Global Ecological Land Units	250 m ecophysiographic stratification based on bioclimate, landform, lithology and land cover data – https://catalog.data.gov/dataset/global-ecological-land-units-elus
IUCN Red List species distributions	Distribution polygons for some 60,000 species – http://www.iucnredlist.org/technical-documents/spatial-data
Map of Life	Species richness and diversity information – https://mol.org/
World Database on Protected Areas	Boundaries and attributes for marine and terrestrial protected areas – http://www.protectedplanet.net/
European coverage	
CORINE Land Cover	100 m land cover classifications for 1990, 2000, 2006 and 2012 – https://www.eea.europa.eu/publications/COR0-landcover
Ecosystem Types of Europe	Combines spatially-explicit land cover information with non-spatially referenced habitat information to improve the representation of European ecosystems – https://www.eea.europa.eu/data-and-maps/data/ecosystem-types-of-europe
NATURA 2000 sites	Boundaries and properties of sites designated under EU directives – https://www.eea.europa.eu/data-and-maps/data/natura-8

Note: Compiled from information available in August 2017

Land cover is another key variable where data provision has been transformed by advances in satellite sensor technology. Table 5.3 provides details of relevant data sources. At present the global standard is set by the European Space Agency for 300 m resolution classifications, derived from ENVISAT MERIS imagery. These now include an annual time series from 1992–2015 which provides the basis for many forms of change assessment. It also seems likely that with current developments in the Sentinel satellites a 10 m land cover map could exist in the foreseeable future (http://www.esa.int/Our_Activities/Observing_the_Earth/Improving_land_cover_mapping_with_Sentinel-2).

At the European scale the CORINE land cover database is a core resource for many types of environmental assessments, particularly because it provides information at several points in time. In addition, land cover databases have been combined with other environmental layers to generate maps of ecosystem or habitat types at both European and global scales. There is also detailed mapping of protected areas. Various international databases exist on species distributions and aspects of biodiversity, and their quality is improving all the time. However, for many such purposes it is necessary to consult national-scale resources (e.g. https://nbnatlas.org/).

Table 5.4 Sources of administrative, population and socio-economic data

Name	Content and Internet address
Global coverage	
Global Administrative Areas	Administrative boundaries for countries and lower level subdivisions such as provinces, departments or counties – http://gadm.org/
NASA Socioeconomic Data and Applications Center (SEDAC)	Extensive range of data including urban settlements and gridded population distributions – http://sedac.ciesin.columbia.edu/
Global Human Settlement Layer	Settlement layers and gridded population distributions for 1975, 1990, 2000 and 2015 at 250 m and 1 km resolutions – http://ghsl.jrc.ec.europa.eu/
European coverage	
Eurostat GISCO	Boundaries of administrative units and population distributions – http://ec.europa.eu/eurostat/web/gisco/geodata/reference-data
Eurostat	Extensive source of statistical information for European administrative areas – http://ec.europa.eu/eurostat/web/main/home

Note: Compiled from information available in August 2017

Many forms of socio-economic data are collected through censuses or surveys and published as statistics for administrative areas. To incorporate such data into GIS analyses it is necessary to obtain digital versions of the boundaries. These are normally straightforward to obtain from some of the sources listed in Table 5.4 or national census websites. A recent trend has been to combine census data with details of settlements derived from satellite imagery to derive gridded population distributions and there are now several examples of such products at 250 m or 1 km resolutions. The NASA Socioeconomic Data and Applications Centre is particularly recommended as a useful resource. Other sources mentioned earlier in Table 5.1, such as the EEA Data Centres and OpenStreetMap, can also provide valuable socio-economic information.

5.4 Data Integration and Uncertainty Considerations

It should be evident from the previous section that a wide variety of spatial data sets are now readily available to help support the integration of ES assessments into landscape planning. In many respects, the challenge now is more one of ensuring that the available data are integrated in appropriate ways to ensure that the resulting database is as *fit for purpose* as possible. It is also important to appreciate that any representation of real-world features in a GIS database is likely to involve some *uncertainty*, both in terms of positional information (e.g. the boundaries of an area) and attributes (e.g. the characteristics that distinguish a zone). This is particularly true of many phenomena important in landscape planning where there is *inherent uncertainty* arising from gradual transitions (e.g. between biomes) rather than crisp boundaries or some ambiguity in different taxonomies (e.g. how features such as 'wetlands' or 'woodlands' are defined) (Longley et al. 2015, p 99–127). Consulting the *metadata* (i.e. data about data) associated with a data set is often a useful initial step in appreciating how it should be interpreted and the following are suggested as a checklist of some other key issues to consider.

1. Visually checking data layers for completeness or consistency is a good starting point. This is particularly important when using crowdsourced data such as OpenStreetMap where the level of mapping detail can change from one neighbourhood to another and the classification of the same type of feature (e.g. woodland) may vary between contributors (Taigel et al. 2017). Another example occurs in some gridded population distributions where although the cell size is uniform the level of detail varies between countries or regions because of changes in the number and size of administrative units that the cell population totals were interpolated from.
2. If data are obtained from several geoportals then it is quite likely that they will be in different spatial *coordinate systems*. Global data layers tend to be supplied in coordinates such as degrees longitude and latitude, while European or national data are typically in a map projection with units in meters. For instance, much of the data supplied by European agencies uses the ETRS89 datum and a Lambert Azimuthal Equal Area projection (*ETRS89_LAEA*). The disadvantage of degrees as a coordinate system is that the distance represented by a degree longitude varies with latitude so for any analytical purposes it is usually best to convert the data to a meters-based map projection. The specific choice of projection may well depend on individual circumstances, but the advantage of having all data layers in the same coordinate system is considerable.
3. It is also common that the data obtained are at several different source scales (e.g. 1:10,000, 1:50,000 or 1:100,000 for vector outlines) or cell resolutions (e.g. 100 m or 1 km for raster grids). Since a GIS makes it straightforward to zoom in on part of map it is quite easy for data (especially vector layers) to appear more accurate than they really are. The source scale or resolution always has consequences for the extent of *generalisation* in feature representation and it is important not to overlook this. For instance, since a line cannot be drawn much narrower than about 0.5 mm on a 1:10,000 scale map, the minimum distance which can be represented is about 5 m. On a 1:100,000 scale map the resolution is 50 m. This also has consequences for the minimum size of area that can be shown (Goodchild 1993) with, for example, the CORINE land cover data using a minimum mapping unit of 25 ha for areal phenomena. It is very important to be aware of such characteristics, especially if the analysis involves small or irregularly shaped features.
4. A related issue is the need for caution when overlaying or otherwise integrating a set of data layers with contrasting resolutions. There is a general tendency for a *lowest common denominator* rule to apply so that, for example, the poorest resolution input data layer has by far the greatest influence on the accuracy of an overlay operation result (Burrough et al. 2015, p 261–265). This is an example of *operational uncertainty* introduced into analysis outputs through data

processing operations. If quantitative measures of accuracy are included in metadata then there are also techniques that can be used to model the implications of these for results (e.g. Heuvelink 1998; Zhang and Goodchild 2002; Rae et al. 2007). In practice, it may be worth questioning whether an appreciably coarser data layer is really needed or explicitly choosing to generalise more detailed data to a coarser resolution to improve the comparability of a set of layers. This is quite a common issue in ES assessments where, for example, terrain data tend to be at the finest resolution, land cover occupy an intermediate position and socio-economic data is the coarsest. As a consequence, raster grid sizes such as 100 m, 500 m or 1 km cells are quite widely used as intermediate resolutions for integration purposes.

5. Connected to these issues of different sources and scales is the matter of temporal consistency. While some environmental parameters (such as aspects of geology and soils) are essentially stable on annual or decadal timescales, others such as land cover, species distributions, biogeochemical properties or policy designations may be much more dynamic. When assessing such variables in a planning process it is therefore important to check that the measurements refer to dates and times which are sufficiently close to enable a meaningful comparison.
6. Following on from these points it is worth emphasising that the most detailed data are not what is always required – it should always be a matter of thinking about fitness for purpose. This is particularly true with themes (such as elevation data) where a range of resolutions are readily accessible. Fig. 5.1 shows six examples of different resolutions of elevation data for an area around the city of Oxford, UK. The top four examples all cover a 20 km by 20 km area, while the bottom two are enlargements of a 5 km by 5 km box focused on the city itself. The 1000 m data are probably too blocky as a representation of the regional terrain but the 500 m or 50 m data could suffice for this purpose. Similarly, there are no especially pronounced differences between the 5 m elevation model and 1 m LiDAR, so unless there was a particular need for the detail of the latter (e.g. for local flood risk assessment) the former could be used for many landscape planning purposes. A key matter is to think about what is really needed, rather than just selecting the most detailed data for the sake of it.
7. Lastly, it is also relevant to consider how to communicate the uncertainty associated with different data set and analysis outputs (e.g. by using differences in hue, forms of shading or animation to represent such properties). Examples of different approaches are discussed by Hunter and Goodchild (1996), Appleton et al. (2004) and Beven et al. (2015).

5.5 Conclusions

This chapter has discussed the factors that have resulted in substantial changes in the extent and availability of spatial data over the past two decades. The overview of global and European sources indicates that there is now a broad baseline of information to support the needs of landscape planning that can be supplemented by other

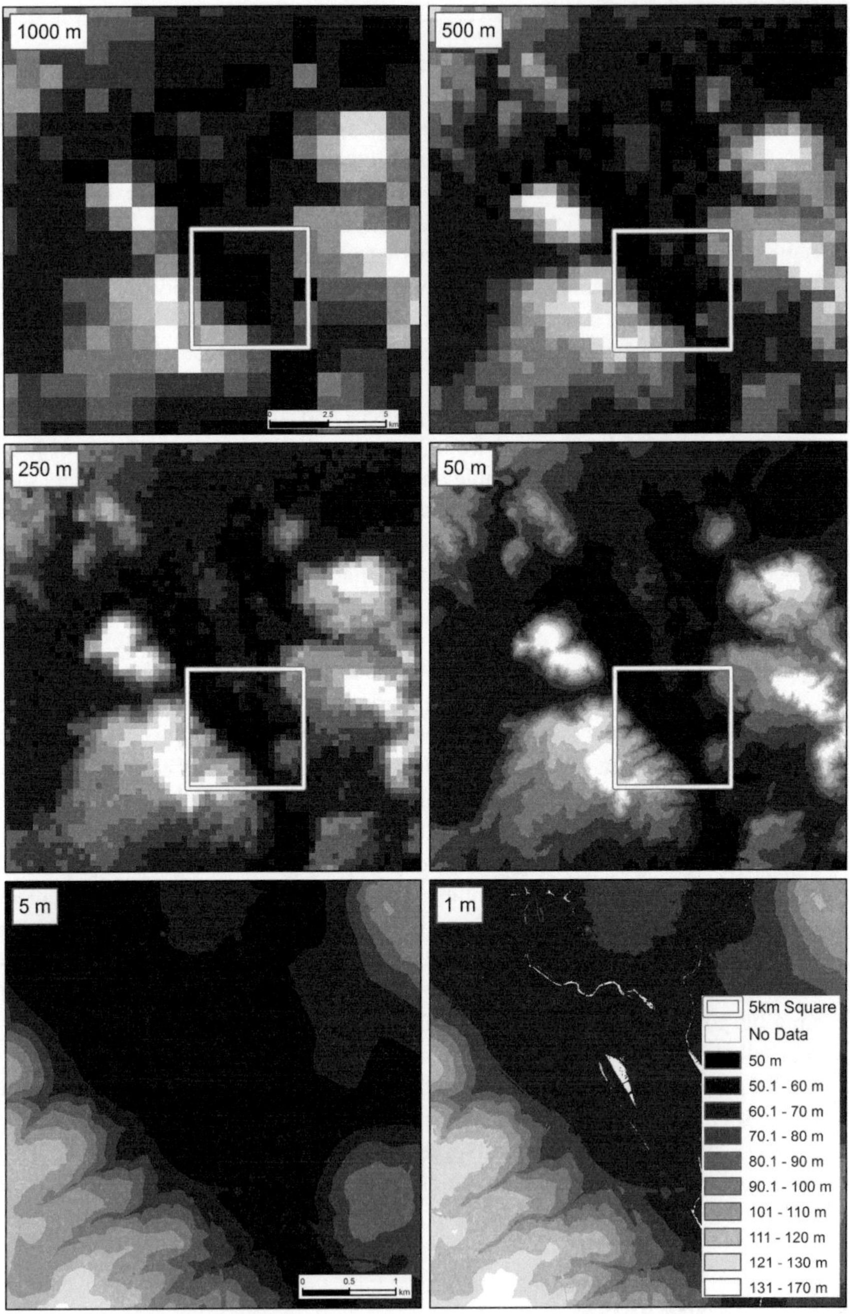

Fig. 5.1 Different resolutions of elevation data for the area around the city of Oxford, UK. (Sources: 1000 m, 500 m and 250 m from http://srtm.csi.cgiar.org/; 50 m from https://www.ordnancesurvey.co.uk/opendatadownload/products.html; 5 m from http://digimap.edina.ac.uk/ and 1 m from https://data.gov.uk/dataset/lidar-composite-dtm-1m1). Acknowledgements: Contains OS data © Crown copyright and database right (2017). Contains public sector information licensed under the Open Government Licence v3.0

data layers obtained at the national or regional scales. Indeed, it is arguable that the main challenge now is one of integrating the available data in appropriate ways and a checklist of issues has been presented to assist in this process.

Acknowledgements Work on this chapter was supported by grant number ES/LO11859/1 from the Business and Local Government Data Research Centre, funded by the UK Economic and Social Research Council to improve access to data for researchers and analysts.

References

Appleton, K., Lovett, A., Dockerty, T. et al. (2004). *Representing uncertainty in visualisations of future landscapes*. Paper presented at ISPRS Congress, Istanbul. http://www.cartesia.org/geodoc/isprs2004/comm4/papers/385.pdf. Accessed 14 June 2018.

Bateman, I. J., Harwood, A. R., Mace, G. M., et al. (2013). Bringing ecosystem services into economic decision-making: Land use in the United Kingdom. *Science, 341*, 45–50. https://doi.org/10.1126/science.1234379.

Beven, K., Lamb, R., Leedal, D., et al. (2015). Communicating uncertainty in flood inundation mapping: A case study. *International Journal of River Basin Management, 13*, 285–295.

Burrough, P. A., McDonnell, R. A., & Lloyd, C. M. (2015). *Principles of geographical information systems* (3rd ed.). Oxford: Oxford University Press.

European Commission. (2007). *INSPIRE Directive 2007/2/EC Infrastructure for spatial information in the european community*. https://inspire.ec.europa.eu/inspire-legislation/26. Accessed 14 Aug 2017.

European Data Portal. (2016). *Open data maturity in Europe 2016 – Insights into the European state of play*. https://www.europeandataportal.eu/sites/default/files/edp_landscaping_insight_report_n2_2016.pdf. Accessed 14 Aug 2017.

Goodchild, M. F. (1993). Data models and data quality: Problems and prospects. In M. F. Goodchild, B. O. Parks, & L. T. Steyaert (Eds.), *Environmental modeling with GIS* (pp. 94–103). Oxford/New York: Oxford University Press.

Heuvelink, G. B. M. (1998). *Error propagation in environmental modelling with GIS*. London: Taylor & Francis.

Hunter, G. J., & Goodchild, M. F. (1996). Communicating uncertainty in spatial databases. *Transactions in GIS, 1*, 13–24.

Lillesand, T. M., Kiefer, R. W., & Chipman, J. W. (2015). *Remote sensing and image interpretation* (7th ed.). Hoboken: Wiley.

Longley, P. A., Goodchild, M. F., Maguire, D. J., et al. (2015). *Geographic information science and systems* (4th ed.). Hoboken: Wiley.

Maune, D. F. (Ed.). (2007). *Digital elevation model technologies and applications: The DEM users manual* (2nd ed.). Bethesda: American Society for Photogrammetry and Remote Sensing.

Milner, G. (2016). *Pinpoint: How GPS is changing our world*. London: Granta Books.

Rae, C., Rothley, K., & Dragicevic, S. (2007). Implications of error and uncertainty for an environmental planning scenario: A sensitivity analysis of GIS-based variables in a reserve design exercise. *Landscape and Urban Planning, 79*, 210–217.

Sui, D. Z., Elwood, S., & Goodchild, M. (Eds.). (2013). *Crowdsourcing geographic knowledge – Volunteered Geographic Information (VGI) in theory and practice*. Dordrecht: Springer.

Taigel, S., Lovett, A. A., & Sünnenberg, G. (2017). *Integrating spatial data sources to develop a representation of green infrastructure for local government* (ESRC Business and local government data research centre working paper WPS2016–19). http://www.blgdataresearch.org/portfolio-item/integrating-spatial-data-sources-to-develop-a-representation-of-green-infrastructure-for-local-government/. Accessed 14 Aug 2017.

Zhang, J., & Goodchild, M. (2002). *Uncertainty in geographical information*. London: Taylor & Francis.

Using GIS in Landscape Planning

6

Daniela Kempa and Andrew A. Lovett

Abstract
Geographical Information Systems (GIS) are effective tools for data integration, analysis and display that are central to many aspects of landscape planning. This chapter provides an overview of some key GIS concepts and developments regarding landscape planning. We also provide possible sources of software and some examples for different applications where GIS has been employed. The final section discusses the strengths, weaknesses, opportunities and threats associated with GIS applications in landscape planning.

Keywords
Geographical Information Systems · Computer software · Data integration · Spatial analysis · Decision support

6.1 Introduction and Definitions

Spatial coordination and policy integration are central to many tasks in modern landscape planning. Assessment of development options, habitat connectivity, the valuation of ecosystem services and enhancement of multifunctional land use, to name just a few, all benefit from the strengths of technologies for spatial data processing. These technologies include geographical information systems (GIS), global

D. Kempa (✉)
Institute of Environmental Planning, Leibniz University Hannover, Hannover, Germany
e-mail: kempa@umwelt.uni-hannover.de

A. A. Lovett
School of Environmental Sciences, University of East Anglia, Norwich, UK
e-mail: a.lovett@uea.ac.uk

C. von Haaren et al. (eds.), *Landscape Planning with Ecosystem Services*,
Landscape Series 24, https://doi.org/10.1007/978-94-024-1681-7_6

positioning systems (GPS) and software for digital image processing. Not surprisingly, GIS and associated technologies have become an integral part of landscape planning practice today.

GIS are computerized information systems, containing hardware, software and data. A GIS provides structures and tools for capturing, storing, analysing, modelling, transforming and presenting spatial and attribute data referenced to the Earth (Department of the Environment 1987; Longley et al. 2015). This ability to link spatial and attribute data is central to many of the needs of landscape planning, especially because it also offers scope for integrating information collected for different geographical units. Modern GIS provide tools for many types of two- or three-dimensional data analysis and display. Whether planning questions are about site suitability, change detection, connectivity of features or assessment of management options, by skillful combination of tools it is possible to efficiently generate results and present them in a variety of visual formats.

This chapter discusses some key conceptual, technical and practical issues associated with GIS applications in landscape planning. It does not seek to provide a 'how to' guide to GIS analysis and for such information the reader is directed to sources such as Kennedy (2013) and de Smith et al. (2015).

6.2 Data and Software for Landscape Planning

The data used in a GIS are essentially of two main types. These are:

- Spatial data describing the geographical locations of features
- Attribute data describing the characteristics of those features in numerical or text formats

The way these types of data are connected varies between different types of *data structure*. Vector structures represent features as points, lines or polygons. The geographical positions of these features are defined by sets of x, y coordinates and the attribute information is stored in separate tables and linked to the spatial data through common feature ID codes. Raster structures represent phenomena as sets of attribute values in regularly-spaced grid cells. This approach is particularly suitable for continuously varying phenomena (e.g. elevation) or where data are collected through such a regularly-spaced monitoring framework (e.g. many types of airborne or satellite sensors). It is also efficient in terms of processing speed for many analysis purposes, but the representation of features can become quite distorted depending on how their size compares with the resolution of the grid cells. Further details of such structures and more complex extensions are given by Zeiler and Murphy (2010).

Features such as road networks, property boundaries and administrative data are usually stored in vector structures, while digital elevation models and land cover are most commonly provided in raster formats. As discussed in Chap. 5 there are now a wide diversity of spatial data sources, with much information available through online geoportals. Consequently, it is important to ensure that the available data are integrated in appropriate ways so that the resulting database is as '*fit for purpose*' as possible. Section 5.3 provides guidance on this matter.

The range of software available to GIS users is now extensive. Table 6.1 provides details of some commonly used products. The market for proprietary GIS is still dominated by ArcGIS from ESRI, widely used at many universities, planning authorities and consultancies. However, ArcGIS is now facing increasing competition from free and open source GIS, such as QGIS, OpenJUMP and many others. In addition, the capabilities of software such as ArcGIS and QGIS can be extended and customised by using programming languages such as Python and R (Brunsdon and Comber 2015; Yang 2017).

Furthermore, there is rapid development of applications using online platforms such as ArcGIS Online and Google Earth (e.g. Harwood et al. 2015) and smartphone apps for navigation, data collection and data display (e.g. Bishop 2015; Gill and Lange 2015). These innovations offer enormous opportunities for activities such as crowdsourcing of data or communication of information and their use in landscape planning that will become much more widespread in coming years.

Table 6.1 GIS software and apps for landscape planning purposes

Name	Details and Internet address
ArcGIS	Very widely used proprietary GIS software – http://www.esri.com/arcgis/about-arcgis
Terrset/ IDRISI	Offers an extensive range of raster analysis, image processing and modelling tools – https://clarklabs.org/
QGIS	Widely used free and open-source cross-platform desktop GIS – https://qgis.org/
OpenJUMP	An open source GIS written in the Java programming language – http://www.openjump.org/
OSGeo	The Open Source Geospatial Foundation (OSGeo) is a not-for-profit organization whose mission is to foster global adoption of open geospatial technology. The website has links to many different software products and resources – https://www.osgeo.org/
Python	Programming language widely used for spatial data manipulation and analysis – https://www.python.org/
R	R is a free software environment for statistical computing and graphics – https://www.r-project.org/
ArcGIS Online	A cloud-based online mapping service – https://www.arcgis.com
Google Earth	Virtual globe with extensive imagery resources and customisation capabilities – https://www.google.com/earth/
OsmAnd	An open source GPS Navigation and map application that runs on many Android and iOS devices – http://osmand.net/
Collector for ArcGIS	Smartphone app for field data collection – http://www.esri.com/products/collector-for-arcgis

Note: Compiled from information available in May 2018

6.3 GIS in Landscape Analysis and Presentation

GIS can be used at many different stages and scales of the landscape planning process. It is also worth noting that landscape planning has had an important influence on the development of analysis tools within GIS since one of the key concepts, polygon overlay, was first introduced by McHarg (1969) in his book 'Design with Nature'.

Following the framework of the DPSIR model (see Chap. 3), GIS can be used to record, assess and display the state of a landscape and to assess pressures by recalculating or updating input data. Impacts, such as e.g. changes in the delivery of ecosystem services, can then be evaluated by models based on the overlay of spatial data (e.g. Lee and Thompson 2005) or the assessment of connectivity between locations. One common example of the latter is the assessment of habitat connectivity and the ecological implications of this for areas to support individual species or wider biodiversity (Nikolakaki 2004; Marulli and Mallarach 2005). Drivers (e.g. agricultural policy) and responses (e.g. new regulations) are, as a rule, more difficult to include in GIS analysis unless they can be translated into spatially explicit statements.

Similarly, GIS is applicable in all six themes of the widely-used geodesign framework for landscape planning proposed by Steinitz (1990, 2012) (see also Chap. 27). Within the themes of Representation, Process, Evaluation, Change, Impact and Decision, the GIS functions of data integration, analysis and display are particularly important and helpful. Related to the Change and Impact themes many GIS-based analyses are concerned with 'what if?' questions. Typically, these are either based on expert judgement to overlay different factors to identify priority sites or incorporate stakeholder opinion through techniques such as multi-criteria evaluation to assess the suitability of land for particular purposes (Malczewski 2004; Bailey et al. 2006; Watson and Hudson 2015).

The geodesign framework also emphasises landscape design as process-oriented planning, one that informs and convinces not only by scientific statements but also by participation. Working with planning scenarios in GIS, stakeholder participation can be supported and benefit from spatial functionalities, such as overlay, turn off/on of information layers, simulation of time series for landscape changes, display of different options for future development etc. (Kwartler 2005; Walker and Daniels 2011). Such tools make it easier to handle complex decision processes for stakeholders and support communication of decisions. Today, decision processes are characterised by higher requirements for spatial coordination, not least because economic development policies are increasingly changing from a sectoral focus to one that is place-based (e.g. Organisation for Economic Co-operation and Development 2006; Department for Business, Energy & Industrial Strategy 2017).

Producing landscape plans in a digital format from a GIS is advantageous because of the opportunities to publish and disseminate by web portals and social media. Consequently, landscape plans can reach a wider audience and contribute to better public information than if they were in an analog format. Adding interactive and multimedia techniques, GIS-supported public participation can also lead to greater interest in planning results and improvement of implementation (von Haaren and Warren-Kretzschmar 2006; Galler et al. 2014). Some municipalities are already testing these possibilities by presenting their plans on the web with interactive tools enabling georeferenced comments and annotation of landscape information (http://koenigslutter.entera-online.com/entera/mapserv.phtml). Other projects use GIS data to underpin virtual reality 3D-tools to discuss landscape changes due to renewable energy projects (e.g. http://lenne3d.com/category/referenzen/energie/). New forms of interactive commenting in webGIS are increasingly used to retrieve landscape knowledge from the public. This kind of crowd sourcing is used by informal types of georeferenced data mining (e.g. google earth with google pictures; flickr; panoramio) but also by authorities for formal applications, e.g. georeferenced data on species occurrence (e.g. Rüter et al. 2010).

Another current area of GIS application is the integration of data from the natural environment and socio-economic realms to develop spatially explicit inventories of ecosystem services (Maes et al. 2013; Norton et al. 2018). GIS has become a particularly important tool for initiatives to apply monetary values to such natural resources through techniques such as travel cost and hedonic pricing, or to transfer benefit estimates from one location to another (Bateman et al. 2002; Troy and Wilson 2006). These techniques are further discussed in Chap. 20.

In the next section two examples of GIS application for landscape planning are described at greater length.

6.4 Examples of GIS Applications in Landscape Planning

6.4.1 Identifying Possible Sites for Biomass Crops in England

The first example is based on a process of expert judgement and overlay to identify priority sites. Lovett et al. (2009) examined the scope for increased planting of Miscanthus (a type of perennial biomass crop) in England. Such crops have the potential to offer more sustainable sources of energy, but there have been concerns about their possible impacts on other ecosystem services such as food production, landscape amenity and water resources. A GIS was first used to estimate spatial variations in Miscanthus yield and then a range of possible constraints were overlaid to mask out unsuitable areas due to factors such as natural habitats, cultural heritage and landscape sensitivity. As shown in Fig. 6.1, the resulting maps were then further analysed to avoid planting on the best quality (Grades 1 and 2) agricultural land and assess the implications for energy generation and possible displacement of food crops. The findings suggested that it would be possible to meet current planting targets without adverse impacts and maps such as those in Fig. 6.2 were used to discuss possible sites with communities where there was interest in developing biomass heat and power generation. In this instance, therefore, the use of GIS allowed potential land-use conflicts to be evaluated and provided a basis for targeting of funding initiatives.

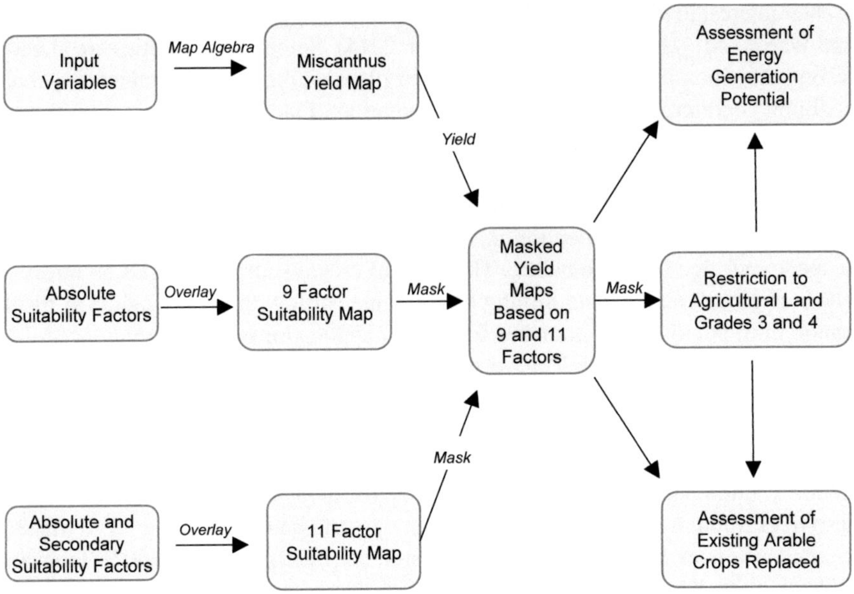

Fig. 6.1 Steps in the assessment of land suitable for Miscanthus planting in England. (Source: adapted from Lovett et al. 2009: 19)

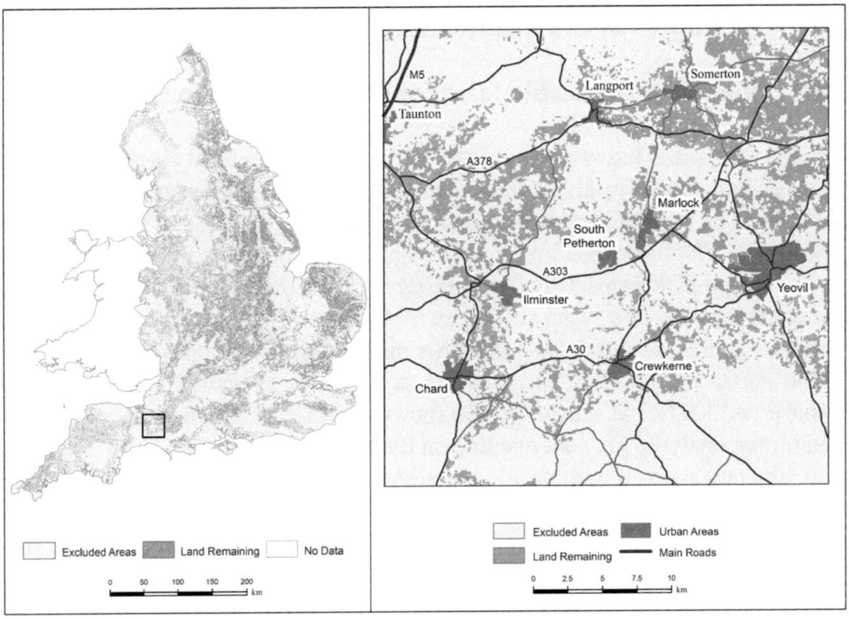

Fig. 6.2 Grade 3 or 4 land outside the 11 constraints considered in the overlay analysis. National results for England with an enlargement of a region in Somerset and Dorset. (Source: partially redrawn from Lovett et al. 2009: 23)

Lovett et al. (2014) subsequently extended the approach to consider landscape naturalness as a site suitability factor and compared the results with those based on a simple exclusion of designated areas such as National Parks and Areas of Outstanding Natural Beauty. The results were also validated against the distribution of biomass crop planting supported by a government initiative and 83% of this was found to be in identified suitable areas and a further 12% on other arable land or improved grassland. These results therefore gave confidence in the robustness of the approach.

6.4.2 Biodiversity Assessment and Decision Support for Farmers

As a second example a tool for farm biodiversity assessment is presented. It combines expert assessment based on data overlay and information for stakeholders to provide decision support.

Farmers are paid by EU agri-environmental programs if compliance with national environmental protection or nature conservation aims is achieved. A major aim of these programs is the maintenance and promotion of agrobiodiversity and ecosystem services. To measure and compare on-farm biodiversity a standardised system is necessary, one which can assess spatial information about state and pressures and identify suitable management options.

To respond to such needs, the Institute of Environmental Planning at Leibniz University Hannover developed a farm assessment system (MANUELA, cf. von Haaren et al. 2012) as a software PlugIn for the open source GIS OpenJump. MANUELA provides a toolset to produce an assessment of biodiversity including habitats, species, habitat development potential, habitat connectivity (see Chaps. 17 and 18) and the influence of land use. Nearly all these assessments depend on site properties, such as soil types, crop choices and other cultivation options. Therefore, the role of GIS is central in providing IT support for biodiversity assessment on farms. Additional advantages of a digital GIS-supported system include the scope to combine the biodiversity assessment with precision farming applications (e.g. GPS-guided fertilizer applications) and to ensure standardised automated processing. The latter includes repeatability, modelling of alternatives and updating of data as well as options for data import or export. Figure 6.3 shows an annotated view of the software user interface. Data import/export and assessment modules can be activated by a submenu and results are presented in a map window, in table or graph format. Results, such as habitat values or connectivity are presented on the map display with graded colours and additional information in tables can be accessed by a click on a field or element. This specific information can be used by the farmer or farm advisor to plan different management options or to locate agri-environmental measures.

The expert underpinnings of the biodiversity assessment are derived from ecological risk analysis theory, literature reviews and statistical analyses of recorded farm data. With the help of GIS these complex methods are automated and designed

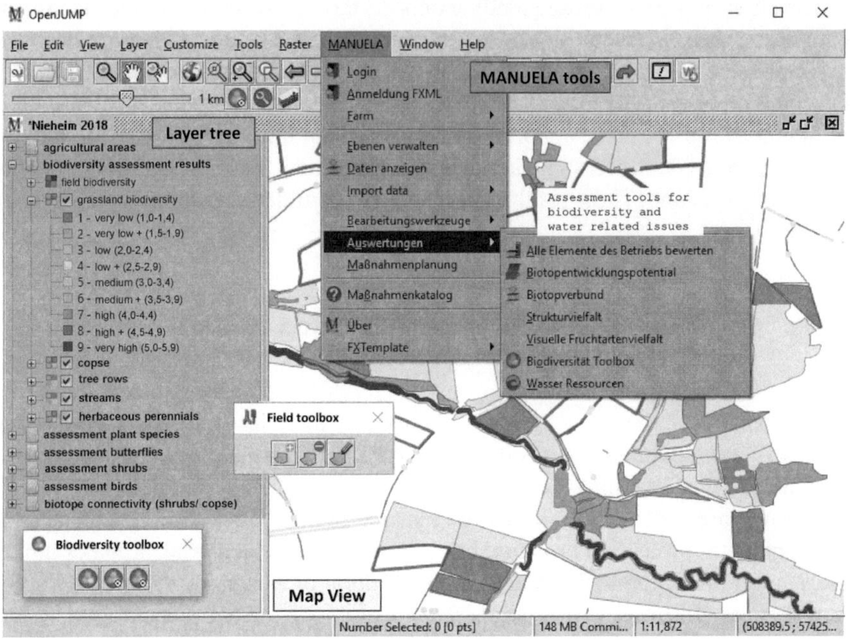

Fig. 6.3 Annotated user interface of the farm management software MANUELA

to provide simple but well-founded assessment results. The outputs can be used to determine and locate farm management options to enhance habitat quality or connectivity, and thus help to improve biodiversity on the farm. However, interpretation of the results is necessary to reach an appropriate decision and to choose suitable agri-environmental measures. These decisions can, in turn, be digitally recorded, to document targeted, performance-based agri-environmental measures and biodiversity efforts. The latter can also play an important role for food companies interested in environmental documentation, as they can use it to satisfy consumer demands for goods produced by environmentally-friendly means (Kempa 2012, 2013).

6.5 Strengths, Weaknesses, Opportunities and Threats of GIS Applications

From the previous discussion and examples, it is apparent that there are many strengths that can help address stages of the DPSIR model or the key landscape planning questions outlined by Steinitz (1990, 2012). These are evident in the wide range of applications and stem particularly from the ability of GIS for place- based data integration, as well as capabilities to investigate 'what if?' issues and present information in forms that support communication and participation in decision making.

GIS-based processing is indispensable for planning and analyses purposes, particularly those that due their complexity would not be feasible with analogue working methods. For example, land use classification based on satellite images can only be conducted with GI systems, because images are provided as digital data and classification algorithms need computer processing. Similarly, the analysis of high-resolution digital elevation models, e.g. for erosion modelling or visibility studies, requires GIS due to its high calculation requirements.

Other strengths of GIS are the ability to manage and merge different kind of data, methods and forms of presentation in one system. Information is often provided as either text, vector-based, in a raster grid format or in numbers. With the help of GIS these datasets can be imported and managed in one database and spatial (vector or raster) data can be related to textual or statistical information. Information from the database then can also be depicted in maps, tables, statistics, diagrams etc. and (if necessary) exported for further use in other software.

Due to standardisation and automation of analysis processes in GIS, regular updating for monitoring purposes (e.g. distributions of high nature value species) or forward projection of landscape plans is facilitated. By using models and scripts (e.g. Model Builder in ArcGIS, Python or R) an automated sequence of tools can be created to minimize errors in repeated processes. This has potential in many types of applications, including recording and assessment of ecosystem services in EU member states, as specified by the EU Biodiversity Strategy (cf. European Commission 2011).

However, standardization assumes that comparable data sources in terms of scale and classification are available and accessible for the areas considered (e.g. CORINE for EU-wide land cover information). In terms of weaknesses, it is important to recognise that the analysis options within GIS are always only as good as underlying data. Therefore, significant conceptual and practical challenges often need to be addressed. This includes feature representation since many elements of a landscape are inherently 'fuzzy' but are often simplified to be depicted on a precise 'hard' basis (Fisher 2009). These issues are further discussed in Sect. 5.4.

Restricted data availability and simplification of 'real' landscapes are also reasons for uncertainty in ecosystem service assessment results. The extent to which uncertainty is quantified or explicitly modelled is often limited. Sensitivity analysis or land use scenario modelling are first steps and opportunities to measure uncertainty but are still not common and face considerable challenges (cf. Lowell 2008). The need to represent and communicate such uncertainties is especially important where policy options are contested and is likely to become increasingly significant as landscape planning seeks to address the challenges posed by climate change (Sheppard 2015).

Opportunities for future enhancement of spatial data can be found in improvement of remote sensing data in terms of costs, coverage and higher resolution of images. This includes radar and multispectral products from the Sentinel satellites https://sentinel.esa.int/web/sentinel/home) and daily 3 m resolution imagery from the Planet satellite constellation (https://www.planet.com/). Another promising development is the extraction of data from crowd sourcing. Information on landscape characteristics can be collected by the lay public and volunteers, then shared with wider communities via GPS-enabled mobile apps and web portals. These huge amounts of data can be used by experts to cover gaps in more specialist databases. However, attention must be paid to data quality as standardisation of this data is often on a low level and copyrights need to be clarified.

Crowd sourcing mechanism can nonetheless be very powerful in terms of facilitating community input and engagement in landscape planning and conservation matters. With continued enhancements of internet bandwidth it will also become feasible to not just derive and analyse community information online, but to also provide real time visualisation and modelling of scenarios for community engagement in planning processes. Figure 6.4 gives an overview of strengths, weaknesses, opportunities and threats of GIS applications in landscape planning.

As a final point it is essential to recognise that a technology such as GIS is used within societal and political context. The above discussion has focused on what is technically possible with GIS, but many types of more innovative applications may not progress far without active interest from the relevant agencies, data availability, legislative support for whole landscape planning, financial incentives for co-operation between land owners etc. In the current financial climate of austerity there may well be resourcing constraints and the extent to which there are governance frameworks in place to support landscape-scale collaboration also varies between countries (see Chap. 30).

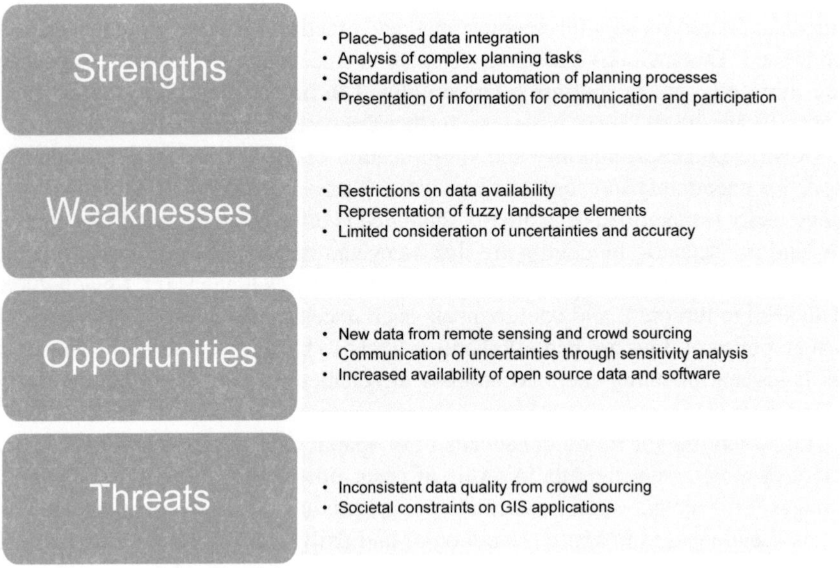

Fig. 6.4 Strengths, weaknesses, opportunities and threats for GIS in landscape planning

Acknowledgements Work on this chapter was supported by grant number ES/LO11859/1 from the Business and Local Government Data Research Centre, funded by the UK Economic and Social Research Council to improve access to data for researchers and analysts.

References

Bailey, N., Lee, J. T., & Thompson, S. (2006). Maximising the natural capital benefits of habitat creation: Spatially targeting native woodland using GIS. *Landscape and Urban Planning, 75*, 227–243.

Bateman, I. J., Jones, A. P., Lovett, A. A., et al. (2002). Applying geographical information systems (GIS) to environmental and resource economics. *Environmental and Resource Economics, 22*, 219–269.

Bishop, I. D. (2015). Location based information to support understanding of landscape futures. *Landscape and Urban Planning, 142*, 120–131.

Brunsdon, C., & Comber, L. (2015). *An introduction to R for spatial analysis and mapping*. London: Sage.

de Smith, M. J., Goodchild, M. J., & Longley, P. A. (2015). *Geospatial analysis – A comprehensive guide* (5th ed.). http://www.spatialanalysisonline.com/HTML/index.html. Accessed July 2018.

Department for Business, Energy & Industrial Strategy. (2017). *Industrial strategy: Building a Britain fit for the future*. London: Department for Business, Energy & Industrial Strategy.

Department of the Environment. (1987). *Handling geographic information: Report of the committee of enquiry chaired by Lord Chorley*. London: HMSO.

European Commission. (2011). *The EU biodiversity strategy to 2020*. Luxembourg: Publications Office of the European Union.

Fisher, P. (2009). The representation of uncertain geographic information. In M. Madden (Ed.), *Manual of geographic information systems* (pp. 235–264). Bethesda: ASPRS.

Galler, C., Krätzig, S., Warren-Kretzschmar, B., et al. (2014). Integrated approaches in digital/interactive landscape planning. In U. Wissen Hayek, P. Fricker, & E. Buhmann (Eds.), *Peer reviewed proceedings of digital landscape architecture* (pp. 70–83). Berlin: Wichmann.

Gill, L., & Lange, E. (2015). Getting virtual 3D landscapes out of the lab. *Computers, Environment and Urban Systems, 54*, 356–362.

Harwood, A. R., Lovett, A. A., & Turner, J. A. (2015). Customising virtual globe tours to enhance community awareness of local landscape benefits. *Landscape and Urban Planning, 142*, 106–119.

Kempa, D. (2012). *Bedingungen für den Einsatz einer Software für Naturschutzberatung und -dokumentation auf landwirtschaftlichen Betrieben* PhD dissertation. Leibniz Universität Hannover.

Kempa, D. (2013). Environmental services coupled to food products and brands: Food companies interests and on-farm accounting. *Journal of Environmental Management, 127*, 184–190.

Kennedy, M. (2013). *Introducing geographic information systems with ArcGIS: A workbook approach to learning GIS* (3rd ed.). Hoboken: Wiley.

Kwartler, M. (2005). Visualization in support of public participation. In I. Bishop & E. Lange (Eds.), *Visualization in landscape and environmental planning* (pp. 251–260). London: Taylor & Francis.

Lee, J. T., & Thompson, S. (2005). Targeting sites for habitat creation: An investigation into alternative scenarios. *Landscape and Urban Planning, 71*, 17–28.

Longley, P. A., Goodchild, M. F., Maguire, D. J., et al. (2015). *Geographic information science and systems* (4th ed.). London: Wiley.

Lovett, A. A., Sünnenberg, G. M., Richter, G. M., et al. (2009). Land use implications of increased biomass production identified by GIS-based suitability and yield mapping for Miscanthus in England. *Bioenergy Research, 2*, 17–28.

Lovett, A. A., Sünnenberg, G. M., & Dockerty, T. L. (2014). The availability of land for perennial energy crops in Great Britain. *Global Change Biology Bioenergy, 6*, 99–107.

Lowell, K. E. (2008). Uncertainty in landscape models: Sources, impacts and decision making. In C. Pettit, W. Cartwright, I. Bishop, et al. (Eds.), *Landscape analysis and visualisation* (pp. 367–382). Berlin: Springer.

Maes, J., Teller, A., Erhard, M., et al. (2013). *An analytical framework for ecosystem assessments under action 5 of the EU biodiversity strategy to 2020* (Mapping and assessment of ecosystems and their services). Luxembourg: Publications Office of the European Union.

Malczewski, J. (2004). GIS-based land-use suitability analysis: A critical overview. *Progress in Planning, 62*, 3–65.

Marulli, J., & Mallarach, J. P. (2005). A GIS methodology for assessing ecological connectivity: Application to the Barcelona metropolitan area. *Landscape and Urban Planning, 71*, 243–262.

McHarg, I. L. (1969). *Design with nature*. New York: Natural History Press.

Nikolakaki, P. (2004). A GIS site-selection process for habitat creation: Estimating connectivity of habitat patches. *Landscape and Urban Planning, 68*, 77–94.

Norton, L. R., Smart, S. M., Maskell, L. C., et al. (2018). Identifying effective approaches for monitoring national natural capital for policy use. *Ecosystem Services, 30*, 98–106.

Organisation for Economic Co-operation and Development. (2006). *The new rural paradigm: Policies and governance*. Paris: OECD Publications.

Rüter, S., Hachmann, R., Krohn-Grimberghe, S., et al. (2010). *GIS-gestütztes Gebietsmonitoring im ehrenamtlichen Naturschutz. (inkl. DVD-Beilage)*. Stuttgart: ibidem-Verlag.

Sheppard, S. R. J. (2015). Making climate change visible: A critical role for landscape professionals. *Landscape and Urban Planning, 142*, 95–105.

Steinitz, C. (1990). A framework for the theory applicable to the education of landscape architects. *Landscape Journal, 9*, 136–143.

Steinitz, C. (2012). *A framework for Geodesign: Changing geography by design*. Redlands: ESRI Press.

Troy, A., & Wilson, M. A. (2006). Mapping ecosystem services: Practical challenges and opportunities in linking GIS and value transfer. *Ecological Economics, 60*, 435–449.

von Haaren, C., & Warren-Kretzschmar, B. (2006). The interactive landscape plan: Use and benefits of new technologies in landscape planning and discussion of the interactive landscape plan in Koenigslutter am Elm. Germany. *Landscape Research, 31*, 83–105.

von Haaren, C., Kempa, D., Vogel, K., et al. (2012). Assessing biodiversity on the farm scale as basis for ecosystem service payments. *Journal of Environmental Management, 113*, 40–50.

Walker, D., & Daniels, T. (2011). *The planners guide to communityviz: The essential tool for a new generation of planning*. Chicago: Orton Family Foundation Books/American Planning Association/Planners Press.

Watson, J. J. W., & Hudson, M. H. (2015). Regional scale wind farm and solar farm suitability assessment using GIS-assisted multi-criteria evaluation. *Landscape and Urban Planning, 138*, 20–31.

Yang, C. (2017). *Introduction to GIS programming and fundamentals with Python and ArcGIS®*. Boca Raton: CRC Press.

Zeiler, M., & Murphy, J. (2010). *Modeling our world* (2nd ed.). Redlands: ESRI Press.

Part II
Sources of Drivers and Pressures

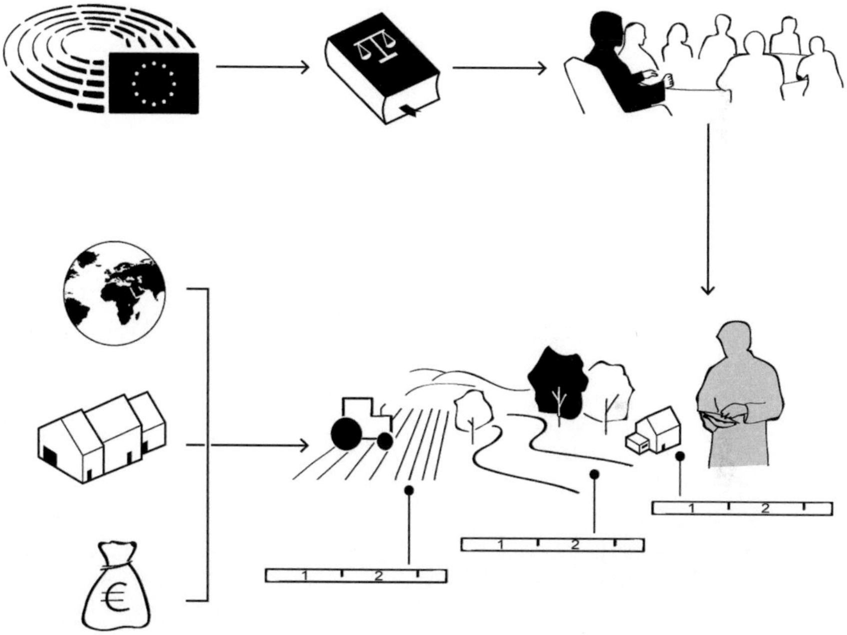

An overview of the state of ecosystem services in Europe introduces this part of the book. Relevant European Union policies and national legislative frameworks are presented as drivers which influence the state of ecosystem services. Other drivers such as global change, market forces and the ubiquitous interests of people (i.e. settlement development) also generate pressures on biodiversity and ecosystem services. Knowledge about these framing conditions enables landscape planners to understand what can (or cannot) be changed in specific cases. Methods for the assessment of pressures are introduced as a first step of the planning process.

Ecosystem Services Under Pressure

7

Felix Kienast, Julian Helfenstein, Adrienne Grêt-Regamey, Roy Haines-Young, and Marion Potschin

Abstract

Three 'megatrends' have had a particular influence on ecosystem service provision and trade-offs in Europe. These are (i) the globalization of industrial and agricultural production, (ii) the changing lifestyles of a post-industrial society and (iii) the production of green energy. This chapter discusses the pressures arising from these trends and employs land use/cover data to cluster European landscapes into regions with similar changes in land use and associated ecosystem services.

Keywords

Ecosystem services · Megatrends · Land use change · Meta-analysis · Trade-offs

---

F. Kienast (✉)
Swiss Federal Institute for Forest, Snow and Landscape Research WSL, Birmensdorf, Switzerland
e-mail: felix.kienast@wsl.ch

J. Helfenstein
Department of Environmental Systems Science, Swiss Federal Institute of Technology (ETH) Zurich, Zürich, Switzerland
e-mail: julian.helfenstein@usys.ethz.ch

A. Grêt-Regamey
Swiss Federal Institute of Technology (ETH) Zurich, Institute for Spatial and Landscape Development, Zurich, Switzerland
e-mail: gret@ethz.ch

R. Haines-Young · M. Potschin
Fabis Consulting Ltd, Nottingham, UK
e-mail: Roy.Haines-Young@Nottingham.ac.uk; marion.potschin@fabisconsulting.com

C. von Haaren et al. (eds.), *Landscape Planning with Ecosystem Services*, Landscape Series 24, https://doi.org/10.1007/978-94-024-1681-7_7

7.1 Introduction

Landscape change accelerated dramatically in the twentieth century and continues unabatedly today. This has led to a substantial loss of ecosystem goods and services[1] that are crucial for human well-being, such as the pollination of crops, flood protection, soil formation, and bio-remediation (MA 2005). Prime causes of these changes are land use alterations associated with technological development, changing lifestyles, and energy and agricultural production. In the case of some land uses associated with traditional low-impact management, some services were transformed into dis-services by modern use (Swinton et al. 2007). This is the case in many traditional agricultural landscapes that under current intensive practices contribute substantially to air and water pollution, greenhouse gas emissions, erosion, and biodiversity loss (Helfenstein and Kienast 2014). The interest in these historical land use trajectories and how they have influenced ecosystem service provision has recently gained attention (Bürgi et al. 2012, 2015). This complements current mainstream research on ecosystem services, which has tended to overlook the temporal dynamics of ecosystem service provision.

In this chapter we focus on three 'megatrends' and how they have altered ecosystem service provision and trade-offs in Europe in some selected cases. These megatrends are; (a) the globalization of industrial and agricultural production; (b) the changing lifestyles of a post-industrial society; and (c) the production of green energy. Conceptually, ecosystem service trade-offs can be assessed through supply and demand for ecosystem services as suggested by Burkhard et al. (2012) and Nedkov and Burkhard (2012). For a detailed debate on the ecosystem service trade-off paradigm see Seppelt et al. (2011) and Potschin and Haines-Young (2011).

To better illustrate our assessment of the impacts of the mentioned megatrends, we employ a land use intensity gradient ranging from urban to natural, characterised by decreasing human disturbance (Fig. 7.1). Most of the observed ecosystem service trade-offs in Europe and elsewhere can be attributed to land use shifts along this gradient (Braat and ten Brink 2008; Haines-Young 2009; Potschin and

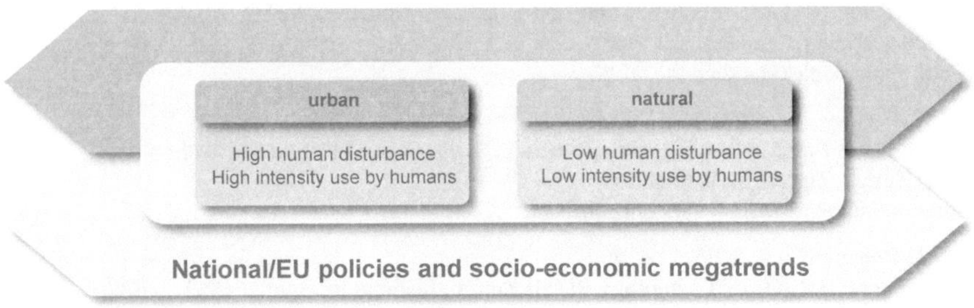

Fig. 7.1 The urban-natural gradient used to illustrate trade-offs in ecosystem service provision. (Modified after Haines-Young 2009)

---

[1] All ecosystem services mentioned in this chapter are named after the CICES 4.3 classification, see Haines-Young and Potschin (2013).

Haines-Young 2013). Currently policy instruments in Europe, such as the Natura 2000 or the Common Agricultural Policy (CAP), try to control the shifts along the gradients exhibited in Fig. 7.1 with regulatory measures or economic incentives. We emphasize, however, that less human disturbance does not implicitly mean that more or more adequate ecosystem services are provided. On the contrary, many services that are important for example for biodiversity or cultural heritage are only brought about by human disturbance. During the course of European history, massive land use shifts were experienced along the urban to natural gradient. For example, the founding of German cities in the middle ages led to a considerable increase in the amount of 'urban' areas. Only after about 1870, with the industrial revolution, did urban populations start to boom. The urban population in Germany increased from 12% in the 1870s to 22% in 1890 (Brophy 2011). By the turn of the twenty-first century the share of urban population reached 70–95% in many parts of Europe. The subsequent marginalization of less favorable land has been a major cause of the abandonment in rural areas (Modica et al. 2012). These significant changes along the gradient, shown in Fig. 7.1, were catalysed by the three megatrends discussed here.

7.2 Megatrends and Their Effects on Ecosystem Services in Europe

7.2.1 Megatrend 1: Globalization of Industrial and Agricultural Production

This megatrend is – irrespective of any regulatory framework – responsible for major landscape changes in Europe and corresponding trade-offs in ecosystem service provisioning. After the 1970s many labour-intensive businesses relocated to emerging economies with low unit labour costs and frequently low emission standards. This trend seemed to flatten out at the turn of the twenty-first century as labour costs and production standards increased in emerging economies, and wage costs became less important compared to technical innovation and distribution. The departure of many environmentally critical businesses led to the improvement of several ecosystem services in Europe, as polluting practices were 'out-sourced' (Lin et al. 2014). Other aspects of public life however, have been negatively influenced by the loss of small to medium manufacturing businesses, and it is undisputed that this loss contributed considerably to the decline of many small-to-medium sized municipalities.

Globalisation also left its footprint on provisioning services, such as cultivated crops. As prices for many agricultural goods decreased on the global market, farmers tried to maintain or increase profitability by intensive farming techniques, which often included removing small landscape features and the abandonment of low-yielding land. Thus, production has progressively increased since 1960 despite a decrease in farmed area. However, direct payments and other government support programs were able to slow down the abandonment process, and thanks to an increasingly wealthy urban population, agricultural niche products can increasingly be sold with relatively high net margins. This has increased farm incomes in many parts of Europe. The geographical distribution of the net loss of arable land, permanent crops, and heterogeneous agricultural areas, since 1990, is strongly dependent on landscape type (Haines-Young et al. 2012). Highest losses are found in marginal areas. Lowest rates of loss occur on highly fertile soils, such as the hotspots of arable production in the eastern part of the UK, northern France, parts of Belgium and the Netherlands, and Denmark, together with a broad sweep of land in the northern part of Germany and Poland. Areas with low potential are mountainous areas and the Nordic regions. Land use projections, such as those by Meijl et al. (2006), Verburg et al. (2006), Westhoek et al. (2006) and Verburg et al. (2009), suggest that this trend may continue to 2030, at least for a 'global market' scenario. This scenario envisages rapid economic growth, the global population peaking at around nine billion in 2050, a rapid uptake of new technologies, and globalised societies.

As human activities across the world become increasingly interconnected, the analysis of land use change and ecosystem service provision requires understanding land systems as open systems encompassing large flows of goods, people, capital, and information that connects localities across great distances and at global scales (Liu et al. 2013). While environmental interactions at broad spatial scales are nothing new, and socio-economic interactions across great distances have occurred since the beginnings of human history, external drivers of land change and thus ecosystem service provision; such as trade, global capital allocation, transnational land deals, technology transfer, policies, etc., have become more widespread and intense and rather 'placeless' (Kienast et al. 2007). As a result, land-relevant decisions are increasingly made outside national governance systems, leading to new trade-offs between ecosystem services supplied locally and demanded globally (Adger et al. 2009).

7.2.2 Megatrend 2: Changing Lifestyles of a Post-industrial Society

The changing lifestyle of an increasingly urban and mobile population is another megatrend that leaves its footprint on ecosystem services. As shown in a report by the EEA (2009), this transformation is driven by demographic changes such as migration, increased numbers of elderly people, and deferred parenthood. Further drivers of change include concurrent multiple careers, increased mobility, smaller households, and higher average floor-space requirements per capita. The economic crisis of 2009 might have slowed down some of these trends, but as a general megatrend it is likely to continue affecting ecosystem service supply and demand in the future. As a result, a number of ecosystem services – primarily regulation services – are under pressure. Increased water consumption and soil sealing are examples of some of the consequences of urbanisation and the densification of road networks. Fragmentation (see Fig. 7.2) and urban sprawl have been identified as major challenges for regulation and cultural ecosystem services in and around mid- to large-sized cites. Such impacts are often even more pronounced in cities that are shrinking.

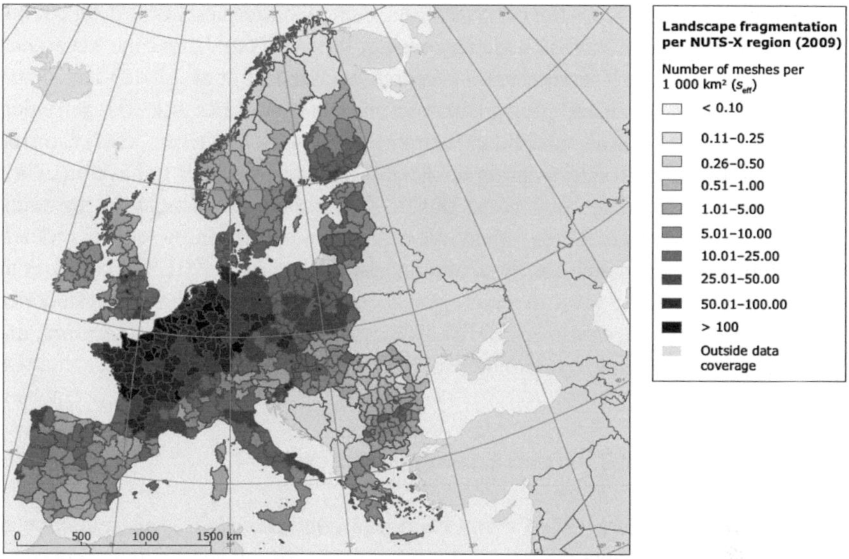

Fig. 7.2 Landscape fragmentation in Europe (2009) measured with effective mesh density. Mesh density is high when many roads, railways or settlements subdivide the land into small patches. Mesh density is low when large contiguous patches of land cover dominate. (Adapted from Jäger et al. (2011). Data basis of original file: EEA/FOEN 2011)

A recent report estimates that the negative externalities of urban sprawl in the US amount to circa $ (US) 1000 billion a year (Litman 2015). A number of measures and incentives such as defining maximum land uptake per person, zoning regulations, densification of existing built-up areas, or tax release for smart city approaches, will help to control these negative impacts.

Other sectors such as forestry have also adopted innovative management practices to meet the demand for ecosystem services. A prime example is European forest management. Central European forests are gradually being managed for multi-functionality (Farrell et al. 2000; Hahn and Knoke 2010). The realisation that forests provide other important services (notably recreation, protection from natural hazards, water filtration, and biodiversity conservation) and that these services are not guaranteed if management only concentrates on timber production, led to e.g. a new Swiss forest law in 1991. This law centred on multi-functionality (WaG 1991). In forests close to urban centres, management conducive to recreation overrides timber production objectives. Only small disturbances are allowed, tree species and age diversity is retained, and paths are maintained to suit visitors. This legislative framework sets the stage for a multifunctional forest management, where recreation, biodiversity, timber production, natural hazard protection, and other services are provided within the same landscape.

Meeting the demand for ecosystem services not only requires a supply, but also access. The increasingly sedentary and leisure-oriented urban society is a challenge for both public health and city planners. There are indications that society – and especially young people – have fewer chances to experience nature and are increasingly alienated from the everyday landscape (Charles and Louv 2009). While long distance travel and tourism are increasing, the everyday landscape experience is frequently missing. The recreation research by Kienast et al. (2012) has shown that inhabitants in urban and peri-urban environments should be able to reach easily accessible natural or semi-natural landscape structures such as rivers, creeks, or forest patches within a 10 min walking or driving distance, in order to have an everyday landscape experience and some physical exercise. Landscapes in the urban fringe are, however, frequently abandoned and need to be improved in order to enable a nature experience for their inhabitants (Home et al. 2012; Martens et al. 2011). We increasingly need to also regard urban areas as potential ecosystem service providers (Helfenstein et al. 2014). Green walls and roofs, urban farming, and other such developments, can provide such ecosystem services in a city.

7.2.3 Megatrend 3: Green Energy Production

The future energy strategies of many European countries envisage a reduction of fossil and nuclear energy sources and an increase in renewables (wind, hydropower, photovoltaic, geothermal). In the case of photo voltaics (PV) we have seen an increase in the European energy market from 0.3 TWh in 2002 to 71.0 TWh in 2012. Germany, Italy, and Greece were the major producers. In regard to wind, there was an increase from circa 75 TWh to circa 205 TWh between 2005 and 2012 (Eurostat 2014). The main producers are currently Spain, the UK, Germany, and Denmark.

While renewables reduce greenhouse gas emissions, they are also a driver of land use change because they compete with other uses and hence impact on the delivery of ecosystem services at the local scale. Wind turbines, for example, have an impact on ecosystem services at the landscape scale, effecting scenic beauty and sense of place. Since the best wind conditions usually occur on ridges or off-shore, wind turbines can also lead to conflicts with wildlife, for example birds, bats or fish (Kienast et al. 2014; Göke and Lamp 2012).

In principle, there are two options for ecosystem service management when dealing with renewables; (a) 'hide' the power generators in order to avoid conflicts with cultural ecosystem services; or (b) 'promote' – in a participatory way – energy regions where renewable energy becomes a symbol for the region and its economic prosperity. Both approaches can currently be observed, depending on the planning culture of the country concerned.

Whatever management option is adopted, however, strategies for minimising conflict with other ecosystem services are required. GIS and satellite remote-sensing technologies enable us to perform large-scale assessments and select those sites with the least impacts on other ecosystem services. Figure 7.3 illustrates such a spatial conflict analysis for photovoltaics for an area of roughly 40,000 km^2 of mountain and hilly terrain in Switzerland. PV on roofs has the least conflict with other ecosystem service unless the PV panels are placed in villages with cultural heritage sites (assuming the technology of ca. 2014). PV on semi-natural or abandoned agricultural sites has a higher conflict potential. First, there is an issue with the regulating service

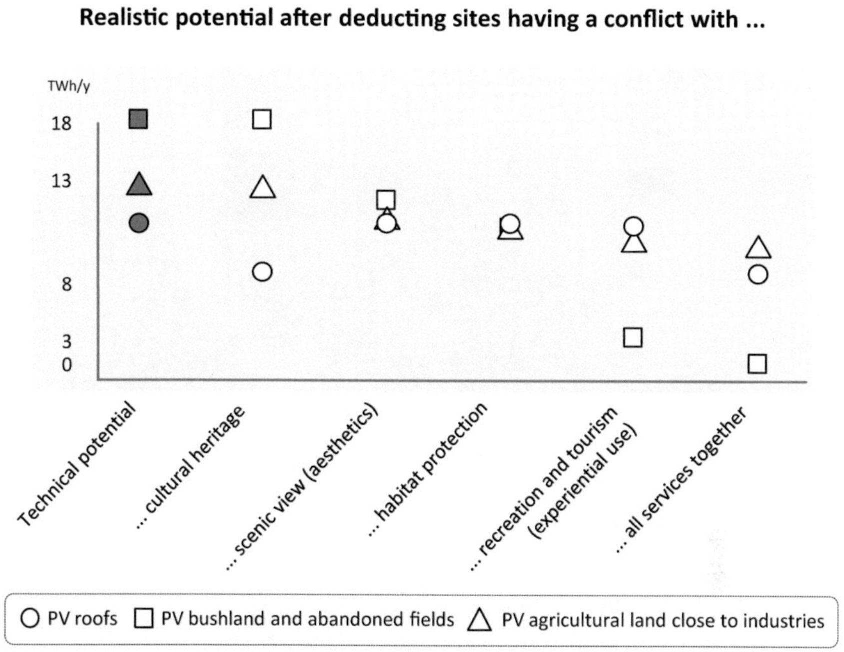

Fig. 7.3 Energy potential of photovoltaic panels before (very left, solid symbols) and after deducting all sites that have a conflict with an ecosystem service. The results stem from a spatially explicit GIS study involving a roughly 40,000 km^2 sized area of Switzerland. (Adapted from Hersperger et al. 2014)

'lifecycle maintenance, habitat and gene pool protection', since these areas are often important for biodiversity. Second, there is a conflict between PV and the scenic beauty and landscape character of these areas. PV on arable land close to industries and other infrastructure has the least conflicts with other services. As a rough estimate, PV production is reduced by circa 50% (100% = full technical potential) if all sites showing a high to medium conflict with other ecosystem services are excluded. This reduction holds true for the Swiss context, i.e. a relatively high population and municipality density, and relatively strict landscape conservation legislation.

7.3 Where it Happens

Finding *general* links between land use transitions and megatrends is easy compared to assessing *specifically* where the corresponding ecosystem service trade-offs have already taken place or will do so in the future. Haines-Young et al. (2012) undertook such a broad-scale analysis to gain an idea of the geographical extent of such trade-offs in Europe. The authors used land use/cover data between 1990 and 2006 and land use projections up to 2030 to cluster Europe's landscapes into regions with similar changes in land use and associated ecosystem services (see Fig. 7.4 and Table 7.1).

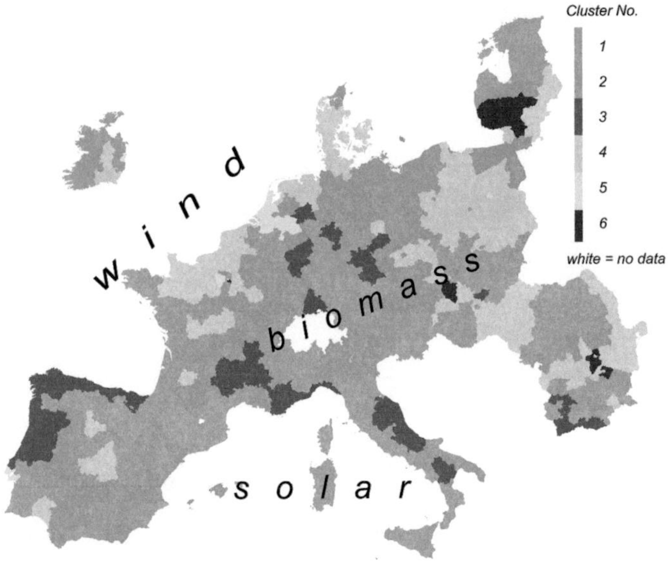

Fig. 7.4 Clusters of similar land use and ecosystem service trajectories from 1990 to 2030 (modified after Haines-Young et al. 2012). The period between 1990 and 2006 is based on real land cover data (LEAC see EEA 2002, 2006); the period 2006–2030 stems from a land use model using the A1 story lines (Meijl et al. 2006; Verburg et al. 2006; Westhoek et al. 2006; Verburg et al. 2009). Administrative boundaries are NUTS-X regions. Areas that do not have LEAC data in any time steps are excluded. Areas with potentials for renewable energy are highlighted

Table 7.1 Cluster descriptions

Number	Description
1	Areas with moderate to low crop potential and decreasing to stable provisioning services since 1990, high nature and recreation potential and trend towards rewilding areas by 2030
2	Areas with stable mixed ecosystem and medium potential for crop production; low potential for wildlife products, little changes expected by 2030
3	Nature-dominated areas with a trade-off between crop production and rewilding by 2030; high potential for nature tourism
4	Crop zones with a low nature potential. Slightly reduced provisioning services since 1990 but very positive perspectives for food production by 2030. Habitat services will not be affected by increased production capacities
5	Highly populated areas with moderate crop production potential but high recreation needs. Thus experiential services (recreation etc.) will increase at the expense of habitat, crop and wildlife services
6	Intensive crop zones where crop potential decreased since 1990 but will strongly increase by 2030

The spatial extent of megatrend 1 (globalization) is clearly visible in the trade-off between the potential for the services related to habitat maintenance and cultivated crops. There are at least two clusters (~ geographical areas) where this trade-off is pronounced, i.e. Clusters 1 and 3. Cluster 1 areas show early declines in agricultural production which will most likely continue in a moderate way until 2030 with modest rewilding or extensification. Two clusters show a high potential for crop-related services in the future (Clusters 4 and 6). Only one cluster (Cluster 2) has a fairly stable mix of ecosystem service provision. Cluster 2 shows a medium potential for crop production, low potentials for wildlife products and only small changes are expected by 2030.

For Cluster 3 'nature-dominated areas', the authors predict an increasing importance of ecosystem services that favour nature experience and nature-oriented tourism, services that are increasingly important for modern society. If megatrend 2 (lifestyle) persists over the coming decades, the areas of Cluster 3 will be indispensable for the physical and mental health of an increasingly urban dominated society (Hartig et al. 2011).

In addition to the ecosystem service trade-offs exhibited in Fig. 7.4, we show very generalised hotspots for megatrend 3, the emerging abiotic output 'renewable energy sources' (not an ecosystem service according to CICES V4.3, Haines-Young and Potschin 2013). At the European level these hotspots follow the wind potential (Atlantic regions), the solar potential (Mediterranean regions), and for biomass the regions with highest yields in both crop and forest production (Central Europe). Regional land use planning will be necessary for realisation of these ecosystem service gains and to generate synergies, such as those between PV and extensive pastures or between wind and agricultural production.

References

Adger, W., Eakin, H., & Winkels, A. (2009). Nested and teleconnected vulnerabilities to environmental change. *Frontiers in Ecology and the Environment, 7*, 150–157.

Braat, L., & ten Brink, P. (Eds.). (2008). *The cost of policy inaction. The case of Not meeting the 2010 biodiversity target*. Study/Report for the European Commission, DG Environment Under Contract ENV.G.1/ETU/2007/0044 (Official Journal Reference: 2007/S95–116033).

Brophy, J. (2011). The end of the economic old order: The great transition, 1750–1860. In H. Walser Smith (Ed.), *The Oxford handbook of modern German history* (pp. 169–194). Oxford: Oxford University Press.

Bürgi, M., Kienast, F., & Hersperger, A. M. (2012). Chapter 7: In search of resilient behavior using the driving forces framework to study cultural landscape. In T. Plieninger & C. Bieling (Eds.), *Resilience and the cultural landscape. Understanding and managing change in human-shaped environments* (pp. 113–124). Cambridge: Cambridge University Press.

Bürgi, M., Silbernagel, J., Wu, J., et al. (2015). Linking ecosystem services with landscape history. *Landscape Ecology, 30*, 11–20.

Burkhard, B., Kroll, F., Nedkov, S., et al. (2012). Mapping ecosystem service supply demand and budgets. *Ecological Indicators, 21*(Special Issue), 17–29.

Charles, C., & Louv, R. (2009, September). Children's nature deficit: What we know – And don't know. *Children and Nature Network*.

European Environmental Agency (EEA). (2002). *Corine land cover update 2000: Technical guidelines*. EEA: Copenhagen.

European Environmental Agency (EEA). (2006). *Land accounts for Europe 1990–2000. Towards integrated land and ecosystem accounting*. EEA Report 11/2006, EEA Copenhagen, p 107.

European Environmental Agency (EEA). (2009). *Ensuring quality of life in Europe's cities and towns. Tackling the environmental challenges driven by European and global change*. EEA Report 5/2009, EEA Copenhagen, p 108.

Eurostat. (2014). *Europe 2020 indicators – Climate change and energy*. Online resource https://ec.europa.eu/eurostat/statistics-explained/index.php/Europe_2020_indicators_-_climate_change_and_energy. Accessed 23 Mar 2015.

Farrell, E., Führer, E., & Ryan, D. (2000). European forest ecosystems: Building the future on the legacy of the past. *Forest Ecology and Management, 132*, 5–20.

Göke C, Lamp J (2012) *Case study Systematic site selection for offshore wind power with Marxan in the pilot area Pomeranian Bight*. Balt Sea Plan Report 29.

Hahn, W., & Knoke, T. (2010). Sustainable development and sustainable forestry: Analogies, differences, and the role of flexibility. *European Journal of Forest Research, 129*, 787–801.

Haines-Young, R. (2009). Land use and biodiversity relationships. *Land Use Policy, 26*(1), 178–186.

Haines-Young, R., & Potschin, M. (2013). Common *International Classification of Ecosystem Services (CICES)*. Report to the European Environment Agency (download: www.cices.eu).

Haines-Young, R., Potschin, M., & Kienast, F. (2012). Indicators of ecosystem service potential at European scales mapping marginal changes and trade-offs. *Ecological Indicators, 21*, 39–53.

Hartig, T., van den Berg, A. E., Hagerhall, C. M., et al. (2011). Health benefits of nature experience: Psychological, social and cultural processes. In K. Nilsson, M. Sangster, C. Gallis, et al. (Eds.), *Forests, trees and human health* (pp. 127–163). New York/Dordrecht/Heidelberg: Springer.

Helfenstein, J., & Kienast, F. (2014). Ecosystem service state and trends at the regional to national level a rapid assessment. *Ecological Indicators, 36*, 11–18.

Helfenstein, J., Bauer, L., Clalüna, A., et al. (2014). Landscape ecology meets landscape science. *Landscape Ecology, 29*, 1109–1113.

Hersperger, A., Segura Moran, L., & Kienast, F. (2014). Nationale Planung: Räumliche Modelle für die Schweiz. *Hotspot, 29*, 16–17.

Home, R., Hunziker, M., & Bauer, N. (2012). Psychosocial outcomes as motivations for visiting nearby urban green spaces. *Leisure Sciences, 34*, 350–365.

Jäger, J. A. G., Soukup, T., Schwick, C., et al. (2011). Landscape fragmentation in Europe. In J. Feranec, T. Soukup, G. Hazeu, & G. Jaffrain (Eds.), *European landscape dynamics – CORINE land cover data* (pp. 157–198). Boca Raton/London/New York: CRC Press.

Kienast, F., Ghosh, R., & Wildi, O. (Eds.). (2007). *A changing world: Challenges for landscape research* (Landscape Series). Dordrecht: Springer.

Kienast, F., Degenhardt, B., Weilenmann, B., et al. (2012). GIS-assisted mapping of landscape suitability for nearby recreation. *Landsc Urban Plann, 105*, 385–399.

Kienast, F., Hersperger, A., Hergert, R., et al. (2014). Landschaftskonflikte durch erneuerbare Energien. *WSL Ber, 21*, 69–74.

Lin, J. T., Pan, D., Davis, S. J., et al. (2014). China's international trade and air pollution in the United States. *PNAS, 111*, 1736–1741.

Litman, T. (2015). *Analysis of public policies that unintentionally encourage and subsidize Urban Sprawl*. Victoria Transport Policy Institute Supporting paper commissioned by LSE Cities at the London School of Economics and Political Science on behalf of the Global Commission on the Economy and Climate (www.newclimateeconomy.net) for the New Climate Economy Cities Program.

Liu, J., Hull, V., Batistella, M., et al. (2013). Framing sustainability in a telecoupled world. *Ecology and Society, 18*, 26.

Martens, D., Gutscher, H., & Bauer, N. (2011). Walking in the "wild" and "tended" urban forests the impact on psychological well-being. *Journal of Environmental Psychology, 31*(1), 36–44.

Meijl, H. V., van Rheenen, T., Tabeau, A., et al. (2006). The impact of different policy environments on agricultural land use in Europe. *Agriculture, Ecosystems and Environment, 114*, 21–38.

Millennium Ecosystem Assessment. (2005). *Ecosystems and human well-being: Biodiversity synthesis*. Washington, DC, World Resources Institute.

Modica, G., Vizzari, M., Pollino, M., et al. (2012). Spatio-temporal analysis of the urban-rural gradient structure: An application in a Mediterranean mountainous landscape (Serra San Bruno Italy). *Earth System Dynamics, 3*, 263–279.

Nedkov, S., & Burkhard, B. (2012). Flood regulating ecosystem services – Mapping supply and demand in the Etropole municipality Bulgaria. *Ecological Indicators, 21*(Special Issue), 67–79.

Potschin, M., & Haines-Young, R. (2011). Introduction to the special issue: Ecosystem services. *Progress in Physical Geography, 35*, 571–574.

Potschin, M., & Haines-Young, R. (2013). Landscapes, sustainability and the place-based analysis of ecosystem services. *Landscape Ecology, 28*, 1053–1065.

Seppelt, R., Dormann, C. F., Eppink, F. V., et al. (2011). A qualitative review of ecosystem service studies: Approaches, shortcomings and the road ahead. *Journal of Applied Ecology, 48*, 630–636.

Swinton, S. M., Lupi, F., Robertson, G. P., et al. (2007). Ecosystem services and agriculture: Cultivating agricultural ecosystems for diverse benefits. *Ecological Economics, 64*(2), 245–252.

Verburg, P. H., Veldkamp, A., & Rounsevell, M. D. A. (2006). Scenario-based studies of future land use in Europe. *Agriculture, Ecosystems and Environment, 114*(1), 1–6.

Verburg, P. H., van de Steeg, J., Veldkamp, A., et al. (2009). From land cover change to land function dynamics: A major challenge to improve land characterization. *Journal of Environmental Management, 90*(3), 1327–1335.

Waldgesetz (WaG). (1991). Bundesgesetz über den Wald. 4. Oct 1991 as at 1 July 2013.

Westhoek, H. J., van den Berg, M., & Bakkes, J. A. (2006). Scenario development to explore the future of Europe's rural areas. *Agriculture, Ecosystems and Environment, 114*(1), 7–20.

European Union Policies and Standards as Drivers for Ecosystem Service Provision and Impairment

8

Christian Schleyer, Sylvia Herrmann, Irene Bouwma, Eeva Primmer, and Klara J. Winkler

Abstract

This chapter describes the potential of different EU policies for implementing the concept of ecosystem services (ES) at different scales. We provide an overview of a broad range of EU policies and their role as drivers for ecosystem services provision and impairment on the regional and local scale and present the consideration of the ES concept within these policies. This is based on research conducted in the EU FP7 OpenNESS project. Additionally, we provide examples of the implementation of selected policies in EU Member States: the EU Water Framework Directive and Public Policy Appraisals allowing for a more in-depth evaluation of opportunities and obstacles of the implementation processes.

Keywords

Water framework directive · Public policy appraisal · Pressures

C. Schleyer (✉)
University of Innsbruck, Institute of Geography, Innsbruck, Austria
e-mail: christian.schleyer@uibk.ac.at

S. Herrmann
Institute of Environmental Planning, Leibniz University Hannover, Hannover, Germany
e-mail: herrmann@umwelt.uni-hannover.de

I. Bouwma
Alterra Wageningen UR, Wageningen, The Netherlands
e-mail: irene.bouwma@wur.nl

E. Primmer
Environmental Policy Centre, Finnish Environment Institute (SYKE), Helsinki, Finland
e-mail: Eeva.Primmer@ymparisto.fi

K. J. Winkler
Natural Resource Sciences, McGill University, Ste-Anne-de-Bellevue, Canada

C. von Haaren et al. (eds.), *Landscape Planning with Ecosystem Services*, Landscape Series 24, https://doi.org/10.1007/978-94-024-1681-7_8

8.1 Introduction

This chapter describes the potential of different EU policies for implementing the concept of ecosystem services (ES)[1] at different scales. We assume that by including the ES concept in different policy frameworks a positive contribution for the provision and protection of ES can be reached. Identifying the ES impacted by specific policies, plans, and projects might lead to a better balancing of different interests. However, a fully-fledged empirically-based assessment of the concrete effects of EU and national policies, as drivers for ES provision and impairment on the regional and local scale, is an enormous methodological challenge beyond the reach of this analysis. Instead, we analyse existing EU policies and regulations with respect to their potential for mainstreaming the ES concept.

In the first part, we provide an overview of a broad range of EU policies and their role as drivers and conclude with remarks on the consideration of the ES concept within these policies. These findings are based on research conducted in the EU FP7 OpenNESS project (http://www.openness-project.eu/), which focused on the conceptual challenges involved in using the ES concept in decision-making at EU and also at regional and local levels (Schleyer et al. 2015a, b; Bouwma et al. 2017).

In the second part, we provide examples for the implementation of selected policies in EU Member States (MS): the EU Water Framework Directive and Public Policy Appraisals. These examples allow for a more in-depth evaluation of the opportunities and obstacles linked with the implementation processes and their implications for mainstreaming the ES concept on the ground.

8.2 Analysis of Selected EU Policies

8.2.1 The Operationalization of the Ecosystem Services Concept in Policies

An important issue for the operationalization of ES is the way in which, and the extent to which, policies enable or impede the mainstreaming of this relatively new concept. However, 'mainstreaming' itself has different meanings, and is connected to several expectations and challenges. There is a broad range of interpretations among policymakers and other concerned stakeholders as to what 'mainstreaming' actually means, or what it should mean. These interpretations range from pure 'window-dressing' to its comprehensive and compulsory incorporation or application. 'Window-dressing' in this context might simply mean adding ES references or terminology to preambles or general parts of existing or emerging policies, with only indirect, non-binding implications for the recipients and implementers of the policies. At the other end of this continuum, a concrete, comprehensive, and compulsory incorporation or application of the ES concept could encompass all relevant and available tools and methods for the assessment and valuation of ES at all stages of the policy process.

[1] For a definition of ecosystem services, see Chap. 1.

Mainstreaming the ES concept in EU policies allows, in principle, the adoption of a systematic and an integrative perspective on linkages between ecosystems and society. In the policy context, the concept of ES is seen as a process through which scientific evidence about the causes of ecosystem change and their consequences for human well-being are identified, so that appropriate management and policy options can be developed to support the needs of decision-makers (MEA 2005a; Jordan and Russell 2014). Here, the ES concept enables policy makers to exploit synergies, in addition to considering and managing trade-offs between different categories of ES (i.e. provisioning, regulatory, and cultural ES) and between individual ES within these categories (see, for example, Maes et al. 2013; Hauck et al. 2013a, b; Schleyer et al. 2015a, b and Matzdorf and Meyer 2014). Thus, effective mainstreaming of the ES concept in EU policymaking, i.e. introducing it into a variety of policy fields, goes well beyond simply introducing ES-related terminology. Consequently, if the ES concept is introduced into EU policymaking in a comprehensive and explicit way, regional policy makers and other implementers would, to some extent, be compelled to adopt an ES perspective. They might, for example, have to address the ecosystem impacts of policies systematically and improve coherence between different policy fields (Hauck et al. 2014).

To gain an overview of the EU policies in which the ES concept is already addressed a policy analysis was carried out. This is done either explicitly, i.e. actually using ES-related terminology, or implicitly, i.e. by referring to particular services or by containing terms referring to ecosystems as complex systems or ecosystem functions. In order to identify policies relevant for an in-depth analysis a three-step selection process was applied. First, a literature and document review resulted in an initial list of 53 EU policies. Second, the policies were prioritised based on their relevance for the OpenNESS case studies. Thirdly, EU policy makers identified key policies at a focus group workshop in Brussels in January 2014. The 11 EU policies selected through this process cover policy fields ranging from biodiversity, forest, and water policies to climate policies and policies for rural and urban areas as well as a mobility and infrastructure-related policy (see Table 8.1) (Schleyer et al. 2015a).

Table 8.1 Policies analysed (year of publication)

Policy Field	EU Policy
Biodiversity policies	Green Infrastructure Strategy (2013) Habitats Directive (1992) Biodiversity Strategy to 2020 (2012)
Water policies	Water Framework Directive (2000) Marine Strategy Framework Directive (2008)
Forest policy	Forest Strategy (2013)
Policies for rural and urban areas	Common Agricultural Policy (2013) incl. Rural Development Regulation Thematic Strategy on the Urban Environment (2006)
Climate policies	Renewable Energy Directive (2009) Climate Change Adaptation Strategy (2013)
Mobility and infrastructure-related policy	Trans-European Network – Transport (2014)

Source: Schleyer et al. (2015a)

8.2.2 Analytical Approach

Policy coherence refers to the degree to which policy goals and associated instruments of different laws or policies are in line with one another or form a meaningful ensemble (May et al. 2006; Mickwitz 2003). Coherence can be measured at the level of definitions, objectives, instruments, or in implementation practices (Deloitte Consulting 2011; Nilsson et al. 2012). *Internal* coherence relates to the logic between goals, objectives, instruments, and the implementation processes within a particular policy field. In our case, this also relates to the coherence between the ES concept and the respective EU policy within this policy field. *External* coherence relates to the overlap or alignment of definitions, objectives, instruments, or implementation processes across different policy fields. In this analysis, the term coherence is used to measure the extent to which the different EU policies already address or can incorporate the ES concept to 'produce' a meaningful and integrated policy at the level of definitions, objectives, and implementation (financing/monitoring/sanctioning instruments). In other words, the analysis seeks to answer the question: what is the level of internal coherence between the ES concept and the various dimensions of a particular EU policy?

The dimensions of coherence covered in the analysis are:

1. *Coherence at the level of definitions*. Concepts and ideas develop and change over time and often become embedded in policymaking and politics (Schmidt 2008). This can be seen in the way legal frameworks, strategies, and other policies reflect new ideas and concepts, and in the extent to which they have incorporated them. The concept of ES has been used in scientific circles since the beginning of the 1990s (e.g. de Groot 1992; Daily 1997). Since then, the number of scientific publications addressing this issue has substantially increased (e.g. Potschin and Haines-Young 2011). In the political domain, the ES concept gained momentum after the publication of the Millennium Ecosystem Assessment (MEA) in 2005 and the TEEB reports (The Economics of Ecosystem Services and Biodiversity) (TEEB 2008). In the analysis, attention was paid to the date the policy came into force and any revisions that were made since.
2. *Coherence of aims or objectives*. Another important issue in the analysis was the degree with which the policy objectives were, in principle, compatible with the 'spirit' of the ES concept. Here, the analysis focused on the aims and objectives and to what extent the policies distinguished between categories of ES; with or without mentioning these terms explicitly. Based on the extent to which the ES concept was reflected in definitions, objectives, or instruments, the policies were assessed using a scoring mechanism that was adapted from a characterisation of impact assessments (Helming et al. 2013; Table 8.2). The analysis also looked at the drivers, i.e. the developments or trends at the global to regional scale that the respective EU policy set out to tackle. These often determine the type and range of objectives of policies and usually have substantial repercussions for the choice of measures and instruments suggested or introduced. Tackling the loss of biodiversity resulting from deforestation might bring different policy measures

Table 8.2 Typology of regulatory frameworks with respect to references to the environment and/or (the) ES (concept) based on Helming et al. (2013)

Type	Description
Type 0	No ecological or environmental issues mentioned or referred to
Type 1	Environment mentioned, but neither a prominent objective nor relevant for/mirrored in policy measure design or monitoring
Type 2	Environment mentioned and/or relevant for/mirrored in policy measure design or evaluation
Type 3	Strong environmental framing and evaluation, but ecosystems or ES not explicitly mentioned
Type 4	Contains framing around ES or use of terminology, but is hardly relevant for/mirrored in policy measure design or evaluation
Type 5	ES fully embedded throughout the regulatory framework, including objectives and policy measure design and monitoring

Source: Schleyer et al. (2015a)

to the fore, compared to tackling loss resulting from the pollution of rivers. In line with the MEA (2005a), the analysis distinguished between direct and indirect drivers. Direct drivers are physical or biological drivers that influence ecosystem processes, including changes in climate or land use. Indirect drivers operate more diffusely by altering one or more direct drivers, including demographic, economic, socio-political, science and technology, and cultural drivers.

3. *Coherence at the level of implementation (degrees of freedom).* EU policy is considered to be regulatory in nature (Jordan and Adelle 2012), usually leading to a top-down implementation process. Nevertheless, in many cases attention is paid to processes taking place at different governance levels in a non-hierarchic fashion (Hooghe and Marks 2001; Wurzel et al. 2013). In practice, however, many directives and regulations define ambitions, goals, instruments, and settings as well as the policy targets to be achieved. This leaves little room for 'freedom' of implementation. Instead, the interest lies in the dominance of goals, ambitions, and instruments formulated by the government and how they are designed, as well as the 'coherence' of policies and policy instruments across policy fields. Another approach argues that the meanings of the policies are constantly reframed in various debates at all levels of implementation (i.e. EU, national, regional, or local level) (e.g. Hajer and Wagenaar 2003). This reframing can lead to situations in which the original policy intent deviates considerably from what is happening 'on the ground'. Attention is primarily on implementation practices and the degree of freedom inherent in the design of a particular policy: how policies are (and can be) interpreted and modified at the various implementation levels, how they play out in real life, and how this varies in different settings. Due to the set-up of many directives, regulations, and policies such a 'reframing' is often problematic.

Focusing on EU policies, the analysis adopted both these approaches by, on the one hand, paying attention to the high-level obligations or expectations and the established systems of reporting and monitoring, and on the other hand, by recognising the practical implications and room for local interpretation. The analysis covered the policies' inherent degrees of freedom for implementation provided to MS and local stakeholders. This is important as it could have considerable bearing on the implementation practices that they can develop. It defines the flexibility they have to address issues that are of particular importance at the national or regional level and to incorporate new emerging ideas and concepts. In this analysis, the type of the policy under scrutiny (e.g. Strategy or Directive) and the dominant mode of steering (e.g. command-and-control, advisory, or economic) were assessed through reporting, monitoring, and evaluation duties and (usually EU-funded) financing mechanisms defined in the policy.

8.2.3 Results

The analysis shows that the ES concept has barely been introduced, remaining confined to the policy arena that addresses natural ecosystems, forestry, or agriculture. Only five EU policies refer explicitly to ES:

- Biodiversity Strategy
- Green Infrastructure Strategy
- Forest Strategy
- Common Agricultural Policy
- Marine Strategy Framework Directive

The first three policies reflect the ES concept in their design of measures which may be relevant for the implementation of landscape planning. Other more recent policies (e.g. Climate Change Adaptation Strategy) do not take up the ES concept at all. Policies that explicitly mention ES usually refer to all three ES categories and biodiversity. In all policies, regulating ES are mentioned in much greater detail than the other ES categories. Those policies, however, that only mention ES indirectly tend to refer just to a small selection of regulating ES, such as carbon storage or water quality. The relatively few references to cultural ES often focus on tourism and recreation.

There is a broad and diverse range of drivers, both from the natural and the social spheres, mentioned in the policies. These range from the overexploitation of natural resources and climate change to changes in lifestyle, education, and demographic change. Most direct drivers relate to the main objectives pursued, such as maintenance of biodiversity (Biodiversity Strategy) or improvement of water quality (Water Framework Directive), reflecting the sectoral nature of most policies. All three policies in which the ES concept is fully embedded are strategies featuring an advisory (even symbolic) mode of steering, reflecting the novelty of the ES approach and the reluctance of MS to adhere to strict regulation across different ES categories. This is also in line with the current trend in the EU to reduce direct regulation and simplify procedures. Very few policies require MS to report on the stock/flow of a particular ES. Also, accounting for the environmental impacts is not standard for all policies, or only very specific ecosystems are focused on (e.g. Water Framework Directive). While for most policies there are various EU funds available to finance measures, there is, however, a small set of EU funds that feature prominently. These include the EU Cohesion Fund, the Life/Life+ Programme, and the EU Agricultural Fund for the Rural Development (Schleyer et al. 2015a).

8.2.4 Conclusions Regarding EU Policies

The analysis shows that there is considerable scope to improve the mainstreaming of the ES concept, in order to achieve coherence between the aims of different policy sectors. Measures should be taken to actively promote uptake of the ES concept in new EU policies as well as in revisions of existing ones. However, this will require a deeper understanding of the factors affecting uptake, including communication barriers, stakeholder attitudes to the ES concept, and tensions between policy sectors. There is also scope to improve the uptake of the ES concept through dedicated financing mechanisms, common methods for monitoring and evaluation of ES (especially cultural ES), and better tools to help policy makers exploit synergies and manage trade-offs between ES (Schleyer et al. 2015a).

8.3 Implementation of the ES Concept in EU Member States

The use of knowledge on ecosystem functions and services as components of policy making is increasing in different contexts (e.g. Water Framework Directive, Habitat Directive, agri-environmental measures). This has occurred ever since the Millennium Ecosystem Assessment Report criticised the fact that "decision-makers […] not use all of the relevant information that is available" (MEA 2005b, p. 23). However, implementing the ES approach in a more holistic sense, i.e. including all ES categories, plus the respective trade-offs, while concurrently focusing on the integration between environmental, social, and economic aspects of the utilisation of ecosystems, is still lacking. This section explores whether the implementation and consideration of ES in policies and planning on the MS level is more or perhaps less widely spread than the results of the EU level have shown. We use two examples, the EU Water Framework Directive and Public Policy Appraisals, to describe the current situation.

8.3.1 The Water Framework Directive and the ES Concept

In Germany and other MS a lively discourse has emerged, both among academics and (local) water management practitioners, regarding if and to what extent the ES concept can contribute to and complement the implementation of the Water Framework Directive (WFD), in particular with respect to the River Basin Management Plans and the related Programs of Measures (e.g. BfG 2015; Hansjürgens and Herkle 2012; Vlachopoulou et al. 2014; Wallis et al. 2011).

Above all, a holistic interpretation of the ES concept that looks beyond water-related ES is hoped to overcome or at least mitigate some of the weak spots of the WFD. These are, in particular, technical and conceptual problems of the WFD such as the consideration of social aspects, or the communication of the purpose and objectives of the WFD to a wide range of stakeholders across policy fields, including to specialists but also the general public (Spray and Blackstock 2016; Vlachopoulou et al. 2014). Here, the ES concept could be a powerful communication tool enabling a dialogue between different stakeholders who can use it as a common language (Martin-Ortega 2012). It can also serve to facilitate horizontal policy integration, linking, for example, water management, nature protection, and energy, as well as local residents with non-local stakeholders. This would also induce a shift from the currently rather narrow technical perspective, that aims to comply with the quality standards, to a broader view considering societal costs and benefits (Everard 2011; Wallis et al. 2011; Reyjol et al. 2014) that goes well beyond the WFDs objective of reaching a 'good (ecological) status' of all waters. A more holistic perspective on ecosystems, in terms of services they deliver, would further allow for the identification and assessment of a broader range of ES affected by water policy measures (see Fig. 8.1). Thus, links and interdependencies between different (sectoral) policies are becoming more obvious and understandable (Martin-Ortega 2012). Integrating landscape planning analysis and objectives for the whole spectrum of ES into management plans could serve this purpose. On the other hand, information about water body related ES can be integrated into landscape planning.

Furthermore, some researchers argue that the ES concept with its broad understanding of human well-being might help to specifically consider non-monetary values of nature, which are not explicitly included in the economic valuation toolbox favoured by WFD implementers (Seeconsult GmbH and InterSus 2012). In practice, however, it is the development or adaptation of economic valuation approaches that are hoped to act as 'catalysers of innovation' for the implementation of WFD (DESSIN 2014; Koundouri et al. 2016).

Looking at the practical implementation, however, it is interesting to note that the ES concept is not mentioned in the current legal text of the Water Framework Directive (Type 3) (EC 2000). Nevertheless, the design and the recent implementation of the WFD seem to follow some of the basic ideas of the ES concept, for example by adapting an integrative perspective on ecosystems following the catchment principle, by making public participation and involvement of stakeholders systematic and compulsory, and by assessing and (economically) valuing water-related services. With respect to the latter, some successful applications already exist such as a web-based tool to support the valuation of ES in Flanders, Belgium, or a tool for valuing the impacts of ES interactions for policy effectiveness in the UK (Wallis et al. 2011).

Moreover, there are also cases that show how the ES approach could be combined with the implementation of the WFD in a more holistic way, i.e. going beyond

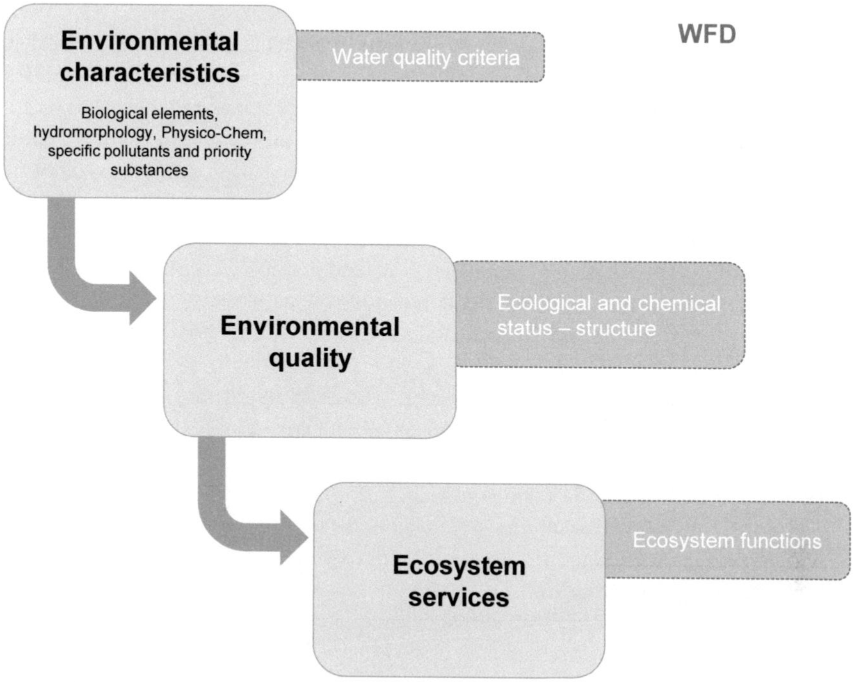

Fig. 8.1 The link between environmental characteristics of surface waters and ecosystem services in the WFD. (Adapted from Vlachopoulou et al. 2014: p. 686)

economic valuation. For example, focus groups organised by the Centre of Expertise for Waters (CREW), were used to co-construct a common understanding between scientists, practitioners, land managers, and other stakeholders and to develop an ecosystem service vision for the Lunan Catchment in Eastern Scotland (see Fig. 8.2; James Hutton Institute 2012).

To our knowledge, there are no documented cases in Germany of where the ES concept has actually been used in a systematic and holistic way in the context of WFD implementation. Some research has even revealed that stakeholders involved in drafting the River Basin Management Plans, at the regional level, were often not even aware of the ES concept (see, for example, Krüger (2016) for the case of the River Ems, Lower Saxony). However, some elements of the larger ES toolbox, in particular various forms of economic valuation, have already been employed by regional water managers for quite some time; though mostly without explicitly referring to the ES concept. In addition, there are a number of pilot projects exploring the options of taking an ES perspective for improving the cost-effectiveness of river restoration. However, yet again, the projects highlight economic aspects by focusing on the development of a 'Payments for Ecosystem Services Scheme' (Borowski-Maaser et al. 2014). In Germany, with area-covering landscape planning, the most efficient way of integrating a broader perspective on water-related ES

Box 8.1: Why the Integration of the ES Concept in the WFD is Challenging

First, there is the conceptual relation between the aim of achieving a 'good status' for all waters and the provision of related ES. Does reaching a 'good ecological and chemical status' automatically lead to a desirable status of all related ES (Hartje and Klaphake 2006; Interwies 2011; Tolonen et al. 2014)? Does 'good status' need to be redefined? Is it necessary to identify and assess all related ES?

Second, if there is such a relation, it is argued that a coherent and sufficient quantification of most relevant ES is not possible especially as the required data are not available and/or insufficient (Seeconsult GmbH and InterSus 2012).

Third, the spatial scales of affected ecosystems might not fit or coincide with the scales of river basins, which are the main administrative units of water management in the WFD. Also, time scales might be out of sync (Bastian et al. 2012; Spray and Blackstock 2016).

Fourth, the latter two challenges make a systematic (economic) valuation and assessment of related ES an almost impossible endeavour. Nevertheless, is it necessary to embark on an (economic) valuation of all relevant ES in order to achieve the WFD's objective of adequate water pricing, i.e. reflecting the 'true' costs?

Fifth, the integration of the ES concept in the WFD might be another burden (actual and/or perceived) for water administrations in all EU MS. They are already confronted with a variety of WFD-related policy innovations, such as the combination of pollution prevention with economic analyses of water use, the provision of active involvement rights to the general public, and a detailed and comprehensive system of monitoring and reporting that needs to be established (EC 2000). Is the adoption of the ES concept on top of this indeed feasible for water managers or administrations?

into the River Basin Management Plans would be to use landscape planning as a source and information system (Galler et al. 2013).

In conclusion, the experiences with the WFD illustrate the typical expectations and challenges faced when mainstreaming the ES concept in EU policies (see Box 8.1). On the one hand, it fosters a more holistic approach to water management, helps to communicate the benefits of the WFD to other policy fields and the wider public, and facilitates an even more comprehensive economic approach. On the other hand, however, there are tensions between the WFDs objectives and maximising ES. There are also methodical and practical questions related to the substantial vagueness of the ES concept when it comes to appropriate and meaningful assessments and (economic) valuations of ES.

It is important to note that the few success cases above are dependent on the willingness of stakeholders, scientists, and policy makers to 'think out of the box'

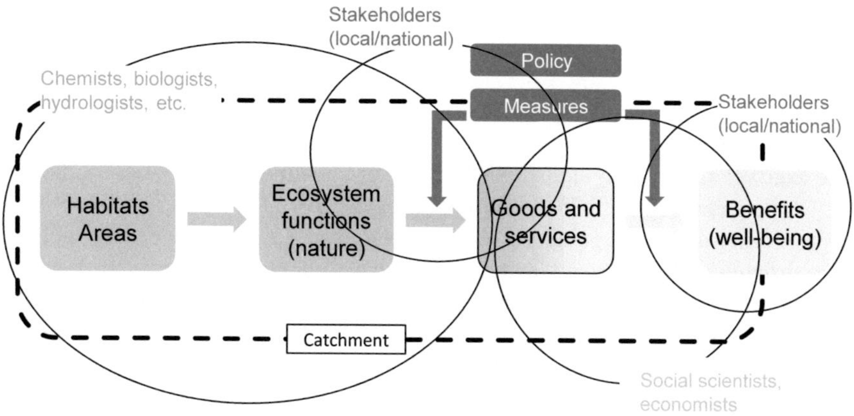

Fig. 8.2 Co-constructed framework for application of an ES approach for the Lunan Catchment. Local Stakeholders Focus Group, Tannadice 14.3.2012. (Adapted from James Hutton Institute 2012: n.p.)

and to not solely focus on their individual interests, but to act in favour of the collective outcome. Furthermore, these processes are time consuming and necessitate long-term collaboration between the different stakeholders and decision-makers (Waterton and Tsouvalis 2015). Thus, it seems judicious to carefully reflect on the aims and purposes for incorporating the ES concept in any EU policy. The Blueprint to Safeguard Europe's Water Resources stresses the need to improve EU water policies and to include cross-cutting problem solving, which might trigger in the future a more pronounced application of the ES concept for sustainable water management in the EU (Reyjol et al. 2014).

8.3.2 Implementation of the ES Concept in Public Policy Appraisals

Public Policy Appraisals are one type of context in which knowledge about ecosystems can be embedded into policy. As a central policy making tool, Environmental Impact Assessments[2] are particularly promising for embedding the ES concept into policy making. What makes them promising is, unlike other venues where knowledge flows from knowledge generators into the decision-making processes, the policy maker has to search for the knowledge that is useful for the assessment (Turnpenny et al. 2014). The following example from the UK examines if and how

[2] Environmental Impact Assessment (EIA) is a process of evaluating the likely environmental impacts of a proposed project or development, taking into account inter-related socio-economic, cultural, and human-health impacts, both beneficial and adverse (Convention of Biological Diversity – CBD).

the ES concept is utilised in public policy appraisals, more specifically in Impact Assessments (IA), which are seen as central venues for embedding ecosystem information.

8.3.2.1 Method Used

The example is based on an extended review of 75 national-level IA carried out in the UK between 2008 and 2012 (Turnpenny et al. 2014). Here, 17 (22.7%) of the IA targeted environmental policies, 36 (48%) targeted environment-related policies (e.g. agriculture, housing and land, energy and natural resources, transport), and 22 (29.3%) targeted non-environmental policies (e.g. social security, sport, criminal law). The study used document analysis for the assessment of the IA, but did not specifically examine the influences of the ES concept on policy outputs and longer term outcomes (Turnpenny et al. 2014). The typology of Impact Assessments developed by Helming et al. (2013) formed the basis for classifying the IA. They were investigated according to the degree in which environmental considerations or the ES concept are embedded in the appraisals. The following criteria in Table 8.3 were used to assess the IA.

8.3.2.2 Results

The study found that only 12% of the IA included the ES concept, for example, Type 4 (framing around an ES concept, but does not carry out much impact analysis) or Type 5 (fully embedding ES throughout the process including long term impacts) (see Fig. 8.3). Here, the most prominent policy fields covered by these IA were climate change, energy, and nature conservation. However, the vast majority of IA (88%) did not refer to the ES concept at all.

Fig. 8.4 presents the same data from an administrative, sectoral perspective. It distinguishes between IA led by the Department for Environment, Food and Rural Affairs (DEFRA) and those from other ministries. Interestingly enough, only 20% of the DEFRA-led IA showed clear ES framing. Furthermore, the ES concept was also used by non-environmental departments such as the Department for Culture, Media and Sport, though this is the exception.

These findings are also visible when an assessment of the impacts of policies is performed. The cases showing evidence of framing of the ES concept

Table 8.3 Criteria for the assessment of the IA

Type	Description
Type 0	No ecological or environmental knowledge referred to
Type 1	Environment mentioned, but not evaluated at all
Type 2	Environment mentioned, but some elements are missing, and only weakly evaluated
Type 3	Strong environment framing and evaluation, but ecosystems not explicitly mentioned
Type 4	Contains framing around an ES concept, but does not carry out much impact analysis.
Type 5	ES concept fully embedded throughout (explicitly referring to one or more of the ecosystem services; examining long-term and indirect impacts; taking an integrative approach).

Source: Adapted from Turnpenny et al. (2014)

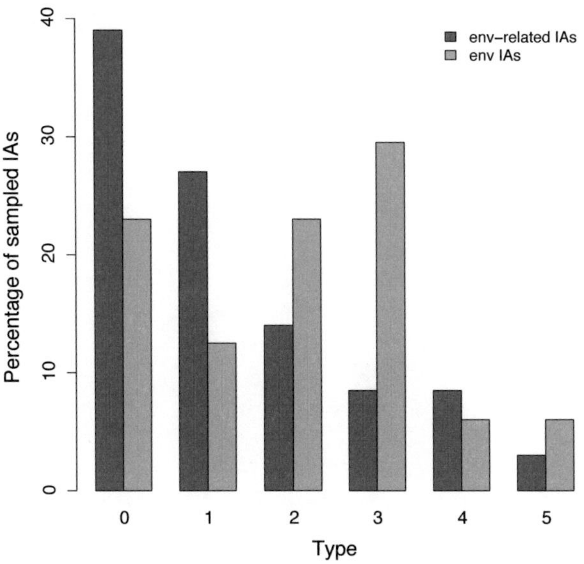

Fig. 8.3 Percentage of sample IA with different types of ES *framing*. (Adapted from Turnpenny et al. 2014)

Fig. 8.4 Percentage of sampled IA with different types of ES *framing*: Defra vs. non-Defra IA. (Adapted from Turnpenny et al. 2014)

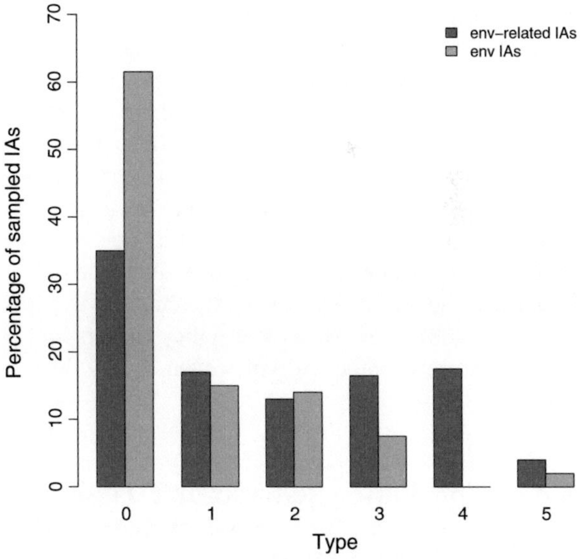

(Fig. 8.3 and 8.4) also developed their impact analysis around an ES concept. All cases classified as Types 3, 4, or 5 were analysed with respect to the types of ES they include. Fig. 8.5 shows regulating services as the most prominent ES element in the IA, while supporting and cultural services were relatively scarce.

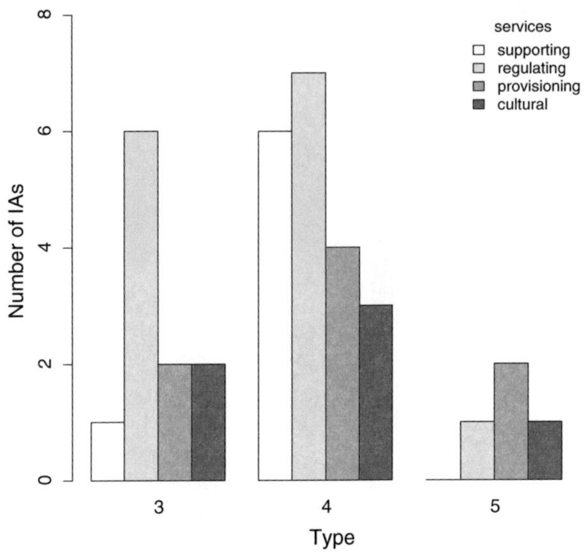

Fig. 8.5 Number of the subsample of IA displaying different elements of an ES concept. (Adapted from Turnpenny et al. 2014)

In conclusion, it can be said that policy appraisals indeed seem to be a crucial venue to embed the ES concept within policy and practice, or at least to implement ES thinking. However, despite the long experience of UK policy makers and practitioners with applying the ES concept, there are still significant obstacles for a systematic embedding of ES in practice. Most of the time only environmental terms are mentioned while ecosystem services only appear in a few cases. One of the reasons for this might be that the ES concept has been predominately well known mainly by scientists since the 1990s. Yet, the diffusion of scientific knowledge into policy making and practice can be extremely slow due to institutional inertia, policy constraints, or the time needed by the stakeholders to adjust to new ideas (Owens 2012; Sabatier 1988). This means that policy makers and practitioners at least seem to think about environmental issues even if they do not use the ES concept for the analysis.

8.4 Concluding Remarks on EU Policies and Standards as Drivers for Ecosystem Service Provision and Impairment

Our analysis shows that the ES concept has been mainstreamed in policies to varying degrees, depending on the timing and specificity of the policy, the sector, and the level at which the policy operates. As the example of the WFD illustrates, the consideration and implementation of ES in policy and planning is not very visible, particularly at the local or regional level. Possible reasons for limited consideration and implementation include:

- well-functioning, existing implementation procedures
- lack of a clear national or European framework forcing the stakeholders at the local or regional level to take ES in consideration
- too general a description of the ES concept by scientists, which does not support operationalization
- no clear empirical evidence on the advantages of using the ES concept in policy and planning
- specifically in Germany, a long tradition of using landscape functions as guidelines for spatial and landscape planning
- due to the separation of spatial (economic) planning and landscape (ecosystem-related) planning, integrative effects of the ES concept are not taken into account

Mainstreaming the ES concept into EU policy-making is no 'silver bullet' – some expectations may be met, but others may be disappointed. For example, some scientists are rather sceptical when it comes to the usefulness of the ES concept for biodiversity conservation (e.g. McShane et al. 2011; Turnhout et al. 2013) or are concerned about issues of (environmental) justice (Glotzbach 2013; Hauck et al. 2013b; Jax et al. 2013). Thus, 'expectations management' for policy-makers as well as practitioners and NGOs, is necessary to avoid frustration and backlash.

There are several challenges for both horizontal and vertical policy integration of the ES concept. Mainstreaming requires substantial capacity building, and the consideration of various policy fields and decision-making levels. Participatory approaches are a must for horizontal and vertical integration and may be helpful at least for local policy integration and balancing trade-offs. They may fail due to administrative challenges facing vertical policy integration or opposing agendas for horizontal policy integration, which are additionally combined with imbalanced power relations. Thereby, the mainstreaming must be seen rather as a means to an end. More precisely, the sustainable use and management of ecosystems should not become a top-down and sector-based form of management but an opportunity to make environmental aspects an integral part of management practices.

Clearly, mainstreaming the ES concept cannot resolve all of the challenges connected with biodiversity loss, ecosystem degradation, and risks for human well-being. The ambiguity of the concept should be taken seriously. Sometimes, there is a misleading belief and blind trust in the communicative potential of the concept, which may conflict with its shortcomings. However, a well-facilitated and careful process of reflection may improve the potential of mainstreaming the ES concept and may significantly improve the governance of ecosystems and natural resources. Ultimately, the usefulness of mainstreaming the ES concept depends on its impact on decision-making on the ground (Schleyer et al. 2015b).

The different challenges of mainstreaming the ES concept are not easy to address simultaneously. Rather, policy integration and inclusive participation sometimes pull in opposite directions (Green and Penning-Rowsell 1999). Participatory approaches are a must but at the same time limited in their contribution to more effective complex policies. To become trustworthy and effective organisations, new participatory elements need to be linked up with the existing administrations and democratically legitimised decision-making structures (Theesfeld and Schleyer 2013). Beside the need to balance their usefulness for horizontal and vertical policy integration, they may stimulate processes of public reasoning to deal more sustainably with natural resources and societal dependencies on functioning ecosystems.

To consider the ES concept in policy and planning the following policy mechanisms should be considered:

- It seems necessary to establish incentives to encourage consideration of ES in policy formulation and implementation in planning by formulating a clear framework for the European and national levels. For example, payments in the context of the EU Common Agricultural Policy (CAP) could be used in such a way
- The competencies and the responsibility of different political levels for protection and restoration of ES should be discussed with respect to the spatial extent of ecosystems and the value of their ES
- The integration of agricultural, regional, and landscape policies has to be intensified. The meaning of ES, underlining the contribution of nature for human well-being, can support this integration
- Landscape planning, as integrative environmental planning, covers multiple ES and tries to develop integrated concepts for different types of landscapes. This can be used as a first tool to show how an integration of environmental and economic objectives can be reached by using an ES concept
- To establish the ES concept at the local or regional levels, shared projects with scientists and practitioners are necessary to collectively explore the pros and cons of such an approach. So far, the ES approach seems to be more a concept developed by scientists without a strong connection to practice
- The newly introduced greening component of the CAP, and its possible extension in the future, might be conducive for fostering the ES perspective here

References

Bastian, O., Grunewald, K., & Syrbe, R. (2012). Space and time aspects of ecosystem services, using the example of the EU Water Framework Directive. *International Journal of Biodiversity Science, Ecosystem Services and Management, 8*(1–2), 5–16. https://doi.org/10.1080/21513732.2011.631941.

BfG (Bundesanstalt für Gewässerkunde). (Ed.). (2015). Ökosystemleistungen – Herausforderungen und Chancen im Management von Fließgewässern. 5. Ökologisches Kolloquium am 5./6. Mai 2015 in Koblenz. Veranstaltungen 3/2015, Koblenz, Oktober 2015, p 100.

Borowski-Maaser, I., Sauer, U., & Cortekar, J., et al. (2014). *Final Report (DII.6–V4) on Phase II of an ecosystem services project in the Vecht basin: Developing a proposal for a regional scheme on payments for ecosystem services.* (VechtPES Project). http://www.interessen-im-fluss.de/wp-content/uploads/2014/09/VechtPES_FinalReport_09September2014_small.pdf. Accessed 14 Mar 2016.

Bouwma, I., Schleyer, C., Primmer, E., et al. (2017). Adoption of the ecosystem services concept in EU policies. *Ecosystem Services, 29*(Part B), 213–222. https://doi.org/10.1016/j.ecoser.2017.02.014.

Daily, G. C. (1997). *Nature's services*. Washington, DC: Island Press.

de Groot, R. S. (1992). *Functions of nature: Evaluation of nature in environmental planning, management and decision-making*. Groningen: Wolters Noordhoff BV.

Deloitte Consulting. (2011). *Support to fitness check water policy*. European Commission – General Directorate Environment.

DESSIN (Demonstrate Ecosystem Services Enabling Innovation in the Water Sector). (2014). *About Dessin: What do we do?* https://dessin-project.eu/?page_id=16. Accessed 14 Mar 2016.

European Commission (EC). (2000). *Establishing a framework for Community action in the field of water policy* (Water Framework Directive – WFD). Directive 2000/60/EC, Brussels.

Everard, M. (2011). Introduction on ecosystem services and water management: A manager's point of view. In C. Wallis, N. Séon-Massin, F. Martini, et al. (Eds.), *Implementation of the water framework directive: When ecosystem services come into play, 2nd "water science meets policy" event, Brussels*. 29-30th September 2011; http://www.onema.fr/EN/EV/meetings/ecosystem-services.pdf (pp. 108–112).

Galler, C., von Haaren, C., & Albert, C. (2013). Planning multifunctional measures for efficient landscape management: Quantifying and comparing the added value of integrated and segregated management concepts. In B. Fu & K. B. Jones (Eds.), *Landscape ecology for sustainable environment and culture* (pp. 249–284). Berlin: Springer. ISBN:978-94-007-6529-0.

Glotzbach, S. (2013). Ecosystem services and distributive justice: Considering access rights to ecosystem services in theories of distributive justice. *Ethics, Policy & Environment, 16*(2), 162–176. https://doi.org/10.1080/21550085.2013.801203.

Green, C., & Penning-Rowsell, E. (1999). Inherent conflicts at the coast. *Journal of Coastal Conservation, 5*, 153–162. https://doi.org/10.1007/BF02802753.

Hajer, M., & Wagenaar, H. (Eds.). (2003). *Deliberative policy analysis: Understanding governance in the network society*. Cambridge: Cambridge University Press.

Hansjürgens, B., & Herkle, S. (Eds.). (2012). Der Nutzen von Ökonomie und Ökosystemleistungen für die Naturschutzpraxis. Workshop II: Gewässer, Auen und Moore. BfN-Skripten 319.

Hartje, V., & Klaphake, A. (2006). *Implementing the Ecosystem Approach for Freshwater Ecosystems – A case study on the Water Framework Directive of the European Union*. Bonn.

Hauck, J., Schweppe-Kraft, B., Albert, C., et al. (2013a). The promise of the ecosystem services concept for planning and decision-making. *Gaia, 22*, 232–236.

Hauck, J., Görg, C., Varjopuro, R., et al. (2013b). Benefits and limitations of the ecosystem services concept in environmental policy and decision making: Some stakeholder perspectives. *Environmental Science & Policy, 25*, 13–21.

Hauck, J., Schleyer, C., Winkler, K. J., et al. (2014). Shades of greening: Reviewing the impact of the new EU Agricultural Policy on ecosystem services. *Change and Adaptation in Socio-Ecological Systems, 1*, 51–62.

Helming, K., Diehl, K., Geneletti, D., et al. (2013). Mainstreaming ecosystem services in European policy impact assessment. *Environmental Impact Assessment Review, 40*, 82–87.

Hooghe, L., & Marks, G. (2001). *Multilevel governance and European integration*. Oxford: Rowman and Littlefield.

Interwies, E. (2011). Utilizing the esa for WFD implementation between theory and practice: Opportunities and challenges. In C. Wallis, N. Séon-Massin, F. Martini, et al. (Eds.), *Implementation of the water framework directive: When ecosystem services come into play. 2nd "water science meets policy" event, Brussels* (pp. 127–128). 29-30th September 2011; http://www.onema.fr/EN/EV/meetings/ecosystem-services.pdf.

James Hutton Institute. (2012). Exploring ways for the application of ecosystem services approaches at the catchment level. www.hutton.ac.uk/research/themes/managing-catchments-and-coasts/ecosystem-services/catchment-level. Accessed 26 Mar 2019.

Jax, K., Barton, D. N., Chan, K. M. A., et al. (2013). Ecosystem services and ethics. *Ecological Economics, 93*, 260–268. https://doi.org/10.1016/j.ecolecon.2013.06.008.

Jordan, A., & Adelle, C. (Eds.). (2012). *Environmental policy in the EU: Actors institutions and processes* (3rd ed.). New York: Routledge/Earthscan. ISBN: 978-1849714693.
Jordan, A., & Russel, D. (2014). Embedding the concept of ecosystem services? The utilisation of ecological knowledge in different policy venues. *Environment and Planning C: Government and Policy, 32*, 192–207.
Koundouri, P., Ker Rault, P., Pergamalis, V., et al. (2016). Development of an integrated methodology for the sustainable environmental and socio-economic management of river ecosystems. *Science of the Total Environment, 540*, 90–100.
Krüger, A-L. (2016). Institutional analysis of the river management contract 'Masterplan Ems 2050'. Master thesis, Humboldt-Universität zu Berlin.
Maes, J., Hauck, J., Paracchini, M. L., et al. (2013). Mainstreaming ecosystem services into EU policy. *Current Opinion in Environment Sustainability, 5*, 128–134.
Martin-Ortega, J. (2012). Economic prescriptions and policy applications in the implementation of the European Water Framework Directive. *Environmental Science & Policy, 24*, 83–91. https://doi.org/10.1016/j.envsci.2012.06.002.
Matzdorf, B., & Meyer, C. (2014). The relevance of the ecosystem services framework for developed countries' environmental policies: A comparative case study of the US and EU. *Land Use Policy, 38*, 509–521.
May, P. J., Sapotichne, J., & Workman, S. (2006). Policy coherence and policy domains. *Policy Studies Journal, 34*(3), 381–403.
McShane, T., Hirsch, P., Trung, T., et al. (2011). Hard choices: Making trade-offs between biodiversity conservation and human well-being. *Biological Conservation, 144*(3), 966–972. https://doi.org/10.1016/j.biocon.2010.04.038.
Mickwitz, P. (2003). A framework for evaluating environmental policy instruments: Context and key concepts. *Evaluation, 9*(4), 415–436.
Millennium Ecosystem Assessment (MEA). (2005a). *Ecosystems and human wellbeing: Current state and trends, volume 1*. Washington, DC: Island Press.
Millennium Ecosystem Assessment (MEA). (2005b). *Ecosystems and human wellbeing: Synthesis*. Washington, DC: Island Press.
Nilsson, M., Zamparutti, T., Petersen, J. E., et al. (2012). Understanding policy coherence: Analytical framework and examples of sector-environment policy interactions in the EU. *Environmental Policy and Governance, 22*(6), 395–423.
Owens, S. (2012). Experts and the environment – The UK Royal Commission on the environmental pollution 1970–2011. *Journal of Environmental Law, 24*(1), 1–22.
Potschin, M., & Haines-Young, R. (2011). Ecosystem services – Exploring a geographical perspective. *Progress in Physical Geography, 38*(5), 575–594.
Reyjol, Y., Argillier, C., Bonne, W., et al. (2014). Assessing the ecological status in the context of the European Water Framework Directive: Where do we go now? *Science of the Total Environment, 497–498*, 332–344.
Sabatier, P. (1988). An advocacy coalition framework of policy change and the role of policy-oriented learning therein. *Policy Sciences, 21*, 29–168.
Schleyer, C., Bouwma, IM., & Primmer, E., et al. (2015a). *Paper on the policy analysis*. European Commission FP7, 2015.
Schleyer, C., Görg, C., Hauck, J., et al. (2015b). Opportunities and challenges for mainstreaming the ecosystem services concept in the multilevel policy making within the EU. *Ecosystem Services, 16*, 174–181. https://doi.org/10.1016/j.ecoser.2015.10.014.
Schmidt, V. A. (2008). Discursive institutionalism: The explanatory power of ideas and discourse. *Annual Review of Political Science, 11*, 303–326.
Seeconsult GmbH, InterSus. (2012). German Case Study Report.
Spray, C., & Blackstock, K. (2016). *Optimising water framework Directive River basin management planning using an ecosystem services approach*. Aberdeen: Centre of Expertise for Waters (CREW) Facilitation Team, James Hutton Institute.
TEEB. (2008). *The economics of ecosystems and biodiversity: Ecological and economic foundation*. Cambridge: Earthscan.

Theesfeld, I., & Schleyer, C. (2013). Germany's light version of integrated water resources management. *Environmental Policy and Governance, 23*, 130–144. https://doi.org/10.1002/eet.1602.

Tolonen, K. T., Hämäläinen, H., Lensu, A., et al. (2014). The relevance of ecological status to ecosystem functions and services in a large boreal lake. *Journal of Applied Ecology, 51*, 560–571. https://doi.org/10.1111/1365-2664.12245.

Turnhout, E., Waterton, C., Neves, K., et al. (2013). Rethinking biodiversity: From goods and services to "living with". *Conservation Letters, 6*, 154–161. https://doi.org/10.1111/j.1755-263X.2012.00307.x.

Turnpenny, J., Russel, D., & Jordan, A. (2014). The challenge of embedding an ecosystem services approach: Patterns of knowledge utilisation in public policy appraisal. *Environment and Planning C: Government and Policy, 32*, 247–262.

Vlachopoulou, M., Coughlin, D., Forrow, D., et al. (2014). The potential of using the ecosystem approach in the implementation of the EU water framework directive. *Science of the Total Environment, 470–471*, 684–694.

Wallis, C., Séon-Massin, N., Martini, F., et al. (2011). Implementation of the water framework directive – When ecosystem services come into play. In *2nd "water science meets policy" event, Brussels*. 29–30th September 2011; http://www.onema.fr/EN/EV/meetings/ecosystem-services.pdf.

Waterton, C., & Tsouvalis, J. (2015). On the political nature of cyanobacteria: Intra-active collective politics in Loweswater, the English Lake District. *Environment and Planning D: Society and Space, 33*, 477–493.

Wurzel, R., Zito, A. R., & Jordan, A. J. (2013). *Environmental governance in Europe: A comparative analysis of the use of new environmental policy instruments*. Cheltenham: Edward Elgar Publishing.

9 Assessing Pressures in Landscape Planning

Christian Albert, Rodrigo Caracciolo Martins, and Andrew A. Lovett

Abstract

This chapter provides an overview of relevant pressures, their more specific effects in different types of ecosystems, some guidelines for assessing pressures, and an illustrative case study. It provides insights concerning which pressures need to receive particular consideration in different types of ecosystems, and how they can be evaluated. An example is provided through a case study of pressures on natural capital in East Anglia, UK.

Keywords

Pressures · Landscape impacts · Ecosystem impacts · Risk register

9.1 Introduction

The process of landscape planning, in general, follows the well-established Driving Forces, Pressures, State, Impacts and Reponses (DPSIR) framework put forward by Smeets and Weterings (1999). As outlined in detail in Chap. 3, landscape planning according to the DPSIR model identifies driving forces (D) leading to specific pressures (P) which in turn have an effect on the state (S) of landscapes and ecosystems.

C. Albert (✉) · R. C. Martins
Institute of Environmental Planning, Leibniz University Hannover, Hannover, Germany
e-mail: albert@umwelt.uni-hannover.de

A. A. Lovett
School of Environmental Sciences, University of East Anglia, Norwich, UK
e-mail: a.lovett@uea.ac.uk

C. von Haaren et al. (eds.), *Landscape Planning with Ecosystem Services*, Landscape Series 24, https://doi.org/10.1007/978-94-024-1681-7_9

The changes of the state can be interpreted as impacts (I). If those impacts are perceived as undesirable, landscape planners can craft responses (R) in order to alleviate driving forces, minimise pressures, safeguard the state, or decrease the impacts.

Information on current and future pressures is of great importance for the assessments of ecosystems (European Commission 2016; Rounsevell and Harrison 2016), and for landscape planning more specifically (von Haaren 2004). On the one hand, information on pressures needs to be considered in evaluating the current state of the landscape. For example, assessments of the current delivery of ecosystem services in a particular landscape need to recognise the extent to which this current provision is already impacted by human pressures. The degree and spatial location of pressures already provides some proxy indication of the probable state of the landscape. On the other hand, knowledge on current and likely future pressures is needed to identify priorities for action (i.e. where current pressures exceed standards) and to craft appropriate response strategies that avoid, minimise, or compensate for these pressures.

9.2 Five Major Groups of Pressures on Landscapes

Building on previous work by the Millennium Ecosystem Assessment (2005), Maes et al. (2014) proposed five major groups of pressures causing changes in ecosystems and landscapes. These major groups include habitat change, climate change, overexploitation, invasive alien species, and pollution and nutrient enrichment (Table 9.1).

Habitat change arises both from direct changes in landscapes (e.g. of site characteristics) or indirect pressures such as fragmentation. The main driver influencing habitat change is alterations in land use, with about half of Europe's land area being used for agriculture, forests being intensively used and natural areas being increasingly fragmented (EEA 2010). Another important habitat change is abandonment of fields and grasslands, which may have both negative and positive effects for nature conservation.

Climate change has always influenced landscapes across Europe, but is expected to become even more important over future decades. The types and impacts of climate changes differ substantially across Europe (for an overview see EEA 2012). In north-western Europe, for example, increases in winter precipitation and river flow are expected, combined with a northward movement of species, and an increasing risk of river and coastal flooding. In contrast, northern Europe is expected to see temperature increases larger than the global average, a decrease in snow and ice cover, an increase in crop yields, hydropower potentials and summer tourism. Furthermore, climate change may lead to a shift in habitats so that they encounter new pressures in terms of land use changes or other processes (Haslett et al. 2010).

Overexploitation refers to the unsustainable use and management of landscapes (European Commission 2016). It has mainly occurred due to two simultaneous pressures. On the one hand, settlement and infrastructure development has decreased areas available for farming, on the other hand the demands for agricultural products

Table 9.1 Groups of pressures on ecosystems

Group of pressures	Description
Habitat change	The main pressure causing habitat change in terrestrial ecosystems is land take. This causes impacts, such as fragmentation, soil sealing, soil erosion and soil degradation that can cause direct degradation of a habitat or its loss and replacement by another habitat type. For some areas, abandonment of farmland leading to replacement by shrub or forest is also significant. For marine and coastal ecosystems, the main pressures are destructive fishing techniques and coastal development, and, for freshwater ecosystems, they are human modifications such as the creation of dams and diversion of rivers
Climate change	Anthropogenic climate change causes fluctuations in the life cycles of plants and animals and extreme events such as floods, droughts and fires change the health and characteristics of habitats and the species present
Overexploitation (unsustainable land or water use or management)	Pressures arise from the use of ecosystems for production of food, fuel and fibre. Intensive land management and overexploitation of natural resources, including overfishing and over-extraction of water, has already seriously reduced habitat quality and biodiversity in Europe
Invasive alien species	Invasive alien species can replace native species, occupying their habitats, reducing their survival and abundance and leading to loss of biodiversity
Pollution and nutrient enrichment	Pollution and nutrient enrichment occur when excessive harmful components such as pesticides, fertilisers and industrial chemicals are introduced into an ecosystem, exceeding its capacity to maintain their natural balance and resulting in their ending up in the soil, groundwater, surface water and seas, leading to ecosystem changes

Source: European Commission (2016). Adapted from EEA (2015)

have been growing, for example through increased use of bio-energy. Overexploitation means maximising the delivery of one ecosystem services, with negative effects on several others which could be simultaneously delivered in less intensive management regimes.

Invasive alien species are described by the European Commission (2016) as plants, animals, pathogens and other organisms which originate from other ecosystems and may cause negative effects on ecosystems, ecosystem services delivery, and human well-being. For example, invasive alien species can act as vectors for new diseases, alter ecosystem processes, and change species compositions in landscapes. Furthermore, they may influence landscape aesthetic values and can be a threat for urban ecosystems regarding biological diversity (Zisenis 2015).

Pollution and nutrient enrichment can be understood as the introduction of quantities of substances that exceed the ecological balance of the respective landscapes (European Commission 2016). Pollution and nutrient enrichment can result in substantial threats for biodiversity, ecosystem processes and services, and humans. For example, the excessive nutrient inputs to rivers cause substantial impacts for both the river and marine ecosystems (see Chap. 11).

9.3 Effects of Pressures on Ecosystems

The third report of the Mapping and Assessment of Ecosystem and their Service initiative (European Commission 2016) assessed the possible effects of the five groups of pressures introduced above on different types of ecosystems. Examples of these relationships are provided in Table 9.2 in terms of the specific pressures applicable to selected types of ecosystems.

Urban ecosystems experience demands for new building and infrastructure developments, as well as other pressures such as landscape fragmentation, soil sealing and resource extraction. Population migration towards major cities as favoured residences increases the contemporary demand for mobility and transportation of products as well. Hence, shipping and airports services cause high amounts of air, noise, water pollution and GHG emissions that are directly and indirectly burden ecosystems (EEA 2017). The pressures originating from urban areas affect both local and peri-urban lands, but also more distant rural areas. Climate changes affect the health of urban dwellers.

Cropland ecosystems face threats from overexploitation due to harmful farming practices, leading to water quality deficits and soil degradation. Arable land undergoes a path of slow shrinkage but its environmental impacts of cropland expand (Tilman et al. 2001). Increased agricultural intensity through technical applications leads to a loss of biodiversity (Reidsma et al. 2006). Habitats, pollinators and biological pest control services have also deteriorated.

The extent of grassland ecosystems has declined substantially over recent decades and many of the remaining grasslands are in unfavourable condition from a nature-conservation perspective due to abandonment that is often caused by a decrease in the number of livestock (Stoate et al. 2009). The loss and inappropriate management of many grasslands has led to species extinctions and effects on landscape aesthetics.

The size of woodland and forest ecosystems is slowly increasing across Europe, but many of them are quite evenly aged and have limited diversity in species composition. Demands for wood and timber products are expected to increase in the future, making higher levels of intensification likely.

Heathland, shrub and sparsely vegetated land systems cover only a small part of the EU, but are often of high ecological and cultural value. Nevertheless, many are in unfavourable condition due to fragmentation, overgrazing, or the abandonment of traditional grazing systems. Biodiversity and cultural values, among others, are threatened.

Wetland ecosystems in Europe are mostly located within intensively managed land and thus impacted by pollution and nutrient enrichment from surrounding catchments. Wetlands are lost due to conversion into agricultural use, afforestation, peat extraction or through natural succession after changes in the water regime. Overexploitation of the local aquifer through intensive irrigation causes degradation of wetland ecosystems and leads to salinisation of soil and groundwater (Kløve et al. 2011).

Table 9.2 Major pressures in selected ecosystem types

	Habitat change	Climate change	Overexploitation	Invasive alien species	Pollution and nutrient enrichment
Urban ecosystems	Land take	Extreme events droughts, floods, fires, heatwaves	Gravel extraction around cities	Expansion of alien species	Soil contamination by heavy metals due to industrial activities
	Landscape fragmentation due to urban sprawl and roads around urban areas	Rise in sea level for coastal cities	Overexploitation of groundwater and freshwater	Introduction of exotic species in gardens	Air pollution and critical levels of ozone
					Water pollution caused by poor wastewater management
					Sludge, waste
Cropland ecosystems	Land take	Changes in temperature and precipitation	Agricultural intensification: intensive cultivation and overharvesting	Expansion of invasive alien species	Pesticide use
	Landscape fragmentation	Extreme events (floods, droughts, heatwaves, fires)	Groundwater over-extraction		Critical levels of ozone
	Agricultural specialisation				Nutrient enrichment
	Intensification and abandonment				Soil salinisation

(continued)

Table 9.2 (continued)

	Habitat change	Climate change	Overexploitation	Invasive alien species	Pollution and nutrient enrichment
Grassland ecosystems	Landscape fragmentation	Changes in temperature and precipitation	Agriculture intensification	Expansion of invasive alien species	Fertilisers
	Abandonment of grazing or mowing	Extreme events	Overgrazing		Nutrient run-off
	Land take	Fires	Groundwater extraction		Critical levels of ozone
	Habitat loss				Heavy metals
Woodlands and forests	Land use change: urbanisation, conversion to agriculture	Changes in temperature and precipitation	Unsustainable exploitation of timber and non-wood products	Fast-growing invasion alien species	Nitrogen enrichment
	Changes in forest pattern	Fires	Recreation and tourism	Pests and disease agents, e.g. *Phytophthora*	Acidification
	Fragmentation due to roads, forests isolation	Extreme events (droughts, frost, floods, storms)	Game hunting		Heavy metals
		Pests and diseases	Overgrazing		Air pollution
					Critical levels of ozone
Heathland, shrub and sparsely vegetated land ecosystems	Land use change	Extreme events	Lack of appropriate site management	Invasive plants, e.g. rhododendron, water fungus, and disease agents, e.g. *Phytophthora*	Nitrogen enrichment
	Landscape fragmentation	Fires	Recreational and urban disturbance		Critical levels of ozone
	Land take				Heavy metals
	Land abandonment				

Wetland ecosystems	Land take	Drought	Blocking and extraction of the water inflow	Introduction of invasive fish	Eutrophication
	Fragmentation	Changes in rainfall	Overexploitation of groundwater resources	Plant species such as *Hydrocotyle ranunculoides* (floating pennywort) and *Azolla filiculoides* (water fern)	Pesticides
	Drainage for agriculture		Water abstraction		Acid rain
			Reed harvesting, also for biofuels		Litter (e.g. plastic)
Freshwater (rivers and lakes) ecosystems	Modification of watercourses	Changes in temperature, precipitation and average river flows	Water abstraction	Invasive plants, fish, mammals (e.g. American mink), molluscs and crustaceans (e.g. American signal crayfish)	Nutrient intake from diffuse and point sources (agriculture, wastewater, aquaculture)
	Channelling	Droughts	Gravel extraction		Pesticides
	River regulation				Deposition of acid and nitrifying substances
	Fragmentation (dams)				Heavy metals; households and industrial chemicals
	Soil erosion from agriculture leading to sedimentation of gravel riverbed habitats				Endocrine disruptors
					Sediment transport from soil erosion

Taken from European Commission (2016)

Europe's freshwater ecosystems, including rivers and lakes, are experiencing pressures from diverse sources such as urbanisation, intensive agriculture, hydropower generation, inland water navigation and flood protection schemes. Climate change is becoming an additional source of pressure, with increased water temperatures, risks of flooding and more severe droughts. Water quality has improved, but indicators such as ecological status are still frequently below desired standards.

9.4 Assessing Pressures in Landscape Planning

Assessing pressures in landscape planning follows a problem-oriented approach including an assessment of current land uses types and intensities with relevance for the conservation and sustainable use of biodiversity and ecosystem services (von Haaren 2004). The scale of the assessment relates to the study area and the issues at stake.

The source data usually includes existing land use information such as topographic data, habitat maps, and forestry management plans. Satellite imagery and aerial photographs are an important and powerful source of information, as they can cover large areas and are increasingly frequently updated. Additional field studies are useful for amending, updating and ground-truthing available data. Many relevant data sets can now be found online (see Chap. 5 for examples and details).

The assessment of pressures first identifies and spatially represents the sources of relevant pressures, and/or the locations where these sources exert pressure on the landscape. The results of the assessment of pressures are presented in maps and described in accompanying text.

Once relevant pressures have been identified and spatially described, their level of severity and importance can be evaluated. The evaluation process can draw on legal limits and standards. For example, legal standards may serve as the benchmark against which the type and level of existing pressures can be evaluated. If such minimum standards are not met, the need for response measures becomes immediately evident. Legal standards can be found in EU or national environmental legislation (e.g. regarding water quality or waste disposal), but also in documents such as the cross-compliance regulations linked to EU support payments for farmers. Monitoring compliance with such regulations is increasingly based on remote sensing, but in some cases there is no substitute for a field visit.

9.5 Case Studies of Pressures on Natural Assets in the UK

One approach to evaluating pressures is to compile a natural capital asset statement and risk register (Natural Capital Committee 2017). The asset statement is an inventory of the natural assets in an area and their condition, while the risk register identifies the likelihood and scale of changes to the natural assets which could impact upon their delivery of benefits. In order to construct a risk register likely drivers and pressures which may stimulate socio-economic and environmental change need to be identified and the DPSIR framework can be useful in this respect. Analysing changes in land cover is often a useful first step in identifying the key features of an area and therefore the likely importance of different pressures such as housing and urban development, agricultural change and climate change.

Mace et al. (2015) used a combination of existing data and expert judgement to construct a natural capital risk register for the UK. The register used eight habitat types (e.g. semi-natural grasslands, enclosed farmlands, woodlands) and ten major benefits (e.g. food, clean water, recreation, hazard protection). For each habitat-benefit relationship, Mace et al. (2015) explored the influence and modification of quantity, quality or spatial configuration of habitat on the identified benefit (i.e. the provision of a usable service or good to human populations). The results were summarised in a matrix, with relationships classed as high, medium or low risk. Another feature of the analysis was the substantial degree of uncertainty, either because of substantial gaps in the knowledge base (e.g. regarding marine habitats) or low confidence in assessments.

Lovett et al. (2018) applied a similar approach to compile a risk register for eastern England. The resulting register highlighted the pressures on water-related benefits, particularly within farmland and urban habitats in this relatively dry and fast-growing region of the UK. To extend the analysis, two indicators of pressures (projected population growth and restrictions on water abstraction) were mapped for the 63 local authorities in the region and then compared with distributions of five natural capital assets (high quality agricultural land, carbon storage in soil and vegetation, habitat suitability for pollinators, amenity and recreation areas and priority habitats and sites for nature conservation). The two dimensions were then used to produce a classification of the local authorities as shown in Fig. 9.1. This map provides an indication of the different combinations of relative pressures and presence of natural capital assets that exist within the region, with the darker shadings highlighting authorities which score relatively highly on both dimensions, emphasising the need for particularly careful spatial and resource planning in these areas.

9.6 Conclusions

This chapter has provided a brief overview of relevant pressures, their more specific effects in different types of ecosystems, some guidelines for assessing pressures, and an illustrative case study. In each practical application, landscape planners need to consider the specific context, the issues at stake, and the data and resources available in order to choose the most appropriate approach. Although important data on

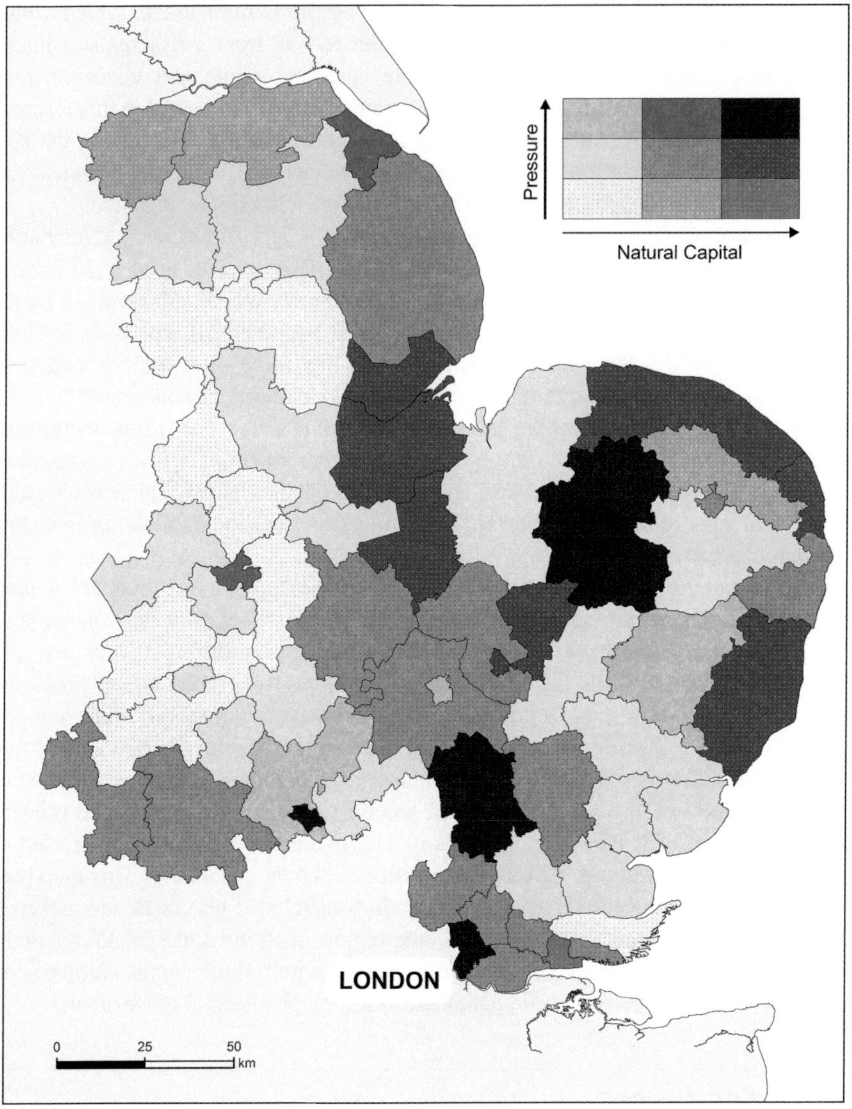

Fig. 9.1 A classification of local authorities in eastern England based on pressures and natural capital assets

pressures can be readily derived from existing land use information, more quantitative data on how particular pressures play out in the landscape at particular points in time are still needed.

References

EEA. (2010). *The European environment: State and outlook 2010, land use*. http://www.eea.europa.eu/soer/europe/land-use/at_download/file. Accessed 18 June 2018.

EEA. (2012). *Climate change, impacts and vulnerability in Europe 2012: An indicator-based report* (EEA Report No 12/2012). http://www.eea.europa.eu/publications/climate-impacts-and-vulnerability-2012. Accessed 4 July 2018.

EEA. (2015). *European ecosystem assessment: Concept, data, and implementation* (EEA Technical Report No 6/2015). http://www.eea.europa.eu/publications/europeanecosystem-assessment. Accessed 18 June 2018.

EEA. (2017) *Aviation and shipping – impacts on Europe's environment* (EEA Report No 22/2017). https://www.eea.europa.eu/publications/term-report/#content-1. Accessed 18 June 2018.

European Commission. (2016). *Mapping and assessment of ecosystems and their services, mapping and assessing the condition of Europe's ecosystems: Progress and challenges*. Luxembourg: Publications office of the European Union.

Haslett, J. R., Berry, P. M., Bela, G., et al. (2010). Changing conservation strategies in Europe: A framework integrating ecosystem services and dynamics. *Biodiversity and Conservation, 19*, 2963–2977. https://doi.org/10.1007/s10531-009-9743-y.

Kløve, B., Allan, A., Bertrand, G., et al. (2011). Groundwater dependent ecosystems. Part II. Ecosystem services and management in Europe under risk of climate change and land use intensification. *Environmental Science & Policy, 14*, 782–793. https://doi.org/10.1016/j.envsci.2011.04.005.

Lovett, A. A., Turner, R. K., & Sünnenberg, G., et al. (2018). *A natural capital asset check and risk register for the Anglian Water Combined Services Area*. CSERGE and The Anglian Centre for Water Studies, School of Environmental Sciences, University of East Anglia.

Mace, G. M., Hails, R. S., Cryle, P., et al. (2015). Towards a risk register for natural capital. *Journal of Applied Ecology, 52*, 641–653.

Maes, J., Teller, A., Erhard, M., et al. (2014). *Mapping and assessment of ecosystems and their services. Indicators for ecosystem assessments under action 5 of the EU biodiversity strategy to 2020, 2nd report*. Luxembourg: Publications office of the European Union.

Millennium Ecosystem Assessment. (2005). *Ecosystems and human well-being: Synthesis*. Washington, DC: Island Press.

Natural Capital Committee. (2017). *How to do it: A natural capital workbook Version 1*. https://www.gov.uk/government/groups/natural-capital-committee. Accessed 18 June 2018.

Reidsma, P., Tekelenburg, T., van den Berg, et al. (2006). Impacts of land-use change on biodiversity: An assessment of agricultural biodiversity in the European Union. *Agriculture, Ecosystems and Environment, 114*, 86–102. https://doi.org/10.1016/j.agee.2005.11.026.

Rounsevell, M. D. A., & Harrison, P. A. (2016). Drivers of change for ecosystem services. In M. Potschin, R. Haines-Young, R. Fish, & R. K. Turner (Eds.), *Routledge handbook of ecosystem services* (pp. 94–98). London: Routledge.

Smeets, E., & Weterings, R. (1999). *Environmental indicators: Typology and overview* (Technical Report No. 25). Copenhagen: EEA.

Stoate, C., Baldi, A., Beja, P., et al. (2009). Ecological impacts of early 21st century agricultural change in Europe – A review. *Journal of Environmental Management, 91*, 22–46. https://doi.org/10.1016/j.jenvman.2009.07.005.

Tilman, D., Fargione, J., Wolff, B., et al. (2001). Forecasting agriculturally driven global environmental change. *Science, 292*(5515), 281–284. https://doi.org/10.1126/science.1057544.

von Haaren, C. (Ed.). (2004). *Landschaftsplanung*. Stuttgart: Eugen Ulmer.

Zisenis, M. (2015). Alien plant species: A real fear for urban ecosystems in Europe? *Urban Ecosystems, 18*, 355–370. https://doi.org/10.1007/s11252-014-0400-1.

Part III
Methods for Assessing State and Impacts

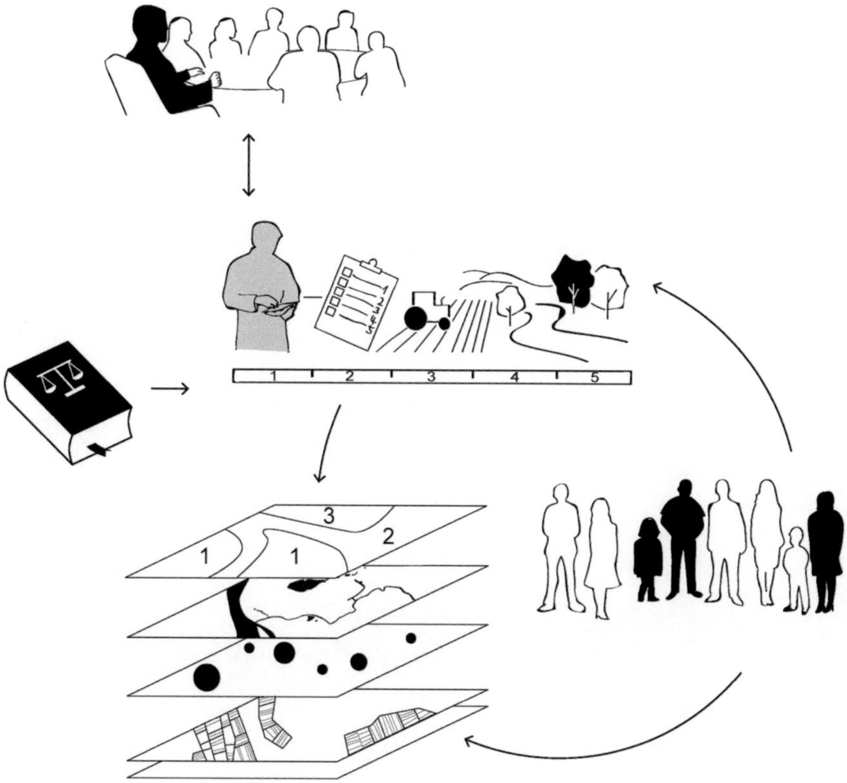

This core part of the book provides a set of methods for assessing the state and value of biodiversity, geodiversity and ecosystem services. With adaptations, these should be applicable in landscape planning throughout Europe. The scope for landscapes to provide multiple ecosystem services is reviewed, as are methods for the economic valuation of the benefits to people. Local knowledge from the public and interest groups, as well as local and regional priorities are integrated into the assessment.

Assessing Productive Capacities of Agro-Ecosystems

10

Jan Bug

Abstract

The ability of an ecosystem to produce the raw materials and food necessary for human economic activities is termed 'productive capacity'. The productive capacity of an ecosystem sets the physical limits for the provision of ecosystem services (ES), in turn defined as the benefits that humans gain from the natural environment. The productive capacity can be evaluated to help landscape planning in the allocation of land use. But ES are under the constant threat to lose their productive capacity. Land degradation processes such as soil erosion, contamination, compaction and the sealing of land can cause a massive decline in production of raw materials. Landscape planning must evaluated these threats with proper tools to develop measures for a sustainable and regional differentiated land use.

Keywords

Productive capacity · Food · Biomass · Soil quality

10.1 Introduction

The ability of an ecosystem to maintain its natural, original, or current condition and to produce goods and services is termed its 'ecological capacity' (Fig. 10.1). Ecological capacity is, in turn, determined by soil, air and water limitations and by geo and bio diversity.

J. Bug (✉)
Landesamt für Bergbau, Energie und Geologie, Hannover, Germany

Institute for Physical Geography and Landscape Ecology, Leibniz University Hannover, Hannover, Germany
e-mail: Jan.Bug@lbeg.niedersachsen.de

C. von Haaren et al. (eds.), *Landscape Planning with Ecosystem Services*, Landscape Series 24, https://doi.org/10.1007/978-94-024-1681-7_10

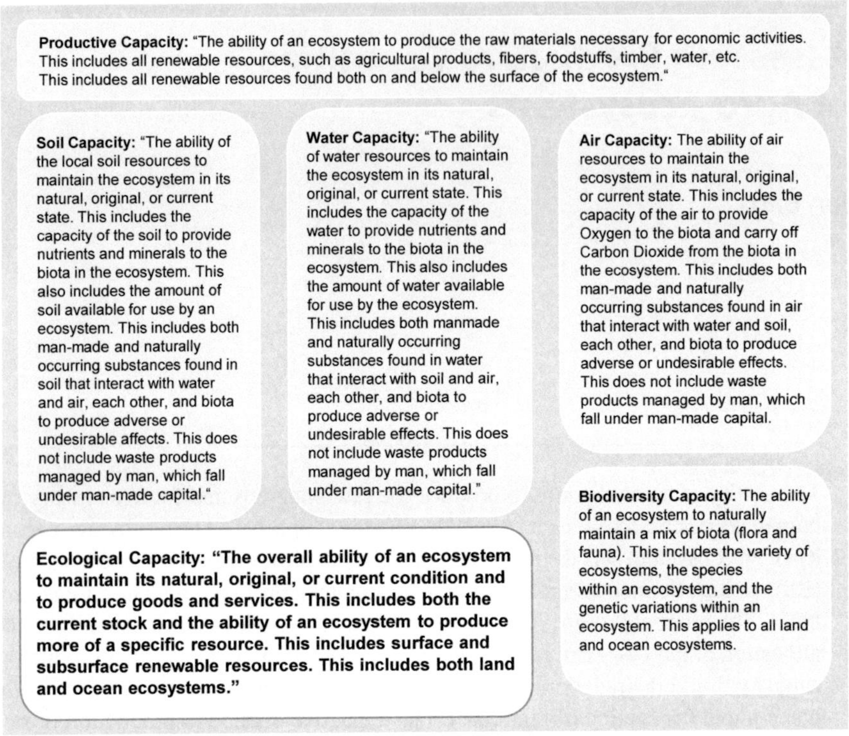

Fig. 10.1 Productive capacity (delivered ES) in the context of overall ecological capacity. (Source: compiled from https://www.hq.nasa.gov/iwgsdi/Ecologicai_Capacity.html)

From a human perspective, the ability of an ecosystem to produce the raw materials necessary for human economic activities is termed its 'productive capacity'. The productive capacity of an ecosystem sets the physical limits for the provision of ecosystem services (ES), in turn defined as the benefits that humans gain from the natural environment.

Chapter 3 introduced a further refinement in the definition of ES differentiating between the overall total amount of an ES that is available (its productive capacity), which, in this book, is termed 'delivered ES', and the amount that humans actually make use of, termed 'utilized ES'.

The productive capacity of an ecosystem is not static. It changes over time, through both natural processes and the influences of humans upon it. All societies rely heavily on the productive capacity of their environment and have often adapted their way of life to suit it. Since the first hunter-gatherer societies, people have sought to take what they need from the environment. With the development of agriculture and domestication of wild animals, humans have progressively shaped the environment to increase the production of food and goods, altering a natural ecosystem to an agro-ecosystem. This can increase the production of goods for human use enormously. However, these changes may also cause a reduction of other benefits from the ecosystem. For example, a drained peatland can have a much higher initial **food production** output than a natural wetland, but the carbon stock and therefore the **climate regulation** function it has may be severely damaged.

In Europe, one of the most intensively exploited continents on earth, over 80% of the area is used for settlement, production systems (including agriculture and forestry) and infrastructure (EEA 2016). Humans have shaped both terrestrial and marine ecosystems to increase ES production. For instance, marine ecosystems provide fish for food and are a location for the production of energy (e.g. offshore wind farms).

The productive capacity of Europe's ecosystems is under constant threat. The intensification of agriculture in the last century resulted in a phenomenal increase in food production but also caused many problems. Examples are soil erosion, soil compaction, the destruction of habitats and attendant loss of biodiversity, pollution, and eutrophication of rivers and other water bodies. Global climate change is likely to accentuate some of these problems and may further impact productive capacity both positive and negative. Climate change can lead to accelerated soil degradation as well as to new opportunities to grow new things or to extend range of some crops e.g. vines and horticulture in regions where climate has not allowed it.

Estimation of productive capacity in agro-ecosystems and potential impacts on it is an important task in landscape planning. Due to the growing world population, production is still intensifying in already altered ecosystems and additional land is being brought into cultivation for production with serious consequences for biodiversity.

Landscape planning can be used to spatially assess productive capacity and develop targets and measures to protect important natural resources, through spatial targeting of land uses. For example, the protection of fertile soils from pressures such as residential development.

This chapter examines factors influencing productive capacity in agro-ecosystems, and, using soil fertility as an example, highlights models used to assess and evaluate soil fertility and the major threats to this ES which underpins the actual provision of food and fibre for materials and biomass energy.

10.2 Factors Influencing Productive Capacity in an Agro-Ecosystem

The assessment of the productive capacity of an agro-ecosystem will concern the **actual state** of the ecosystem with all long-term anthropogenic influences and changes, like draining or other land improvements.

The most important natural factors which influence production capacity are climate (water budget) and soil. Topography can be a factor as well, usually influencing microclimate and water management. The slope of land clearly affects its suitability for agricultural production, mainly through the restrictions steeper slopes impose on mechanization of crop management, and therefore, the potential yield.

Soil improvement from the past, (for example the introduction of organic material (humus) imported from nearby sites), must be taken into account together with other factors such as artificial lowering of the groundwater table. Actual land use or vegetation cover can give an indication of soil quality but usually any evaluation of the individual factors affecting productive capacity will be blurred by practices to increase crop yield (e.g. fertilization, liming or tillage methods).

Hence, an assessment of productive capacity is not a simple task and requires the use of indirect methods.

Data from a combination of environmental indicators which influence plant growth, can be used to evaluate productive capacity. This enables comparison between different sites and the causes of a lower productive capacity to be more easily detectable.

As previously mentioned, the most important environmental variables that influence productive capacity in an agroecosystem are climate, soil and topography.

Climate In relation to climate, total precipitation, as well as the distribution of rainfall in a growing period, are important parameters in the assessment. Dryness, which is defined as the result of a permanent imbalance in water availability due to low precipitation and high evaporative water demand, can limit plant growth. Conditions that are too wet can also hinder plant growth due to limited air capacity in the soil.

Temperature limits growth in two ways: First, those that are too low compromise the crop performance or survival. Second, temperatures that are too high can affect plant growth due to limited water availability. The mean, maximum and minimum temperatures in the growing period, as well as the length of the growing period itself, are important indicators in the evaluation of productive capacity.

Soil Soil influences productive capacity in many different ways. Soil has the ability to allow passage of air and water, withstand erosive forces, and provide a medium for plant roots. Soils offer plants physical support, air, water, temperature moderation, nutrients, and protection from toxins. Furthermore, soils provide readily available nutrients to plants by converting dead organic matter into various nutrient forms. Soils act as the main storage for water and nutrients in an ecosystem. A lack of both can limit plant growth substantially.

The ability of a soil to store water for plants is dependent on texture, organic matter content, density, and porosity. The available water capacity (AWC) is the most common indicator used to evaluate water storage capacity. It defines the amount of water in a soil that is available for plant growth. The upper limit is set by the field capacity and the lower limit is defined as the permanent wilting point. Clay loam soils with a large rooting depth have the highest available water capacity, while sandy soils or soil with a reduced rooting depth possess low values.

Soil stores, releases and recycles nutrients and other elements. By biogeochemical processes, nutrients can be transformed into plant available forms, which are stored in the soil, or emitted into air or water. The ability of soils to store plant available nutrients can be described by the anion- and cation-exchange capacity of soils. Both depend on the texture as well as on the organic matter content and the soil pH value. Humus and clays have the highest cation exchange capacities. Higher soil pH values lead to a higher cation-exchange capacity. Conversely, the anion exchange capacity will generally increase when pH drops and decrease when pH rises.

Topography Topography also influences soil formation, as well as microclimate, and is therefore a factor in the evaluation of productive capacity. In the most common approaches for assessing soil quality, topography, or more precisely slope, is used to estimate the operability of tillage practices at specific sites.

All of these individual parameters and indicators are also determinants for other ES. Consequently, some sites with a high productive capacity can also be important for other ES. Sites with low productive capacity are usually more interesting as habitats for biodiversity (see Chap. 17). It is advisable to perform separate assessments for the different ES and compare the value in an overall landscape plan.

10.3 Socio-Economic Valuation of a Market or Potential Use Value

According to FAO (2009) the world population is expected to increase by over a third, or 2.3 billion people, between 2009 and 2050. At the same time, the projected annual global economic growth rate of about 2.9 percent will lead to increasing demands for both food and animal fodder. The growing demand for biofuels, as energy, may lead to an even higher need for the production of biomass. The FAO conclude that the feeding of a world population of over 9 billion requires a growth in production of over 70% by 2050. 90% of the growth has to be realized from already exploited ecosystems as land expansion can only contribute a further 10%.

The productive capacity of agro-ecosystems, however, is not growing. Instead, due to climate change and other pressures, a deterioration is more than possible in many regions of the world. For instance, in regions that currently have high yields like Central Europe, rainfall is projected to be more irregular (Kovats et al. 2014). To secure the benefits of productive capacity for societies, particularly the production of food, landscape planning has a role to guide land users towards more sustainable growth.

In some regions of the world, extension of arable land area can still be considered in order to secure food distribution and food quality. However, the demand for non-food biomass from other stakeholders is also growing. Biomass from dedicated energy crops such as *Miscanthus* grasses or Short Rotation Coppice is used, for instance, as a renewable energy source. Natural woodlands and plantation forestry supply the need for natural building materials, but in some parts of the world demand is higher than the actual supply or a failed management leads to a decline in the area used as forests (FAO 2016). It is then the spatial distribution and management of land use that is key for the economic and sustainable use of ecosystems and, simultaneously, the protection of the environment. Here landscape planners also have a key role.

For these reasons landscape planners need tools to assess the potential quality of a particular site for the production of different kinds of biomass in a sustainable way. For economic valuation, many indicators can be measured/estimated. One common indicator is biophysical production (yield) that is measured, for instance, in kilograms of maize per hectare or tonnes of tuna landings. The provisioning of ecological goods such as food, fuelwood, or fibre, depends both on the flow and the stock of the good, just as is the case with manufactured goods.

Pagiola et al. (2004) suggest two different methods to assess the value of produced goods. The actual **market price** of products, observed directly in markets can provide a quite straightforward overview of the value of an environmental good and service. A problem, however, is that many other ES cannot be monetarized, and often subsidies distort the actual prices. An alternative method is the estimation of the value of an environmental product as an **input in the production of marketed goods**. However, the estimation is technically difficult and has high data requirements. Notwithstanding these difficulties, both methods are rated as highly amenable and highly transferable, as discussed by Farber et al. (2006) (see also Chap. 20).

Whichever approach is taken, the ***valuation*** of productive capacity is essential to enable a comparison of people's perspectives on the benefits of ES.

10.4 Practical Relevance of the Assessment of Productive Capacity and Resulting Demands for Its Representation in Planning

A main goal of landscape planning is the protection of the ability of an ecosystem to regenerate and to be used as a resource in the future. The estimation of the productive capacity of ecosystems, is crucial for the overall planning of a landscape. Assessment of productive capacity supports the following planning tasks:

- Evaluation of different uses of a landscape.
- Protection of sites with medium to high productive capacity against destruction (e.g. by urbanization), in order to safeguard their capacity for future ES provision.
- Planning of agricultural measures for soil protection to secure the productive capacity of agro-ecosystems.
- Enhancement of agricultural measures to increase biomass production in a sustainable way (sustainable intensification).
- Allocation of agri-environmental measures to sites with low productive capacity, for example to develop habitats.
- Selection of sites for compensatory measures in the context of impact mitigation regulations.

Assessment of productive capacity will also support the allocation of measures to boost other ES. It can help detect and protect low nutrient sites, groundwater-influenced soils, soils with high salinity and organic soils with a high greenhouse gas (GHG) storage. A land-use dependent evaluation of productive capacity can also help to allocate different land uses to gain a higher overall yield from a landscape.

Tools to assess productive capacity can be useful in two different ways: Firstly, they highlight sites where food or fibre production is most efficient. Here landscape planning can help to intensify production in a sustainable way to attain food and non-food biomass output targets. It is important that the usage of the site is sustainable and intensified production is not allowed to diminish productive capacity. Sites with a high productive capacity should not be destroyed by development of any kind. Where a change in land use is being considered, estimation of the impacts on productive capacity can help optimize land use. Such tools reveal sites that have a low or very low productive capacity. Here, the production of biomass may not be very efficient. Other usages for the site, for example for recreation or as a habitat, should be considered. Information about the productive capacity is especially applicable for the planning of compensation measures and the conversation of specific sites. Ecosystems with a low productive capacity might therefore be considered as sites for habitat development (Chap. 17) or for other uses, e.g. environmental protection. That is, the overall benefit may be higher if such sites are used as natural ecosystems rather than as arable land with low or very low yields or with need for a very high input of fertilizer, water, etc.

In order to spatially allocate the different human and natural activities most effectively, it is important to assess the value of sites for different purposes. The assessment of the **production-supporting features** of a site is key to understanding the differences between the potentials. This includes evaluation of the abiotic features of the environment such as soil, geology, climate, and hydrological properties.

10.5 Methods for the Assessment of Productive Capacity

As previously discussed, the most valuable sites for food and fibre production must be protected from threats to secure the present productive capacity for future use. Landscape planning therefore needs both tools to assess ES and methods to describe potential threats to productive capacity from the site scale to the ecosystem scale.

An assessment of productive capacity requires a combined rating of climate, soil, topographical and land use data. One example of such an approach is the Müncheberger Soil Quality Rating (SQR).

10.5.1 Site Scale Evaluation: The Müncheberger Soil Quality Rating

The SQR (Mueller et al. 2007) is a means of evaluating the ***potential*** yield of arable land and grassland. It is a comparative on-site assessment, adapted for the application of soil maps. The SQR refers to the current condition of a soil including the hydrological, thermal, geological, and terrain conditions as well as human impact. The focus of the method is on rain-fed cropping in temperate zones and rotations with a dominance of cereals, mainly wheat. However, the SQR method is not restricted to such sites.

The result of the method is a soil quality score, which ranges between 0 and 100. The score is a measure of the long-term soil quality and will provide a rough estimate of the local crop yield potential. The first step in the method is the valuation of basic soil indicators, for example the field capacity of a soil. The second step is the assessment of potential hazards to the yield, for example soil salinity or drought. Fig. 10.2 shows the eight basic indicators and 13 hazard indicators.

All indicators are classified and their scores range between 0–2 or 0–3. The eight basic soil indicators are added to define the basic soil score, which is then multiplied by the lowest of all 13 hazard indicators. The results are two scores for the site, one

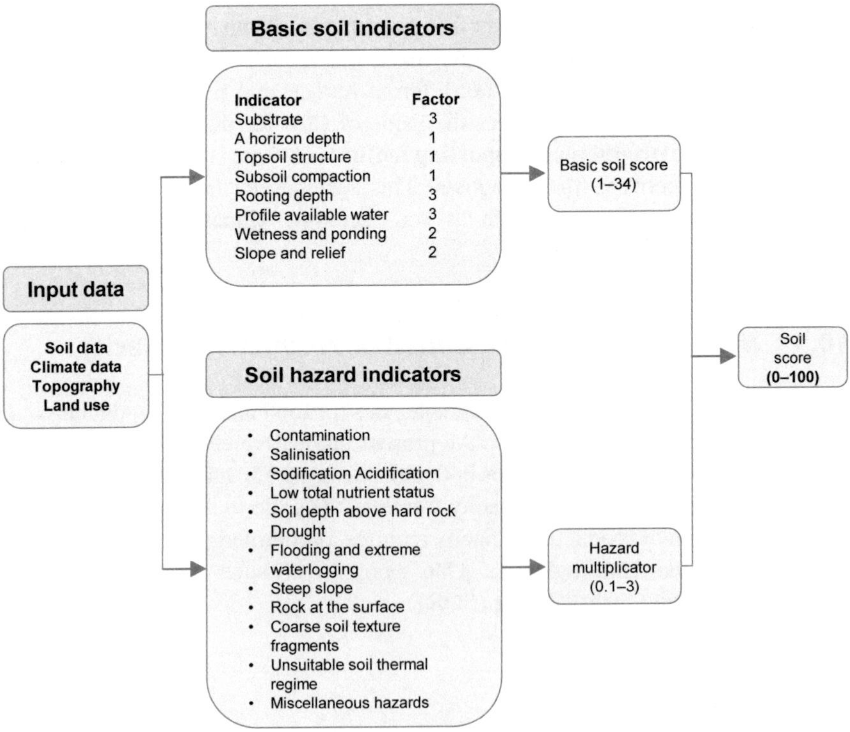

Fig. 10.2 Soil quality indicators and soil hazard indicators (risk of impairment)

for suitability as grassland and one for arable land. The SQR is not appropriate to assess production capacity for forestry.

The SQR includes all main indicators for any soil quality assessment: texture, organic matter, soil depth, density, slope, available water in the soil, pH, and salinization (exchangeable sodium percentage). The SQR also uses further indicators and can be used in both field surveys (Mueller et al. 2010) and as a pedotransfer function (PTF) (Richter et al. 2009). The US Department of Agriculture (2001) has introduced an approach quite similar to the SQR. It is also based on the assessment of soil and climate parameters.

10.5.2 Evaluation of Productive Capacity at Larger Scales

Whilst the SQR is useful for evaluation of productive capacity at a site (e.g. farm) scale, different methods are needed for the evaluation of the productive capacity at a larger scale (e.g. catchment or region). These need to parameterize the factors discussed earlier, namely climate, soil characteristics and topography. As described before, the interpretation of *climate data* is often the first step in a regional evaluation. Water and warmth are crucial factors for plant development. Climate data, most commonly *mean temperature* and *mean precipitation*, are used on a regional and global scale. Productive capacity is positively correlated with precipitation and temperature. However, a lack of water cannot be compensated for by more warmth.

A very common indicator for describing the water regime of an ecosystem is the calculation of the climatic *water balance.* This shows the plant-available water in a defined region for period of time, usually a growing season. A general water balance equation is:

$$Aw = P - Q - E + S$$

where

Aw is plant available water
P is precipitation
Q is runoff
E is evapotranspiration (potential evaporation)
S is the capacity of a storage (in soil or bedrock).

A high value for Aw indicates a good prospect for plant growth and therefore a high productive capacity. However, other factors influence plant growth as well. For the tasks of landscape planning and assessments on a more local scale, more complex methods are needed.

A simple extension of the Aw method is the *available water capacity* (AWC) of the soil. Depending on properties, like depth, texture and organic matter a soil can hold water after any excess has drained away. This stored water can be used in periods with insufficient precipitation until the suction pressure of plants is lower than the soil water potential. The indicator 'effective water balance of the growing season', therefore, integrates soil and climate data.

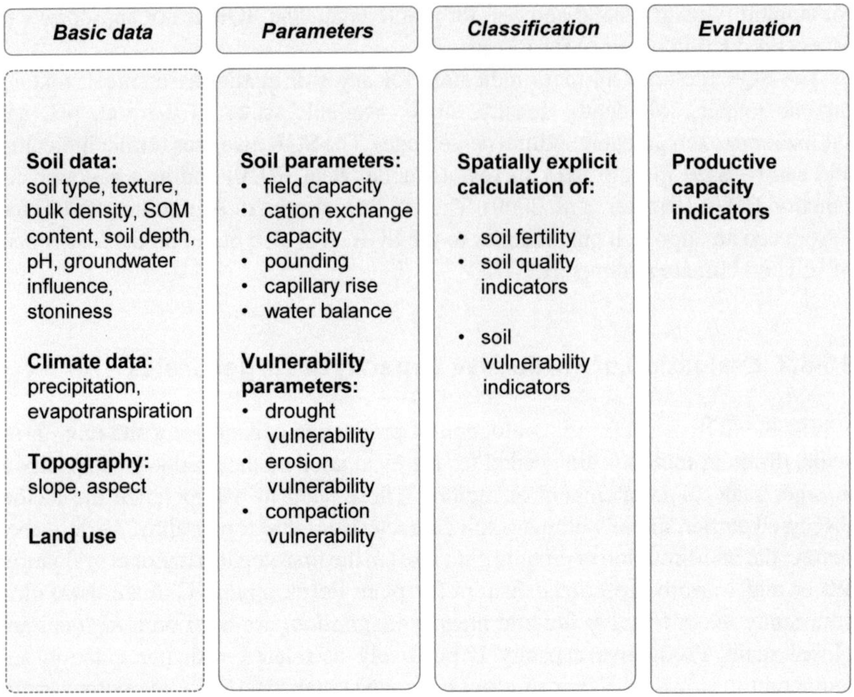

Fig. 10.3 Workflow for the evaluation of productive capacity

More complex methods to assess productive capacity often use soil parameters as their basis. Climatic and topographic data improves the assessment. However, the usage of more data does not automatically mean a better prediction of productive capacity. The availability and quality of data determine the choice of method and scale of the prediction. The most common parameters and methodological workflows that enables assessment of productive capacity are shown in Fig. 10.3.

From the basic data sets a number of parameters of varying complexity can be derived. Typical soil parameters for the evaluation of productive capacity are 'field capacity', 'rooting depth', 'capillary rise', ' water balance of the growing season', 'soil pH' or 'cation exchange capacity (CEC)'. For a more accurate assessment a combined system is needed. The availability of suitable parameters may differ depending on scale. Soil parameters are normally available from regional or local soil maps with scales ranging from 1:5000 to 1:50,000 in many European member states. Climate data is provided by the national meteorological services. Topography data, such as digital elevation models (DEMs) are often available from state geodetic agencies. For details of pan European sources see Chap. 5.

The components of a soil quality assessment include the evaluation of the ability of soil to enhance crop production (productivity component) or mitigate risks to yield such as contaminants or pathogens, as well as evaluating the linkage between soil quality and plant, animal and human health (Arshad and Martin 2002; US

Table 10.1 Example tools for the evaluation of productivity and other ES

Name	Data requirements/platform	Output of productive capacity
InVEST (Integrated Valuation of Environmental Services and Tradeoffs, Bagstad et al. 2013)	Spatial explicit GIS-tools	Managed timber production
	Climate data	Crop production
	Observed yields	Marine Finfish Aquacultural Production
TIM (The Integrated Model, Bateman et al. 2014)	Terrain, soil, climate, land use datasets	Agricultural production module:
	Prices of inputs and outputs	Stocking intensities
		Agricultural land uses
MAES (Mapping and Assessment of Ecosystems and their Services, Maes et al. 2018)	Biodiversity centered approach to show value of ES	Biomass:
		Food (fodder)
		Fibre, timber

Department of Agriculture 2018). For landscape planning, simple indexing approaches are more relevant than complex models to predict actual yield. There are numerous other spatially-explicit models that can evaluate productive capacity. Table 10.1 lists some examples and outlines their data requirements, outputs and approaches.

The most important indicator for all respective assessment methodologies is the observed yield. Economic information e.g. about the prices of in- and output help to monetize the benefit, which is achieved in economic terms.

10.6 Pressures on Agricultural Production Capacity: Impact Assessment

As highlighted above, the productive capacity is under constant pressure. An unsustainable utilization of the ecosystem can cause a deterioration of the quality. The section describes the pressures and threats to productive capacity and therefore to the provisioning of food und biomass.

10.6.1 Structure and Components of Soil Production Capacity Impact Assessment

As previously discussed, humans have modified their environment to boost the production of food and fuel since the invention of domestication and agriculture. However, the changes have negative consequences as well (see Fig. 10.4) with the growing population and economic development being the main human-induced drivers. During the last century, the demand for biomass as a renewable energy and as building material has grown and so has the cultivation of formerly unused ecosystems. The often volatile agricultural markets can cause a radical change in the usage of ecosystems within years and with often new and sometimes inadequate

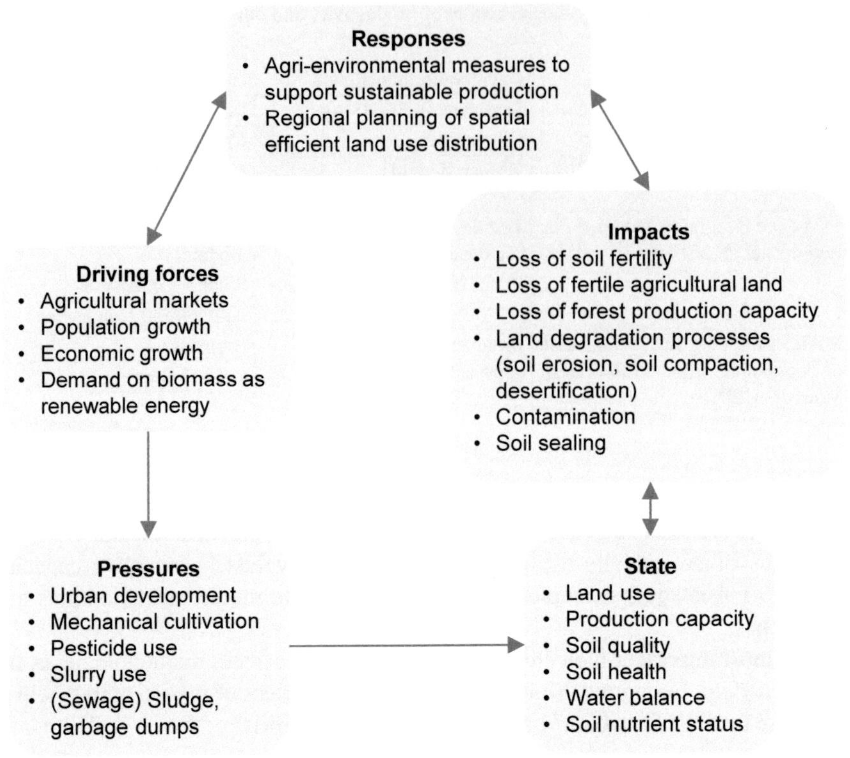

Fig. 10.4 Soil related drivers, pressures and impacts

land-use measures. So many cultivation measures trigger new pressures on the ecosystem, for example land degradation processes. These processes often lead to loss of soil depth, organic matter or nutrient leaching, leading to a negative impact on the soil quality. The total loss of fertile ground for the production of goods is the most severe impact on the productive capacity.

The main pressures on production capacity and soil properties are summarized in Fig. 10.5.

Sealing leads to a loss of area available and therefore of the productive capacity of the ecosystem. Land use practices can cause a decline because the trigger soil degradation process indirectly.

10.6.2 Methods to Determine Impairment to Soil Quality, Production Capacity, and Biomass Yield

Impairments in soil function may be caused by human interference as well as by natural processes. Land degradation processes such as wind and water erosion are natural processes, which are accelerated by the human use of ecosystems. The

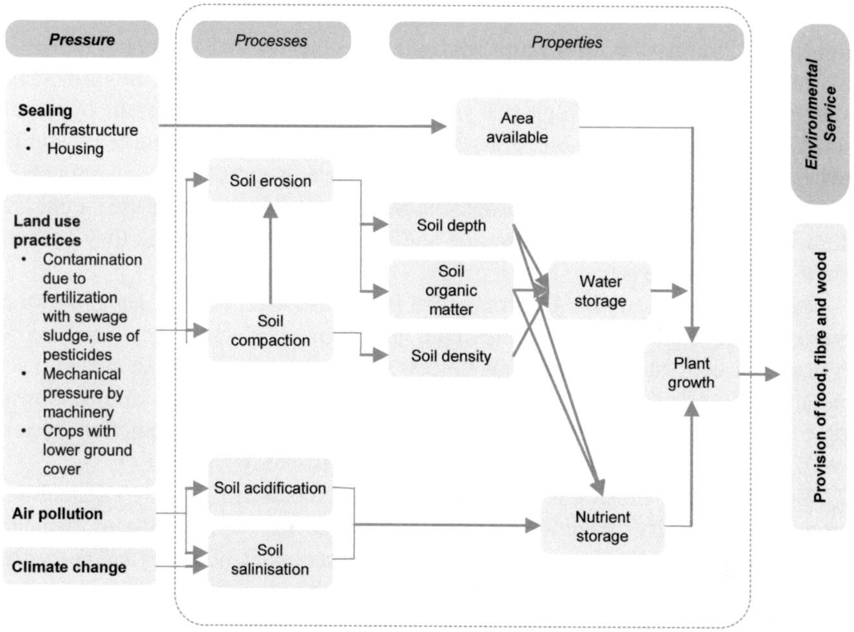

Fig. 10.5 Pressures and process changes in production capacity

processes may lead to an actual decline in the yield of food and raw material, but a more severe problem is the gradual loss of soil fertility over years and decades. Methods are needed to show the effects of the process. But actual the link between land and soil degradation models and the decline in ES is missing. A monetarization of the processes is not yet established. Nevertheless a more detailed description of the processes and the models is appropriate to show the threats to the productive capacity of ecosystems.

Water erosion is a naturally occurring process in which soil particles become detached (usually from the soil surface) by wind or water. Erosion rates can be increased as a result of human activities such as the removal of protective vegetation cover by farming, over-grazing, down-hill ploughing, and soil compaction. Soil erosion by water depends on the potential of rain to erode, (rainfall erosivity) and the susceptibility of soil to erosion (erodibility). The erodibility of a soil decreases with increasing vegetation cover. Soil erosion leads to a reduction in soil fertility due to a loss of nutrient-rich topsoil, the loss of carbon stored in the soil, the diffuse pollution of surface watercourses with nutrients, and contaminants such as pesticides and fertilizers. Overland flow and soil erosion can trigger severe floods in downstream catchments.

Modelling water erosion can therefore be very helpful in landscape planning td for avoiding offsite damage such as eutrophication of water bodies. The most common model is the USLE (Wischmeyer and Smith 1978), which has been adapted and revised many times (e.g. Renard et al. 1991). Recent work by the European Commission Joint Research Centre has created an assessment for the 28 EU member states at 100 m resolution (Panagos et al. 2015). On the farm scale, simplified assessments of high-risk sites can be conducted with details of slopes, soil type and land use (arable, grassland forest etc.). For more detailed quantitative evaluations of the actual soil erosion risk, farm or remote sensing data about the crop rotation are required. Crops can be classified in terms of the extent and duration of ground cover they provide and consequently how much they contribute to erosion processes.

Wind erosion has similar environmental impacts to that of water. The removal of protective vegetation cover is the main cause of accelerated wind erosion in agricultural ecosystems. Numerous models of wind erosion exist. A widely used model is the empirical Wind Erosion Equation (WEQ, Woodruff and Siddoway 1965). A continuous, process-based model, the Wind Erosion Prediction System (WEPS, Wagner et al. 2007), has now been developed to replace WEQ.

Wind and water erosion not only cause localized on-site damage to agricultural production capacity. Additionally they can produce severe off-site damage. Examples include the pollution of water bodies by sediment, nutrients and pesticides (see Chap. 11), traffic hazards arising from eroded material on roads or by sand storms limiting visibility.

Compaction is the increase of bulk density and an associated decrease in porosity of soil. It is caused by heavy machinery traffic and, to a lesser extent, by animals trampling on wet soils. Compaction leads to reduced numbers and sizes of pores within soil, especially larger pores which are needed to circulate air. This results in a reduction in the AWC, an increase in anaerobic subsurface conditions which, in turn, reduces the amount of oxygen available to organisms, increases the risk of nitrogen dioxide and methane production, limits root growth and therefore plant development, and enhances run-off and flooding risk. So soil compaction diminishes the productive capacity of a site and the actual production of food and raw materials. Modelling soil compaction is complex and no widespread model can be recommended at present. However, a very general rule is that clay soils are more susceptible to compaction than sandy ones and that use of heavy machinery should adapt to the different sensitivity of the soils. Water-saturated soils are highly susceptible, so traffic on the land should be limited to days with drier conditions.

Soil acidification and contamination with chemicals or heavy metals are other sources of soil impairments. It can lead to a total loss of productive capacity of a site, because of missing measures or very high costs of them. Modelling the processes is a complex, because data on the actual state and the input load are needed to get good results.

An assessment of impacts like compaction or acidification needs to take account of local characteristics and susceptibility. Regional maps and models can provide some guidance, but an assessment in the field is still the key to understand the threats to productive capacity it.

10.7 Conclusion

The production of food, raw materials and energy is probably one of the most important utilized ES. The ability of an ecosystem to provide goods is based on the local site characteristics, most important climate, soil and topography. Methods to evaluate the ES and parts of it, like the soil quality, are available. A sustainable use of an ecosystem is not always possible due to many pressures, which are mainly driven by population growth and increasing welfare. An overuse can cause severe land degradation process and lead to impacts like the loss of soil quality. To counteract these impacts, landscape planning has to supply suitable responses to enable sustainable production of biomass. Land and agro-environmental management systems can help to mitigate the negative impacts. To identify priority areas for soil conservation where land use should be changed, or where agri-environmental measures to alter land management may be most beneficial, an assessment of negative pressures and impacts is advisable. Future impacts and their consequences can be evaluated by comparing the current situation and the projected state.

References

Arshad, M. A., & Martin, S. (2002). Identifying critical limits for soil quality indicators in agro-ecosystems. *Agriculture, Ecosystems and Environment, 88*, 153–160.

Bagstad, K. J., Semmens, D. J., & Winthrop, R. (2013). Comparing approaches to spatially explicit ecosystem service modeling: A case study from the San Pedro River, Arizona. *Ecosystem Services, 5*, 40–50.

Bateman, I., Day, B., Agarwala, M. et al. (2014). *UK national ecosystem assessment follow-on* (Work Package Report 3: Economic value of ecosystem services). UNEP-WCMC, LWEC, UK.

EEA, European Environment Agency. (2016). *Land use*. http://www.eea.europa.eu/themes/landuse. Accessed 20 June 2018.

FAO, Food and Agriculture Organization. (2009, October 12–13). *Global agriculture towards 2050. High-level expert forum "How to feed the world 2050"*. Rome.

FAO, Food and Agriculture Organization. (2016). *State of the World's Forests 2016. Forests and agriculture: land-use challenges and opportunities*. Rome.

Farber, S., Costanza, R., Childers, D. L., et al. (2006). Linking ecology and economics for ecosystem management. *Bioscience, 56*, 117–129.

Kovats, R. S., Valentini, R., Bouwer, L. M., et al. (2014). Europe. In V. R. Barros, C. B. Field, D. J. Dokken, et al. (Eds.), *Climate change 2014: Impacts, adaptation, and vulnerability. Part B: Regional aspects. Contribution of working group II to the fifth assessment report of the intergovernmental panel on climate change* (pp. 1267–1326). Cambridge, NY: Cambridge University Press.

Maes, J., Teller, A., Erhard, M. et al. (2018). *Mapping and assessment of ecosystems and their services: An analytical framework for ecosystem condition*. Luxembourg: Publications Office of the European Union. http://ec.europa.eu/environment/nature/knowledge/ecosystem_assessment. Accessed 26 June 2018.

Mueller, L., Schindler, U., Behrendt, A., et al. (2007). *The Muencheberg Soil Quality Rating (SQR)*. Müncheberg: Zalf.

Mueller, L., Schindler, U., Shepherd, T. et al. (2010, August 1–6). Assessing agricultural soil quality on a global scale. 9th World congress of soil science, soil solutions for a changing world. Brisbane.

Pagiola, S., von Ritter, K., & Bishop, J. (2004). *Assessing the economic value of ecosystem conservation* (Environment department paper no. 10). Washington.

Panagos, P., Borrelli, P., Poesen, J., et al. (2015). The new assessment of soil loss by water erosion in Europe. *Environmental Science & Policy, 54*, 438–447.

Renard, K., Foster, G. R., Weesies, G. A., et al. (1991). RUSLE revised universal soil loss equation. *Journal of Soil and Water Conservation, 46*, 30–33.

Richter, A., Hennings, V., & Müller, L. (2009, September). Anwendung des Müncheberger Soil Quality Ratings (SQR) auf bodenkundliche Grundlagenkarten. Jahrestagung der DBG, Kommission V DBG. Bonn.

US Department of Agriculture. (2001). *Guidelines for soil quality assessment in conservation planning*. Natural Resources Conservation Service.

US Department of Agriculture. (2018). *Soil quality indicator sheets*. https://www.nrcs.usda.gov/wps/portal/nrcs/main/soils/health/assessment/. Accessed 26 June 2018.

Wagner, L. et al. (2007). *WEPS 1.0 User manual*. https://infosys.ars.usda.gov/WindErosion/weps/wepshome.html. Accessed 26 June 2018.

Wischmeier, W. H., & Smith, D. D. (1978). *Predicting rainfall erosion losses: A guide to conservation planning* (Agriculture handbook No. 537). Washington, DC: USDA/Science and Education Administration/Government Printing Office.

Woodruff, N. P., & Siddoway, F. H. (1965). A wind erosion equation. *Soil Science Society of America Journal, 29*(5), 602–608.

Catchment Water Resources

11

Richard J. Cooper and Kevin M. Hiscock

Abstract

This chapter provides an introduction to the ecosystem services and assessment methods associated with catchment water resources. It considers the main pressures on such resources and the different techniques that can be used to monitor and evaluate the state of water quality and quantity in a catchment. Issues associated with the design of monitoring programmes and tools for modelling water resources are also reviewed.

Keywords

Water cycle · Catchment pressures · Water quality · Water resources · Catchment monitoring

11.1 Introduction

Since the beginning of the twentieth century, a rapidly growing human population coupled with the widespread expansion of industrial and agricultural activities has led to an increasing demand being placed upon the provisioning services provided by catchment water resources. Across the planet, water resources have been so intensively modified (biologically, chemically and physically) to meet human needs that many of the natural functioning systems and services provided by these water resources have been severely degraded or lost entirely. Globally, this has resulted in 1.1 billion people lacking access to safe drinking water, 2.6 billion lacking access to proper sanitation and 20% of the world's population living in regions which produce

R. J. Cooper (✉) · K. M. Hiscock
School of Environmental Sciences, University of East Anglia, Norwich, UK
e-mail: richard.j.cooper@uea.ac.uk; k.hiscock@uea.ac.uk

C. von Haaren et al. (eds.), *Landscape Planning with Ecosystem Services*,
Landscape Series 24, https://doi.org/10.1007/978-94-024-1681-7_11

no appreciable supply of renewable freshwater (WHO 2004). Within the European Union there are more than 100,000 surface waterbodies of which 80% are rivers, 15% lakes and 5% coastal and transitional (estuary) waters. As of 2015, only half of these waterbodies were meeting targets to achieve *'good'* ecological and chemical status, with rivers and estuaries being in particularly poor condition. The poor ecological status of freshwater bodies is most pronounced in regions of intensive agriculture and high human population density in central and north-western Europe where hydromorphological pressures couple with pollution from agrochemicals, sewage effluent and industry to severely threaten sustainable ecosystem functioning. Currently more than 40% of European waterbodies are affected by diffuse agricultural pollution and 25% are affected by point source pollution from wastewater treatment plants and industrial facilities. However, since the implementation of a comprehensive framework of water resources legislation in the early 2000s, the situation is improving, with average nutrient concentrations in Europe's rivers declining by 38% between 1992 and 2011 (EEA 2015), mainly because point sources have been reduced. Nevertheless, threats to long-term water quality still remain and these will have major implications for the future delivery of ecosystem services without significant changes in agriculture. Landscape planning could have a significant role in multi-criteria assessment of water resources and efficiently allocating mitigation measures to reduce anthropogenic pressures on catchment water resources (Box 11.1).

Box 11.1: Definitions and Concepts

Catchment: An area of land onto which precipitation falls and ultimately drains to a common location, typically the ocean. A catchment consists of both *surface* (streams, rivers, lakes, estuaries) and *subsurface* (soil water, groundwater) waterbodies and is separated from neighbouring catchments by a region of higher topography known as the *catchment boundary*. Large-scale water movement through a catchment is *gravity driven*, draining from higher (headwaters) to lower (lake/sea) elevation. Alternative terms for a catchment include *drainage basin* and *watershed* (American).

Sustainable catchment management: The practice of integrating all *environmental*, *economic* and *social* issues associated with anthropogenic activity within a catchment into a holistic management strategy to improve water quality. Practitioners of sustainable catchment management typically adhere to the *adaptive management cycle* as a means of incorporating an appropriate combination of regulation, advice, land use measures, incentives and voluntary action for a collaborative and integrated approach to protect water resources (Macleod et al. 2007; Smith et al. 2015).

Source-pathway-receptor paradigm: A concept applied to water resources management which involves identifying the origin of water pollution (*source*), determining the method of action to transport the pollutant from source to the waterbody (*pathway*), and understanding the impact upon water quality of the pollutant once it has entered the waterbody (*receptor*) (Beven et al. 2005).

11.2 The Water Cycle

There is an estimated 1.39 billion km^3 of water on the Earth. The world's oceans contain the largest volume (96.5%), but its high salinity renders it largely unsuitable for human consumption without expensive desalination treatment. Of the remaining 3.5%, just 2.5%, or 34.8 million km^3, is considered freshwater and 1% is contained in saline or brackish groundwater. The majority of this freshwater (69.6%) is locked away as ice, snow and permafrost in polar and glacial regions and is therefore largely inaccessible for human exploitation (Fig. 11.1). The largest available freshwater resource is groundwater (30.1%) stored within the void space of porous and permeable bedrocks and superficial deposits beneath the Earth's surface, where it exists as modern (shallow) and ancient (deep) reserves. The ultimate source of all freshwater is precipitation from the atmosphere which infiltrates down through the soil profile and into the underlying rocks. Water held within the soil is absorbed by plant roots to support the growth of primary producers, the dominant biota of the Earth's ecosystems that are heavily exploited by humans for food and materials consumption. Soil water not taken up by the biota, or which does not infiltrate down into groundwater, runs off the land surface as streams and rivers where it can enter lakes, reservoirs and swamps before ultimately discharging into the ocean at estuaries. These surface water resources, which represent just 0.3% of the total freshwater on the planet, are heavily exploited by humans for the ecosystem services they provide to domestic, agricultural and industrial sectors. Across the planet, both surface water and groundwater have become severely degraded due to overexploitation by humans, thus threatening the long-term sustainability as both a resource and the ecosystem services which these freshwater bodies provide.

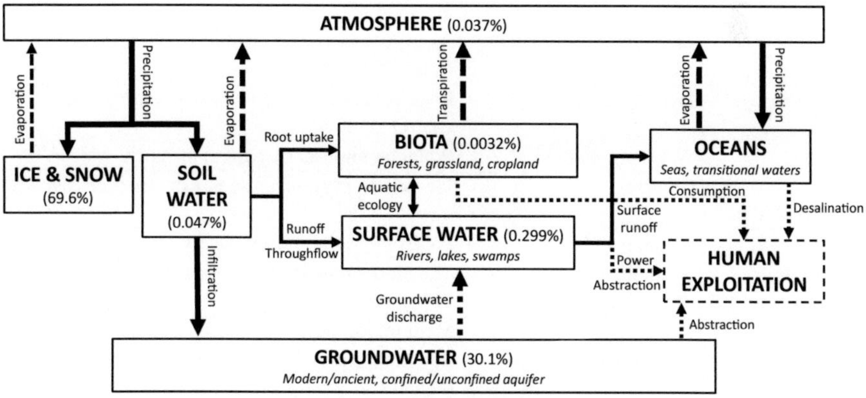

Fig. 11.1 Schematic of the global water cycle. Percentages refer to the distribution of global freshwater within each environment. (Shiklomanov and Rodda 2003)

11.3 Ecosystem Services Linked to Catchment Water Resources

Water is the physiological basis of all life on Earth and as such provides a wide range of *provisioning*, *regulatory*, *cultural* and *supporting* ecosystem services collectively valued at $6.6 trillion per year globally (Costanza et al. 1997; Finlayson and D'Cruz 2005; Grizzetti et al. 2016). A list of common ecosystem services provided by catchment water resources which are essential for the landscape planner to consider within the environmental decision-making process are presented in Table 11.1.

Provisioning services of water resources include *consumptive services*, such as the supply of potable domestic drinking water, harvesting of aquatic organisms for food (e.g. fisheries) and water used during the food manufacturing process. *Non-consumptive services* include water for crop irrigation and production (particularly

Table 11.1 Ecosystem services of catchment water resources

Ecosystem Service	Type	Examples
Provisioning	Freshwater (consumptive)	Drinking water (domestic); food manufacturing
	Freshwater (non-consumptive)	Agricultural production, irrigation; industry; power generation; cooling
	Food	Fisheries; aquaculture; irrigated and rain-fed crops; game
	Transport	Shipping; canals
	Biodiversity	Aquatic habitats; riverine and lacustrine ecosystems
	Forestry	Timber production; paper manufacturing; firewood
	Medicines	Species, genes or biochemicals extracted from freshwater
Regulatory	Hydrological	River baseflows; groundwater storage; flood control
	Water purification	Pollutant removal by natural filtration systems
	Climate	Precipitation; greenhouse gas storage; thermal mass
	Erosion	Preserving soils and sediments
Supporting	Nutrient cycling	Transformation, movement and supply of nutrients
	Solvent	Medium for dissolving chemicals and nutrients
	Soil formation	Retention and accumulation of sediment and organics
Cultural	Tourism	Cruises, boating holidays, adventure activities
	Recreation	Fishing, sailing, wildfowling
	Education	Outdoor learning
	Aesthetic	Enjoyment of natural environment
	Spiritual	Health and wellbeing

high water dependency crops such as rice), power generation (e.g. hydropower stations, cooling water) and heavy industry. Across the 28 EU member states, approximately 50 km^3 of water is abstracted from the environment annually for domestic use, with a further 140 km^3 abstracted for industry and power generation, and 154 km^3 abstracted for agriculture. In addition to these services, water provides the essential basis for all aquatic ecosystems and the biodiversity that exists within them which can be exploited for food, medicines or other consumable products. Water is essential for supporting commercial forestry operations which provide timber and paper based products as well as firewood for fuel. Water resources also provide transport services (e.g. shipping) associated with large-scale movement of material goods via container.

Regulatory services include the supply of groundwater to maintain *river baseflow* during periods of drought and to sustain freshwater-dependent ecosystems (e.g. ponds, estuaries and mangroves). Through the storage of precipitation and surface runoff, lakes and floodplains are able to attenuate high river flows and thus regulate *flood risk*. Aquatic environments such as marshes and reed beds play an important role in regulating water quality by acting as *natural filtration* systems to remove particulate and dissolved contaminants, thus reducing incidences of sedimentation and eutrophication (see Sect. 11.5). Water resources also help regulate *climate* through the transfer of thermal energy between the atmosphere and waterbodies, and through the storage of carbon in wetlands acting as a sink for carbon dioxide and thereby mitigating climate change (Chap. 14).

Supporting services of water resources include the role of riverine and lacustrine waters in *nutrient cycling*, specifically moving, transforming and supplying the essential nutrients which support the existence of primary producers in aquatic habitats. As a *solvent*, water provides a medium for the dissolution and transport of materials both within aquatic environments (e.g. oxygen, nutrients, ions) and within organisms (e.g. food, enzymes). River flooding and the associated deposition of sediments and organic material onto the floodplain also supports the formation of fertile *soils* which can be exploited for crop production.

Cultural services include *tourism* related activities such as river cruises, boating holidays and adventure activities, as well as other *recreational pursuits* such as fishing, sailing, canoeing and wildfowling. Across the EU, more than ten million people are thought to be actively involved in recreational fishing with an annual expenditure of over €25 billion (Pawson et al. 2007). Similarly, in the UK, eight million tourists visit the 300 km^2 freshwater Broads National Park each year generating ~€670 million for the regional economy (Broads Authority 2017). Freshwater bodies such as lakes, rivers and wetlands also provide opportunities for *educational development* for both children and adults through outdoor learning courses and workshop programmes. Furthermore, the spiritual, tranquil and visually spectacular nature of waterbodies can contribute to improved *mental health* and *wellbeing* of people who visit them through increased engagement with, and enjoyment of, the natural environment (Chap. 15).

In the following pages we focus on methods that can be used for assessing the fresh water provisioning service including pollutant removal from water bodies and the flood control service. These are final services which are not discussed elsewhere in this section of the book. The task of landscape planning in this context is to identify pressures which can possibly affect water resources, analyse their influence, find, and then prioritise areas for protecting and restoring the good status of water bodies in the context of other ecosystem services. Increasingly this means that water resources need to be managed in the context of multifunctional landscape benefits (Chap. 19) and that those agencies responsible for catchment management need to collaborate with a range of other organisations.

11.4 The Main Pressures on Catchment Water Resources

11.4.1 Nutrient Enrichment

Diffuse nutrient pollution from agriculture and point source pollution from sewage treatment works (STWs) represent two major drivers behind the *eutrophication* of freshwater systems which causes an array of detrimental economic and environmental impacts which threaten the ability of these systems to provide ecosystem services (see Fig. 11.2) (Smith et al. 1999; Némery and Garnier 2016). As naturally limiting nutrients of plant growth in aquatic systems, the enhanced transfer of fertiliser or sewage derived nitrogen (N) and phosphorus (P) from land into waterbodies fuels blooms of phytoplankton, periphyton and neuro-toxin secreting cyanobacteria colonies which can dramatically lower species diversity and lead to a fundamental breakdown of ecosystem functioning and the services that are provided. Globally, the greatest source of nitrogen enrichment is from the application of nitrate (NO_3) fertilisers to arable crops to facilitate increased food production. Nitrate is highly

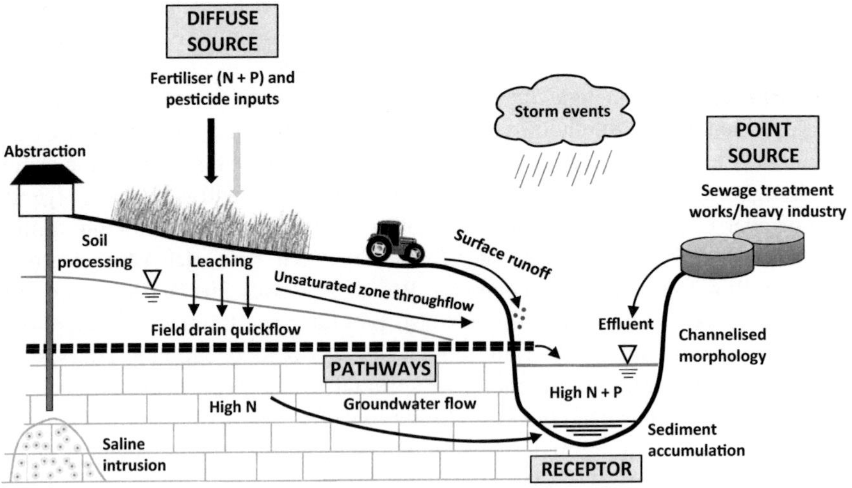

Fig. 11.2 Conceptual model of the threats to catchment water resources expressed through the source-pathway-receptor paradigm

soluble in water and will readily *leach* down through the soil matrix (*throughflow*) during precipitation events, eventually entering groundwater before upwelling into surface waterbodies through the riverbed. N can also enter subsurface agricultural field drain networks which act as *preferential pathways* for the direct discharge of N enriched water into the river system (*quickflow*).

Conversely, P exists in various forms, which affect both its mobility within the environment and its delivery pathway from land to river. Phosphorus binds strongly with clay minerals and metal oxides in soils via a process called *sorption* to form comparatively low mobility particulate phosphorus (PP) compounds which are transported via erosive surface runoff of soils during heavy precipitation events. However, P also exists in highly mobile dissolved forms, such as phosphate (PO_4^{3-}), enabling it to leach through the soil and enter watercourses via throughflow-pathways. The reactivity of P means it is subject to immobilisation and re-mobilisation processes, such as sorption and desorption, as it moves along delivery pathways linking land to water.

Once instream, nitrate concentrations are commonly diluted by fresh water input during precipitation events, with concentrations peaking several hours/days post-event as nitrate slowly leaches through the soil profile into the river (Fig. 11.3). On the other hand, P concentrations commonly exhibit a *flashy* response to storm events with little lag between peak discharge and the highest P concentration, a characteristic linked to the rapid activation of surface runoff pathways. The provisioning services of drinking water, fisheries, and freshwater habitats are all threatened by eutrophication, as are the cultural services of tourism and recreation which are

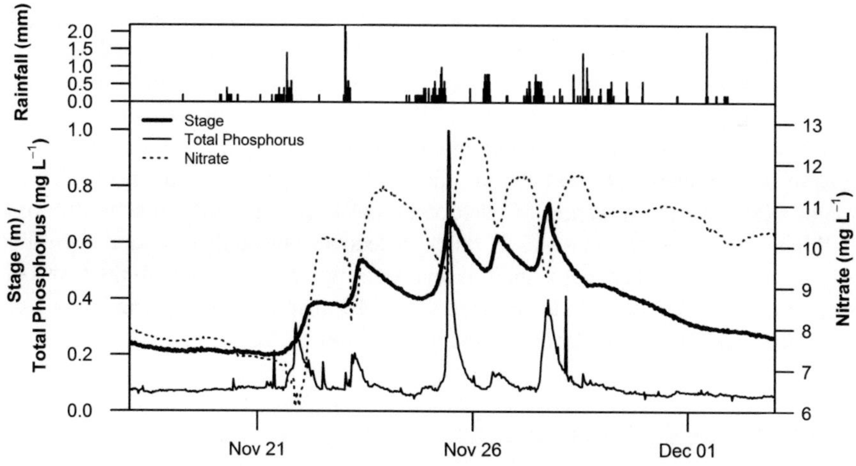

Fig. 11.3 Impact of precipitation events on river stage (i.e. depth) and the concentrations of nutrients being transported out of the catchment. Heavy rainfall quickly initiates a rise in water level and a sharp increase in phosphorus concentration as phosphate-rich sediments are transported by surface runoff into and within the river channel. Conversely, nitrate concentration are initially diluted by fresh rainwater before concentrations slowly increase hours-to-days later as nitrate is leached out of the soil profile across the catchment

impacted by the reduced aesthetic appearance of green, phytoplankton dominated water. Treating eutrophic water incurs significant economic costs, with water companies having to remediate problems with taste, colour and odour whilst lowering concentrations of contaminants in order to make the water potable for human consumption. In the United Kingdom, the total costs of eutrophication have been estimated at £75–114 million per year (Pretty et al. 2003). Despite high nutrient loadings into waterbodies across the EU, between 1992 and 2011 concentrations of N and P declined by 20% and 57%, respectively, thanks largely to improvements in nutrient stripping at wastewater treatment plants (EEA 2015).

11.4.2 Sediment Enrichment

Extensive anthropogenic modification of natural landscapes, in particular the intensification of agriculture, has resulted in the widespread sediment enrichment of environmentally-sensitive freshwater environments (Wilkinson 2005; Quinton et al. 2010). Clearance of permanent natural vegetation (e.g. forests, grasslands) and its replacement with cultivated crops has resulted in hugely accelerated rates of soil erosion (cf. Chap. 10). Cultivating the land disturbs the upper 0.5 m of the soil profile, reducing structural stability and thereby increasing the likelihood of erosion. Many commercial crops are seasonal, meaning that soils are left fallow over winter. Without the protection of above ground vegetation or stabilising subsurface root networks, agricultural soils are exposed to enhanced rainsplash and erosive surface runoff during precipitation events, carrying sediment-laden water off the land and into nearby watercourses. Surface waterbodies affected by elevated sediment volumes experience an array of detrimental impacts which threaten sustainable ecosystem functioning (Bilotta and Brazier 2008). Fine clay and silt sized fractions increase *turbidity*, clog fish gills, smother gravel salmonid spawning grounds and benthic habitats, reduce oxygen circulation through the streambed, and abrasively scour macrophytes, periphyton and small invertebrates. The high specific surface area of fine-grained material (<63 μm diameter) also enables sediments to act as a major vector for the transport of adsorbed phosphorus and other potentially toxic pollutants through stream systems that can lead to eutrophication and fish kills. Alongside ecological concerns there is an economic impact, with high rates of sedimentation reducing the provisioning services of transport and shipping, enhancing flood risk, and increasing dredging and water treatment costs.

11.4.3 Pesticide Contamination

The widespread use of pesticides, also known as *plant protection products* (PPP), in agriculture to kill plant and animal pests, which would otherwise reduce crop yields, has been instrumental in enhancing global agricultural productivity since the mid-twentieth century (Popp et al. 2013). There are six main classes of pesticides, each designed to kill a specific type of pest; *acaricides* (ticks/mites), *fungicides* (fungi), *herbicides* (plants), *insecticides* (insects), *molluscicides* (slugs/snails) and *nematicides* (nematodes). The vast majority of these pesticides can be classified into four groups based on their chemical composition (*organohalogen*, *organophosphorus*, *organonitrogen* and *organosulphur*), which in turn determines how the chemical affects both the target species and interacts within the wider environment. Global pesticide usage stood at ~2.4 Megatonnes in 2015. However, the harmful environmental impacts of applying toxic chemicals across large areas of the planet's surface, particularly on the aquatic environment, are coming under increasing scrutiny. Pesticide pollution can either arise from *diffuse sources*, such as airborne spray drift, leaching and overland flow, or from *point sources*, such as accidental spillages, leakages from equipment or from contaminated machinery washings. High profile cases of pesticide pollution, such as the effect of the insecticide DDT on the hatching success of raptors in the 1960s and 1970s, brought into focus the potential for pesticides to bio-accumulate within organisms and bio-magnify up the food chain and negatively impact upon non-target species (Ames 1966). Similarly, recent research has linked the use of neonicotinoid insecticides to the decline of bee populations in Europe and North America (Whitehorn et al. 2012). Because most pesticides are water soluble, many have a high mobility within the environment and can readily enter catchment waterbodies, threatening drinking water provision and aquatic habitats and resulting in significant economic costs associated with removing these chemicals. Between 1991 and 2000, water companies in the United Kingdom spent £2 billion treating pesticide contaminated water supplies (Jess et al. 2014).

11.4.4 Organic Pollutants

Faecal contamination of waterbodies represents a major cause of failure of the EU Bathing Water Directive and is a direct threat to domestic water supplies. Faeces contains a wide diversity of protozoa (e.g. *Amoeba*, *Giardia*, *Cryptosporidium parvum*, *Toxoplasm gondii*) and bacteria (e.g. *coliforms*, *Enterobacteriaceae*, *streptococci*) which can cause a variety of mild to serious intestinal disorders if ingested by humans or other mammals. The major sources of faecal contamination of waterbodies in the EU include sewage treatment works, leaking septic tanks, pastured livestock and concentrated animal feeding operations. In order to determine whether a waterbody is polluted with faeces, the abundance of a subset of commonly detected species, known as *faecal indicator organisms* (FIOs), is used as a proxy for contamination (Crowther et al. 2002). However, it is not just microbial life that threatens water resources, excessive quantities of *organic matter* from eroding organic-rich soils (diffuse) and sewage treatment works effluent (point) can also directly impair water quality. Organic matter (OM) is any material ultimately derived from photosynthesis or chemosynthesis and can exist in either dissolved (DOM) or particulate (POM) phases. Organic matter is essential to the fertility and structural stability of soils, decomposing to release nutrients for plant growth and binding soil aggregates together to reduce the risk of soil erosion (Lal et al. 2004). It also helps regulate soil water content, holding onto water during prolonged periods without rainfall and thereby reducing soil desiccation. However, excessive organic matter inputs into waterbodies rapidly accelerate the rate of biodegradation by aerobic heterotrophic detritivores which increases the biological oxygen demand (BOD) and reduces dissolved oxygen concentrations. Zones of hypoxia can form, particularly downstream of sewage effluent discharge points, resulting in the suffocation and death of sensitive aerobic organisms, especially fish. The provisioning services of fisheries and aquatic habitats are subsequently degraded and the recreational value is diminished through the reduction in fishing opportunities and production of foul smelling hydrogen sulphide gas in areas of anoxia.

11.4.5 Industrial Pollutants

Waterbodies draining areas of heavy industry, manufacturing and large municipal centres can become contaminated by a diverse range of hazardous substances, many of which are non-biodegradable and therefore have high persistence in the environment (Duruibe et al. 2007). These include heavy metals, such as lead, arsenic, mercury and cadmium, released from metal workings, coal mines and scrap yards. Chlorinated solvents, such as trichloroethylene, can be discharged into rivers and groundwater from facilities handling paints, resins, and cleaning solutions, whilst landfill sites, petrol stations and asphalt manufacturing plants are sources of aromatic hydrocarbon pollution. Crucially, because many industrial pollutants are toxic to both humans and aquatic species, contamination of waterbodies can lead to the loss of drinking and domestic water provisioning services as well as the loss of ecological diversity. Industry, particularly power generation, is also responsible for thermal pollution of waterbodies. The abstraction of cold water from the natural environment to cool reactors or steam pipes risks discharging water back into the environment at temperatures elevated above the tolerance threshold of aquatic organisms and decreasing dissolved oxygen concentrations.

11.4.6 Morphological Changes

Across large parts of the EU, streams, rivers and wetlands have been extensively modified from their natural form in order to satisfy the needs of agricultural intensification. Wetlands have been routinely drained and converted into arable cultivation or livestock pasture. Rivers have been widened, straightened and deepened to quickly move water out of catchments and away from valuable commercial crops sensitive to waterlogged soils. Regular *dredging* and riparian vegetation clearance increase the water holding capacity of the river channel and reduce the incidences of overbank flow. Whilst these morphological modifications have helped to increase agricultural productivity, such aggressive disturbance of the natural functioning of freshwater environments has had severe impacts upon the ability of these environments to provide ecosystem services. The loss of physical features such as *meanders*, *riffles* and *pools* mean rivers now lack the structural diversity to support a diverse array of aquatic ecosystems. This homogeneity of channel morphology directly contributes to a lack of flow diversity, meaning there are fewer areas of fast flow needed for flushing pollutants and deposited riverbed sediments out of the catchment. The removal of wetland vegetation such as reed beds has resulted in the loss of water purification regulatory services. The deepening of rivers and the construction of elevated riparian *levees* results in the loss of floodplain connection meaning they no longer provide the flood alleviation regulatory service that would be provided under natural, unmodified conditions.

11.4.7 Invasive Alien Species

An *invasive alien species* is any species that exists outside of its natural distribution (i.e. is *non-native*) which poses a threat to biological diversity or to ecosystem functioning (Chap. 18). This threat is usually associated with the tendency of an invasive species to spread to such a degree that it inflicts damage upon an ecosystem, typically because of either favourable growing conditions or a lack of natural competitors to keep the population under control. Under such circumstances, invasive species are able to outcompete native species which can, in severe cases, lead to the extinction of native flora and fauna and the breakdown of sustainable ecosystem functioning. The primary mechanism behind the spread of non-native aquatic species is international trade where species are transported across the planet by humans both intentionally (e.g. the pet trade, food supply or horticulture), or unintentionally (e.g. on ship hulls or in ballast water). Around 140 invasive species have currently been identified residing in freshwater habitats across the EU out of a total of 12,000 invasive species recorded in total across all European terrestrial and marine environments (Sundseth 2014). Common examples of invasive species in EU waterbodies include water hyacinth (native to the Amazon), zebra mussel (native to the Black Sea and Caspian Sea), and Canadian pondweed, signal crayfish and American mink (all native to North America).

11.4.8 Overexploitation

The global renewable water resource, which is ultimately determined by the amount of precipitation entering surface water or groundwater bodies, is estimated to total between 33,500 and 47,000 km^3 per year (Shiklomanov and Rodda 2003). Importantly, this renewable resource is not evenly distributed across the planet and does not coincide with the areas of greatest human exploitation. Consequently, even though global freshwater abstraction rates total ~4000 km^3 per year (just 10% of the renewable resource), local human exploitation of groundwater and surface water is occurring at rates greater than the renewable yield, resulting in the unsustainable depletion of catchment water resources. Overexploitation by abstracting too much water from a river or groundwater aquifer can result in water table drawdown and river levels dropping below environmentally acceptable limits required to maintain the functioning of aquatic ecosystems and the services which they provide. In coastal areas, particularly around the Mediterranean where effective precipitation is low and water demand is high, over-abstraction of groundwater leads to saline intrusions, which render the aquifer resources unsuitable for domestic consumption and crop irrigation. Eight EU member states (Germany, UK, Italy, Spain, Belgium, Bulgaria, Cyprus and Malta) are classified as water-stressed where total water abstraction is more than 20% of the long-term freshwater resource (EEA 2007). Globally, this issue is worsening due to rising demand for freshwater abstraction which increased by 20% per decade between 1960 and 2000.

11.5 Assessing the State of Water Quality

According to Article 8 of the EU WFD (Box 11.2), member states must establish monitoring programmes to provide a coherent and comprehensive overview of water status within each river basin district. The purpose of doing so is to be able to identify waterbodies suffering from severe environmental degradation, to understand the main threats at local and national scales, and to be able to effectively target mitigation measures and policy actions aimed at improving the environmental state. More specifically, the stated monitoring requirements include:

- Classification status of surface water and quantification of reference conditions
- Chemical and quantitative status of all groundwater bodies
- Estimates of the direction and rate of flow in groundwater bodies
- Estimates of pollutant load transfers across international boundaries or discharged to sea
- Assessment of changes in waterbody status
- Causes of waterbodies failing to achieve environmental objectives
- The magnitude and impacts of accidental pollution
- Compliance assessments with the standards and objectives of protected areas

Addressing these objectives across the EU requires the collection and assimilation of large quantities of monitoring data at a sufficiently detailed spatial and temporal resolution to enable accurate characterisation. *Eionet Water* is a European Environment Agency (EEA) monitoring network established to assess the status of EU water resources and to understand how they respond to environmental pressures through the amalgamation of European-wide water quality data. The type of monitoring conducted is classified as either *surveillance*, *operational* or *investigative*. The exact suite of water quality parameters monitored will ultimately depend upon the purpose of the monitoring programme, but a list of commonly analysed physicochemical determinands is presented in Table 11.3.

Chemical – as drivers of eutrophication, *nitrogen* and *phosphorus species* are commonly monitored in waterbodies threatened by nutrient enrichment and exist in a variety of dissolved/particulate and organic/inorganic forms due to their highly

Box 11.2: Legislation to Protect Water Resources: Water Framework Directive (2000/60/EC)

The European Union Water Framework Directive (officially Directive 2000/60/EC of the European Parliament and of the Council of 23 October 2000 establishing a framework for Community action in the field of water policy; abbreviated to WFD) was implemented in December 2003 and required all member states to achieve *good qualitative* and *quantitative status* of all waterbodies (surface, subsurface and marine up to 1 nautical mile off-shore) by 2015. There are five levels of *status* referring to ecological and chemical conditions, ranging from high (no or very low human pressures) to bad (severely impacted by human pressures). The assessment criteria are sub-divided into four categories:

- *Biological quality* determined by populations of fish, benthic invertebrates, plankton and macrophytes;
- *Chemical quality* determined by exceedance of maximum concentration standards for specific pollutants (Table 11.2);
- *Hydromorphological quality* determined by river profile, connectivity, bed substrate, and water depth/flow;
- *Physicochemical quality* determined by parameters such as temperature, pH, salinity and dissolved oxygen concentration.

Under the WFD, waterbodies in EU member states are divided into *River Basin Districts* based on the river catchment area rather than political/national boundaries. A total of 160 river basin districts have been designated, of which 40 are international across borders and cover 60% of EU territory. National governments within each river basin district have to produce a *River Basin Management Plan* which provides a clear strategy and timeframe for how the status of waterbodies within the river basin district are to be improved, with this plan updated every 6 years. Plans are based on integrated river basin management and adopt a holistic approach to protecting the whole waterbody from its source to its mouth. Public participation is a fundamental principle to enable EU citizens to play an influential role in planning and implementing WFD measures, without which regulatory measures will not succeed. The WFD serves as an umbrella directive for a plethora of related legislation, such as the Groundwater (2006/118/EC) and Urban Wastewater Treatment (91/271/EEC) directives. The current 6-year cycle runs from 2016–2022. Targets for the previous cycle (2009–2015) were missed with >40% of waterbodies across the EU failing to achieve good ecological and chemical status.

reactive nature. For N, analysis tends to focus on the concentrations of ammonium (NH_4^+), nitrite (NO_2^-) and nitrate (NO_3^-) which form during the oxidation of urea based synthetic fertilisers and organic manures and therefore serve as an indicator

Table 11.2 Water quality guidelines for specific pollutants as defined under the EU Water Framework Directive (WFD)

Pollutant	Legislation	Guideline
Nitrate (NO_3)	Drinking Water Directive(98/83/EC)	11.3 mg/L of NO_3-N at tap.
Phosphorus (P)	Habitats Directive(92/43/EC)	20–60 μg/L for headwater streams
		40–100 μg/L for moderate rivers
		60–100 μg/L for large rivers
Sediment	Freshwater Fisheries Directive (2004/44/EC)	25 mg/L in waters suitable for salmonid and cyprinid fish.
Pesticides	Drinking Water Directive (98/83/EC) & Groundwater Directive (2006/118/EC)	Individual pesticide = 0.1 μg/L
		Total pesticides = 0.5 μg/L

Table 11.3 Common physicochemical and meteorological parameters used in the assessment of catchment water resources

Physical	Chemical	Meteorological
Water temperature (°C)	Nitrogen species: nitrate, nitrite, ammonium, total N (mg L^{-1})	Precipitation (mm)
Turbidity (NTU)	Phosphorus species: phosphate, dissolved, particulate and total P (mg L^{-1})	Air temperature (°C)
Suspended sediment (mg L^{-1})	Pesticides: organo-phosphorus, –halogen, –sulphur (μg L^{-1})	Wind speed (ms^{-1})
Electrical conductivity (μS cm^{-1})	Organic matter (mg L^{-1})	Radiation (Wm^{-2})
Stage (m)	Anions/cations (meq)	Relative humidity (%)
Discharge (m^3 s^{-1})	Chloride (mg L^{-1})	Wind direction (degrees)
Soil moisture (%)	Alkalinity (mg $CaCO_3$ L^{-1})	
Groundwater level (m)	pH	
	Dissolved oxygen (mg L^{-1})	

of diffuse agricultural pollution. Similarly, highly bioavailable phosphate (PO_4^{3-}) concentration is a commonly monitored P species derived from both organic and inorganic fertiliser and from sewage treatment works effluent. Measurement of total nitrogen (TN) and total phosphorus (TP) concentrations provide a good indication of the overall nutrient status of the waterbody as they incorporate all forms of the nutrients (i.e. dissolved/particulate and organic/inorganic). Other important chemical parameters to monitor include *chloride* which occurs naturally in freshwater bodies but can be enriched in wastewater from sewage works, in agricultural runoff and in runoff from salted (gritted) roads; *pH* to which many aquatic organisms are sensitive and typically prefer a range of 6–9; *alkalinity* which reflects the ability of a waterbody to resist changes in pH due to the presence of carbonate (HCO_3) and bicarbonate (H_2CO_3) ions, thus preventing the water becoming more acidic (i.e. the *buffering capacity*); *organic carbon* concentration which has a significant impact upon nutrient cycling and biological oxygen demand; *dissolved oxygen* concentration which is an indicator of pollution by oxygen consuming substances (e.g. organic matter); and the balance of *anions* and *cations* which is a measure of the amount of negatively and positively charged ions, respectively, dissolved within a waterbody.

Physical – *suspended sediment* concentration is an important physical parameter to measure as it concerns the finest particulate fraction (typically <63 μm in diameter), which has the greatest impact upon aquatic systems due to its elevated mobility and high sorption capacity. It can either be determined directly, by filtering a known volume of water and weighing the mass of sediment removed, or indirectly using *turbidity* as a proxy where an optical sensor measures the degree of light scattering within a water sample due to the presence of suspended material. Other important physical parameters for assessing water quality include *electrical conductivity* (EC), which is a measure of the dissolved ion content of the water and therefore serves as an effective indicator of pollution by fertilisers, pesticides and heavy metals; and *water temperature* which is important due to its inverse relationship with dissolved oxygen content and can be an indicator of increased microbial activity in areas polluted with excessive organic matter (e.g. near sewage treatment works outflows).

11.6 Assessing the State of Water Quantity

To accurately determine the quantity of water resources held within a catchment requires an assessment of both surface and subsurface environments and will be heavily impacted by meteorological factors. The amount of surface water runoff is a direct measure of the *flood control regulation ecosystem service***.** A small surface water runoff is an indicator of high flood control capacities of the landscape in a catchment area. The subsurface water is an indicator for the *water provisioning service*. The more water that is transported to the aquifer the better the availability of water for drinking and other uses. The capacity of the ecosystem (mainly the soil) to store and transport water is also linked to the flood control functions. A high infiltration will reduce surface runoff. Based on a sound assessment of surface and subsurface water the landscape planner will need to identify areas of hydrological importance, which are especially relevant for flood control and for water provisioning.

Meteorological parameters – precipitation ultimately determines the amount of water available for supporting ecosystem services and is therefore an essential parameter to monitor. Precipitation totals can be measured either via a standard rain gauge, which accumulates precipitation over a number of days and the volume is measured manually, or via a tipping bucket rain gauge which, if telemetered, has the potential to record real-time high-resolution (minutes) precipitation totals. Precipitation records are complimented by monitoring of air temperature, wind speed, relative humidity and solar radiation which all impact upon evapotranspiration rates and thus determine the flux of water between catchment water resources and the atmosphere.

Surface water – the amount of surface runoff out of a catchment is determined by measuring the river *flow* or *discharge*, for which a number of techniques are available. (1) The *velocity-area* method involves measuring the cross-sectional area (width x depth) of the river and then measuring the water velocity at numerous points across the width of the channel using either an impeller device, electromagnetic (EM) meter or acoustic Doppler current profiler. The velocity-area method can be labour-intensive due to the relatively large number of measuring points required to achieve an accurate measurement and is therefore not always suitable for regular river flow monitoring. (2) A more convenient alternative is the *stage-discharge relationship* method which initially involves calibrating the river depth ('*stage*') against river discharge (Fig. 11.4). Once a robust regression has been established, known as the '*rating curve*', only river stage needs to be monitored regularly in order to produce a high temporal resolution river flow record. (3) Another option is the *slope-area* method based on the Manning's equation – an empirical formula for estimating water velocity in a river channel:

$$v = \frac{R^{\frac{2}{3}} \times S^{\frac{1}{2}}}{n}$$

where v is velocity; R is the hydraulic radius (cross-sectional area of the channel divided by the wetted perimeter); S is the channel bed slope; and n is the Manning roughness coefficient. Velocity, v, is then multiplied by cross-sectional area to obtain river discharge. (4) *Fluorescent tracers*, such as fluorescein and rhodamine WT, applied to rivers in either by *continuous* or *gulp* injection provide a visual means of assessing water velocity by recording the time it takes the tracer to travel a known distance downstream. Lastly, (5) *weirs* and *flumes* in rivers are fixed gauging

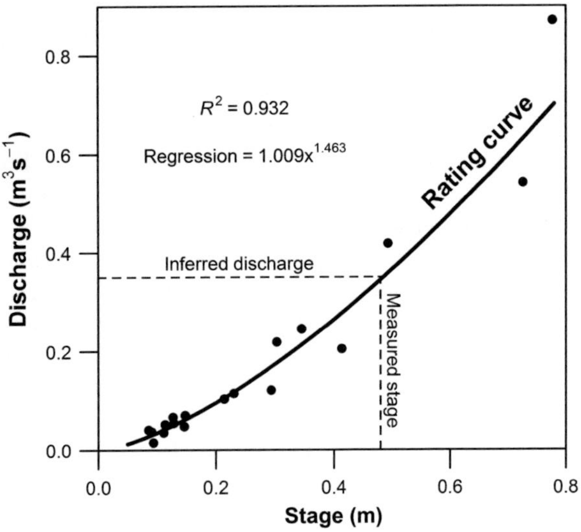

Fig. 11.4 A stage-discharge rating curve can be used to estimate river flow from the water depth

structures designed so that stream discharge is made to behave according to well-known hydraulic laws of the general form:

$$Q = KbH^a$$

where Q is river discharge; K and a are coefficients relating to the design of the structure, and b is the width of flow over the weir crest or in the throat (the constricted section) of a flume. Further details of stream gauging methods are presented in Hiscock and Bense (2014).

Subsurface water **–** in the subterranean environment precipitation is stored and transported at shallow depths within the pore spaces of soils (termed *soil water*) and within deeper porous and permeable geological deposits (termed *groundwater*). Monitoring of soil water resources is an important component of catchment management as soil moisture plays a key role in determining how a catchment functions during precipitation events. When heavy rainfall infiltrating into the soil causes soil moisture levels to exceed *field capacity* and become *saturated*, further infiltration is prevented and surface runoff is initiated. It is under these conditions that soil erosion and the transport of pollutants from terrestrial to aquatic environments is most prevalent. Simplified indicator-based methods as explained in Chap. 10, help to define the sites with a high risk of soil erosion and to calculate the amount of soil transported per year. Monitoring of soil moisture content can assist in the determination of *when* pollution risk is greatest and thus also support the adaptation of land use. Soil moisture content can be determined either manually, by collecting and drying soil samples to determine the percentage water content, or automatically using electrical resistivity based soil moisture probes which relate the ability of soil to resist an electrical current to the amount of water present (more water = lower electrical resistivity). Monitoring groundwater levels is equally important, with many water-bearing geological formations (*aquifers*) being commercially exploited to provide drinking water for centres of human population and irrigation water for agriculture. Groundwater levels typically vary seasonally, reducing in height during the summer or dry season when evapotranspiration exceeds infiltration rates, and increasing during the winter or wet season when enhanced *effective precipitation* (precipitation minus evapotranspiration) contributes to aquifer *recharge*. Groundwater levels can be monitored through the drilling of vertical boreholes down into the underlying geological deposits and measuring the height of the *water level* within the borehole using either a manual dipper or an automatic pressure transducer from which water pressure within the borehole is converted to water level elevation.

In landscape planning a simplified approach is sometimes used to define areas which should be protected against building development and associated surface sealing of the ground. For the identification of areas which make a particularly high contribution to groundwater recharge, this involves subtracting from regional average annual rainfall the surface flow (calculated using relief, soil type and geology as indicators) and the evapotranspiration (classified for different types of ground cover and soil types).

11.7 Indicators of the Ecological Status of Freshwater Ecosystems

Across the EU, almost 300 different methods have been implemented to determine the ecological status of surface waterbodies (Birk et al. 2012). Nevertheless, despite such wide ranging approaches, the majority of assessments focus on the routine monitoring of four *biological quality elements* (BQE) which encompass a broad range of trophic levels and thus provide a robust indication of the ecological health of a waterbody. The most commonly assessed BQE are macroinvertebrates (27% of studies), diatoms (21%), fish (14%) and macrophytes (11%). The status of each BQE is assessed using EU WFD compatible methods with the aim of establishing a baseline of ecological status against which any change in response of the system can be monitored. BQE monitoring also helps to identify the causes of failure to achieve good ecological status and links biological response to ecologically-relevant pollutants. Examples of the assessment methods employed in the United Kingdom are presented below:

Diatoms – a statistically robust methodology for assessing freshwater ecological status from diatoms is the *Diatom Assessment of River Ecological Status (DARES)* tool. An estimation of the diatom community is made by collecting representative samples of benthic algae attached to stones (or macrophytes where suitable stones are lacking), randomly selected from the benthos. Attached algae is removed from the surface and preserved in Lugol's iodine solution before being digested with hydrogen peroxide and mounted on microscope slides and examined to identify diatoms at species level. Data are summarised for each sample as a list of species present and their percentage abundance with scores allocated to species using the *trophic diatom index* (TDI), where the total score for the sample is the average weighted by relative abundance.

Macroinvertebrates – the *River Invertebrate Prediction and Classification System (RIVPACS)* is a methodology for assessing freshwater quality based on macroinvertebrate species that provides a biotic response to organic pollutant, flow and sediment pressures. A semi-quantitative kick sample of the benthic material is collected that is representative of the river reach being investigated and which covers all habitats in proportion to their occurrence. Following standard protocol, this involves a 3-minute kick sample of the substrate to entrain any macroinvertebrates into suspension which are then carried downstream into a kick-net. Contents of the net are emptied into a white tray where the relative abundance of taxa resolved to species level can be determined. The physical characteristics of the reach being assessed (e.g. width, depth, discharge) and the substrate composition as percentage cover are also recorded.

Macrophytes – the *Leaf Prediction and Classification System* (*LEAFPACS)* is a biological method to assess the trophic status of streams and rivers based on the presence and abundance of species of aquatic macrophytes, where macrophytes are defined as 'any plant observable with the naked eye and nearly always identifiable when observed'. This definition includes all higher aquatic plants, vascular cryptograms and bryophytes, together with groups of algae which can be seen to be composed predominantly of a single species. A 100 m reach of a watercourse is carefully walked by the surveyor between 1st June and 30th September, avoiding periods during or immediately after high flows which may remove certain taxa. The presence and percentage cover of all macrophyte species are recorded up to the height of the river bank that would normally be submerged for >50% of the year, with the percentage cover converted into a taxon cover value on a scale of 1–9 (where 1 is <0.1% and 9 is >75%). Each taxa identified is assigned a *River Macrophyte Nutrient Index (*RMNI*)* value based on the nutrient tolerance of each species, with lower values indicating reduced nutrient tolerance and therefore reduced nutrient pollution pressures.

Fish **–** the *Fish Classification Tool 2 (FCS2)* can be used to determine the quality of a waterbody based on the composition and abundance of detected fish species. Electric fishing is typically used to capture the fish, whereby a pulsed DC current is passed into a netted section of water which momentarily stuns the fish and draws them toward the anode at the water surface. Stunned fish can then be collected and identified, with scale samples taken to determine the age and health of the individuals within the population. Along a 100 m stretch of river, 3–4 runs of electric fishing would typically be conducted to achieve depletion of the fish stock (i.e. nearly all fish captured and recorded).

Hydrogeomorphology **–** a *River Habitat Survey (RHS)* is a semi-quantitative approach for assessing the physical diversity of rivers based on the presence or absence of more than 200 fluvial geomorphological and ecology characteristics. Typically conducted along a 500 m length of river channel, information is collected on the physical structure of the watercourse, such as the sinuosity, bank height, channel width, riverbed substrate, flow dynamics, presence of riffles and pools, braiding, and anthropogenic modifications. The structural diversity of vegetation growing in and along the banks of the river is also recorded.

Biological data are summarised as indices related to specific pressures or general degradation. Data are assessed using the appropriate WFD-compliant tools to compare against reference conditions (the predicted state of the site in the absence of degradation) and thus provide an *Ecological Quality Ratio* (EQR) for each site. This EQR, which ranges from 0 to 1, is divided into the WFD quality classes of High (>0.8), Good (0.6–0.8), Moderate (0.4–0.6), Poor (0.2–0.4), Bad (<0.2) used to classify the ecological status of a site. Further statistical approaches can then be used to link biological responses to pressures, with the aims of linking stressors to ecological response and of identifying temporal scales of ecological responses.

11.8 Experimental Monitoring Design

The sampling strategy employed in any water quality monitoring programme will be a major determining factor in the usefulness and interpretability of the results obtained. A commonly applied method in catchment science research for assessing the impact of land management interventions to improve water quality is the *Before-After-Control-Impact (BACI)* approach (Fig. 11.5). Here, a manipulated system

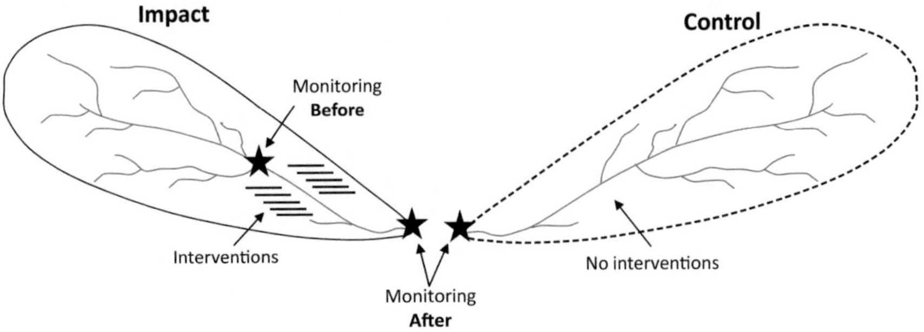

Fig. 11.5 Schematic of the Before-After-Control-Impact (BACI) catchment monitoring design

such as a river catchment (i.e. the impact) is compared with a non-manipulated system (i.e. the control) before and after the implementation of some kind of intervention or mitigation measure aimed at improving water quality. Measurements taken before the intervention provide a baseline against which to compare post-intervention conditions in the focus area. The neighbouring control system provides an additional spatial reference that can be used to factor out the confounding effects of changes in, amongst other things, temperature, precipitation, river flows and land management. Monitoring is typically conducted at the local level and then *upscaled* and reported at the River Basin District (RBD) level in accordance with the agenda given in the Water Framework Directive.

The temporal frequency at which monitoring of water resources is carried out will depend upon a number of factors, including the expected degree of variability in the parameters being monitored, the purpose of the monitoring and the resources available to make quantitative measurements. High-resolution monitoring (e.g. continuous to hourly) can provide important insights into both major and subtle aspects of catchment behaviour which can often be overlooked by lower-resolution (daily to yearly) sampling (Fig. 11.6). High-resolution monitoring can elucidate evidence of *non-stationarity* in water quality parameters, uncover *hysteresis* patterns in sediment and nutrient concentrations, and reveal the intricate dynamics of storm-dependent pollutant transfers by capturing a broader range of concentrations (Halliday et al. 2012). High-resolution monitoring is essential at water treatment plants where public water suppliers need to ensure the water is potable and free of harmful levels of pollutants before distribution to consumers. It is also very important for studying high-flow storm event conditions when approximately 90% of nutrients, sediments and other pollutants are moved through the catchment during less than 10% of the year. However, high-resolution monitoring is usually expensive to conduct due to either the high capital costs of installing, maintaining and running suitable instrumentation or due to high labour costs from time-consuming sample collection and processing. These costs typically render high-resolution analysis unsuitable for many water resource monitoring programmes where the need for it is not essential. Low-resolution monitoring is cheaper, quicker to conduct, and easier to deploy over a wider geographical area. Although low-resolution monitoring will fail to capture

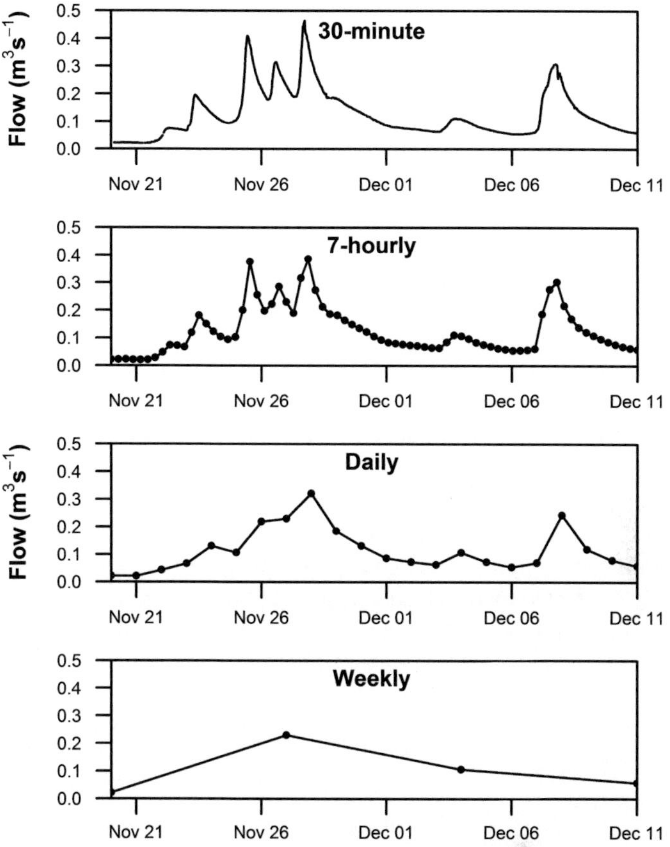

Fig. 11.6 Effect of the temporal monitoring resolution on recorded river flows over a 21-day period

the full range of pollutant concentrations and waterbody conditions, it can elucidate important information on the underlying background conditions and thereby provide a useful benchmark with which to compare between sites monitored at comparable resolutions.

11.9 Assessing Water Resource Abstraction

In order to ensure that the ecosystem services provided by catchment water resources are maintained, it is imperative that resources are sustainably managed and not overexploited. This is particularly important when it comes to abstracting water from rivers and aquifers to satisfy domestic, industrial or agricultural requirements. Collectively, EU member states abstract approximately 350 km^3 of water per year, representing 10% of the total global abstracted freshwater resource (EEA 2007). On average, agriculture accounts for 44% of abstracted reserves, industry and energy production 40%, and 15% is used for domestic water supply. To prevent environmental degradation from occurring it is necessary to calculate the abstraction *safe yield*, the level above which resource mining and environmental degradation occur. For river abstractions, the safe yield is equivalent to the effective precipitation minus the surface runoff and groundwater discharge needed to support sustainable river flows whilst balancing the change in groundwater storage (Fig. 11.7):

$$Abstraction = (P - ET) - (S_R + G_R) \pm \Delta S$$

where P is precipitation, ET is evapotranspiration, S_R is surface runoff, G_R is groundwater discharge, and ΔS is the change (gain or release) in groundwater storage. Data to support this calculation is obtained by monitoring of the physical and meteorological parameters listed in Table 11.3, principally river flows, groundwater levels and precipitation totals. For groundwater abstractions, a safe yield is where abstraction is less than or equal to groundwater discharge into the river. The long-term sustainability of water extraction from aquifers is dependent upon the amount of water extracted being less than the aquifer recharge, otherwise groundwater levels will decrease, and so threaten baseflow supply to surface waterbodies or wetlands. The *borehole hydrograph* method can be used to determine groundwater recharge levels and is calculated by the change in height of the water table (determined from boreholes drilled into the aquifer) multiplied by the aquifer *storage coefficient*. An alternative is the *chloride budget method* where recharge is estimated based on the

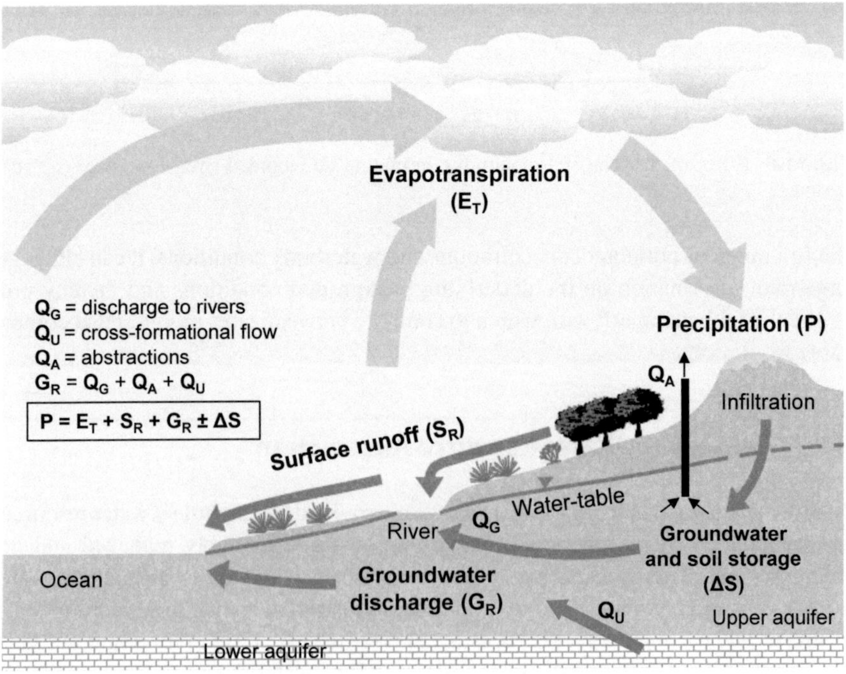

Fig. 11.7 Conceptual model of the components of the catchment water balance equation

chloride concentrations in rainfall and groundwater assuming that all chloride is from atmospheric sources:

$$R_d = P\left(\frac{C_p}{C_g}\right)$$

where R_d is direct aquifer recharge, P is mean annual precipitation (mm), C_p is mean chloride concentration in rainfall, and C_g is mean chloride concertation in groundwater. Lastly, the conventional *Penman-Grindley soil moisture balance* method can be used to determine groundwater recharge based on the balance between precipitation and evapotranspiration, with recharge assumed to occur when precipitation exceeds evapotranspiration. The most difficult aspect is to calculate actual evapotranspiration (*AE*). In general, the potential evapotranspiration (*PE*) is first defined as the maximum rate of evapotranspiration under prevailing meteorological conditions over short-rooted vegetation with a limitless water supply. A budgeting procedure is used to convert *PE* to *AE* with the degree to which potential and actual evapotranspiration rates diverge being controlled by a root constant, a function of soil and vegetation characteristics and a measure of readily available water within the root range. For further details, including a discussion of the estimation of *PE*, the reader is directed to Hiscock and Bense (2014) (Box 11.3).

Box 11.3: Modelling Water Resources

There are numerous catchment science tools available for modelling water resources, a selection of which are described below.

Soil & Water Assessment Tool (SWAT) – a continuous time, catchment-scale hydrological transport model used to quantify the impact of land management practices on water resources on a daily time-step. The primary objective of SWAT is to predict the long-term impacts of land management actions, principally related to agricultural activities, on water quality and water quantity in order to identify best management practices. To do this, SWAT incorporates meteorological, surface runoff, groundwater flow, evapotranspiration, reservoir storage, irrigation, crop growth, water transfer, and nutrient, pesticide and sediment loading components at the catchment-scale (Arnold et al. 2012).

SCIMAP – a spatially-distributed model for mapping diffuse pollution risk across a catchment. Pollution risk mapping in SCIMAP is performed by calculating the spatial pattern of erosion risk based on land cover information, rainfall patterns, terrain analysis and hydrological connectivity. The information is combined to map the location of the critical source areas where there is both a source of pollution and a connection to the river channel (Reaney et al. 2011).

TOPography based hydrological MODEL (TOPMODEL) – a distributed rainfall-runoff model in which predictions of catchment response to precipitation events are made based on the theory of hydrological similarity, which in turn is based on the topographic index (Beven and Freer 2001).

(continued)

Box 11.3 (continued)

Mesoscale Hydrologic Model (mHM) – a spatially-explicit distributed hydrologic model that uses grid cells as a primary hydrologic unit, and accounts for the following processes: canopy interception, snow accumulation and melting, soil moisture dynamics, infiltration and surface runoff, evapotranspiration, subsurface storage and discharge generation, deep percolation and baseflow, and discharge attenuation and flood routing (Samaniego et al. 2010).

Groundwater modelling – Numerical modelling of groundwater flow is an indispensable tool for managing local and regional groundwater resources and enables the prediction of groundwater flow patterns, for example the effects of different groundwater abstraction patterns on sensitive aquatic systems, or the shape of wellhead capture zones for protecting groundwater quality, or future aquifer response to changing recharge amounts under climate change. The primary aim of a groundwater model is to represent adequately the different features of groundwater flow through the aquifer within the model area or domain. In this respect, the important features to consider in governing the response of an aquifer to a change in hydrogeological conditions include: aquifer inflows (recharge, leakage and cross-formational flows), aquifer outflows (abstractions, spring flows and river baseflows), aquifer properties (hydraulic conductivity and storage coefficient), and aquifer boundaries (constant or fixed head, constant flow or variable head, and no-flow boundaries). A popular finite-difference model for application in two- and three-dimensional groundwater flow problems is the United States Geological Survey's code MODFLOW (McDonald and Harbaugh 1988), with demonstrations of this model presented by Chiang and Kinzelbach (2001) and Anderson et al. (2015).

Sediment fingerprinting – a catchment science tool for estimating the sediment contributions from various eroding terrestrial sources to fluvial sediment load via a mixing model approach. The technique relies on selecting appropriate markers or *'fingerprints'* that are transported from eroding source areas to the river *'target'* in a reliable manner through well understood biotic or abiotic pathways. A variety of fingerprints can be used to help differentiate potential sediment source areas, including major and trace elements, colour coefficients, fallout radionuclides, mineral magnetism and compound-specific stable isotopes. These can help to identify sediment contributions from sources such as arable topsoils, stream channel banks, forests, grassland, road verges, urban areas, wildfire-burned land and contrasting geological provinces. There are also a variety of mixing model approaches which can be employed to quantitatively assess sediment volumes derived from each source area, with models ranging in complexity from simple optimisation routines to more comprehensive Bayesian uncertainty assessments (Cooper et al. 2015).

11.10 Conclusion

An important task in landscape planning is to review existing information and results about ES assessment of water resources and integrate them in a multifunctional assessment (Galler et al. 2015). Due particularly to the WFD, existing assessments of water resources across Europe tend to be of a higher standard than those for many other ES. Nevertheless, locally specific results may be missing and it will be the task of the planner in cooperation with hydrologists, regulators and administrators to decide whether, when and at what cost the methods described in this chapter should be applied. Some of the methods have considerable resource requirements, but depending on the data situation and consequences of the decision, more simplified approaches (as noted for assessing groundwater recharge rates) may be quite sufficient.

References

Ames, P. L. (1966). DDT residues in the eggs of the osprey in the north-eastern United States and their relation to nesting success. *Journal of Applied Ecology, 3*, 87e97.

Anderson, M. P., Woessner, W. W., & Hunt, R. J. (2015). *Applied groundwater modeling: Simulation of flow and advective transport* (2nd ed.). San Diego: Academic.

Arnold, J. G., Moriasi, D. N., Gassman, P. W., et al. (2012). SWAT: Model use, calibration, and validation. *Transactions of the ASABE, 55*, 1491–1508.

Beven, K., & Freer, J. (2001). A dynamic TOPMODEL. *Hydrological Processes, 15*, 1993–2011.

Beven, K., Heathwaite, L., Haygarth, P., et al. (2005). On the concept of delivery of sediment and nutrients to stream channels. *Hydrological Processes, 19*, 551–556.

Bilotta, G. S., & Brazier, R. E. (2008). Understanding the influence of suspended solids on water quality and aquatic biota. *Water Research, 42*, 2849–2861.

Birk, S., Bonne, W., Borja, A., et al. (2012). Three hundred ways to assess Europe's surface waters: An almost complete overview of biological methods to implement the Water Framework Directive. *Ecological Indicators, 18*, 31–41.

Broads Authority. (2017). *Facts & figures*. http://www.broads-authority.gov.uk/learning/facts-and-figures. Accessed 14 June 2018.

Chiang, W.-H., & Kinzelbach, W. (2001). *3D-Groundwater modeling with PMWIN: A simulation system for modeling groundwater flow and pollution*. Berlin: Springer.

Cooper, R. J., Krueger, T., Hiscock, K. M., et al. (2015). High-temporal resolution fluvial sediment source fingerprinting with uncertainty: A Bayesian approach. *Earth Surface Processes and Landforms, 40*, 78–92.

Costanza, R., d'Arge, R., de Groot, R., et al. (1997). The value of the world's ecosystem services and natural capital. *Nature, 387*, 253–260.

Crowther, J., Kay, D., & Wyer, M. D. (2002). Faecal-indicator concentrations in waters draining lowland pastoral catchments in the UK: Relationships with land use and farming practices. *Water Research, 36*, 1725–1734.

Duruibe, J. O., Ogwuegbu, M. O. C., & Egwurugwu, J. N. (2007). Heavy metal pollution and human biotoxic effects. *International Journal of Physical Sciences, 2*, 112–118.

EEA. (2007). *Water abstraction*. https://www.eea.europa.eu/themes/water/water-resources/water-abstraction. Accessed 14 June 2018.

EEA. (2015). *The European environment state and outlook 2015: Protecting, conserving and enhancing natural capital*. Copenhagen: European Environment Agency.

Finlayson, C. M., & D'Cruz, R. (2005). Chapter 20: Inland water systems. In R. Hassan, R. Scholes, & N. Ash (Eds.), *Ecosystems and human well-being: Current state and trends* (Millennium ecosystem assessment) (pp. 553–583). Washington, DC: Island Press.
Galler, C., von Haaren, C., & Albert, C. (2015). Optimizing environmental measures for landscape multifunctionality: Effectiveness, efficiency and recommendations for agri-environmental programs. *Journal of Environmental Management, 151*, 243–257.
Grizzetti, B., Lanzanova, D., Liquete, C., et al. (2016). Assessing water ecosystem services for water resource management. *Environmental Science & Policy, 61*, 194–203.
Halliday, S. J., Wade, A. J., Skeffington, R. A., et al. (2012). An analysis of long-term trends, seasonality and short-term dynamics in water quality data from Plynlimon, Wales. *The Science of the Total Environment, 434*, 186–200.
Hiscock, K. M., & Bense, V. F. (2014). *Hydrogeology: Principles and practice* (2nd ed.). Chichester: Wiley.
Jess, S., Kildea, S., Moody, A., et al. (2014). European Union policy on pesticides: Implications for agriculture in Ireland. *Pest Management Science, 70*, 1646–1654.
Lal, R., Griffin, M., Apt, J., et al. (2004). Managing soil carbon. *Science, 304*, 393.
Macleod, C. J. A., Scholefield, D., & Haygarth, P. M. (2007). Integration for sustainable catchment management. *The Science of the Total Environment, 373*, 591–602.
McDonald, M. G., & Harbaugh, A. W. (1988). *A modular three-dimensional finite-difference ground-water flow model* (Techniques of water-resources investigations of the United States Geological Survey. Book 6, Chapter A1). Washington, DC: Scientific Software Group.
Némery, J., & Garnier, J. (2016). Biogeochemistry: The fate of phosphorus. *Nature Geoscience, 9*, 343–344.
Pawson, M. G., Tingley, D., Padda, G., et al. (2007). *EU contract FISH/2004/011 on "sport fisheries" (or marine recreational fisheries) in the EU* (p. 242). Lowestoft: Cefas.
Popp, J., Pető, K., & Nagy, J. (2013). Pesticide productivity and food security: A review. *Agronomy for Sustainable Development, 33*, 243–255.
Pretty, J. N., Mason, C. F., Nedwell, D. B., et al. (2003). Environmental costs of freshwater eutrophication in England and Wales. *Environmental Science & Technology, 37*(2), 201–208. https://doi.org/10.1021/es020793k.
Quinton, J. N., Govers, G., Oost, K. V., et al. (2010). The impact of agricultural soil erosion on biogeochemical cycling. *Nature Geoscience, 3*, 311–314.
Reaney, S. M., Lane, S. N., Heathwaite, A. L., et al. (2011). Risk-based modelling of diffuse land use impacts from rural landscapes upon salmonid fry abundance. *Ecological Modelling, 222*, 1016–1029.
Samaniego, L., Kumar, R., & Attinger, S. (2010). Multiscale parameter regionalization of a grid-based hydrologic model at the mesoscale. *Water Resources Research, 46*(5), W05523. https://doi.org/10.1029/2008WR007327.
Shiklomanov, I. A., & Rodda, J. (2003). *World water resources at the beginning of the 21st century*. Cambridge, UK: Cambridge University Press.
Smith, L., Porter, K., Hiscock, K., et al. (2015). *Catchment and river basin management: Integrating science and governance*. Oxford: Routledge.
Smith, V. H., Tilman, G. D., & Nekola, J. C. (1999). Eutrophication: Impacts of excess nutrient inputs on freshwater, marine, and terrestrial ecosystems. *Environmental Pollution, 100*, 179–196.
Sundseth, K. (2014). *Invasive alien species: A European Union response* (European Commission report). Luxemburg: Publications Office.
Whitehorn, P. R., O'Connor, S., Wackers, F. L., et al. (2012). Neonicotinoid pesticide reduces bumble bee colony growth and queen production. *Science, 336*, 351–352.
WHO. (2004). *Meeting the MDG drinking water and sanitation target: A mid-term assessment of progress*. Geneva: World Health Organisation.
Wilkinson, B. H. (2005). Humans as geologic agents: A deep-time perspective. *Geology, 33*, 161–164.

Renewable Energy Production Capacities and Goods

12

Claudia Palmas, Michael Rode, and Andrew A. Lovett

Abstract

The contributions of landscapes to produce renewable energy from sources such as wind, solar and biomass has recently attracted enhanced interest from policy and business stakeholders. At the same time, potential conflicts with nature conservation, tourism interests and the delivery of other ecosystem services have become apparent, originating from both increased pressures for land use intensification and changes in the energy grid. The objective of this chapter is to present a method for estimating sustainable renewable energy potentials and exploitable energy yields for wind and solar energy taking account of other ecosystem services. The method first spatially assesses energy potentials for each source. It then identifies the most suitable areas for decentralized renewable energy generation, considering both spatial efficiency and environmental trade-offs. A case study application in the Hanover region, Northern Germany, demonstrates the applicability of the method and the outputs that can be generated. The information generated by our method can usefully enhance landscape and spatial planning with important information on renewable energy potentials, and it can help to identify where investment in electricity grid infrastructure appropriate for harnessing these potentials might be required. Last but not least, the method can identify potential opportunities and conflicts in advance of developments, help alleviate conflicts and harness synergies between diverging interests.

Keywords

Wind energy · Solar energy · Energy potentials · Trade-offs

C. Palmas · M. Rode (✉)
Institute of Environmental Planning, Leibniz University Hannover, Hannover, Germany
e-mail: palmas@umwelt.uni-hannover.de; rode@umwelt.uni-hannover.de

A. A. Lovett
School of Environmental Sciences, University of East Anglia, Norwich, UK
e-mail: a.lovett@uea.ac.uk

C. von Haaren et al. (eds.), *Landscape Planning with Ecosystem Services*,
Landscape Series 24, https://doi.org/10.1007/978-94-024-1681-7_12

12.1 Introduction

European energy transition objectives are for at least 40% cuts in greenhouse gas emissions (from 1990 levels) and at least 27% share in energy provision for renewable energy (European Commission 2014). In 2050, a low carbon economy with probably 100% energy from renewables will need to be implemented if the climate goals of the Paris agreement are to be fulfilled (United Nations 2018). Several member states have set themselves corresponding targets. For Germany to achieve these objectives a striking increase in both photovoltaic generation and installed wind power capacity will be needed. Wind energy technology has become increasingly efficient recently (Sathaye et al. 2011). However, there can be negative on-site effects on humans, species and landscape aesthetics (Kienast et al. 2017). How significant these impacts are, depends on the size and technology of the wind turbine and location of the wind farm. The installation of wind power plants in proximity to urban areas can cause negative human health effects, due to noise, light emissions and shadowing. Noise levels, together with landscape transformation and concerns regarding impacts on nature protection and tourism, have led to opposition against on-shore wind power developments (Klinski et al. 2007; Krekel and Zerrahn 2017). Comparable conflicts with other types of land-uses arise with the increased use of ground-sited photovoltaics (PV) (Marcheggiani et al. 2013) and bioenergy production (Buhr et al. 2013). As a direct consequence, public protests against wind farms, ground-sited PV arrays, electricity distribution infrastructure and maize cultivation (for anaerobic digesters) have occurred in many regions (Gailing and Röhring 2016; Kemfert and Horne 2013). Integrated planning approaches, which combine spatial with energy planning, are needed in order to facilitate the development of diverse renewable energy facilities in sympathy with nature conservation, tourism and food production interests (von Haaren et al. 2013). When replacing fossil fuel sources with renewable energy, land availability becomes an important consideration, particularly in terms of competition with residential, commercial, agricultural, industrial and transport uses (Stoeglehner and Narodoslawsky 2009; Huber et al. 2017). However, current planning practices in Europe do not yet include an integrated analysis of renewable energy production or an effective evaluation of energy related land-use issues. Sectorial approaches are not sufficient when dealing with the above mentioned issues (Peters 2013; Palmas et al. 2015). Instead, it is necessary to estimate GIS-based renewable energy potentials as an input for scenario development (Palmas et al. 2012) and to minimize the land consumption for renewable energy production on both the local and regional scales (Diefenbacher 2009). This can be achieved by the optimal exploitation of combined renewable energy potentials together with a simultaneous minimization of environmental trade-offs (Palmas et al. 2015).

This chapter introduces a method for minimizing negative environmental impacts from renewable energy production. The method has been developed for the German legislative context (e.g. concerning noise protection standards in relation to onshore wind developments) and data availability, and it has been tested in the Hanover region. However, its methodological steps can be applied in other EU member states. Energy crop production is not considered further because the efficiency of the energy yield per hectare is very low in comparison to wind and solar power (Palmas et al. 2015). For 2050 we project that only residual and waste biomass will be used to satisfy the remaining demand for fuels in certain processes (Walter et al. in print).

12.2 Methodology and Data

12.2.1 Overview of Scenario Modeling

The methodological steps followed for developing renewable energy scenarios taking account of other ecosystem services are described below. This example is for wind energy (cf. BWE 2012):

1. exclusion of locations with low physical energy potentials, e.g. low-wind locations with speeds <3.2 m/s at 100 m);
2. overlay of the technical renewable energy potentials with hard and soft restriction areas (spatial, legal and practical planning restrictions);
3. allocation of turbines on the suitable areas (in the case of wind the minimum distance between two turbines in all directions is 300 m);
4. calculation of the renewable power capacity and energy production.

Based on this approach the methodology consists of three elements:

1. a raster data-set of GIS-based theoretical energy potentials (wind and solar), where the estimate depend only on geophysical properties of the landscape;
2. a refinement of the first raster layer where the technical requirements of the different power plants are included;
3. a map of hard and soft restriction areas (representing different environmental trade-offs) to allow exclusion of unsuitable areas.

Different scenarios can be developed by overlaying the GIS layers to identify suitable sites and potential conflict areas for renewable energy production. The outcomes can be compared with the actual decision-space for renewable energy development on the regional level and politically-defined generation targets.

The renewable energy potentials are calculated using established methods (Palmas et al. 2012; Calvert et al. 2013) which are adapted to the information available at the regional scale (Palmas et al. 2015). Geographical variations in the potentials are modelled on the basis of geophysical properties i.e. solar irradiation [kWh/m^2*d] and wind speeds [m/s]. These variables are site specific and are related to land use, latitude, altitude, climate, and topography. For instance, slope and aspect are important in identifying suitable locations for ground-based solar arrays (Watson and Hudson 2015). The resulting technical potentials are expressed as the energy

Table 12.1 Input data sources used for renewable energy potential estimation

	Inputs	Scale/Unit/ Format	Example sources
Solar	Digital Terrain Model (DTM)	50 m×50 m	State Department for Geographic Information and Land Development ("Landesamt für Geoinformation und Landentwicklung", LGLN 2014)
	Aspect (azimuth of solar panel)	180°	Joint Research Center of the European Commission, JRC 2015
	Slope (solar panel inclination)	37°	
	Linke atmospheric turbidity coefficient	3,0–4,2	
	Albedo coefficient	0,2	
	Real-sky beam radiation coefficient	0,118–0,62	
	Real-sky diffuse radiation coefficient	0,86–5,49	
	Day of the year	1–365 [d]	
	PVGIS Database	[kWh/m^2]	Photovoltaic Geographical Information System Interactive Maps (Joint Research Center of the European Commission, JRC 2010)
Wind	Digital Terrain Model (DTM)	50 m×50 m	State Department for Geographic Information and Land Development ("Landesamt für Geoinformation und Landentwicklung", LGLN 2014)
	Wind speeds	[dm/s]	German Weather Service ("Deutscher Wetterdienst", DWD 2013), 1981–2000

yield per unit area, calculated for a year [MWh/ha*a]. Table 12.1 provides an overview of the main data sources used for energy potential estimation.

In addition to geographical and technical constraints, legal and practical planning restrictions need to be considered in order to prevent an overestimation of energy potentials. In this example the restrictions for renewable energy development are derived from either legally-binding regulations (e.g. the German Federal Nature Conservation Act, BNatSchG) which are referred to as 'hard' restriction areas, or from recommendations by official planning institutions that represent 'soft' restrictions (Table 12.2) When including legal and/or policy requirements it is also important to refer to the most current regulations in each case.

If no such particular planning recommendations are in place, EU directives, national and regional legislation and strategies referring to species and habitats, landscape amenities and noise pollution should be interpreted with respect to their current and future spatial implications (Walter et al. in print). Other examples of such siting constraints elsewhere in Europe are discussed by Watson and Hudson (2015), Gove et al. (2016), and Sánchez Lozano et al. (2013, 2016).

Table 12.2 Overview of the types of restrictions for renewable energy developments, – indicates no restrictions

Area Categories		Wind power	Ground PV array	Legal basis
Built-up areas	Residential areas	Hard	–	§5 BlmSchG[a]; §35 BauGB[b]
	400 m residential buffer zone	Hard	–	§5 BlmSchG[a]; §35 BauGB[b]
	Commercial/industrial areas	Hard	–	§5 BlmSchG[a]; §35 BauGB[b]
	400 m commercial/industrial buffer zone	Hard	–	§5 BlmSchG[a]; §35 BauGB[b]
Infrastructures	Motorways, 30 m	–	Hard	§35 BauGB[b]
	40 m motorway buffer zone	Hard	–	§9 FStrG[b]; §24 NStrG[d]
	Federal roads, 16 m	–	Hard	§35 BauGB[b]
	20 m federal road buffer zone	Hard	–	§9 FStrG[c]; §24 NStrG[d]
	Overhead power lines, 31 m	Hard	–	DIN EN[e] 50,341-3-5
	82 m overhead power line buffer zone	Hard	–	DIN EN[e] 50,341-3-5
Water and soil	Main watercourses, 50 m buffer zone	Hard	–	§61 BNatSchG[f]
	Flood hazard areas	Hard	Hard	§78 WHG[b]; §35 BauGB[b]
	Slopes >21%	–	Soft	–
Protected areas	Nature conservation areas	Hard	Hard	§23 BNatSchG[f]
	EU-Special Protection Areas (SPA)	Hard	Soft	§31 BNatSchG[f]
	Protected biotopes	Hard	Hard	§30 BNatSchG[f]
	Protected landscape units	Hard	Hard	§28, §29 BNatSchG[f]
	Drinking water protection zones I & II	Hard	–	§51 WHG[b]
	Landscape conservation areas	Soft	Soft	–
	Special areas of conservation (FFH)	Soft	Soft	–

(continued)

Table 12.2 (continued)

Area Categories		Wind power	Ground PV array	Legal basis
Biodiversity, flora and fauna	Faunistic habitats (values 4–5)	–	Soft	–
	Migratory bird resting areas (values 3–5)	Soft	Soft	–
	Wetlands of international importance	Soft	Soft	–
	Breeding bird habitats (values 3–5)	Soft	–	–
	Bat habitats	Soft	–	–
	Areas of special importance for species and ecosystems	–	Soft	–
	Forests	Soft	Soft	–
	Green belts and infrastructure	–	Soft	–
Landscape	Landscape units of special importance	Soft	Soft	–
	Landscape of special structural diversity	–	Soft	–
	Areas of historical/cultural importance	–	Soft	–

Sources: MU 2009; NLT 2014a, b
[a]BImSchG: Federal Control of Pollution Act
[b]BauGB: Federal Building Code
[c]FStrG: Federal Highway Act
[d]NStrG: Lower Saxony road law
[e]DIN EN: German standard for distance of wind energy turbines to high-voltage lines
[f]BNatSchG: Federal Nature Conservation Act
[g]WHG: Water Management Act

12.2.2 Calculating Spatial Renewable Energy Potentials

The theoretical renewable energy potentials for wind and solar power can be estimated based on geophysical properties. Subsequently, by taking into account the technical requirements, different GIS-based raster data-sets can be developed in order to compare different energy yields (Fig. 12.1). These results are then used as inputs to model regional renewable energy development scenarios.

12.2.2.1 Spatial Wind Energy Potential

In order to evaluate wind energy scenarios, potentially suitable areas [ha], their installable power [GW] and possible electricity yields [GWh/a] have to be calculated. As a first step, theoretical wind energy potentials can be estimated by processing data on average annual wind speeds (in the case study: Deutscher Wetterdienst, DWD 2013). The input data for the case study were provided for the reference period 1981–2000, for wind speeds at 10 m height above ground and with a 200 m cell resolution. The data took into account roughness (relief and land

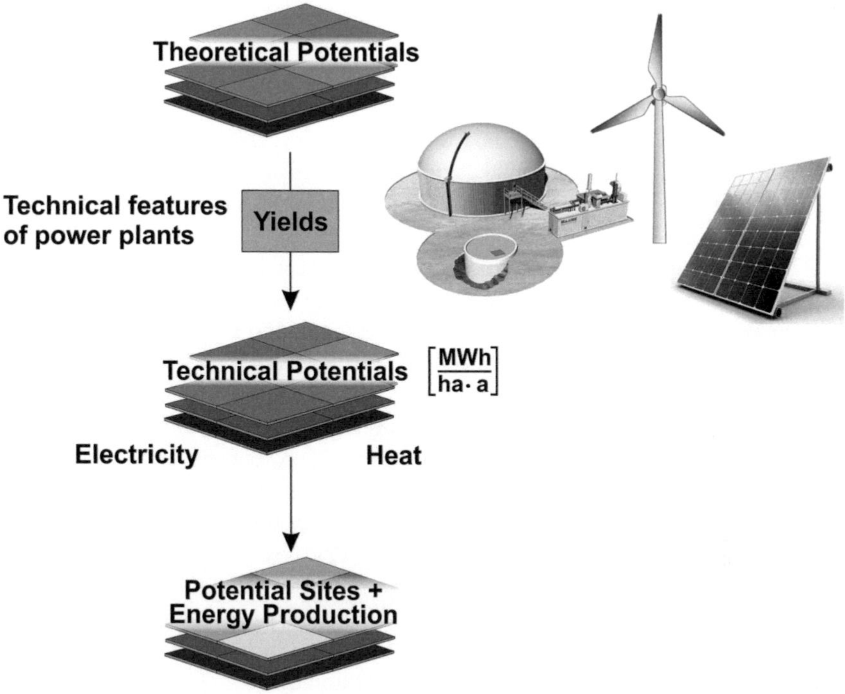

Fig. 12.1 Methodological approach for identifying renewable energy potentials

characteristics), height above sea level, and geographical location. Differences between calculated and measured wind speeds were quoted at ±0.15 m/s. These input data were then downscaled using a Digital Terrain Model (DTM 50). Further processing included calculating wind speeds at the turbine heights given current (or in case of projections – future) technical standards (in the case study 100 m was chosen). The increase in wind speed at different heights can be calculated using Eq. 12.1. The underlying assumption in this equation is that the atmosphere thermic condition is stable, which is expressed by the empirically derived coefficient α at a value of 0.143 (Counihan 1975; Touma 1977).

$$v = v_{\text{ref}}\left(z / z_{\text{ref}}\right)\alpha \tag{12.1}$$

Where:

v: wind speed at height z above ground level;
v_{ref}: reference speed, i.e. a wind speed we already know at height z_{ref};
z: desired turbine height above ground level (e.g. 100 m) for the velocity, v;
z_{ref}: reference height (e.g. 10 m) at which the wind speed is measured v_{ref}.

In order to obtain the technical wind energy potential, expressed in energy yield per unit area and calculated for a year [MWh/ha], a 'standard' wind turbine needs to be defined (in the case study an installable power of 3 MW was chosen). The minimum spacing associated with such turbines should then be identified, which in the case study was based on triple the height of the 100 m mast (BWE 2012). Interpreting this as a 300 m by 300 m grid cell defined an exclusive buffer zone of 90,000 m^2 around each wind turbine. Consequently, the density of turbines will vary according to their height and other technical properties, while energy yields will also depend on the average wind speed distribution across the region.

12.2.2.2 Ground-Based Solar Energy Potential

Theoretical solar energy potentials can be estimated using a raster DTM with the open-source solar radiation model *r.sun*, implemented in GRASS GIS (6.4.2), and the PVGIS CM-SAF estimation utility (JRC 2010). The PVGIS database includes the crucial factor of sky cloud coverage (with a stated overall error of approximately 5% for an entire year, JRC 2015) and assumes regional albedo and Linke turbidity are constant in each month. The output raster map represents the annual average of daily global irradiation totals estimated for an optimized angle [kWh/m^2*d].

When considering large-scale ground-based PV installations (e.g. in agricultural areas), north-facing slopes should be excluded under present technological and market conditions. Furthermore, terrains with an aspect of either 0–59° or 300–360° and a slope of more than 5° are suboptimal. To estimate total regional solar potentials, it may also be relevant to assess the scope for installing roof-mounted PV arrays (see Walter et al. in print for a German example).

12.3 A Case Study of the Hanover Region

The Hanover case study demonstrates the applicability of the method and the results that can be generated. Modelling the geophysical properties of the landscape resulted in the theoretical renewable energy potentials shown in Fig. 12.2. Wind speeds ranged from 2.6 to 6.2 m/s and solar irradiation from 991 to 1349 kWh/m^2*d.

Taking into account the technical requirements (e.g. spacing) of generation equipment, the maps in Fig. 12.3 quantify the different annual energy yields expressed per unit area [MWh/ha*a]. The results show that the technical wind potential reaches higher values than the solar option, but is also more variable across the region.

When calculating the technical potential of solar generation the areas of north-facing slopes and those steeper than 5° are excluded. These amount to about 2.5% of the region. Excluding these areas, the average power density according to PVGIS database is about 22 m^2 to produce 1 kW. Overall, the average solar electricity yields of 420 MWh/ha*a on agricultural areas are similar to the average wind yield (441 MWh/ha*a).

The next step in the assessment was to omit areas excluded by legally-binding regulations (i.e. hard restrictions) or those constrained by planning regulations (i.e. soft restrictions). Fig. 12.4 shows these zones for the Hanover region and indicates

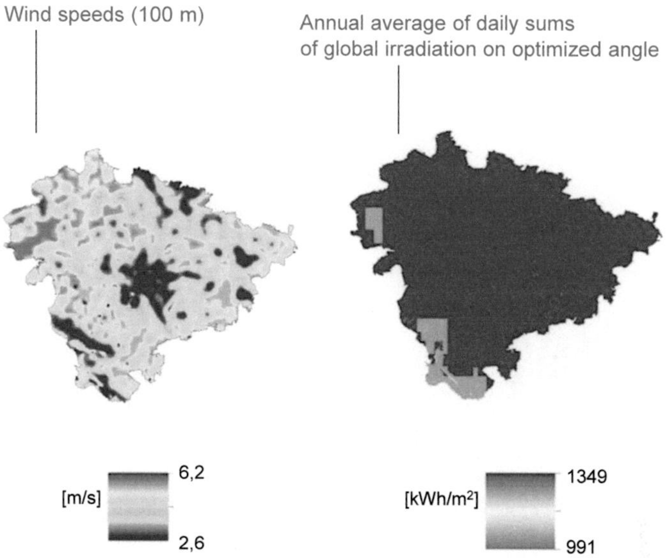

Fig. 12.2 Theoretical wind and solar energy potentials in the Hanover region

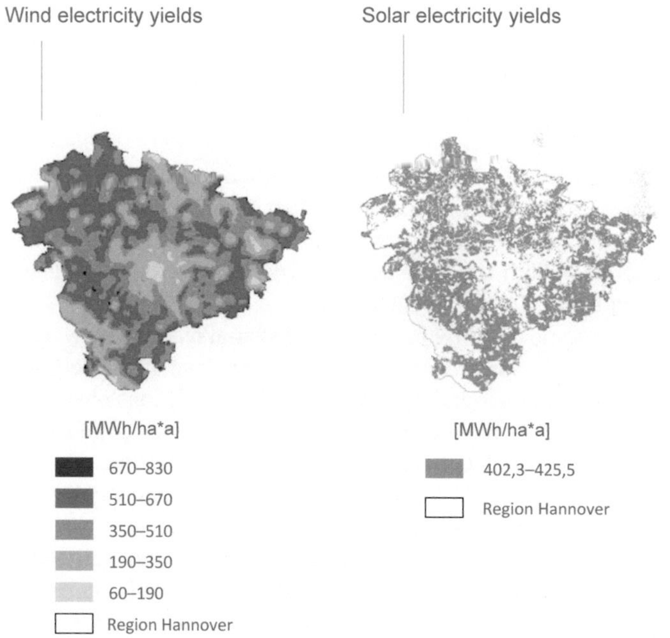

Fig. 12.3 Renewable energy technical potentials in the Hanover region

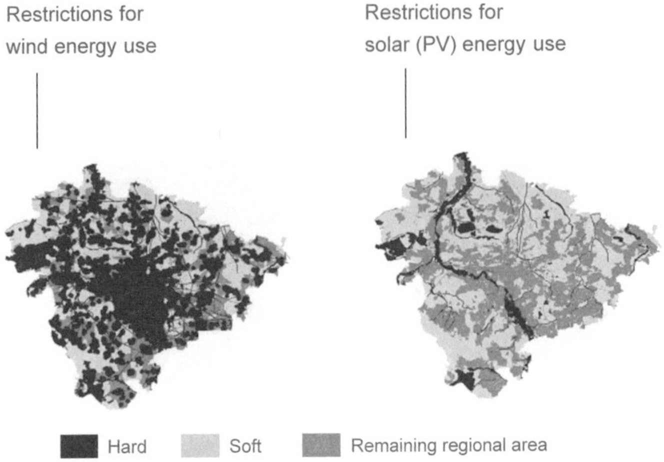

Fig. 12.4 Spatial distribution of hard and soft restriction areas within the Hanover region

Table 12.3 Impact of hard and soft restrictions on electricity yield from wind power

	Hard restrictions	Hard and soft restrictions
Wind energy		
Area [ha]	86,978	25,586
Proportion of region [%]	38	11
Wind turbines on remaining area	10,905	3520
Installable power capacity [GW]	33	11
Energy yields [GWh/a]	4959	1850

a considerable contrast between the two generation technologies. Hard restrictions were much more extensive for wind power than ground-based solar, while the latter had a greater area classed as sensitive under the soft criteria compared to wind. After subtraction of both sets of restrictions, wind power was constrained to 13% of the region and PV to 22% of the total area.

Finally, the technical energy potentials and spatial restrictions were combined to define different generation scenarios. As an example, two wind energy scenarios are presented in Table 12.3. Excluding areas with hard restrictions plus other unsuitable sites, allows for 38% of the region to be used for wind power generation. Here 33 GW can be installed with a total production of 4959 GWh/a. If areas with soft restrictions are added, only 11% remain for unrestricted energy generation where 11 GW can be installed and about 1850 GWh/a can be produced.

These outcomes can be usefully compared against the politically-desired generation targets for the Hanover region. In 2012 two scenarios of future energy production were derived for the Hanover region in order to help increase the sustainable use of renewable energy by 2050 (Region Hannover 2012). 1) The 'trend' energy scenario involved a linear extrapolation of the existing trend in wind energy production through to 2050 resulting in an estimated electricity yield of 1372 GW/a. A second, 'goal path' scenario was based on sustainability objectives: increased efficiency, high implementation of renewable energy, and energy supply self-sufficiency. This estimated a required regional wind energy yield of 3244 GWh/a in 2050.

The GIS-based analysis summarized in Table 12.3 calculated a possible electricity yield of 1850 GWh/a taking into account hard and soft restrictions, increasing to 4959 GWh/a if only the former were applied. This implies that only the 'trend' scenario target could be achieved without adverse effects on other ecosystem services. Moreover, the 'goal path' target would require generation on a large part of the area with soft restrictions. Impairments in other ecosystem services would therefore be likely. As a consequence, careful landscape planning would be needed to minimise negative impacts on other ecosystem services as they may vary from case to case, depending on technical and environmental factors such as geographical location, power plant design or special ecological sensitivities.

12.4 Conclusions and Refinements

The method presented above provides a basis for assessing renewable energy generation potential in particular landscape settings and allowing the possible conflicts or trade-offs with other ecosystem services to be identified. It additionally enables an evaluation of whether current policy targets are feasible. Gaining useful insights from such an approach is, of course, dependent on having reliable input data. For example, technical wind energy potentials can be more precisely estimated if wind speed distributions are available at 100 m above the ground level. Additionally, if Weibull parameters (about scale and shape) are available for the estimation of the Weibull distribution (continuous probability distribution), technical wind energy potentials can be more precisely estimated than by using average wind speeds. Likewise, the area required for the production of solar energy will decrease when the yield of the solar modules is increased through further technical developments.

In some cases, in addition to areas in which no negative effects on other ecosystem services are to be expected and which are thus neither subject to hard nor soft restrictions, parts of soft restricted areas that are already environmentally impacted (e.g. by existing infrastructure projects) may be suitable for renewable energy production as long as they offer satisfactory yields. In order to identify these areas, the interplay between high potential electricity yields and soft restrictions should be analyzed for each landscape individually.

If options that might be less favorable for the environment have to be adopted in order to reach renewable energy production targets, possible mitigation measures should be considered (e.g. inclusion of already impacted areas into development strategies). Indeed, this general approach has already found its way into planning. Evidence of this can be seen with the placement of wind turbines along transport infrastructure (Günnewig et al. 2009) and the siting of ground-based solar panels in former military facilities or in commercial and industrial areas (Staiß 2007).

One way to increase the range of options is to use more differentiated spatial analysis in order to achieve appropriate site selection. To reduce possible land-use conflicts regional land-use planners need spatially-explicit information to help move beyond simple blanket restriction areas and define more specific measures of environmental sensitivities. One example of such an approach is given by Lovett et al. (2014) who used a measure of landscape naturalness, rather than simple exclusion of National Parks and Areas of Outstanding Natural Beauty, in assessing the availability of land for energy crop planting in Great Britain.

The approach presented here can also be extended through the use of multi-criteria evaluation techniques to incorporate policy maker, expert or citizen preferences regarding the relative importance of different siting factors or trade-offs (see Watson and Hudson 2015; Sánchez Lozano et al. 2016). If situated within a wider participatory process (e.g. as part of landscape plan development or updating) involvement in such weighting exercises can be an effective way of engaging citizen interests in the trade-offs that may be necessary. Such an analysis may also help identify investment priorities at a regional scale, for instance where additional electrical grid infrastructure and capacity need to be located in order to allow the efficient utilization of decentralised renewable energy production.

This spatially-explicit approach also needs to include cost and energy engineering considerations, and in the future should aim to link the assessment of renewable energy potentials with operational aspects of power transmission and distribution systems (e.g. see Zeyringer et al. 2018). The trend towards decarbonisation and decentralisation means that geographical perspectives and land use issues are becoming increasingly important in the assessment of energy system options and new approaches to modelling are needed to address such challenges (Price et al. 2018). Developing the interface between energy systems engineering and landscape planning can make an important contribution in this evolving area of research and practice.

Acknowledgement Work on this chapter was supported by grant number NE/M019713/1 for the ADVENT project (Addressing Valuation of Energy and Nature Together) funded by the UK Natural Environment Research Council.

References

Buhr, N., Rode, M. W., & Kanning, H. (2013). Effectiveness of planning instruments for minimizing spatial conflicts of biogas production. *European Planning Studies, 22*(8), 1711–1734.

BWE – Bundesverband Windenergie. (2012). *Potenzial der Windenergienutzung an land.* https://www.wind-energie.de/sites/default/files/download/publication/studie-zum-potenzial-der-windenergienutzung-land/bwe_potenzialstudie_kurzfassung_2012-03.pdf. Accessed 28 May 2018.

Calvert, K., Pearce, J. M., & Mabee, W. E. (2013). Toward renewable energy geo-information infrastructures: Applications of GIScience and remote sensing that build institutional capacity. *Renewable and Sustainable Energy Reviews, 18*, 416–429.

Counihan, J. (1975). Adiabatic atmospheric boundary layers: A review and analysis of data from the period 1880–1972. *Atmospheric Environment, 79*, 871–905.

Deutscher Wetterdienst (DWD). (2013). Wulf-Peter Gerth, Winddaten für Deutschland – Bezugszeitraum 1981–2000. Offenbach.

Diefenbacher, H. (2009). Zum Konfliktpotenzial Erneuerbarer Energien. soFid Sozialwissenschaftlicher Fachinformationsdienst. *Internationale Beziehungen/Friedens- und Konfliktforschung, 2*, 9–18.

European Commission. (2014). *Climate and energy framework for the year 2030*. https://ec.europa.eu/clima/policies/strategies/2030_en. Accessed 28 May 2018.

Gailing, L., & Röhring, A. (2016). Germany's Energiewende and the spatial reconfiguration of an energy system. In L. Gailing & T. Moss (Eds.), *Conceptualizing Germany's energy transition: Institutions, materiality, power, Space* (pp. 11–20). London: Palgrave Pivot.

Gove, B., Williams, L. J., Beresford, A. E., et al. (2016). Reconciling biodiversity conservation andwidespread deployment of renewable energy technologies in the UK. *PLoS One, 11*(5), e0150956. https://doi.org/10.1371/journal.pone.0150956.

Günnewig, D., Wachter, T., Nagel, D., Peters, W., Ahmels, P., Rehfeldt, K., Klinksi, S., Schweizer-Ries, P., & Zoellner, J. (2009). *Abschätzung der Ausbaupotenziale der Windenergie an Infrastrukturachsen und Entwicklung von Kriterien der Zulässigkeit* (Final report), Berlin, client: BMU, Forschungszentrum Jülich PTJ, pp. 1–199.

Huber, N., Hergert, R., Price, B., et al. (2017). Renewable energy sources: Conflicts and opportunities in a changing landscape. *Regional Environmental Change, 17*(4), 1241–1255.

Joint Research Center of the European Commission (JRC). (2010). *Interactive maps*. http://re.jrc.ec.europa.eu/pvgis/apps4/pvest.php. Accessed 10 Dec 2015.

Joint Research Center of the European Commission (JRC). (2015). *Clear sky indexes*. https://ec.europa.eu/jrc/en/about/jrc-site/ispra. Accessed 10 Dec 2015.

Kemfert, C., & Horne, J. (2013). *Good governance of the Energiewende in Germany: Wishful thinking or manageable?* Hertie School of Governance: Hertie school experts on the German federal elections 2013. http://www.hertieschool.org/fileadmin/images/Media_Events/BTW2013/20130820_Good_Governance_of_the_Energiewende_in_Germany_ClaudiaKemfert_Download.pdf. Accessed 31 July 2013.

Kienast, F., Huber, N., Hergert, R., et al. (2017). Conflicts between decentralized renewable electricity production and landscape services – A spatially-explicit quantitative assessment for Switzerland. *Renewable and Sustainable Energy Reviews, 67*, 397–407.

Klinski, S., Buchholz, H., Rehfeldt, K., et al. (2007). Entwicklung einer Umweltstrategie für die Windenergienutzung an Land und auf See. Endbericht. Umweltbundesamt, Berlin. http://tudresden.de/die_tu_dresden/fakultaeten/juristische_fakultaet/jfitur2/forschung/Umweltstrategie%20Wind%20Endbericht%20endg.pdf. Accessed 21 May 2014.

Krekel, C., & Zerrahn, A. (2017). Does the presence of wind turbines have negative externalities for people in their surroundings? Evidence from well-being data. *Journal of Environmental Economics and Management, 82*, 221–238.

Landesamt für Geoinformation und Landentwicklung (LGLN). (2014). *Digitale Geländemodelle*. http://www.lgn.niedersachsen.de/portal/live.php?navigation_id=11080&article_id=51746&_psmand=35. Accessed 9 Dec 2015.

Lovett, A., Sünnenberg, G., & Dockerty, T. (2014). The availability of land for perennial energy crops in Great Britain. *GCB Bioenergy, 6*, 99–107. https://doi.org/10.1111/gcbb.12147.

Marcheggiani, E., Gulinck, H., & Galli, A. (2013). Detection of fast landscape changes: The case of solar modules on agricultural land. In B. Murgante et al. (Eds.), *Computational science and its applications – ICCSA 2013. ICCSA 2013* (Lecture notes in computer science) (Vol. 7974, pp. 315–327). Berlin/Heidelberg: Springer.

Ministerium für Umwelt, Energie und Klimaschutz (MU). (2009). Verordnung über Schutzbestimmungen in Wasserschutzgebieten. SchuVO. http://www.nds-voris.de/jportal/?quelle=jlink&query=WasSchGebV+ND&psml=bsvorisprod.psml&max=true. Accessed 3 May 2018.

Niedersächsischen Landkreistag (NLT). (2014a). Naturschutz und Windenergie. Hinweise zur Berücksichtigung des Naturschutzes und der Landschaftspflege bei Standortplanung und Zulassung von Windenergieanlagen. http://www.nlt.de/pics/medien/1_1414133175/2014_10_01_Arbeitshilfe_Naturschutz_und_Windenergie__5__Auflage__Stand_Oktober_2014_Arbeitshilfe.pdf. Accessed 22 Oct 2014.

Niedersächsischen Landkreistag (NLT). (2014b). Regionalplanung und Windenergie. Empfehlungen des NLT zu den weichen Tabuzonen zur Steuerung der Windenergienutzung mit Ausschlusswirkung in Regionalen Raumordnungsprogrammen. http://www.nlt.de/pics/medien/1_1392281645/2014_02_06_Arbeitshilfe__Ergaenzende_Empfehlungen_NLT.pdf. Accessed 22 Oct 2014.

Palmas, C., Abis, E., von Haaren, C., et al. (2012). Renewables in residential development: An integrated GIS-based multicriteria approach for decentralized micro-renewable energy production in new settlement development: A case study of the eastern metropolitan area of Cagliari, Sardinia, Italy. *Energy, Sustainability and Society, 2*(1), 10.

Palmas, C., Siewert, A., & von Haaren, C. (2015). Exploring the decision-space for renewable energy generation to enhance spatial efficiency. *Environmental Impact Assessment Review, 52*, 9–17.

Peters, W. (2013). Erneuerbare Energien – Strategien für eine naturverträgliche Nutzung. In B. Demuth, S. Heiland, N. Wiersbinski, et al. (Eds.), *Energielandschaften – Energielandschaften der Zukunft? "Energiewende – Fluch oder Segen für unsere Landschaften?"* (BfN-Skripten) (Vol. 337, pp. 122–131). Bonn: Bundesamt für Naturschutz.

Price, J., Zeyringer, M., Konadu, D., et al. (2018). Low carbon electricity systems for Great Britain in 2050: An energy-land-water perspective. *Applied Energy, 228*, 928–941.

Region Hannover. (2012). Masterplan Stadt und Region Hannover. 100% für den Klimaschutz-auf dem Weg zu einer klimaneutralen Region bis 2050. http://www.hannover.de/Leben-in-der-Region-Hannover/Umwelt/Klimaschutz-Energie/Klimaschutzregion-Hannover/Masterplan-100-für-den-Klimaschutz/Ergebnisse. Accessed 8 May 2018.

Sánchez-Lozano, J. M., Teruel-Solano, J., Soto-Elvira, P. L., et al. (2013). Geographical InformationSystems (GIS) and multi-criteria decision making (MCDM) methods for the evaluation of solarfarms locations: Case study in South-Eastern Spain. *Renewable and Sustainable Energy Reviews, 24*, 544–556.

Sánchez-Lozano, J. M., García-Cascales, M. S., & Lamata, M. T. (2016). GIS-based onshore wind farm site selection using fuzzy multi-criteria decision making methods. Evaluating the case of southeastern Spain. *Applied Energy, 171*, 86–102.

Sathaye, J., Lucon, O., Rahman, A., et al. (2011). Renewable energy in the context of sustainable energy. In O. Edenhofer, R. Pichs-Madruga, Y. Sokona, et al. (Eds.), *IPCC special report on renewable energy sources and climate change mitigation* (pp. 707–790). Cambridge/New York: Cambridge University Press.

Staiß, F. (2007). Nutzung der solaren Strahlungsenergie nach §11 EEG. In F. Staiß, M. Schmidt, & F. Musiol (Eds.), *Vorbereitung und Begleitung der Erstellung des Erfahrungsberichtes 2007 gemäß §20 EEG, Report* (pp. 68–77). Stuttgart: ZSW.

Stoeglehner, G., & Narodoslawsky, M. (2009). How sustainable are biofuels? Answers and further questions arising from an ecological footprint perspective. *Bioresource Technology, 100*, 3825–3830.

Touma, J. S. (1977). Dependence of the wind profile power law on stability for various locations. *Journal of the Air Pollution Control Association, 27*, 863–866.

United Nations. (2018). *Climate Change – The Paris agreement*. http://unfccc.int/paris_agreement/items/9485.php. Accessed 3 May 2018.

von Haaren, C., Palmas, C., Boll, T., et al. (2013). Erneuerbare Energien – Zielkonflikte zwischen Natur- und Umweltschutz. In BBN (Ed.), *Neue Energien – Neue Herausforderungen: Naturschutz in Zeiten der Energiewende* (Jahrbuch Naturschutz und Landschaftspflege) (Vol. 59, pp. 18–33). Bonn: BBN.

Walter, A., Wiehe, J., & von Haaren, C. (in print). Naturverträgliche Energieversorgung aus 100% erneuerbaren Energien 2050 "EE100". BMU/BfN.

Watson, J. W. W., & Hudson, M. D. (2015). Regional scale wind farm and solar farm suitability assessment using GIS-assisted multi-criteria evaluation. *Landscape and Urban Planning, 138*, 20–31.

Zeyringer, M., Price, J., Fais, B., et al. (2018). Designing low-carbon power systems for Great Britain in 2050 that are robust to the spatiotemporal and inter-annual variability of weather. *Nature Energy, 3*, 395–403.

Regional Climate Regulation Capacities 13

Hermann Klug and Steffen Reichel

Abstract

In the past decades climate change impacts have become more pronounced, especially in the European Alps. There has been a spatio-temporal diversity of these impacts ranging from east to west, north to south, and with consideration of the orography, from low to high altitudes. The impacts of the present weather conditions and climate change on the environment and on humans show a great variety. As a consequence of dissimilar preconditions, mid-term and short-term adaptation planning approaches cannot simply be transferred from one place to another. They require a local to regional understanding of the drivers, impacts and responses on vulnerable ecosystem services, and regional regulation capacities and adaptation potentials. With respect to climate variables this chapter explores existing climate data repositories and demonstrates how these can be transferred into standardised data offerings, enabling harmonised searching, discovering, assessing, analysing, and processing. Diverse publicly available meteorological datasets, from weather stations in the alpine space, are transferred into an Open Geospatial Consortium Compliant Sensor Observation Service. This standard compliant data repository enables a post-processing with Python scripts which analyses the unified data records and displays them in charts for further interpretation. As a result, readers are able to compile and publicly share standard compliant data repositories and run queries on them for data exploration to estimate impacts on ecosystem services.

Keywords

Climate change · European Alps · Meteorological data · Data synthesis · Python

H. Klug (✉) · S. Reichel
Interfaculty Department of Geoinformatics – Z_GIS, University of Salzburg, Salzburg, Austria
e-mail: hermann.klug@sbg.ac.at; steffen.reichel@sbg.ac.at

C. von Haaren et al. (eds.), *Landscape Planning with Ecosystem Services*, Landscape Series 24, https://doi.org/10.1007/978-94-024-1681-7_13

13.1 Introduction

The Alps have been highlighted as being affected by climate change in a more pronounced way than other European regions (Nogués-Bravo et al. 2007). These changes are not evenly distributed in space (horizontally and vertically) and time. They depend on the specific landscape structures and patterns. Consequently, climate impacts on humans and the environment are different across the Alps (Auer et al. 2007). Moreover, climate changes are altering the functions of ecosystems (Nelson et al. 2013).

The impacts of, for example, flooding (Beniston 2012), storms (Etienne and Beniston 2012), or water scarcity (Hohenwallner et al. 2011), and the related environmental and human consequences from these hazards, are closely watched by local to national stakeholders. Related transdisciplinary efforts and existing non-alpine climate change platforms, that these stakeholders are working with, were documented within the Alpine Space program project C3-Alps (Klug et al. 2014). The produced online repositories were used in iterative stakeholder interactions for designing layout and functionality of a new user-centred knowledge inventory portal on climate change impacts and adaptation planning, tailored to the Alpine space.[1] Such user-adapted and spatially concrete information about climate change and the adaptation capacities of ecosystems and their services is a precondition for integrating climate change adaptation into landscape and spatial planning.

The need for an integrated information platform is illustrated by the fact that more than 350 datasets on climate change impacts and present adaptation strategies on ecosystem services have been developed across the Alps. However, the content focuses on reports and pictures and neglects detailed climate and weather information services on, for example, precipitation, temperature, wind speed, and wind direction. Furthermore, the information is often provided only offline, scattered across websites, or resides in poorly documented reports and in addition, data are often not internationally standardised. An environmental information system – as landscape planning can provide for regions and municipalities – could either import the relevant data from national platforms, or generate the relevant information and process it for use in planning and land use decision.

In this context, this chapter is about an integration and synthesis approach on climate and weather data. A synthesised cross-Alpine standard compliant inventory method is provided as an example. The approach can be adapted to other regions with some changes. It includes scripts and tools to setup a Sensor Observation Service (SOS) (OGC 2012), to insert datasets from available sources into the SOS, and to analyse and visualise the data using Python scripts. This practical approach provides publicly available datasets in the data formats Observations & Measurements (O&M) (OGC 2011; ISO 19156 2011), and Water Markup Language (WaterML) as a hydrological time-series data specialisation of O&M. This is in congruence with the International Organization for Standardization (ISO) and the

[1] http://www.sbg.ac.at/zgis/landscapelab/c3alps

World Meteorological Organization (WMO). Both are closely cooperating in their effort to facilitate a worldwide establishment of weather station networks.

13.2 Material and Methods

Having outlined the spatio-temporal differences of weather phenomena and climate change impacts and the possible alteration of ecosystem services across the Alps, we start with the setup of the spatial data infrastructure (SDI). A SOS from the 52°North repository is setup (Fig. 13.1). Publicly available datasets are automatically (or manually) downloaded from the accessible web resources. Provided XML and Python scripts transfer the data (semi-)operational into the SOS and visualises the datasets for further interpretation.

13.2.1 Setting up the Sensor Observation Service and Data Structures

The Sensor Observation Services have to be set up, in our case, as discussed in the 52°N SOS community.[2] Consequently, the SOS files[3] are downloaded and the

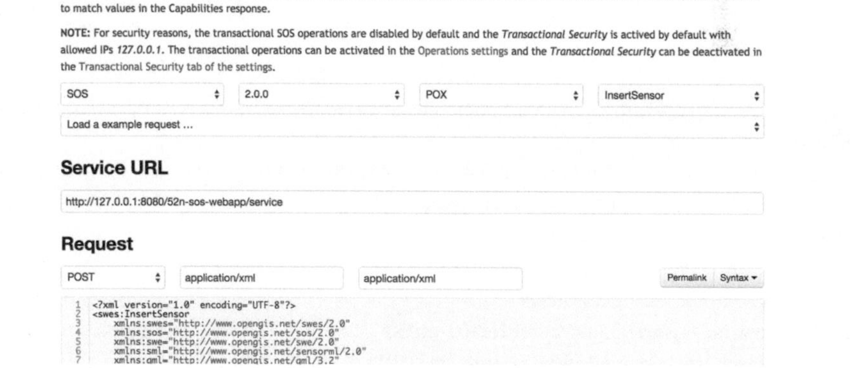

Fig. 13.1 Entering the data structure into the SOS

[2] http://www.52north.org/communities/sensorweb/sos

[3] http://www.52north.org/communities/sensorweb/sos/download.html

'52n-sensorweb-sos-bundle-4.3.0' installed according to the wiki reference.[4] For integration of the data structure into the SOS we set up the sensor metadata using the three examples. Extensible Markup Language (XML) scripts in the 'sensorml' folder.[5] The contents of every file is copy/pasted into input field of the 52°N SOS test client as outlined in Fig. 13.1.[6]

13.2.2 Available Weather Information Services and Datasets

Referring to the flooding, storm, and water scarcity examples in the case study, available precipitation, temperature, and wind speed and wind direction datasets were unified into the O&M 2.0 data format distributed via the SOS. Climate data repositories are organised by diverse data providers and on different organisational levels. National environment agencies, meteorological, and/or hydrological surveys, provide this information. In these repositories, datasets differ from raster, vector to text file representations with a dissimilar monitoring time scale. These may range in time scale from annual to monthly, daily average, or single measurements repeating every 5–15 or 30 min. Further repositories have resulted from joint global (IPCC,[7] NOAA,[8] WorldClim,[9] FAO,[10] NASA,[11] CRU[12]), European (INSPIRE[13]) or Alpine (HISTALP[14]) efforts or have been based on commercial interests (OGIMET,[15] UBIMET,[16] ZAMG[17]). As a result, the selection of data properties, including their temporal resolution, is dependent on the location of the interest and the phenomena under investigation.

[4] https://www.wiki.52north.org/bin/view/SensorWeb/SensorObservationServiceIVDocumentation#Installation

[5] http://www.sbg.ac.at/zgis/landscapelab/downloads/KlugReichel.zip

[6] http://www.YOUR-IP:8080/52n-sos-webapp/client

[7] http://www.ipcc-data.org/index.html

[8] http://www.ncdc.noaa.gov/data-access

[9] http://www.worldclim.org/download

[10] http://www.fao.org/nr/climpag/pub/EN1102_en.asp

[11] http://eosweb.larc.nasa.gov/cgi-bin/sse/sse.cgi?+s01

[12] http://www.cru.uea.ac.uk/data

[13] http://inspire-geoportal.ec.europa.eu

[14] http://www.zamg.ac.at/histalp/dataset/station/csv.php

[15] http://www.ogimet.com

[16] http://www.ubimet.com/en_INT

[17] http://www.zamg.ac.at/cms/en

13.2.3 Importing Climate Datasets into the SOS

This chapter focuses on the datasets, from which respective information should be extracted, in our case exemplary for the European Alps. Python scripts[5] are useful for transferring the below mentioned datasets into the installed SOS.

The National Oceanic and Atmospheric Administration (NOAA) repository is a global data pool. It is intended to be free and have unrestricted access for research purposes, education, and other non-commercial activities, for 18 surface meteorological parameters (including temperature, precipitation, and wind). Historical data from the global summary of the day (GSOD) repository are available for +9000 single stations from 1929 to present, while data from 1973 to present is almost complete for every station.[18] Within the script, daily average temperature values from GSOD are converted from Fahrenheit (°F) to Celsius (°C) and daily precipitation averages from inch to millimetre.

Meteorological Aviation Reports (METAR) are available at airports in encoded format for reporting weather information – mainly to pilots.[19] In contrast to GSOD the publication interval of the parameters is 30 min.

The freely available Historical Instrumental Climatological Surface Time Series of the Greater Alpine Region (HISTALP) database consists of, for example, monthly homogenised temperature (partly since 1760) and monthly precipitation data sums (partly since 1800) (Auer et al. 2007). We manually downloaded the monthly average datasets as homogenised series CSV export.[20]

13.2.4 Visualisation of Data Entries

The open source Python 2D plotting library 'matplotlib' has proven useful for a simple visualisation of data values.[21] We recommend using the Python Anaconda distribution and following their installation instructions[22] or alternatively, our readme instructions in the KlugReichel.zip.[5]

13.3 Results

The automated climate data assembly and processing provides a complex information base on past and present environmental conditions, which in comparison with ecosystem services might expose local regulation capacity to respond to stress

[18] http://www1.ncdc.noaa.gov/pub/data/gsod/, ftp://ftp.ncdc.noaa.gov/pub/data/noaa/isd-history.txt

[19] http://weather.noaa.gov/pub/data/observations/metar/stations/STATION-CODE.TXT (your station code, e.g. LOWS for Salzburg airport).

[20] http://www.zamg.ac.at/histalp/dataset/station/csv.php

[21] http://matplotlib.org

[22] https://store.continuum.io/cshop/anaconda

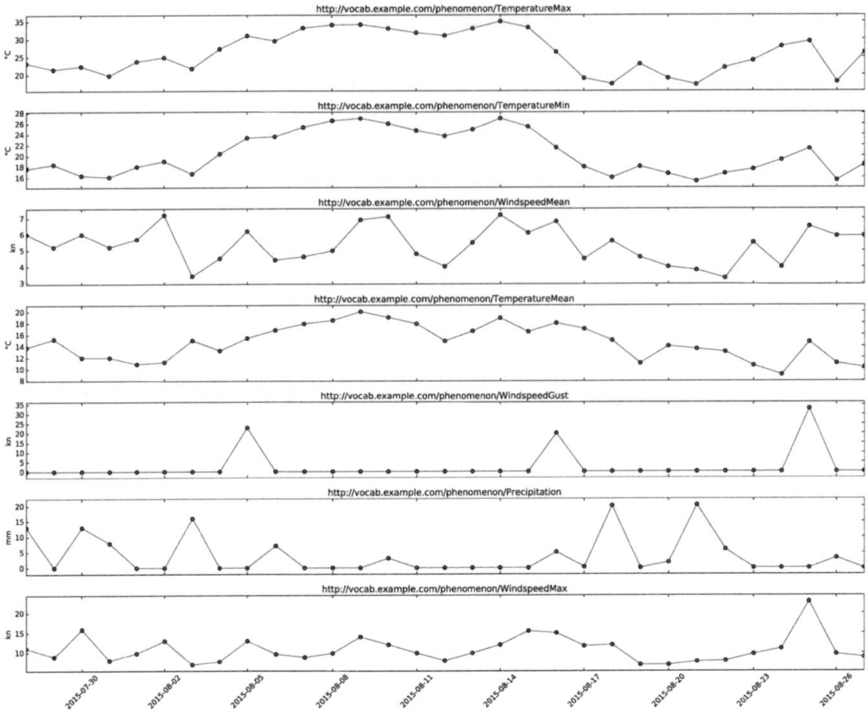

Fig. 13.2 GSOD dataset from Salzburg airport (LOWS)

during flood, storm, and water scarcity times. Targeted temperature, precipitation, wind speed and wind direction dataset are generated. With the unification of the HISTALP data (temperature, precipitation, pressure), GSOD (temperature, precipitation, wind speed), and the METAR datasets (wind speed, wind direction, temperature, and others), a standardised data repository could be established.

There is almost no long-term METAR data repository publicly available but OGIMET hosts information since 2005, which they are willing to share. Actual measurements on the NOAA data repository are discarded after 36 h. Thus, the developed script includes a permanent request on new METRA data every 30 min to set up a larger data repository.

With the 52°N SOS v4.3.0 user interface shown in Fig. 13.1, the values inserted can be retrieved using tailored XML queries. These XML scripts integrated into the provided Python plot script can be used for visualisations in the Python 2D plotting library 'matplotlib' as shown in Figs. 13.2, 13.3, and 13.4.

The visualisations in Figs. 13.2, 13.3, and 13.4 have been kept very simple but can be modified in colour, style, and layout. Especially the representation of the wind direction in Fig. 13.3 is less explorative as it could be with the wind rose extension to 'matplotlib'.[23]

[23] https://pypi.python.org/pypi/windrose/1.5

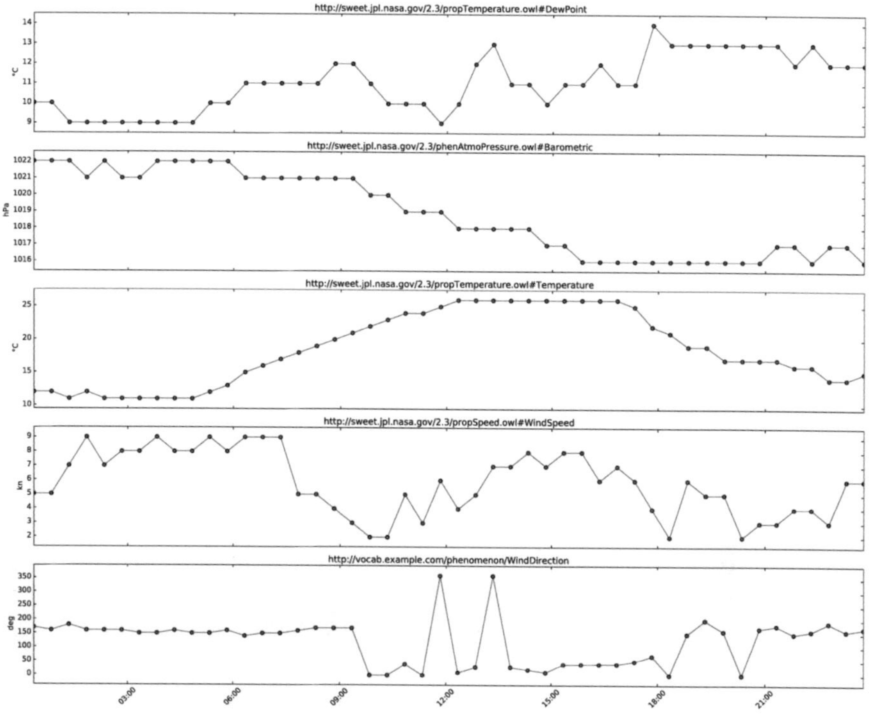

Fig. 13.3 METAR dataset from Salzburg airport (LOWS)

13.4 Discussion and Conclusion

The framework presented here updates existing landscape planning approaches for ecosystem services with an 'Automated Geosynthesis' approach based on climate and weather information. As demonstrated by Kmoch et al. (2015) and Klug and Bretz (2012), other hydrology or soil-related spatio-temporal datasets could also be shared and synthesised in a similar standardised and publicly accessible procedure. This geosynthesis is designed to combine the spatio-temporal components or elements and related processes to form an interconnected whole. As such, it is an entity greater than the sum of its single components. It should provide an improvement in the understanding of the processes and functions operating in landscapes. The automated geosynthesis should deliver locally adapted operational services and products that assemble distributed geospatial information for complex analysis. In turn this will enhance and support the qualified decision making process in landscape planning for ecosystem services.

Critical components in implementing an automated geosynthesis are the many local environmental parameters (e.g. relief, soils, flora, fauna, and hydrology). In the presented case study, only the meteorological information was used. Since neither the local setting nor the impacts and planning approaches are similar across the

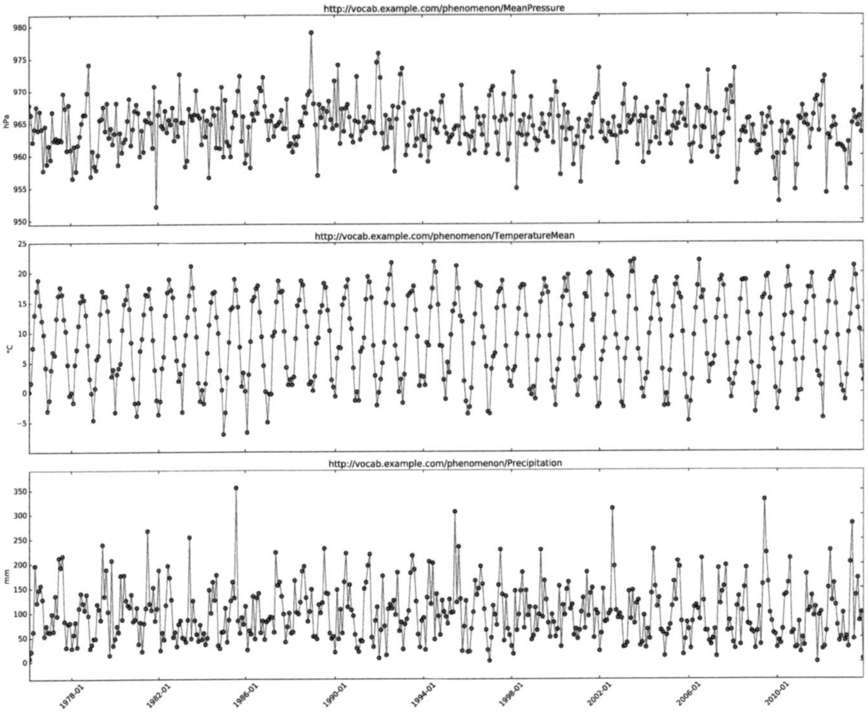

Fig. 13.4 HISTALP dataset from Salzburg airport (LOWS)

Alps, an immediate transfer of the study might be limited. However, the application of a standard compliant concept of connecting geospatial technologies is place independent and has been proven successful. Thus, the described standard compliant geospatial technologies that are independent from computer platforms can be recommended for any further study also outside of the Alps. They enable the interoperable connection of distributed data and metadata services (SensorML; OGC (2007b)) to ensure an exchange of machine- (XML) and human-readable sources (figures) for an increased awareness of past climate change, present weather phenomena, and climate change predictions such as those provided by datasets from EURO-CORDEX or ENSEMBLES (Jacob et al. 2014), which are used for short and mid-term ecosystem service planning approaches. The improved resource allocation and prioritisation, with tailor-made location specific information for mid-term planning support, could then be used as an operational service covering the European Alps.

The HISTALP and GSOD repositories setup include temporal ambiguities. Thus, the interpretation of both datasets is limited in its present form. Post-processing the daily data to average monthly values could help to analyse the datasets on a monthly scale. Nevertheless, temperature measurements are now available in a unique format, not separated into non-comparable degree Fahrenheit (GSOD) and degree Celsius (METAR, HISTALP). The same applies to precipitation information consistently converted from inches (GSOD) into millimetres (HISTALP, METAR). However, the temporal differences of the datasets require post-processing for a consistent daily comparison.

13.5 Outlook

Besides climate datasets, cross-domain environmental monitoring networks across the Alps regularly provide information on the state of the environment (Mirtl et al. 2015). Implemented wireless sensor networks integrating meteorological, hydrological, and pedological observations into a standardised spatial data infrastructure, provide the basis for exposing local to regional vulnerability hot spots in near real-time, for example for flooding (Klug et al. 2015). Immediately available observations from distributed monitoring networks enable forecasting applications incorporating real time *in situ* measurements for validation (Klug and Oana 2015). Coupled with integrated environmental modelling toolkits, installed as a Web Processing Service (WPS) (OGC 2015), they provide information on present and near-future environmental conditions exposing local regulation capacities on stressors such as those mentioned above (Klug and Kmoch 2015; Laniak et al. 2013; Schimak et al. 2010). Hence, the process discussed in this study should be further developed to incorporate spatial data infrastructures and publicly accessible data offerings. This would reduce the number of copied datasets and would enable and strengthen a transparent data holding and sharing within a holistic, transdisciplinary, and integrated analysis. This innovation would further avoid and reduce delay times in data provision and prevent stakeholders and practitioners wasting valuable time and resources in searching, obtaining, and pre-processing of datasets.

For creating the data inventory, services and their access, developers should follow the approach described by Klug and Kmoch (2014), who used a standard compliant web service architecture for inter-operable and thus a multi-purpose use of data and metadata. To ensure the proposed spatial data infrastructure complies with the Infrastructure for Spatial Information in Europe (INSPIRE) further work should follow the INSPIRE theme on Meteorological geographical features.[24] The data repositories should be registered in a Catalogue Service for the Web (OGC 2007a) using the ISO 19115 (2003) metadata standard and possibly extended to the XML schema according to the ISO 19139 (2007) standard. This would enable the transfer of information across platforms; again without copying and pre-processing of datasets but enable an immediate multi-purpose use of available resources.

Acknowledgements The authors wish to acknowledge funding for the SMART project from the Ministry of Business, Innovation, and Employment (MBIE), New Zealand.

[24] http://inspire.ec.europa.eu/theme/mf

References

Auer, I., Böhm, R., Jurkovic, A., et al. (2007). HISTALP—Historical instrumental climatological surface time series of the greater alpine region. *International Journal of Climatology, 27*, 17–46.

Beniston, M. (2012). Impacts of climatic change on water and associated economic activities in the Swiss Alps. *Journal of Hydrology, 412–413*, 291–296.

Etienne, C., & Beniston, M. (2012). Wind storm loss estimations in the Canton of Vaud (Western Switzerland). *Natural Hazards and Earth System Sciences, 12*, 3789–3798.

Hohenwallner, D., Saulnier, G. –M., Castaings, W., et al. (2011). *Water Management in a Changing Environment: Strategies against water scarcity in the Alps. Project outcomes and recommendations*. Interreg IVB, Alpine Space Programme, project 5-1-3-F. University of Savoie, Savoie.

ISO 19115. (2003). *Geographic information – Metadata*, p 140.

ISO 19139. (2007). *Geographic information – Metadata – XML schema implementation (encoding of metadata)*, p 111.

ISO 19156. (2011). Geographic information – Observations and measurements, p 46.

Jacob, D., Petersen, J., Eggert, B., et al. (2014). EURO-CORDEX: New high-resolution climate change projections for European impact research. *Regional Environmental Change, 14*, 563–578.

Klug, H., & Bretz, B. (2012). *Discover INSPIRE compliant harmonised soil data and services. Assessment and strategic development of INSPIRE compliant Geodata-Services for European Soil Data*. GS Soil 2009(06)-2012(05).

Klug, H., & Kmoch, A. (2014). A SMART groundwater portal: An OGC web services framework for hydrology to improve data access and visualisation in New Zealand. *Computational Geosciences, 69*, 78–86.

Klug, H., & Kmoch, A. (2015). Operationalizing environmental indicators for real time multipurpose decision making and action support. *Ecological Modelling, 295*, 66–74.

Klug, H., Kmoch, A., & Reichel, S. (2015). Adjusting the frequency of automated phosphorus measurements to environmental conditions. *Journal for Applied Geoinformatics*. GI_Forum, *2015*, 590–599.

Klug H, Oana L (2015) A multi-purpose weather forecast model for the Mondsee catchment. In: Jekel T, Car A, Strobl J et al (eds) Journal for Applied Geoinformatics. GI_Forum 2015(1):600–609.

Klug, H., Schörghofer, R., & Reichel, S. (2014). A climate change capitalisation knowledge inventory platform. *GI_Forum, 2014*, 57–66. https://doi.org/10.1553/giscience2014s57.

Kmoch, A., Klug, H., Ritchie, A. B. H., et al. (2015). A spatial data infrastructure approach for the characterization of New Zealand's groundwater systems. *Transactions in GIS, 20*(4), 626–641. https://doi.org/10.1111/tgis.12171.

Laniak, G. F., Olchin, G., Goodall, J., et al. (2013). Integrated environmental modeling: A vision and roadmap for the future. *Environmental Modelling and Software, 39*, 3–23.

Mirtl, M., Bahn, M., Battin, T., et al. (2015). *Research for the future – LTER-Austria white paper 2015 – On the status and orientation of process oriented ecosystem research, biodiversity and conservation research and socio-ecological research in Austria*. Vienna: LTER-Austria: Austrian Society for Long-term Ecological Research.

Nelson, E. J., Kareiva, P., Ruckelshaus, M., et al. (2013). Climate change's impact on key ecosystem services and the human Well-being they support in the US. *Frontiers in Ecology and the Environment, 11*, 483–893.

Nogués-Bravo, D., Araújo, M. B., Errea, M. P., et al. (2007). Exposure of global mountain systems to climate warming during the 21st century. *Global Environmental Change, 17*, 420–428.

OGC. (2007a). *OpenGIS Catalogue Service Implementation Specification (ISO 19115), v2.0.2, CSW 2.0.2*. The Open Geospatial Consortium (OGC). http://www.opengeospatial.org/standards/is. Accessed 20 June 2018.

OGC. (2007b). *Sensor Model Language (SensorML) Implementation Specification, 1.0.0.* The Open Geospatial Consortium (OGC). http://www.opengeospatial.org/standards/sensorml. Accessed 20 June 2018.

OGC. (2011). *Observations and Measurements – XML Implementation, O&M v2.0, O&M 2.0.* The Open Geospatial Consortium (OGC). http://www.opengeospatial.org/standards/is. Accessed 20 June 2018.

OGC. (2012). *OGC sensor observation service interface standard, v2.0, SOS 2.0.* The Open Geospatial Consortium (OGC). http://www.opengeospatial.org/standards/is. Accessed 20 June 2018.

OGC. (2015). *OpenGIS web processing service 2.0, WPS.* The Open Geospatial Consortium (OGC). http://www.opengeospatial.org/standards/wps. Accessed 20 June 2018.

Schimak, G., Rizzoli, A. E., & Watson, K. (2010). Sensors and the environment – Modelling & ICT challenges. *Environmental Modelling & Software, 25*, 975–976.

Greenhouse Gas Storage and Sequestration Function

14

Amy Thomas and Catharina Schulp

Abstract

Soil carbon storage and sequestration provides an ecosystem service (ES) through regulation of atmospheric carbon concentration, and associated climate conditions. Landscape planning to enhance this function can be most efficient through spatial assessment, followed by targeting of mitigation efforts to locations where soil carbon (C) stores are high, currently degrading, or there is good potential for sequestration. Direct measurement of C storage in the landscape is complicated by spatial variation in C storage in soil and biomass, hence in order to evaluate GHG storage and sequestration, calculations can be made based on available data for key factors influencing the offering of this ES. Complex process based models are available for prediction, however landscape planning approaches tend to favour simpler approaches utilising inventory figures and emission factors associated with soil type and vegetation. Inventory figures may include significant uncertainty, and represent average values for the relevant land use and soil, however there is an assumption that errors will balance out at larger regional, and certainly at national scale. If soil data on finer scale is available as well, such as in the case of well-defined wetlands, the certainty of results seems to be good enough for deducing measures. This chapter includes a case study on the application of a simple landscape scale model for assessment of potential placement of agri-environmental measures to enhance or protect soil C storage and sequestration in the landscape, as well as description of other models which may be preferred under different application requirements.

A. Thomas (✉)
Centre for Ecology & Hydrology, Environment Centre Wales, Bangor, Gwynedd, UK
e-mail: athomas@ceh.ac.uk

C. Schulp
Department of Earth Sciences, Faculty of Earth and Life Science, Environmental Geography Group, VU University Amsterdam, Amsterdam, The Netherlands
e-mail: nynke.schulp@vu.nl

C. von Haaren et al. (eds.), *Landscape Planning with Ecosystem Services*, Landscape Series 24, https://doi.org/10.1007/978-94-024-1681-7_14

Keywords
Soils · Carbon sequestration · Use values · Modelling tools · Assessment methods

14.1 Introduction

Soil carbon storage is increasingly considered an important factor in landscape planning, since recent global estimates suggest that the soil carbon pool is around three times the size of the vegetation carbon pool and double the atmospheric pool. As such, maintaining or increasing soil carbon storage fulfils an important function of greenhouse gas (GHG) storage and sequestration. Landscape planning to enhance this function can be most efficient through spatial assessment to identify areas with potential for increased storage, or with risk of potential losses. Targeting of mitigation efforts to these locations creates opportunities to avoid changes in land use and management where they may reduce carbon storage in soils, or to encourage change in land use and management at sites where soil carbon stores are currently degrading, or there is good potential for sequestration.

14.2 Definition and Concept (Box 14.1)

Box 14.1: Landscape Planning and Carbon Sequestration

Greenhouse gas (GHG) sequestration and storage refers to the removal of GHG from atmospheric pools and retention in terrestrial and aquatic pools. The sequestration (capturing and long-term storage) of GHG into non atmospheric pools acts to offset some of the increase due to anthropogenic emissions, stabilising atmospheric composition, and associated climate conditions. Storage of GHG also provides an ES through regulation of atmospheric composition, and associated climate conditions. In the context of landscape planning, terrestrial stores are more directly impacted than aquatic stores, although land management may affect transfer of C in sediments to fluvial and coastal stores.

Landscape planning approaches quantify GHG storage and sequestration in terms of quantification of current C storage in soil and vegetation, and quantitative or qualitative assessments of sequestration or loss. Where data are available, assessment may compare C storage in soil and vegetation from historic land use to equilibrium value for current land use to make an evaluation of whether storage is increasing or decreasing. Alternatively where data is unavailable, or to consider the impact of land use changes on the change in potential GHG storage, assessment may assume that C storage in soil and vegetation represents an equilibrium value for the current land use.

Sequestration and storage of GHG provides both a delivered and a utilised ES, which needs no human input to be beneficial for humans: by avoided increase in atmospheric greenhouse gas concentrations, humans benefit from greater global climate stability. Sequestration of carbon and other GHGs may be provided by terrestrial, freshwater or marine ecosystems. For managed terrestrial ecosystems, securing this indirect-use value may require a trade-off with direct-use provisioning services, for example through reduction in agricultural intensity or reduced removal of forestry products. Alternatively, an increase in soil carbon can improve agricultural productivity, through increased soil quality, nutrient availability and water holding capacity.

Since sequestration and storage of GHG is a global non-proximal service, i.e. impacts are felt via a global system and not restricted to local populations (Costanza 2008), it is most often characterised and assessed in terms of the delivered ES. Sequestration and storage of GHG may also be assessed from the perspective of service utilisation, through identification of people or locations which benefit from the service. This includes populations most directly vulnerable to climate instability, for example those in coastal or low-lying areas sensitive to sea level rise, such as parts of the Netherlands, or those in densely populated arid regions at risk of drought under decreasing precipitation, for example areas of southern Europe (Millennium Ecosystem Assessment 2005). Climate change may also influence stability of agricultural yields, making croplands other areas that benefit from climate change regulation. European States with greater utilisation of the ES benefit of climate stability may consider it a greater priority in normative value-based assessments of which ES to prioritise in landscape planning decisions, agri-environmental incentive schemes etc. Thus, divergences between areas which contribute a lot to climate change by GHG emissions and those that suffer from it create spatial problems of fit (Sects. 3.2 and 7.2.1). These can only be solved by multilevel planning and governance, which define the responsibility of each decision level by acknowledging the capacities for climate protection of each spatial unit as well as the global capacities.

Hotspots for carbon storage and sequestration capacity at a global scale are found in the tropics and boreal regions (Serna-Chavez et al. 2014). Measures for GHG storage and sequestration at a global scale may be targeted to these locations via the carbon compliance market clean development mechanism, whereby Annex 1 countries can pay for projects in developing countries and include any storage as part of carbon accounting (Bonn et al. 2014).

To assess the state of ecosystem function for GHG storage and sequestration, soil and biomass samples can be taken to measure directly the storage of carbon in terrestrial soil and vegetation pools. Changes in storage can be monitored over time based on these samples, or exchanges of GHG between terrestrial and atmospheric pools may also be measured. Owing to spatiotemporal variation both in storage in terrestrial pools and exchange with the atmosphere, it is costly to take sufficient samples to be representative, and difficult to upscale findings to calculate totals for flux and storage. Additionally, terrestrial carbon pools are very large compared to annual carbon fluxes from terrestrial ecosystems. Therefore many samples are needed to quantify changes using a sample-based approach. An alternative approach for directly measuring the state of ecosystem function for GHG storage and sequestration is through remote sensing (e.g., Net Primary Productivity as derived from MODIS) or other spectrometry approaches (Stevens et al. 2008).

In order to evaluate GHG storage and sequestration in the absence of direct measurements, calculations can be made based on available data for key factors influencing the delivery of this ES. In general, there is greater availability of data on variables controlling C storage such as soil and vegetation, as compared to direct measurements of soil C. Historic climate data may be available, however limited availability of historic land use and management means that current soil carbon storage is often assessed on the assumption of an equilibrium value.

As well as biotic controls in the form of vegetation, the abiotic environment is significant, since climate, soil and hydrology will act as controls on equilibrium soil C. Vegetation will control amount, type and depth of inputs to soil; these inputs include root exudates, leaf litter, and agricultural residues. Vegetation growth and inputs will themselves be affected by the abiotic controls of soil and climate. More directly, soil depth, texture and mineral composition are significant in determining the C storage capacity of the site. Climate parameters affecting temperature and soil moisture content of the soil will in turn affect the rate of breakdown of C, and thus whether this capacity can be reached.

These controlling variables are also important for assessing provisioning services, as discussed (in Chaps. 10 and 11). Separate assessment must be performed for provisioning services; nevertheless sites with potentially high value for C storage and sequestration are likely to also have high value for some provisioning ES, owing to the relationships between gross primary productivity (GPP) crop or forestry yield and C sequestration in vegetation and soil.

The evaluation of an assessment of the GHG emissions cannot yet draw from legislation and emission standards from the European Union or respective national legislation. But assessment results can be classified on ordinal and cardinal scales. This helps to judge the relevance of a certain case/site in comparison to others and to facilitate spatial priorities and targeting for climate mitigation measures. In addition socio-economic valuations are possible (see Sect. 14.3).

Historic land use and management factors are also significant for assessing C stocks in two ways. First, past land use and management changes influence whether a site is currently at equilibrium soil C, and if not, whether transition to that C storage equilibrium represents sequestration or emissions of GHG. Second, carbon sequestration and especially storage respond slowly to land use and management changes. Therefore, past land use patterns are often still reflected in the present-day spatial patterns of carbon storage and past land use can be a useful factor for calculating carbon stocks (Schulp et al. 2013).

As well as current storage and measurable exchange, longer term trends are significant in terms of the state of ecosystem function for GHG storage and sequestration. This requires comparison between states before and after specific human interventions (which may have already occurred); this may take the form of comparison between equilibrium values for the two different states, and thus consider the change in projected storage. Alternatively if sufficient data are available, comparison may be made between current storage and storage immediately before transition to the new state; however such assessment is likely to be more data intensive, and or require complex modelling.

A range of potential land use or management changes may be considered as part of landscape planning, hence it is useful to be able to quickly calculate the difference between projected C storage under multiple different states, to predict the most advantageous option. Impacts on other ecosystem services must also be considered; co-benefits may be possible, for example, stable and extensive vegetation cover or peat, which are important for C storage, will be likely to have high value for habitat conservation as well.

14.3 Socio-economic Valuation of a Market or Potential Use Value

The Paris climate accord is an agreement within the United Nations Framework Convention on Climate Change (UNFCCC) which deals with greenhouse gas emissions mitigation, adaptation and finance starting in the year 2020. It represents a commitment to keeping global temperature rise this century below 2 °C above pre-industrial levels and to pursue efforts to limit the temperature increase to 1.5 °C. The EU and member states have to take action to achieve these targets. In the context of land use, land use change and forestry (LULUCF), important measures and sources to be considered are: afforestation, reforestation and deforestation occurring since 1990, forest management, revegetation, cropland management, and grazing land management. These are accounted for in terms of change in C storage in soils and vegetation, generally using inventory values of average C for the relevant land use on the relevant soil type, and average rate of change between land uses. Accounting will have significant uncertainty, since these values are often based on expert judgements from limited available data, and average values could not be expected for all sites, however there is an assumption that errors will balance out at national scale.

Historically peatlands in Europe have been drained with the intention of enhancing provisioning services of agriculture and forestry. This was done at the expense of C sequestration and water regulation services provided by wetlands in good condition. Even in the absence of deliberate drainage, soil disturbance and water uptake associated with forestry, or trampling by livestock, can also degrade peat condition. Peat extraction for fuel and other market uses represents an additional provisioning service with direct economic value, in competition with C sequestration; European accounting for peatlands is incomplete (Joosten 2009), and may well represent a significant GHG source. LULUCF emission factors specified for wetlands in the IPCC 2014 inventory guidelines attribute high GHG fluxes to degraded and drained wetlands, largely due to CO_2 emissions, including losses from dissolved organic C, as well as N_2O, and CH_4 from drainage ditches (IPCC 2014). Emissions factors (EFs) vary according to type of wetland, condition and usage, but will generally result in calculation of significant GHG savings for restoration and rewetting;

although CH_4 EF will increase due to anaerobic conditions, this is offset by significant reduction in CO_2 EF. For example, complete GHG balance of a few polders in the Netherlands with different management regimes indicated that the total emission of a field (expressed in CO_2 equivalents) declines with decreasing management (Schrier-Uijl et al. 2014). In practice, emissions savings will be dependent on specific site conditions and the approach to restoration, there may be a large methane peak on rewetting, and it may take 4 years for CO_2 sequestration and decades for true peat formation to resume.

Nevertheless, rewetting of drained wetlands is a measure that could contribute considerably to emissions reduction. Individual EU states may decide whether to include this activity, however for most it would be a beneficial contribution to mitigation calculations, since drainage for the most part was undertaken prior to 1990, hence this category has significant potential to enable the EU to meet mitigation targets (Bonn et al. 2014). This provides a general legitimised goal for member states choosing to include this measure in their inventory to create regulations designating existing wetlands as protected areas and to target agri-environmental measures towards rewetting of drained peats.

The contribution of LULUCF to carbon accounting provides an incentive for the development of agri-environmental schemes which finance measures enhancing C sequestration in the terrestrial environment. This is reflected in the European Agricultural Development Fund. Member states are responsible for specific 'Rural Development Programmes' (RDPs) which must be approved by the commission and meet certain requirements. Supporting regulations explicitly consider carbon storage and sequestration in agriculture and forestry under the core objective of shifting towards a low carbon climate resilient economy. Additionally, cross compliance regulations include minimum requirements for soil cover and land management appropriate to site conditions to minimise erosion. Use of assessment approaches to spatially target agri-environmental funding supports the requirement to designate areas for interventions to ensure efficiency of schemes, and should improve scheme performance at the assessment stage.

An additional economic incentive for landscape planning to maintain and enhance carbon storage and sequestration exists in the form of carbon markets. Carbon storage in terrestrial soil pools is temporary, and will fluctuate due to natural and anthropogenic forcing, hence sequestration is not currently accepted for emissions trading (Lehmann 2007) and although forest-based carbon credits are traded globally, these are not applicable in the current EU Emission Trading Scheme (Ellison et al. 2014). Unofficial 'Voluntary Carbon Markets' may instead be used for financing measures to increase C sequestration in these pools, whereby private companies provide funding in the hope of improving their corporate social image. These are less well regulated, hence values stated for emissions savings may be less reliable, but schemes are easier to set up and implement, and there may be greater consideration for other ES such as biodiversity (Bonn et al. 2014; Newell et al. 2013).

Cumulatively, payments from carbon markets and agri-environmental schemes must offset both the costs of measure implementation, and the income forgone by the landowner, in order to be economically attractive.

14.4 Practical Relevance of Assessment of GHG Storage and Sequestration and Resulting Demands for Its Representation in Planning

The need for economic incentives to implement land use and management approaches which favour sequestration and storage provides an opportunity to spatially target the placement of these measures. Spatial targeting is a requirement of RDP agreements, since optimising location of interventions should improve cost effectiveness. As such, assessment of GHG storage and sequestration is of particular relevance for:

- Areas with currently high C storage or areas currently sequestering C which might otherwise economically favour land use change.
- Areas currently emitting C which would not otherwise economically favour land use change.

The results of GHG storage and sequestration assessment support the allocation of agri-environmental measures to such areas to protect existing C stores or support ongoing sequestration. Large carbon storage losses can occur very quickly after management change, whilst carbon sequestration is a much slower process. Due to this asymmetry, protection of existing stores may be considered more important, particularly where large stores are degrading or at risk of land use change.

From a market perspective, it is also relevant to calculate the potential use value, e.g. provisioning services which may be lost or reduced with changes to agricultural land use or management aimed at increasing C storage. Since payments from agri-enviromental schemes and elsewhere must cover this cost as well as that for implementing the measures, it is useful to target schemes to locations with lower agricultural productivity potential, as well as high GHG storage or sequestration potential. Financial payments may also consider that LULUCF accounting for C sequestration by forestry for EU states is currently capped at 3.5% of 1990 emissions; hence it does not make economic sense to incentivise afforestation or forestry management beyond this threshold, although this may not be reflected in RDP targets.

14.5 Method for Assessment of the Provided ES: GHG Storage and Sequestration

Given the global importance of GHG storage and sequestration ES, and the requirement to include calculation of C storage changes relating to LULUCF, there are many methodologies and models available for assessment of the provided ES. This section will focus on the Land Utilisation and Capability Indicator (LUCI) model and its application in landscape planning decisions, due to its simplicity, and ability to run quickly at national scale. Critically, LUCI is able to aid in landscape planning by identifying locations with existing C stores which may be at risk of degradation, for targeting of agri-environmental measures. The LUCI model was developed from Polyscape, which is described in Jackson et al. (2013).

Other ES models which could be applied for assessment of GHG storage and sequestration will be discussed, including the Total Integrated Model (TIM), Integrated Valuation of Environmental Services and Trade-offs (InVEST), and Artificial Intelligence for Ecosystem Services (ARIES).

The parameters, classification criteria and methodological workflow that guide the assessment of GHG storage and sequestration in LUCI are shown in Fig. 14.1. Within Europe, the LUCI model has currently only been tested for the UK and Greece, but can be applied using datasets available across most of the EU. The model requires spatial datasets in vector format. Parameters used in the classification are: the average soil C storage value expected for the relevant soil type, and the average equilibrium total C value expected for the current land use. A recent refinement of the model enables it to take into account impact of soil type on the average equilibrium total C value expected for the current land use, where data are available. These parameters can be identified from soil and land use maps, which should be used at a scale relevant for the scale of the ES assessment or landscape planning decision. Key datasets compatible with the model include the Corine inventory of

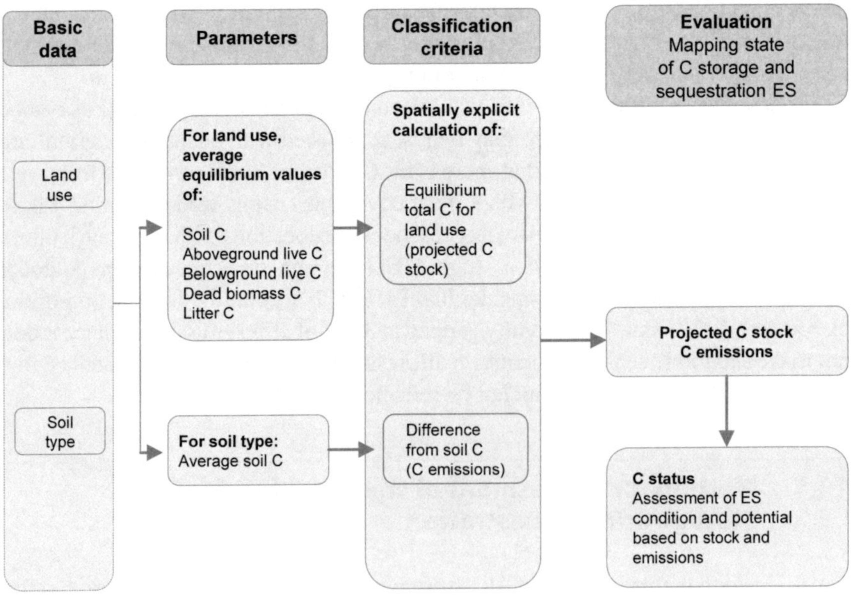

Fig. 14.1 Workflow and parameter criteria for the evaluation of carbon sequestration and storage

land cover, which is split into 44 classes, and available at a scale of 1:100,000; and European JRC soil data, available at a scale of 1:1,000,000. These data are suitable for modelling on state level but will probably be too coarse for regional or local landscape planning. The model can also be run using other datasets available at less coarse scales, and there is an overwrite function for sites where local stakeholder knowledge may contradict these large scale datasets, enabling more accurate assessments at small scale.

A gridded elevation data layer is also required, with calculations for all ES tools made at the spatial scale of these data. A key aspect of the utility of the LUCI model is the ability to operate at relevant spatial scales; the model can be run quickly at the landscape scales relevant to planning decisions, even at a 5 m by 5 m resolution, whilst the fine resolution enables small area modifications such as agri-environmental interventions to be simulated within the landscape.

The data listed in Fig. 14.1 are used for spatially explicit calculation of equilibrium total C for land use (i.e. a projected value for current C stock). Difference from soil C is then calculated (i.e. a value for change in soil C storage resulting from the current land use on that soil type) and mapped as an annual rate for each pixel, assuming a constant rate of transition. In addition to map outputs of numerical values, indicative maps are produced coded as traffic lights, where red (stop) indicates good ES performance so changes should not be made, whilst green (go) indicates an opportunity for ES improvement through interventions. Numerical values may be preferred by those making economic decisions about interventions, whilst the traffic light maps are intuitive and easily interpreted by stakeholders as part of decision making consultations.

Output from other tools within the LUCI model framework can be used to assess where there are trade-offs or win-win situations with other ES. As well as helping to identify where competition with provisioning services with high economic value may prevent uptake of measures to protect or increase soil C, these trade-offs can further optimise targeting of agri-environmental measures to locations where more than one of the core aims of the RDP are met, for example improvements in water quality.

Explicit economic valuation of the ES is not included in this assessment approach; rather, the intention is that the tool may be used to identify locations where investment in this ES would have greatest benefit. Additionally, trade-off assessment against agricultural productivity can be used to guide where greater economic incentives may be required. Ultimately economic assessment should then be based on thresholds for C mitigation, and whether the same mitigation value could be achieved with lower economic incentive requirements at alternative sites.

The LUCI model has been applied to assist the Welsh Government in evaluating spatial placement of agri-environmental payments from the Glastir scheme, along with a suite of other models, and farm specific information gathering. The Glastir scheme is a sustainable land management scheme, funded by the RDP for Wales, aimed at providing value for money on environmental goods and services. The entry level scheme is open to all farmers, whilst the advanced level scheme is targeted to locations best able to meet key objectives.

Advanced level payments will be applicable to farms identified as having good potential to achieve outcomes meeting Welsh Government priorities, which align with the RDP. Sites where the LUCI model identifies high C storage or C stores at risk of depletion may be suitable for advanced scheme measures for C storage and sequestration. The other key Glastir priorities are water management and biodiversity, so sites where trade-off analysis in LUCI identifies opportunities to improve water quality (through nutrient removal), local flood mitigation (through reduced runoff) or potential to enhance biodiversity, will further support evaluation of placement of scheme measures relevant to these objectives. Even greater value for money can be achieved by targeting interventions to farms where these opportunities are co-located with those to protect C stores or enhance C sequestration (for multifunctionality optimisation see Chap. 26). Requirements for farms receiving Glastir payments include:

- Cross compliance with regulations for keeping land in good agricultural and environmental condition, through protection of soil, water and habitats
- Specific Glastir management options designed to specific objectives of the RDP

Contracts for advanced payments will require specific Glastir management options to be applied for delivery of objectives which can most effectively be delivered in that area. In this way, spatial targeting of measures aims to give greater benefits in advanced areas, resulting in good value for money from the scheme. This approach to landscape planning does not require specific economic valuation of ES provided by the interventions, but instead is about identifying locations where maximum benefits can be realised from scheme payments. Alternative approaches to payments for ES such as voluntary and global carbon markets would require explicit economic valuation; for these markets TIM or InVEST would be more appropriate decision making tools. The economic value of C storage and sequestration may be considered equal to the societal cost of the same amount of C; this will vary over time with climate change and market forces (Box 14.2).

Box 14.2: Overview of Spatially Explicit Approaches for Carbon Stock Modelling

Numerous other spatially explicit ES models with carbon modules are available; these vary in terms of data requirements, outputs and approach.

InVEST (Integrated Valuation of Environmental Services and Trade-offs) (Bagstad et al. 2013)

InVEST is similar to LUCI in terms of workflow; a calculation is made by comparing total C for current land use to total C for a future land use, as required by LULUCF. Carbon storage for soil type under natural conditions is not considered, and the model is not intended for identification of sites with degrading C stores. A module for explicit calculation of economic valuation can be included if appropriate data on future carbon prices are available.

TIM (total integrated model) (Bateman et al. 2014)

(continued)

Box 14.2 (continued)

The C modules in TIM are more complex, with separate C approaches for forestry and agricultural C storage and sequestration. Agricultural C is based on the cool farm tool, which performs complete GHG accounting based on land use and management, requiring data on livestock, fertiliser, residue management, transport, energy use etc. the tool does not calculate or identify C losses from drained fens or other organic soils. Forestry C is calculated using the CARBINE model, which simulates tree growth according to species and management, and uses this to calculate C exchanges between pools, including detailed treatment of different harvested wood products. Based on forestry and agricultural C exchanges, TIM then calculates economic values for C sequestration based on a range of estimates for current and future carbon prices.

ARIES (Artificial Intelligence for Ecosystem Services) (Bagstad et al. 2013)

ARIES is a more complex model, which requires a detailed set of input data (including; soil C:N ratio, pH and oxygen conditions, vegetation type composition and stage, and management data on tillage and biomass removal rate) to achieve full potential, although the model can be run with data for only soil type and annual precipitation. C exchanges and uncertainty associated with estimates are modelled using Bayesian networks and probability distributions based on available data or expert opinion, depending on what input data are available.

CLUE-sinks (Schulp et al. 2008)

CLUE-sinks is a bookkeeping model of carbon sequestration in soils (including biomass for forests as well). Emission/sequestration is defined by an emission factor; this is a region-specific, land use type specific amount of sequestration / emission per km^2 per year. The emission for a grid cell is equal to the emission factor. When the land use changes, the emission factor changes to the emission factor of the new land use type. Deforestation causes loss of carbon from biomass. Other factors influencing carbon emission and sequestration are the amount of carbon already present in the soil and the age and management regime of forests. The model is frequently used in EU-scale projects, e.g. Eururalis and VOLANTE, and has recently been applied to estimate the impact of no net loss policy options for the European Commission (Tucker et al. 2013).

MANUELA-climate (Saathoff et al. 2013)

MANUELA-climate is a tool for modelling GHG-storage and emission potentials on the farm scale, using in principle the same indicators as the other models described above. A special feature of the model is that it is designed for the local (implementation) scale in order to supply information on farm-specific climate services or impacts. This information can be used for support

(continued)

Box 14.2 (continued)
of climate-friendly farm management. The model assesses CO_2 emissions from grassland conversion to cropland and peatland cultivation, as well as N_2O emissions from nitrogen fertilisation. As input data, the CO_2 tool requires a classification of soil types according to soil organic carbon storage. The N_2O tool relies on farm data concerning fertilisation. The uncertainties of the results for N_2O are sometimes high, so it is not recommended for use in landscape planning. However, the CO_2-assessment method has been applied via GIS analysis for the landscape plan of Hamburg-Harburg (Germany) and a biosphere reserve nearby (see von Haaren et al. 2012).

Carbon emission and storage values given by LUCI take a Tier 1 IPCC approach similar to the EF approach and as such should not be considered an exact prediction for a specific site. Values are based on expert judgement, and are subject to significant uncertainty, as well as site specific variation with management, land use history, and other factors for which data cannot be entered or may not be available. Given the importance of wetland areas containing significant C stores, it may be appropriate for more in depth assessment than IPCC Tier 1 approaches to be conducted, incorporating data on peat depth, local hydrology and small scale interventions. Many currently available complex soil C models applicable to other areas do not give adequate wetland representation, and this may be considered an important area for future research. Although exact values cannot be calculated, by identifying organic soils currently under land use which is likely to result in C losses, LUCI and similar modelling approaches can highlight sites which are likely to be detrimental to LULUCF inventory and may be deteriorating in terms of ES for C storage and sequestration.

The LUCI model does not include capability to map flow to the end user of ES, however the spatial relationship between the service and beneficiaries may be considered unimportant in the context of GHG sequestration, as a global non-proximal service. TIM and ARIES include capability to map C emitters as beneficiaries of the C sequestration and storage ES, and to calculate the overall local C budget. In some cases, separate modelling approaches may be required for different beneficiary groups, where different variables or perspectives need to be taken into account, and this can significantly increase the complexity of the problem.

14.6 Impact Assessment

Impacts on the ES of GHG storage and sequestration in the landscape are brought about by changes in state, which reflect action from drivers and pressures (from Chap. 3 of the book) as indicated in Fig. 14.2. Drivers and pressures thus influence both the provision of and demand for GHG sequestration and storage ES. For

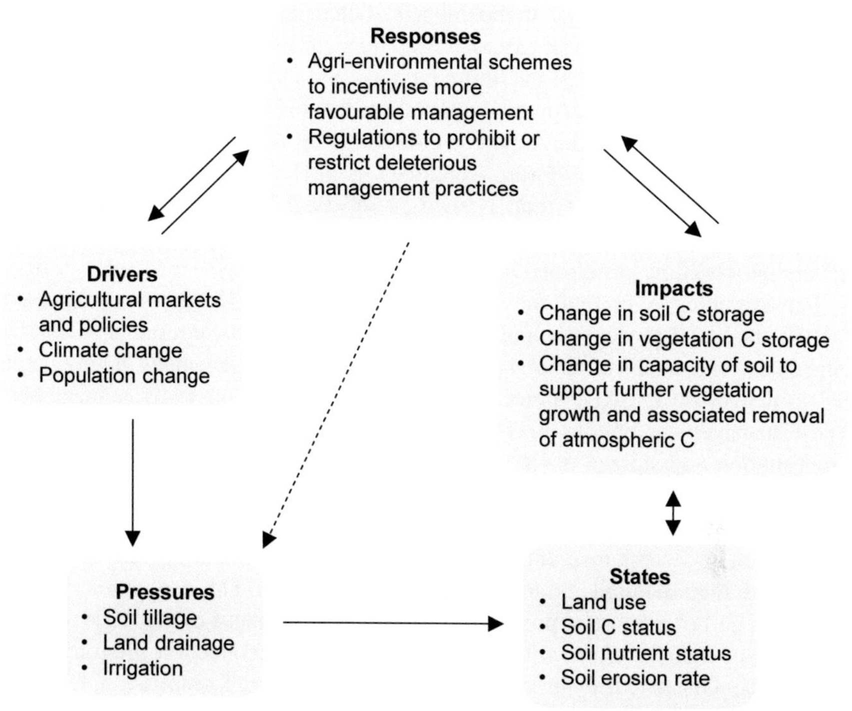

Fig. 14.2 Drivers, pressures, states, impacts and responses as relevant for GHG sequestration and storage in the landscape. Following the Ecosystem Services in Planning (ESIP) Framework

example, potential impacts of climate change on demand (increased global sensitivity to further climate change) and provision (warming may increase rates of soil C decomposition, whilst changes in temperature and precipitation patterns may increase or decrease inputs from vegetation to soil). Drivers of agricultural markets and policies will affect the pressures land is subject to, by dictating which land use and management is most economically beneficial. Land use and management pressures such as vegetation change, changes in stocking density, tillage, drainage, drain blocking or irrigation may then cause a change in state, with impacts relevant for GHG sequestration and storage as indicated in Fig. 14.2. As well as impacts on global climate from change in C sequestration and storage, soil quality in terms of nutrient status and water holding capacity may also be affected, with knock-on effects on agricultural productivity, which may in turn bring about responses which influence the pressure on land.

Impacts may be framed as the differences in the landscape due to human intervention, as compared to a baseline natural state, or as the difference from the current situation to a new scenario following (additional) human intervention. To decide where land use change, or agri-environmental measures to alter land management may be most beneficial, it is useful to assess 'impacts' as the difference due to human intervention, as compared to a baseline natural state (as described in Sect.

14.5). However, to then simulate the impacts of changes being considered as part of landscape planning, it is necessary to assess 'impacts' as the difference between the current situation and (a range of) future land use and management scenarios. LUCI is under development to perform calculations of the difference between equilibrium values for two scenarios internally. The current model version can simulate the current land use, and a range of future scenarios separately, and the user can then compare difference. Alternative models such as INVEST may be applied to directly calculate difference between equilibrium C for different land use scenarios, based on similar workflow as outlined in Fig. 14.1.

For example, as part of the Glastir Monitoring and Evaluation Programme (MEP), the LUCI model was applied to simulate the impacts of a range of the specific Glastir management options identified by Welsh Government as having potential to deliver on the RDP objectives of carbon storage and water management. These management options may be regarded as 'pressures' affecting state of C sequestration and storage in the landscape as indicated in Fig. 14.2. The specific measures simulated by LUCI as part of the evaluation were afforestation in streamside corridor and woodland edge locations; change in C storage was calculated by comparison of C stock map outputs simulated for baseline land use and future scenarios with high medium and low uptake of the measures. This evaluation was performed as part of a range of projections for different elements of the RDP, using an ensemble of models, along with an intensive field data collection programme, as part of the Glastir MEP, in compliance with European Commission Common Monitoring and Evaluation Frame-work (CMEF) for the RDP for Wales.

As the LUCI model is relevant at national, catchment or field scales, and can be run for any time period with available data, the model could equally be applied to identify the impacts of historic or projected future land use change at whatever scale required. Fast computation and easily interpretable output make the model particularly useful for real time interactive scenario analysis, in which farmers and other stakeholders can suggest locations for land use or management changes and receive feedback on potential implications for ES such as carbon storage and sequestration. The model is not currently able to simulate impacts on C sequestration and storage from some small scale landscape interventions such as blocking of peat drains. Additionally, some Glastir interventions such as cessation of nutrient inputs to pasture, which target improvements in other ES, have not been simulated with the LUCI carbon tool, due to a lack of scientific consensus on impacts on soil C storage and sequestration.

References

Bagstad, K. J., Semmens, D. J., & Winthrop, R. (2013). Comparing approaches to spatially explicit ecosystem service modeling: A case study from the San Pedro River, Arizona. *Ecosystem Services, 5*, 40–50.

Bateman, I., Day, B., Agarwala, M., et al. (2014). *UK national ecosystem assessment follow-on. Work package report 3*. Cambridge: Economic value of ecosystem services. UNEP-WCMC, LWEC.

Bonn, A., Reed, M. S., Evans, C. D., et al. (2014). Investing in nature: Developing ecosystem service markets for peatland restoration. *Ecosystem Services, 9*, 54–65.
Costanza, R. (2008). Ecosystem services: Multiple classification systems are needed. *Biological Conservation, 141*(2), 350–352.
Ellison, D., Lundblad, M., & Petersson, H. (2014). Reforming the EU approach to LULUCF and the climate policy framework. *Environmental Science & Policy, 40*, 1–15.
IPCC. (2014). *2013 Supplement to the 2006 Guidelines for National Greenhouse Gas Inventories: Wetlands* (Wetlands Supplement).
Jackson, B., Pagella, T., Sinclair, F., et al. (2013). Polyscape: A GIS mapping framework providing efficient and spatially explicit landscape-scale valuation of multiple ecosystem services. *Landscape and Urban Planning, 112*, 74–88.
Joosten, H. (2009). *The global peatland CO2 picture*. Ede 33: Wetlands International.
Lehmann, J. (2007). A handful of carbon. *Nature Geoscience, 447*, 143–144.
Millennium Ecosystem Assessment. (2005). *Millennium ecosystem assessment synthesis report. Natural assets and human well-being*. Washington, DC: World Resources Institute.
Newell, R. G., Pizer, W. A., & Raimi, D. (2013). Carbon markets 15 years after Kyoto: Lessons learned, new challenges. *The Journal of Economic Perspectives, 27*(1), 123–146.
Saathoff, W., von Haaren, C., Dechow, R., et al. (2013). Farm-level assessment of CO2 and N2O emissions in Lower Saxony and comparison of implementation potentials for mitigation measures in Germany and England. *Regional Environmental Change, 13*(4), 825–841.
Schrier-Uijl, A. P., Kroon, P. S., Hendriks, D. M. D., et al. (2014). Agricultural peatlands: Towards a greenhouse gas sink – A synthesis of a Dutch landscape study. *Biogeosciences, 11*, 4559–4576.
Schulp, C. J. E., Nabuurs, G. J., & Verburg, P. H. (2008). Future carbon sequestration in Europe – Effects of land use change. *Agriculture, Ecosystems and Environment, 127*, 251–264.
Schulp, C. J. E., Verburg, P. H., Kuikman, P. J., et al. (2013). Improving national-scale carbon stock inventories using knowledge on land use history. *Environmental Management, 51*, 709–723.
Serna-Chavez, H. M., Schulp, C. J. E., van Bodegom, P. M., et al. (2014). A quantitative framework for assessing spatial flows of ecosystem services. *Ecological Indicators, 39*, 24–33.
Stevens, A., van Wesemael, B., Bartholomeus, H., et al. (2008). Laboratory, field and airborne spectrometry for monitoring organic carbon content in agricultural soils. *Geoderma, 144*, 395–404.
Tucker, G., Allen, B., Conway, M., et al. (2013). *Policy options for an EU no net loss initiative. Report to the European Commission*. London: Institute for European Environmental Policy.
von Haaren, C., Saathoff, W., & Galler, C. (2012). Integrating climate protection and mitigation functions with other landscape functions in rural areas: A landscape planning approach. *Journal of Environmental Planning and Management, 55*(1), 59–76.

Landscape Aesthetics Capacity as a Cultural Ecosystem Service

15

Sónia Carvalho Ribeiro, Olaf Schroth, Eva Konkoly-Gyuró, Johannes Hermes, Thiemen Boll, and Christina von Haaren

Abstract

This chapter explores established theoretical and empirical work to identify possible indicators to represent landscape aesthetics capacity (LAC) in landscape planning. Throughout this chapter we argue that visual concepts from landscape perception/preferences studies (formed either on an individual or collective basis), together with experiences from implementing Landscape Character Assessments (LCA) throughout Europe, might help in developing frameworks for the assessment of Cultural Ecosystem Services (CES). When compared to provisioning or maintenance/regulating ecosystem services (ES), frameworks for the application of CES are lagging in development. Landscape aesthetics capacity is conceptualized here as delivered ES, which are central to the everyday life of people. The concepts we focus on are derived from landscape preference studies. The empirical cases explored are from LCAs in the United Kingdom

S. Carvalho Ribeiro (✉)
Universidade Federal de Minas Gerais, Belo Horizonte, Brazil
e-mail: soniacarvalhoribeiro@cart.igc.ufmg.br

O. Schroth
Hochschule Weihenstephan-Triesdorf, Freising, Germany
e-mail: olaf.schroth@hswt.de

E. Konkoly-Gyuró
Institute of Forest Resource Management and Rural Development, University of Sopron, Sopron, Hungary
e-mail: konkoly-gyuro.eva@uni-sopron.hu

J. Hermes · C. von Haaren
Institute of Environmental Planning, Leibniz University Hannover, Hannover, Germany
e-mail: hermes@umwelt.uni-hannover.de; haaren@umwelt.uni-hannover.de

T. Boll
Dezernat für Natur- und Landschaftsschutz, Fischerei, Bezirksregierung Münster, Münster, Germany
e-mail: thiemen.boll@bezreg-muenster.nrw.de

C. von Haaren et al. (eds.), *Landscape Planning with Ecosystem Services*, Landscape Series 24, https://doi.org/10.1007/978-94-024-1681-7_15

(UK) and Hungary, from the Landscape Preferences Spatial Framework in Portugal, and from a formal method for mapping and assessing the visual landscape in Germany. There is also a brief overview of current methodological approaches and suggested indicators regarding the utilization of CES. Finally, the chapter emphasises the ways in which landscape aesthetics capacities can be incorporated into planning, by selecting a group of robust indicators (based on theory as well as on our case studies) that could be applied in different European countries.

Keywords
Landscape aesthetics · Cultural ecosystem services · Landscape character assessment

15.1 Introduction

Landscapes, by the simple fact of being out there, provoke feelings in people. The ability of a particular landscape to fulfill human aesthetic needs and desires is described here as Landscape Aesthetics Capacity (LAC). It encompasses a set of immaterial and material landscape characteristics or features that fulfill a common core set of people's aesthetic values (non-use values in the terminology of Chap. 4).

LAC is unarguably one very important aspect of Cultural Ecosystem Services (CES) . Different authors have shown that the perception and appreciation of nature or natural elements in general, and of beautiful landscapes in particular, has positive impacts on human health and well-being (e.g. Russel et al. 2013; Abraham et al. 2009; Kaplan 2001; Kahn et al. 2008). In addition to being regarded as one class of CES, the visual landscape with its aesthetic quality provides the setting for all outdoor recreation activities. It therefore influences the benefits people obtain from it. Furthermore, research has shown that people can highly value a landscape, simply because it exists, even though they never visit, or cannot visit because for example, of restricted access (e.g. Boll et al. 2014).

When addressing LAC we need to deal with two interwoven dimensions. The first is the perspective of the users: the aesthetic perception and preferences of people, which may or may not differ from individual to individual depending on the use they make of the landscape. Individual, subjective preferences may be triggered by different landscape usage for instance by a farmer or a tourist, or by individual dispositions and experience e.g. appreciating landscape from out of an office window. In addition, very similar landscape preferences may be held amongst communities (collective preferences e.g. landscape identity or stewardship cared-for landscapes) and core preferences may even be common to all humans – with cultural variations. For example, Steinitz (2010) shows that residents and tourists in the Valencia region of Spain share common preferences.

The second dimension is the actual physical landscape attributes that may be appreciated or appeal to some people but not to others.

The LAC of physical landscapes can be assessed either by the individual, local users themselves (user-centered evaluation) or modelled, based on knowledge about collective preferences without including the actual users. The best option in landscape planning would is to combine both dimensions and relevant methods are discussed later in this chapter.

Humans need to have the opportunity be in an aesthetically pleasing landscape. This is a necessity for human well-being and health. Green spaces, for example, contribute to a healthy living environment (Croucher et al. 2007; Waltert et al. 2011; Schipperijn et al. 2010). LAC is also a precondition for visual amenities linked to outdoor recreation. In the context of this book, landscape aesthetics capacity is considered as a delivered Ecosystem Service (ES) whereas recreation is considered as a utilized ES.

The multiple relationships between people and their surrounding physical landscape settings, as well as the subjective meanings people associate with these physical settings, can offer valuable knowledge for enhanced planning and management. This, in turn, increases support for including CES such as aesthetics into planning (Opdam et al. 2001; Antrop 2005; von Haaren et al. 2008).

There is a vast body of research addressing landscape preferences, but it is specifically framed at the local scale. This fact raises concerns about the generalizability (Cassatela and Peano 2011) of landscape-based indicators between different scales of analysis. However, a multi-scale assessment of peoples preferences would be challenging (van Zanten et al. 2014). In spite of these limitations, the rich theoretical and empirical work on landscape preferences/perceptions, as well as on the different Landscape Character Assessments (LCAs) throughout Europe, should not be thought of as just a collection of case studies. We argue instead, that exploring the diversity of methods for assessing landscape preferences expressed in the literature, might aid in the development of a suitable framework for assessing the roles and values of landscapes and their ‘material and immaterial’ elements in the provision of CES.

Previous work has shown that visual concepts such as stewardship (Ode Sang and Tveit 2013), historicity, ephemera, coherence/disturbance, diversity (Ode und Miller 2011), and naturalness (Ode et al. 2009), are important drivers of landscape preferences/perceptions (Daniel 2001; Dramstad et al. 2006; Tveit et al. 2006).

The authors of this chapter consider it important to distinguish between perceptions and preferences. Preferences by people, either individual or collective, are based on pre-cognitive responses to landscape features, elements or characteristics which generate feelings of like or dislike (Antrop 2000; Surova and Pinto-Correia 2008; Swanwick 2009; Carvalho Ribeiro et al. 2013). Landscape preference studies throughout Europe have shown that preferences for certain landscape characteristics, or landscape attributes, are likely to differ depending on the use of the landscape (e.g. the preferences of someone picking berries may be different from the same individual's preferences for walking – the functional use shape our preferences) (Tahvanainen et al. 2001).

Perceptions, on the other hand, are cognitively based. Perception is the requirement of aesthetical judgment, which results in decisions about preferences. Antrop

(2000:19) defines perception, “as a complex learning process … [Perception] analyses the observation immediately and interactively and links the results with our knowledge and past experience”. Figure 15.1 schematically shows the processes embedded into perception/preferences by analyzing the mechanism of aesthetical judgement. According to Szerdahelyi (2003), perception builds upon an individual’s sensorial and cognitive system to ‘read’ a certain phenomenon. However, as Fig. 15.1 shows, preference and perception are neither synonyms nor two separate dimensions. Indeed, perception is the cognitive basis that influences our preferences. Both perception and preferences are subjective, mental constructions.

It is important to point out, that perception – the way people perceive landscapes – is influenced but not determined by physical landscape elements (Jacobs 2011). On the contrary, preferences are more straightforward in their relationship to physical landscape characteristics (such as liking or disliking a certain characteristic/attribute of the landscape dependent on the use e.g. sightseeing, hunting, picking mushrooms, bird watching).

However, landscape perception and landscape preferences are not only based on subjective factors. Biological, cultural, and individual factors each influence a person’s landscape preferences and perceptions (Bourassa 1999). Biological factors are intersubjective as they are based on evolutionarily developed innate dispositions, which means they are the same for everybody (e. g. prospect refuge theory by Appleton (1975)). This justifies, for example, human preferences for savannah like landscapes (Wilson 1986). Cultural factors are based on social values and cultural norms, which are likely to differ amongst societal groups. Individual factors are based on previous individual experiences, expectations, needs, hopes, fears, values and moods.

Following on from this therefore, compared to the other ecosystem services which are mostly measurable, CES such as aesthetics are hard to quantify. There are consequently difficulties in applying quantitative, universal assessment standards for CES which would allow assessments based on strong data types such as a ratio scale. This is why aesthetic capacity, contrary to its importance, often lacks attention in ‘formal’ planning as normally the data are ordinal scaled and often only locally valid.

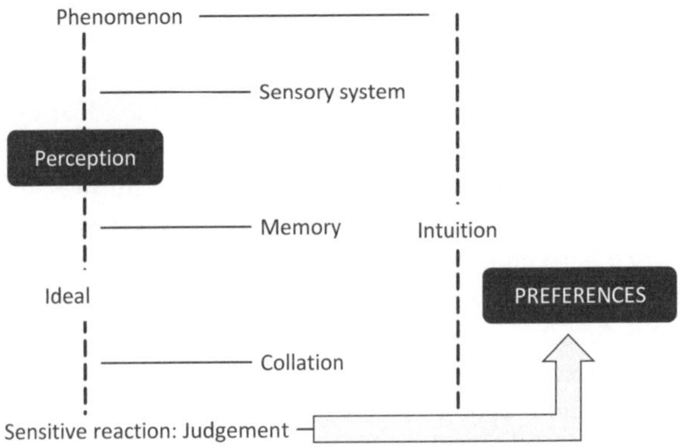

Fig. 15.1 Mechanisms of aesthetical judgement. (Adapted from Szerdahelyi 2003)

15.2 Integrating Landscape Aesthetics Capacity into Multi Scale Planning and Policy

Besides the European Landscape Convention, in the European Union (EU), both the Common Agricultural Policy (CAP Pillar II) and the EU Biodiversity Strategy to 2020 recognise societal demand for CES by calling for the "maintenance, restoration and upgrading of the cultural and natural heritage of villages, rural landscapes and high nature value sites" (CAP Pillar II). However, despite such policy acknowledgment, CES are not explicitly identifiable as policy instruments, but rather, tend to be embedded within the landscape concept (Paracchini et al. 2012). There is no attempt, for example, to link the maintenance of specific CES to landscape payments. There is however a European *landscape state and diversity indicator* framed on the basis of IRENA (Indicator Reporting on the Integration of Environmental Concerns into Agricultural Policy), launched after the publication of the COM (2001) report on 'Statistical Information needed for Indicators to monitor the Integration of Environmental concerns into the Common Agricultural Policy'. The proposal for the *landscape state and diversity indicator* is presented in the Report EUR 25114 (Paracchini and Capitani 2011). The indicator itself is structured in three components: the first concerns the degree of naturalness, the second landscape structure, and thirdly the societal appreciation of the rural landscape (Paracchini et al. 2012; Jones et al. 2016). The third dimension builds on proxy indicators such as protected areas, certified products and farm tourism (Jones et al. 2016).

This limitation certainly contributes to a lack of reliability in assessments of the contributions of 'material' qualities embedded into different CES, such as aesthetics. The failure[1] to agree upon methods for assessing CES and how they can be integrated into planning is a consequence of several interrelated factors (see Warnock and Griffiths 2015). In view of this, it is important to highlight the sharp divide between approaches focusing on the visual interpretation (related to human perceptions/preferences) and the more operational spatial landscape concepts. This split emphasises the lack of accord between the current strong focus on ecological and environmental objectives and human perceptions/preferences.

Such a division does not occur because human values and ecological processes are by nature contradictory. A major cause is the lack of specificity of the EU regulations in the field of ES. Another issue relates to the scale that human preferences/perceptions are framed – the perceptible realm (Gobster et al. 2007) – while ecosystems usually operate at other spatial scales and in other delineations (Carvalho Ribeiro and Lovett 2011). One possible solution is to use different landscape areas for the analysis and present the objectives (responses) at the scale of decision making (administrative spatial units).

The mismatch between spatial scales, at which environmental processes operate and are measured and at which land management operates, also applies to planning and policy institutional scales. Policy and planning framed for one scale of governance may have consequences for the delivery of CES at other scales. Therefore,

[1] In most European countries (exceptions can be found for example in the German Länder).

there have been important calls for the application of multi-scale approaches to policy setting and monitoring (Cash et al. 2006; Dick et al. 2014; Lefebvre et al. 2015).

Several approaches exist to access and map landscapes at different scales of governance (European, national and regional level). To our knowledge there is only one European landscape map named LANMAP 3. This is a pan-European landscape map illustrating the different landscape types across the continent, based on climate, topography, parent material, and land use (Mücher et al. 2010). This overview approach is useful at the European level and has several applications for European projects and policy initiatives. It is a result of the ELCAI project (European landscape character assessment initiative). LANMAP, however, is not suitable for LP on local or regional scale as it does not include subjective landscape dimensions and only works with physical layers (e.g. parental rock, land cover).

In addition to the work conducted at the continental scale, several European countries have implemented methods for identifying landscape quality objectives (under the European Landscape Convention 2000) and to capture the character of their 'local' landscapes. The concept of landscape character has been further developed within the approaches of Landscape Character Assessment (LCA: Swanwick 2002) and Historical Landscape Characterization (Fairclough 2004). Both stress that it is the character which distinguishes landscapes from each other. Since its introduction in the UK in the 70s, LCA has been widely used throughout Europe.

The next section explores the approaches employed in four different European countries to assess landscape qualities. Both the UK (Sect. 15.3.1) and Hungary (15.3.2) have used LCA, although, as shown below, there are differences between them in the application of the method. Section 15.3.3 discusses the Landscape Preferences Spatial Framework developed in Portugal. Finally, in Sect. 15.3.4, the general approach for assessing landscape aesthetic quality (by modeling supra-individual core-preferences) used in Germany is presented. Building on these four case studies and relevant literature, a proposal is then presented for a preliminary set of indicators that can be used to gauge landscape aesthetics capacity on a pan-European basis.

15.3 Methods for the Assessment of Landscape Aesthetics Capacity

15.3.1 Landscape Character Assessment in England, Scotland and Wales

For a variety of reasons, we start with a description of Landscape Character Assessment (LCA) approaches in England, Scotland and Wales. Not only has LCA had a long tradition in the UK, starting in the early 1970s with landscape evaluation, to landscape assessment during the 1980s, and then to the emphasis on landscape character from the mid-1990s onwards (cf. Swanwick 2002 for a detailed overview of the evolution of landscape character assessment in the UK). The UK approach to

LCA has also influenced LCA methods across Europe and can provide a framework for spatial units of cultural services at various scales.

Particularly in Scandinavia and parts of the Baltic, LCA approaches have followed the British example. In Sect. 15.3.2 it is possible to see how LCA practices in Hungary descend from a similar tradition to the British approach. The UK approach also reflects the principles of the European Landscape Convention (ELC) (Council of Europe 2000). It promotes public involvement and acknowledges that the character of an area is the result of interactions between natural and human factors and how the area is perceived by people. Similarly to the ELC, the goal is to describe the distinctiveness of different landscapes rather than categorize landscapes according to their aesthetic quality (for this see the German example in Sect. 15.3.4). Accordingly, the four key principles of the current British LCA approach are (Swanwick 2002:8):

- "the emphasis placed on landscape character;
- the division between the process of characterization and the making of judgements to inform decisions;
- the roles for both objectivity and subjectivity in the process, and
- the potential for application at different scales."

In order to reconcile planning procedures across scales (highlighted as important earlier) the LCA encompasses different scales such as national/regional, local authority and local site scale (see Fig. 15.2). Swanwick (2002) refers to the metaphor of a camera zooming in, from the broad view to the detailed small-scale frame. Natural England, the current non-departmental public body of the UK government responsible for nature conservation and landscapes, has comprehensively characterized all of England into National Character Areas (NCA). NCA number 36 is the 'South Pennines'. The South Pennines area however, encompasses a range of smaller-scale landscape types and areas. 'Moorland Hills' is an example of a character type within the South Pennines and 'Rombalds Hills' is an example of a character area of the Moorland Hills type. If there were to be a development proposal, a new planning policy or any other landscape project within the Rombald Hills (e.g. a wind farm on Romblads Top), the local authority or their planners, the developer, community groups, land owners or private practices/consultants could prepare a LCA at local scale to inform discussion of the proposed activity. Any of these scales could provide the unit for an assessment of CES.

It is important to note that the LCA process distinguishes two stages: the *characterization*, which is supposed to be as value-free and objective as possible (cf. Sect. 15.3.3), and the second stage, which contains the subsequent *judgements* about landscape character and aesthetics. As already implied in the introduction, there has been a long-standing debate about the role of objectivity and subjectivity in the assessment of landscape. It may be argued that this is one of the key distinctions among the various approaches in Europe.

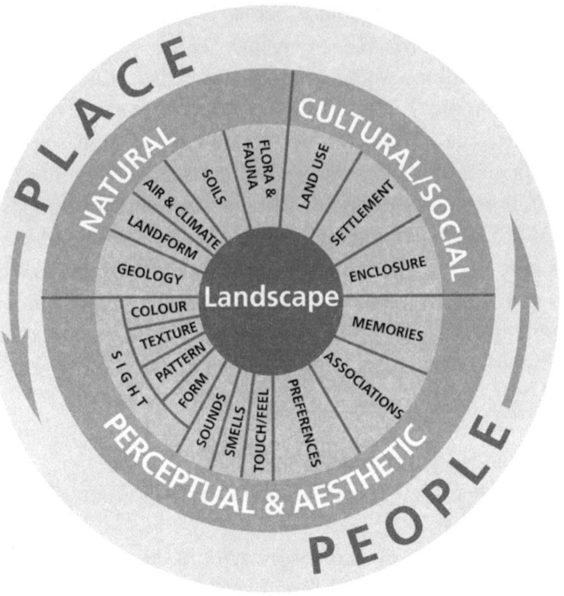

Fig. 15.2 The Landscape Character Assessment spatial hierarchy – an example of the relationship between the different levels. (Swanwick 2002)

Box 15.1 explains the main steps in an LCA based on Tudor (2014) and Swanwick (2002).

Box 15.1: Landscape Character Assessment Method in England, Scotland and Wales

Step 1: Defining the scope. All LCAs need to have a clearly defined purpose as this will critically influence the scale and level of detail of the assessment, the extent to which other subject areas are integrated, resources required, and the ways in which stakeholders can be involved in its preparation. Then, it is suggested to draft a project plan and brief including the nature of the outputs. With increasing LCA coverage, determining how far existing LCAs are up-to-date is also suggested as well as providing an appropriate scale, level of detail and stakeholder involvement to be considered.

Step 2: Desk study. The desk study includes the gathering of baseline data on a) natural factors, b) cultural factors, c) perceptual/aesthetic factors and c) cultural associations. Natural factors benefit significantly from the use of a Geographical Information System (GIS), a computer-based system to manage data input and to map, analyze and present geospatial data. GIS can particularly help in the assessment of natural factors in a LCA but it is also limited by the quality of available data. A GIS database is not a substitute for

(continued)

Box 15.1 (continued)

the LCA because the experiential part is underrepresented in GIS. That said, GIS can be a useful tool to facilitate stakeholder engagement. It is recommended to consider involving the public early on, particularly with regard to the aesthetic and cultural factors.

Step 3: Field survey. Field data are collected in a rigorous way to review the desk study findings but also to record new aesthetic/perceptual and experiential aspects. There is still need for creative methods of collection, e.g. through social media, film or citizen science.

Step 4: Classification and description. The output of the characterization process is refined and finalized by classifying the landscape into landscape character types and/or areas; mapping their extent, based on all the information collected and describing their character.

2. Application of Landscape Character Assessment

Using LCA to inform decisions. The European Landscape Convention (ELC) defines three categories of action, i.e. landscape management, landscape planning and landscape protection. LCA has an important contribution to make in all three of these categories. In addition to landscape character, landscape decision can relate to landscape quality, landscape value, landscape sensitivity, landscape capacity, landscape function and landscape objectives.

Applications. Within the three categories mentioned in the ELC, numerous applications refer to the LCA, for example: landscape strategies and management plans, green infrastructure plans, catchment management plans (landscape management); planning policies, Strategic Environmental Assessment and Environmental Impact Assessments, village design statements and masterplans (landscape planning); landscape designations and National Park management plans (landscape protection).

Step 2, Desk Study, and Step 3, Field Survey, consider not only a wide range of physical landscape characteristics but also perceptual/aesthetic and cultural/social factors and cultural associations (see box above). It has become best practice in the UK to describe these features in short profiles with summaries of the most important characteristics in bullet points, a map of the position and extent of the landscape character type/areas, this illustrated through photos, sketches and diagrams. It may be argued in favour of this LCA approach that these qualitative descriptions are particularly powerful in the description of aesthetic and perceptual factors and also well suited for public involvement.

Although perceptual, cultural and social factors are often easier to collect in a field survey, some relevant datasets do exist at regional and even national scales in England, Scotland and Wales. The Campaign to Protect Rural England (CPRE) compiled so-called 'Dark Skies' maps mapping light saturation for each square kilometer in England. 'Tranquility' maps for England and Wales (Fig. 15.3) and a series of national 'Intrusion' maps for the 1960s, 1990s and 2007 were also created.

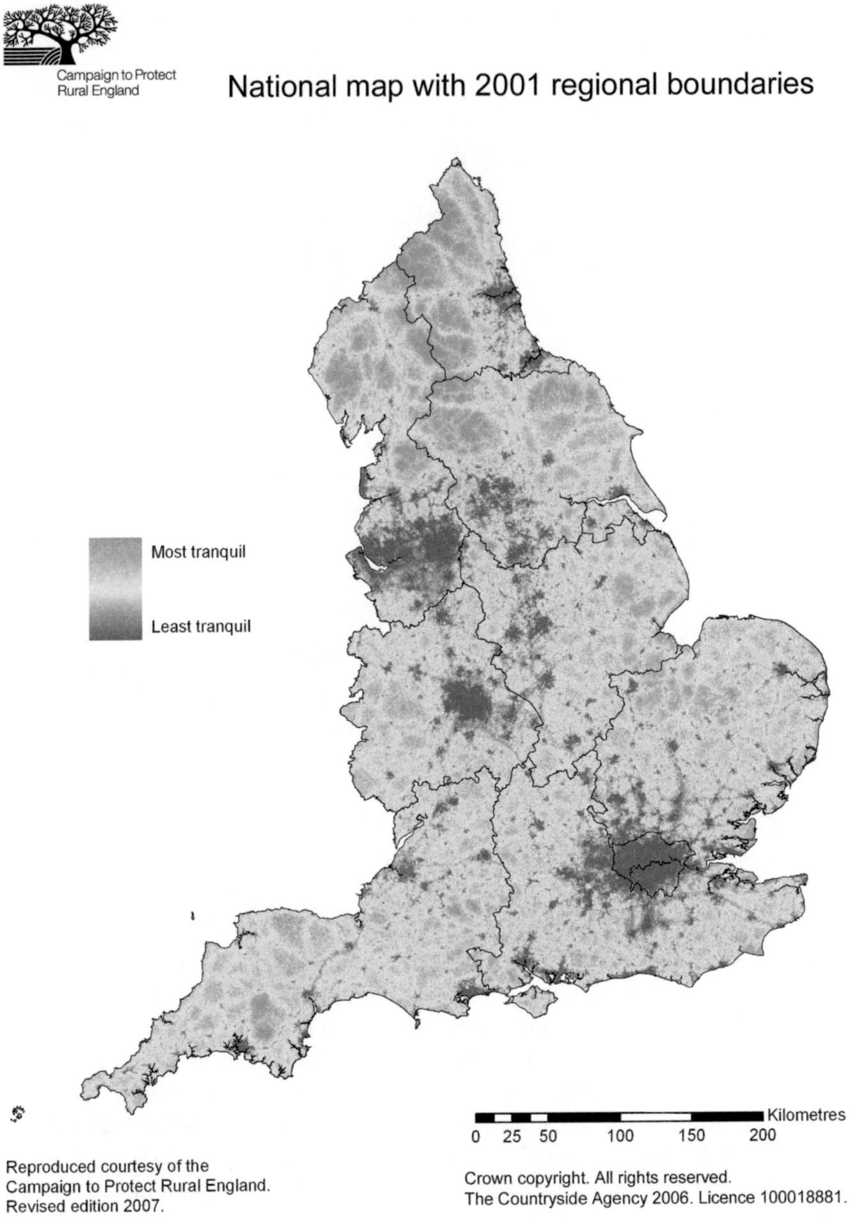

Fig. 15.3 National tranquility map with regional boundaries of 2001. (CPRE 2007)

The national Intrusion maps quantitatively and visually compared the distribution and the amount of land classified as disturbed. Scottish Natural Heritage has mapped areas classified as wild for Scotland. In Wales, LANDMAP contains both visual and sensory information about Welsh landscapes.

Various applications of LCA have evolved. Landscape management strategies, planning policies, design guidance and protecting valued landscapes are common approaches following on from an LCA. Recent LCAs have demonstrated striking versatility. For instance, they informed the development of green infrastructure plans, forestry strategies, river catchment planning and many other policy documents (Land Use Consultants 2011). Considering aesthetic qualities, judgements by both experts and lay people will touch on the concept of landscape value and cultural services (Swanwick 2002). Well-established landscape-related indicators used to determine landscape values are *landscape quality, scenic quality, rarity, representativeness, features of cultural interest, wildness and cultural associations* and *tranquility*, which is a composite indicator. Particularly in large-scale infrastructure projects, such as the new high-speed rail network HS2 and in terms of the cumulative impacts of wind farms, tranquility has received growing interest in the UK.

Additionally, of particular interest in the context of this chapter are capacity studies, i.e. studies of "the degree to which a particular landscape character type or area is able to accommodate change without significant effects on its character, or overall change of landscape character type" (Swanwick 2002:53). A capacity study is specific to a certain type of development, e.g. a wind farm, and interestingly, wind farms have been the main driver for capacity studies in recent years. Capacity studies benefit from public perception studies and some of them have successfully adopted the scenario method to assess landscape capacity under different development options. That said, a debate has started about whether capacity is still an appropriate factor to look at. Land Use Consultants (2011) suggest shifting the emphasis from capacity, as the main criteria, towards sensitivity. On the other hand, it may be argued that examples calculating capacities of natural resources for sustainable development are shifting in the opposite direction, emphasizing capacity even more (see Rockström et al. 2009). This issue cannot be solved here but this book may provide important contributions to the debate.

Several research studies have explored new ways of facilitating public involvement in LCA. These are based on a conceptual framework grounded in perception as a phenomenological experience of landscape, Butler and Berglund (2012) assess 52 British LCAs, dating from 2007 to 2011, to see how public involvement has been considered. They conclude that only a quarter of all assessments involved the public. Butler and Åkerskog (2014) suggest that, despite a lack of participation and the common misconception that awareness-raising about landscape is a top-down process, the LCA method does have the potential for mutual knowledge exchange and collaboration. They conclude that the first step is to acknowledge the values and aspirations attached to a landscape and consequently, mutual public involvement will alter how landscape is perceived. Those who experience a landscape need to be facilitated in expressing their values – which brings us back to the entangled topics of perception and preference.

How can ecosystem services facilitate such a process of two-way public awareness-raising and involvement? It has been argued that ES are an important communication tool for uncovering the benefits of aesthetic capacities of different landscapes. A first step then could be the use of Landscape Description Units (LDU) as suggested in the Living Landscapes approach, which is based on the LCA method for England, Scotland and Wales.

> LDUs are distinct and relatively homogeneous units of land, each defined by a series of definitive attributes, so called because they define the extent of each spatial unit. There are four definitive attributes at Level 1—Physiography and Ground type which describe the underlying natural dimension of the landscape, and Land-cover and Cultural pattern (reflecting settlement pattern and farm structure) which describe the cultural dimension of the landscape (Warnock and Griffiths 2015:265).

According to Warnock and Griffiths (2015), ES can be applied to a variety of geographical units – ecosystems themselves, catchment areas, landscape character areas or types, the particular features and attributes of landscapes, and areas of green infrastructure.

15.3.2 Landscape Character Assessment in Hungary

Arising from the British LCA method (Swanwick 2002), the application in Hungary has considered natural, cultural and aesthetic characteristics. The landscape character types were initially tackled at the micro/regional scale and the assessment placed an even greater emphasis on fieldwork and aspects of perception.

After defining the landscape character types on-site, the detailed micro-regional assessment was adapted for larger areas and the expert judgements validated by factual, mapped information. Furthermore, the relevant and consistent landscape attributes, which represent the uniqueness of the landscape character types, were defined.

An important task is assessing how to map the quantitative and qualitative information, identifying the most appropriate indicators. In Hungary, three main attributes in a hierarchical system were chosen. The first two are complex indicators and the third is a simple factual characteristic.

1. The *relief type* reflecting geomorphology, geology and hydrology that is defined by the physical setting, based on thematic maps.
2. The *human impact* – expressing the intensity and heterogeneity of the land use from the natural state towards the highly transformed urban areas, based mainly on field work. This is the attribute that contains the information on aesthetical/perceptional aspects.
3. *Land cover dominance*, a clear, measurable feature but nevertheless, a very strong characteristic. It is based on CORINE Land Cover maps. It is obviously connected to the first two attributes but also gives highly relevant additional information on the land use, landscape quality, and helps to define the real 'face' of the landscape.

Relief and land cover dominance can be derived from cartographic datasets, but human impact is both a quantitative and qualitative attribute. The definition of human impact has been based mainly on perceptional information gained by field work. This is a crucial part of the method as it allowed the inclusion of qualitative information into a complex attribute and into GIS systems. It needed both expert judgement and accurate knowledge about the individual areas.

When applied in a study area of 2634 km^2 spanning the Austria-Hungary border the combination of the three attributes resulted in the identification of 65 character types. This is an extremely high level of variability. Although these patches are relevant mosaic units of the landscape, and their uniqueness should therefore be taken into account, further aggregation was still considered necessary. The spatial distribution of these mosaic units made possible their aggregation by expert judgement, which resulted in 13 landscape character types in the study area (see Fig. 15.4). Figure 15.5 shows examples of Hungarian character types.

In summary, LCA in Hungary places a greater emphasis on expert judgment, while that in the UK puts more efforts on public engagement. In Hungary more

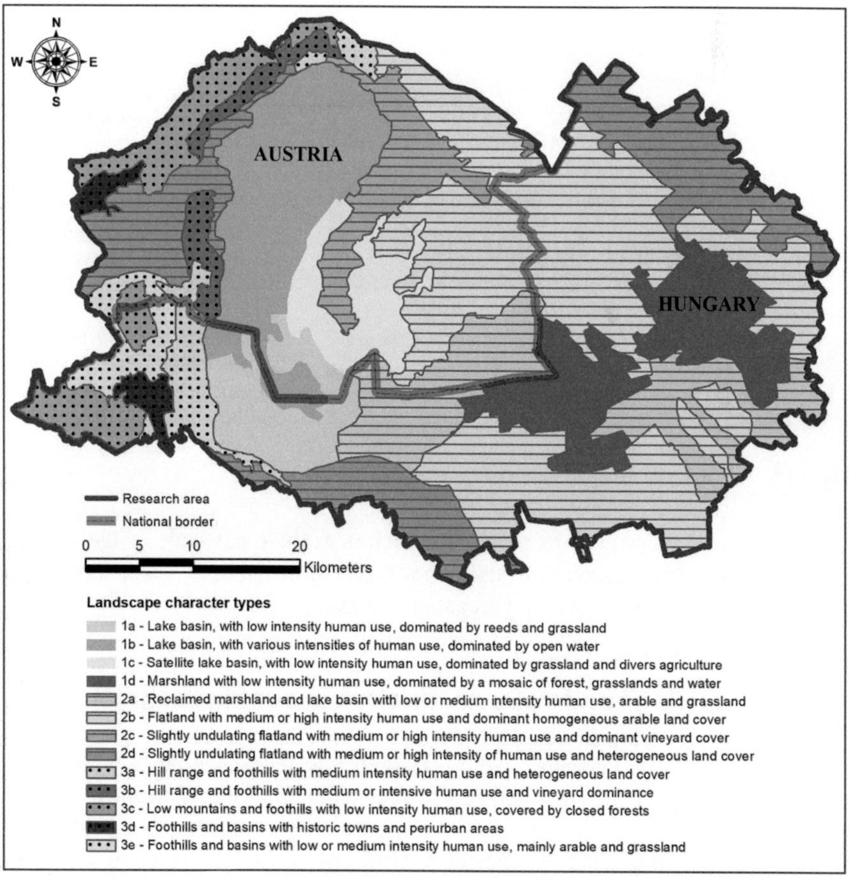

Fig. 15.4 Landscape character types in the Austro-Hungarian study area Fertő-Hanság

Fig. 15.5 Examples of Hungarian landscape character types

emphasis has been put on the perception factors at the local scale of the analysis. Nevertheless, when upscaling to the larger regional or national scale, some of the detailed perception aspects are aggregated and lost.

15.3.3 Assessing Landscape Aesthetics Capacity in Portugal

In 2004 Portuguese landscapes were classified into landscape units 'unidades de paisagem' based on cartographic layers such as relief and vegetation and a considerable field work effort. This project developed through a partnership between the Portuguese planning ministry and the University of Évora (Abreu et al. 2004). Figure 15.6 shows that there are 28 classes, further divided into several subclasses.

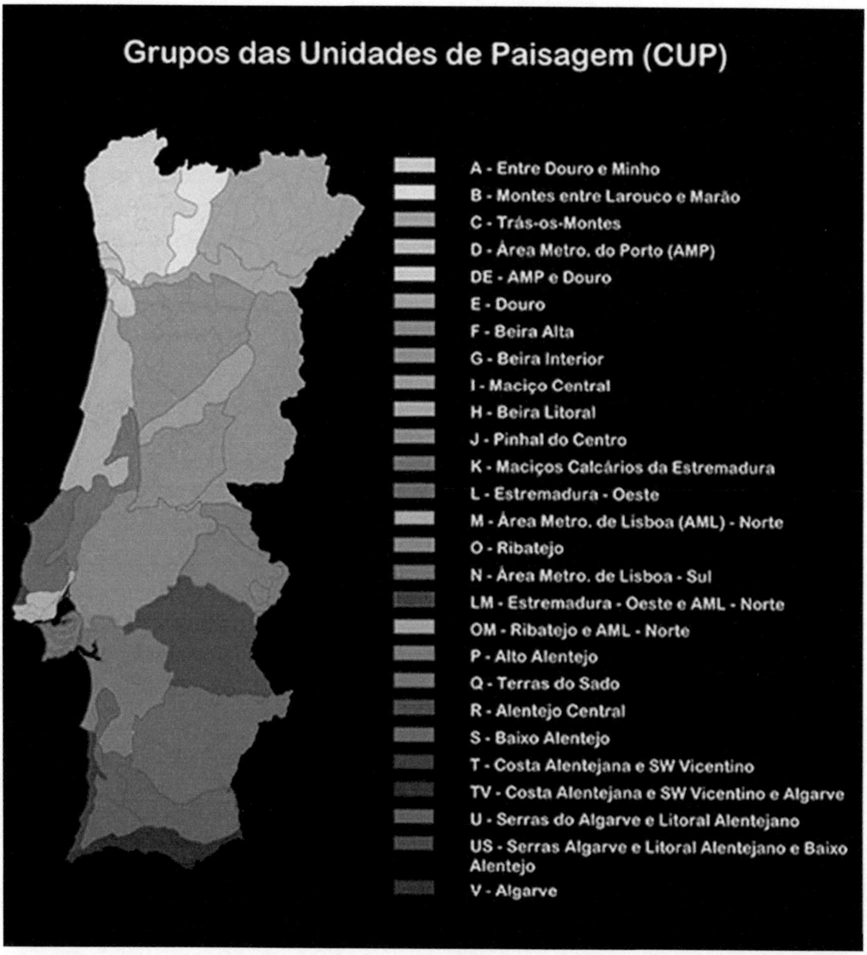

Fig. 15.6 Landscape Units (Carta Unidades de Paisagem) in Portugal. (Source http://www.dgterritorio.pt/sistemas_de_informacao/snit/cup (accessed 20 June 2018))

In Portugal, although the landscape scale is commonly referred to within legislation, the Portuguese planning system is very hierarchical within individual sectors such as forestry and urban development. Therefore, there are as yet no comprehensive landscape scale planning approaches nor are there any formal methods to assess LAC. In 2015 the National Policy in Architecture and Landscape (PNAP Política Nacional Arquitetura e Paisagem) *Diário da República, 1.ª série – N.° 130 – 7 de julho de 2015* was formally enacted. This piece of law was contested by the Portuguese Association of Landscape Ecology (APEP) and other stakeholders. The reason for disagreement was the narrow vision of the policy that were thought to favour the architecture discipline and missed the opportunity to address landscape in a transdisciplinary manner focusing on all landscapes (not only the ones with cultural value) as is the premise of the European Landscape Convention (ELC).

In such an unsettled legislative situation different research projects and initiatives by civil society dealing with landscape issues have developed in a variety of ways. All these initiatives acknowledge that landscapes (all landscapes, not only the ones with cultural and natural value) in Portugal are very important to everyday life. These initiatives and projects, some of them sponsored by the government, address the ways in which Portuguese landscapes satisfy multiple societal demands.

In addition to a national assessment by Abreu et al. (2004), regional partnerships have occurred amongst official planning bodies (particularly in the agriculture and forestry domains) and universities or research centres. In this context, in southern Portugal, the Instituto de Ciências Agrarias e Mediterrânicas (ICAAM) assessed both i) the ways in which different people described their surrounding landscape to others outside the region and ii) which land cover patterns related to landscape aesthetics capacity (see Carvalho Ribeiro et al. 2013).

The study examined which physical landscape components relate to subjective landscape dimensions such as landscape aesthetics and scenic beauty. The physical components with the strongest associations included the so-called montado agro forestry system and heritage sites (castle, churches), in addition to relief and topography (hilly landscapes). There were also other immaterial landscape aspects such as tranquility, smells, and colours that respondents associated with physical landscape attributes so that they could be mapped. In fact, the argument used by different people to justify how and why they composed their preferred land cover patterns, was that of landscape aesthetics capacity (Fig. 15.7).

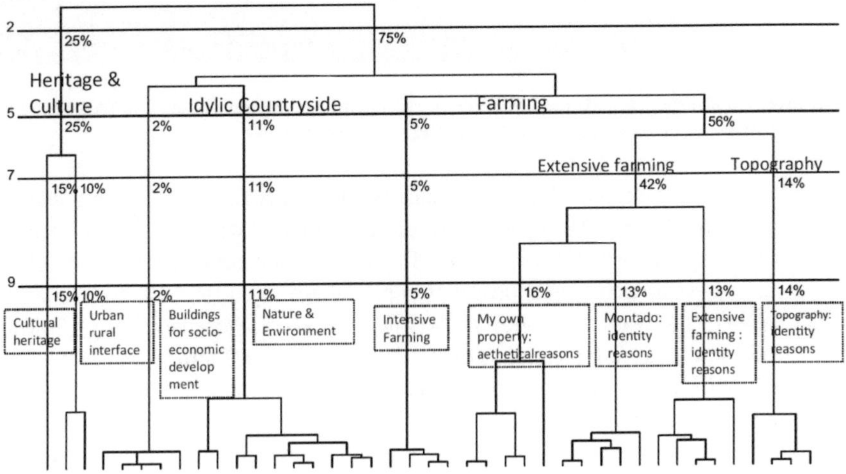

Fig. 15.7 Clusters of responses from Multiple Correspondence Analysis (see Carvalho Ribeiro et al. 2013)

In the majority of landscape preference studies, photo questionnaires are used with real photos or photomontages. In these cases the respondents have to make selections on the basis of the 'existing' landscape. In ICAAM, through the development of the Landscape Preferences Spatial Framework respondents were asked to compose their preferred 3D 'block diagram' of land cover categories and then areas in the real landscape that matched these patterns were mapped (Carvalho Ribeiro et al. 2013). This method was important to explore which land cover patterns were in line with landscape aesthetics capacity. It also created a bridge between personal preferences and mapping which could be used as a basis for landscape planning approaches.

15.3.4 Mapping and Assessing Landscape Aesthetics Capacity in Germany: Formal Methods Adapted for Landscape Planning and Environmental Impact Assessments

15.3.4.1 The Tradition of Landscape Aesthetic Assessment in Germany

Landscape aesthetics assessments have a long tradition in Germany as they are an important part of landscape planning and environmental impact assessments. The German Federal Nature Conservation Act of 1976 earmarked landscape planning as the principal instrument for safeguarding beautiful landscapes and developing the recreational value of landscapes. The purpose of the aesthetic assessments is to identify beautiful landscapes or landscape elements which should be protected, further developed for recreational use or where infrastructure for recreation should be improved. The assessment methodology of landscape planning is also used for comparing the impacts of plans and project variants on landscape aesthetics, in the context of environmental impact assessments, or in order to define landscape adapted compensation needs (as demanded by the German Federal Nature Conservation Act). The first formalized (repeatable) method for transparently and intersubjectively assessing the diversity of the landscape was developed by Kiemstedt in 1967. The method followed a multicriteria benefit analysis approach. In addition, general landscape preferences of the population were also explored (e.g. Hanstein 1967), which could substantiate the formal assessments.

At present there are numerous methods for assessing the visual landscape in Germany. They are usually adapted to different planning scales, from the federal state, to the regional and the communal level. The formal assessments, which are based on general landscape preferences of the German population but do not include the specific perspective of the actual local landscape users, are called user-independent methods. The resulting landscape evaluation is used for designing landscape conservation areas, for considering valuable landscapes and landscape elements in municipal development or infrastructure planning and for improving beautiful landscapes for tourism.

These formal methods can and should be supplemented by user-dependent methods, which capture the actual utilization of the landscape and local landscape preferences. In particular in local scale landscape planning, the specific perception and landscape use of individuals should be included. However, in German landscape planning a systematic user-dependent assessment is not very common because of the time-consuming methodologies. However, the standard procedure does include gathering of user opinions during citizen meetings as part of the participation process. Recently, in interactive or at least web-based landscape planning (e.g. interactive landscape plan Koenigslutter, http://entera-online.com/), it is much easier to collect user preferences, e.g. by participatory GIS tools. Resulting information on how locals see and use the landscape can be used to integrate specific local needs into landscape planning, which may differ from or add to the general landscape valuation.

15.3.4.2 User-Independent, Formal Assessment

In the formalized landscape aesthetics assessment, the landscape is evaluated by three basic criteria using lists of indicators (usually landscape features). The three most widely-used criteria are diversity, naturalness and specific landscape character (uniqueness) (Fig. 15.8). These criteria are supported by the German Federal Nature Conservation Act, which mentions diversity, uniqueness and beauty as integrating properties. The act also mentions the capacities for recreation and natural and

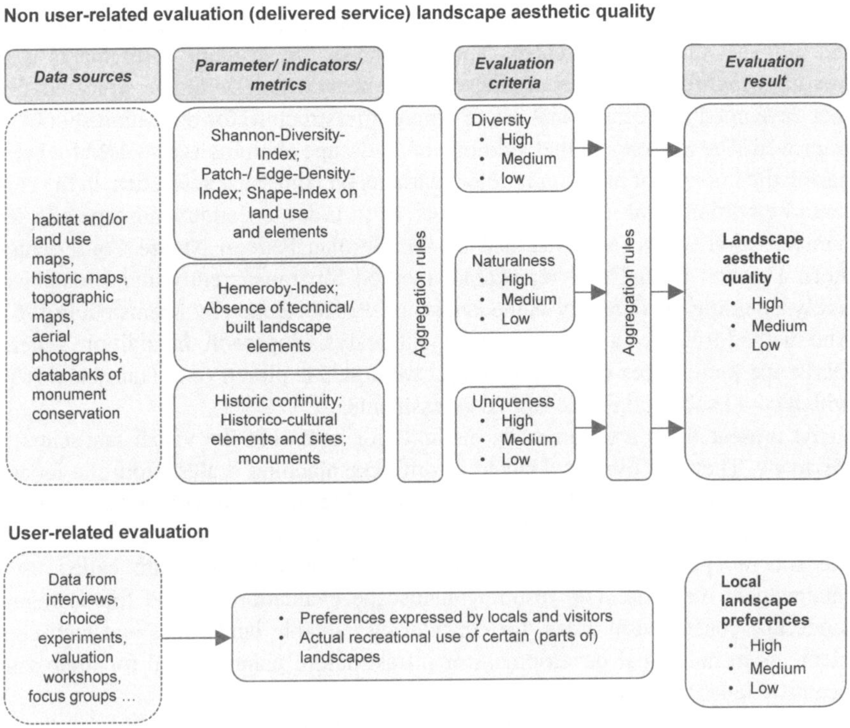

Fig. 15.8 Workflow, criteria and parameters for evaluating landscape aesthetic quality in Germany

historic landscapes as valuable properties and assets to be conserved (BNatSchG §1 (1) 3 and (4)). Furthermore, the criteria are based on preference studies in different landscapes, which have shown that there is a core of basic common landscape preferences. Only this core is measured in the landscape aesthetics assessment. Individual preferences of landscape users, which may be influenced by a diversity of individual preconditions, are neglected in the user-independent methods. The inventory of such individual preferences is left to the user-dependent approaches.

Within the group of user-independent formal assessments, different combinations of objective landscape metrics are used as valuation parameters (see Hermes et al. 2018a). The criterion diversity is understood as the variety and distribution of landscape components (land use-types and landscape elements). It can be measured using a variety of metrics. For example, the Shannon Index represents the number and proportional distribution of different landscape components in a given area. However, it does not indicate their proximity and spatial correlation and therefore, does not indicate the complete composition of the landscape components. For this, the index has to be combined with other indicators such as the Patch Density Index, which calculates the number of single patches per area. Other indices like the Shape Index or Edge Density Index account for the shape of landscape component types, which leads to differences in the perceived structural diversity of a landscape, even though the composition may be the same. Other indices that have been found to be significant when tested with social-empirical methods are the forest/open-landscape ratio and arable/grassland ratio. Next to land cover, relief diversity most influences the visual impression of a landscape. It can be expressed, for example, as roughness or relief energy. Examples of these latter indices can be found in Herbst et al. (2009), Roth and Gruehn (2006), Ode and Miller (2011), Roser (2011), Frank et al. (2013), Hermes et al. (2018b).

Naturalness is measured using the *hemeroby index*, a qualitative scale describing the degree of human impact on the landscape. It expresses the distance of the current state from a constructed potential natural landscape, if all human impact stopped (hemeroby, according to Sukopp 1976; Blume and Sukopp 1976) (see also Walz and Stein 2014; Rüdisser et al. 2012; Kowarik 2006). As this also accounts for irreversible changes, it seems to be more appropriate than using a historic 'natural' state as reference. Such a state is also very difficult to define. Despite its origin in ecology and biodiversity research, the index can be included in assessments of aesthetic values of landscapes, as shown by Frank et al. (2013) and Paracchini et al. (2014). In this context the indicator 'naturalness' should be adapted to better represent the human perception and experience of naturalness (Hermes et al. 2018b). Uniqueness is measured by the occurrence of landscape features that are rare in a nationwide comparison and thus characteristic compared to other landscapes. An analysis of the occurrence of landscape features in different landscapes on a national scale can be used as a basis for judging the degree or rareness of landscape elements. People in local communities, confronted with such uniqueness maps, have been very interested to learn which landscape elements in their municipality are rare elsewhere and thus constitute a uniqueness that they can identify with. However, in Germany the data situation does not yet allow a sound analysis of rareness on national scale because many relevant data sets are available only in the federal states and not in a nationally comparable format. Next to improving the data availability the assessment of uniqueness could be enhanced by extending social-empirical studies to name places or landscape elements that are characteristic for a region or have a high symbolic or recognition value as perceived by people (cf. Steinitz 2010). In the context of uniqueness, historic landscape characteristics and elements/ensembles are also considered. They are included by using, for example, lists of perceptible cultural and natural historic landscape elements such as dolines, 'village lime trees', megalithic tombs, historical fish ponds or sunken roads, which bear testimony to historical landscape continuity. Another way to distinguish historic continuity is to compare current and historic maps (according to Nohl 2001, the historic maps need to be at least 50 year old). All in all, the criterion uniqueness relates to the landscape character in the LCA.

The three criteria, diversity, naturalness and uniqueness, are assessed separately and ultimately combined into the evaluation of landscape aesthetic quality. An example of such output for the Hannover region is shown in Fig. 15.9. Depending on the scale of the analysis (e.g. if the planning area is very small), it might also be relevant to include indicators for sense of taste and the sense of touch (Nohl 2001).

Delineation of spatial valuation units may be performed by identifying landscapes which are perceived to be homogenous in structure and composition of landscape components by the visitor. The differentiation of landscape units is similar to the approach used in LCA. The size of the landscape units is dependent on the scale

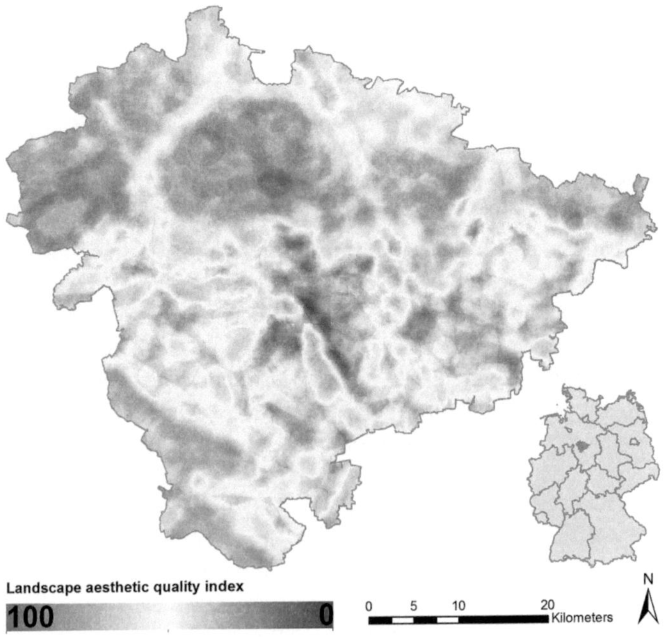

Fig. 15.9 Characterization and assessment of visual landscape in the Hannover region

of the planning area and the character of the landscape itself. However, this method is not very transparent as it greatly depends on individual decisions made by the planner. Furthermore, the method can only be automated with great difficulty. It is much easier to use a GIS to perform an assessment on the basis of grid cells. If needed for planning purposes, grid cells with similar characteristics can be identified and aggregated into more or less homogeneous landscape zones. The problem of very inhomogeneous grid cells can be solved by choosing relatively small grid cell sizes and applying neighborhood analysis tools.

In a reflection of the DPSIR approach (Chap. 3), other existing or proposed landscape disturbances (pressures) and their impact on recreational landscape qualities are also assessed. This includes the impact the disturbances have on the visual landscape as well as disturbances by noise or smell. The disturbances have to be analyzed to determine their intensity and the amount of area affected by them.

The spatial extent of such disturbances is calculated by visibility analysis (e.g. viewshed analysis in GIS software), air pollution and noise propagation models. The affected areas can also be mapped by simple standardized methods such as defining noise bands along streets according to DIN 18005-1 'noise abatement in town planning'. The evaluation of the impact is classified on an ordinal scale and based on German legislation about acceptable emission levels for different user groups. For example, noise standards exist for residential areas or areas with health facilities.

As a basis for mapping the visual landscape, relevant information can be extracted from existing maps such as habitat or land use maps, topographic maps, aerial photographs and other information in order to reduce field mapping efforts. Once in the field, further relevant landscape characteristics and elements need to be mapped. A standard list of landscape elements can be used as the basis for mapping.

15.3.4.3 Landscape Features Required for Specific Recreational Activities

Aesthetic quality is an important component of the natural capacity of a landscape for CES provision. For the more general/unspecific types of landscape enjoyment, the aesthetic quality may even be the only relevant or most important component. However, there are more specific kinds of recreational landscape uses that focus on particular elements and features. These include for example, rivers and lakes, which provide opportunities for all kinds of water-related activities. Additional examples are specific relief forms, which are the precondition for activities such as outdoor-skiing and climbing, or habitats and interesting species that allow for nature observation. Next to natural elements, historic-cultural elements and sites also need to be considered. The density of such features and elements, in a given area, can be used as a metric to map their availability in different landscapes. If possible, it is desirable to include data sources that better represent such assets rather than mere land use/cover data. Such data sources might, for example, indicate the quality of a feature/element, inherent value or suitability for specific recreational use. For nature observation, for example, this includes habitat quality, species richness or population size. Other examples are the quality of bathing waters or the preservation status of certain areas and elements. Such elements should be given specific attention in planning of responses and may lead to an upgrading of the respective patches in the landscape aesthetics assessment results.

15.4 From Delivered CES to Utilization: Determinants for Outdoor Recreation?

For outdoor recreation, not only is the landscape aesthetic quality relevant but there are also large overlaps with other CES. For example, with the group of physical and experiential interactions with ecosystems and their components, as well as some aspects of cultural and natural heritage or education. As the differentiation of these factors is difficult and for some of them no assessment methods are at hand, the following section regarding methods for assessing and mapping CES utilization will focus on outdoor recreation. However, this should not lead to a marginalization of other important CES that are more difficult to assess (Milcu et al. 2013).

The landscape aesthetic qualities and types of properties listed in Table 15.1 are preconditions for activities such as hiking, bike riding, bird watching or rafting. Consequently, these attributes are the most important aspect of a landscape's recreation capacity. Mapping how they are used can be done concurrently. The differentiation between LAC and their actual utilization is a valuable means of deducing response measures for CES (Chap. 24).

Table 15.1 Examples of landscape features and infrastructures relevant for specific activities

	Features	Infrastructure	Data
Suitability for bird watching	Quantity: Area of habitat for interesting species	Observation towers, hides, information boards; guided tours	Official land use data; thematic maps (e.g. bird monitoring)
	Quality: Population size of interesting species; number of different interesting species; rarity of species		
Suitability for hiking/cycling	Scenic views/no noise	Quantity: Density of paths and roads without traffic; benches; number of viewpoints; restaurants and lodging for longer trips	Official land use data; online mapping sources such as OpenStreetMap; hiking maps; tourist information
		Quality: Marked/quality controlled hiking trails/cycling routes; furnishing	
Suitability for rafting	Length of suitable river(s)	Special facilities in the river; rental stations, guided tours	Tourist information

Data to measure and assess actual outdoor recreation is often scarce. Therefore, we need to use a mixture of proxies and scattered direct evidence to model the utilization of the landscape. Important proxies for outdoor recreational activities are the 'human inputs' (i.e. infrastructure) which are needed. Furthermore, demand, accessibility, and information about local and individual preferences are determinants of actual utilization. Empirical data about visitor numbers and actual activities, as well as survey results about the effects of recreation on people, can validate modelling results – if available. Figure 15.10 presents an approach for modeling CES utilization based on capacities, recreational infrastructure and accessibility.

15.4.1 Recreational Infrastructure

A certain amount of recreational infrastructure is needed to harness the natural capacities of a landscape. Recreational infrastructure can be understood as human input that enables, supports, or enhances the utilization of CES and the benefits obtained from them. It includes such things as paths for accessing an area, map and sign trails, furnishing (benches, picnic areas, shelters and information boards), viewpoints, and infrastructure for specific activities such as campsites, restaurants, lodging and other services. The recreational infrastructure determines if and how well the capacity can be utilized. This consequently has a big impact on the material and immaterial benefits that can be obtained from an area. When mapping the availability of recreational infrastructure, their density in a given area is the most significant measure. This can also be combined with additional information regarding the quality of the infrastructure, e.g. to distinguish between marked hiking trails or scenic roads and simple agricultural or logging roads whose primary purpose is not for recreation. In Germany, the Digital Landscape Model (ATKIS Basis-DLM) contains a great deal of information referring to recreational infrastructure. Internationally, data from OpenStreetMap (www.openstreetmap.org) represents a

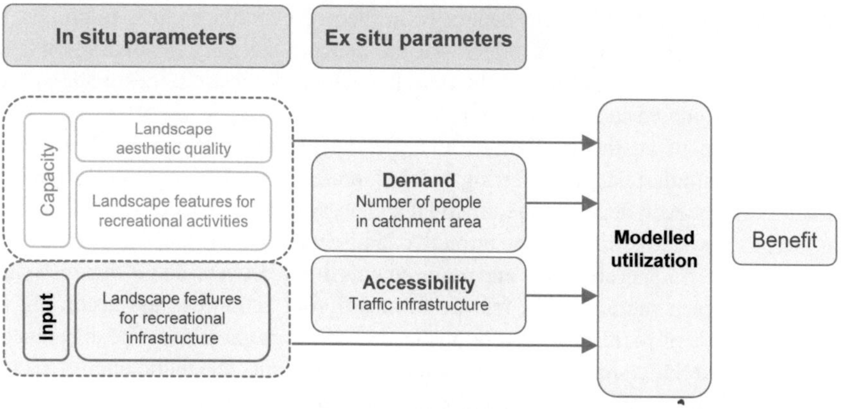

Fig. 15.10 Workflow and parameters for modelling utilized cultural ecosystem services

valuable alternative source, especially when official data are not available. Further information can be gained, for example, from hiking, cycling or tourist information maps. It must also be noted that the role of infrastructure availability varies between different kinds of activities, as some are more dependent on certain infrastructure than others.

15.4.2 Accessibility and Demand

The degree to which potentials are exploited by recreationists depends on their accessibility. Remote areas may be less utilized and thus generate less (economic) benefits from CES than areas with the same or even lower potential, which are located closer to cities and urban areas. This is because the latter areas are accessible to more people. A simple proxy to measure accessibility is the linear distance of an area suitable for recreation to settlement areas (where a demand for recreation opportunities exists). To be more precise, accessibility also relates to travel costs (time and money). Travel costs are determined by travel distances and traffic infrastructure. Travel time/cost maps, for different modes of transport, can be used to characterize recreation sites according to their accessibility (e.g. Sen et al. 2014), where increasing travel time is used as a proxy for decreasing accessibility. Thresholds for travel time are often set to defined areas e.g. for nearby recreation around population centres. Such thresholds need to take into account different kinds of recreation (nearby recreation, day or overnight trips), so that an acceptable relation remains between travel time and duration of the recreational activity. Usually only urban areas of a certain minimum size are considered in such assessments. This is because (a) access to recreation opportunities outside cities is considered to be more relevant in urban than in rural areas (where access is usually assured anyway), and (b) using accessibility from every settlement would not lead to a useful differentiation of landscapes in relatively densely populated countries. Finally, the number of people for whom an area is accessible can be used as a proxy for demand when modelling utilization. The demand relates to the quantity of people likely to utilize the CES in an area and benefit from them. Nevertheless, it is important to note that this proxy does not reflect that the demand can vary depending on user knowledge about the qualities of an area, the extent of substitute destinations that are available and various other factors.

Evaluation of aesthetic qualities and specific activity relevant landscape elements can be undertaken by referring to 'core' preferences. The evaluation of actual utilization however, depends on individual and possibly local collective preferences, as well as the accessibility of the attractive landscape features, and on specific user requirements. Such preferences and requirements differ between socio-demographic groups, between recreationists that perform different activities, and according to different kinds of recreation (nearby visits, day or overnight trips). The differences influence the relative importance of the main determinants (aesthetic quality, recreational infrastructure, and accessibility). In an assessment this can be reflected by using different weightings for the determinants or by adjusting aggregation rules.

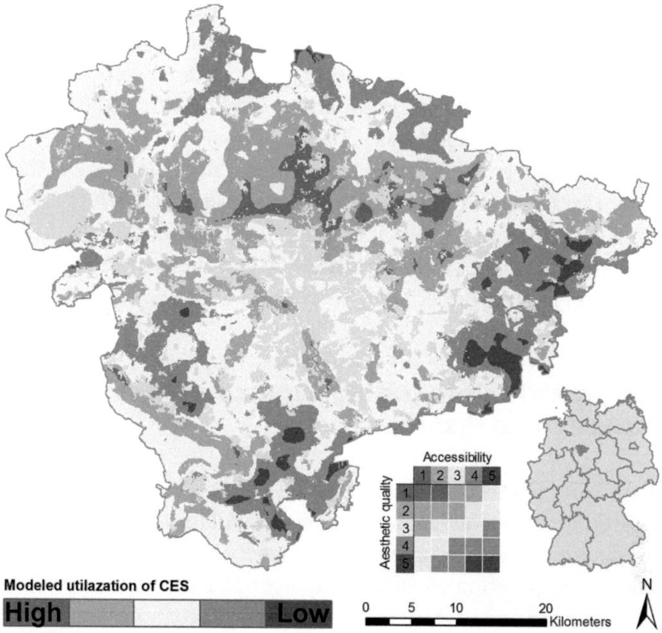

Fig. 15.11 Map of modelled utilization of CES in the Hannover region

Figure 15.11 shows an example of such modelled utilization for the Hannover region. Knowing about these differences and reflecting them in maps for specific types of user, enables planners to respond to them with adequate measures. However, while this is highly desirable it can also be very time consuming to undertake and is not always possible.

Assessments of people's preferences concerning CES require social empirical research such as questionnaire surveys. There are two main approaches that are commonly used, namely revealed preference and stated preferences methods. Revealed preferences methods are based on the observation of actual behavior or individual real life choices. In the case of recreation, the choice to visit a specific site and/or to undertake a specific activity reveals a respondents' preference for certain attributes at the site. Such attributes can include the availability of recreational infrastructure and landscape elements (e.g. visual appearance), as well as, for example, distance from respondents' homes (i.e. the determinants discussed above). A revealed preference approach for economic valuation of recreation benefits is the travel cost method. It is based on the premise that the monetary recreational value of a site is partly expressed through the amount of time and money respondents expend to travel to the site (Whitten and Bennett 2002; Martín-López et al. 2009). In stated preference methods people are asked to rank and/or judge a site's or landscape's attributes or to choose from hypothetical choice sets that are characterized by different combinations of attribute levels, some of which can also be monetary (Adamovicz et al. 1994). As both approaches have their strengths and weaknesses, a combination of both approaches is a desirable option. Finally, the results from such preference analysis can be used to inform the attributes incorporated in spatial analysis to determine and map the aggregated potential for outdoor recreation. For more detailed information on preference analysis and economic valuation see TEEB (2010) and Chap. 20.

While modelling is a valuable and commonly used technique, there are also approaches that focus more directly on the actual use of CES (e.g. Plieninger et al. 2013; Bieling and Plieninger 2013; Wood et al. 2013; Casalegno et al. 2013; Martínez Pastur et al. 2015). Such approaches can be used to supplement and/or validate modelling approaches.

15.5 Indicators for Integrating Landscape Aesthetics Capacity into Planning

As discussed earlier in this chapter, England, Hungary, Germany and Portugal have developed different ways to assess LAC. However, all the described methods include three different components to some degree, namely natural, cultural and perceptual characteristics. Incorporation of these components can provide valuable knowledge for enhanced planning and management, and particularly for including CES such as aesthetics into planning (Opdam et al. 2001; Antrop 2005; von Haaren et al. 2008). Table 15.2 outlines the natural, cultural and perceptual indicators used in the different case studies and presents a summary of other relevant criteria from the literature.

It is apparent that there are differences in how much characteristics other than physical features are considered. In contrast to the other case studies, assessment processes in England, Scotland and Wales usually consider peoples' perception/preferences as at least as important as purely physical aesthetic characteristics. Examples are the concepts of tranquility or wilderness and intrusion. This is also shown in the case of Hungary, where much emphasis is placed on the perceptual dimension. In contrast, with Portuguese landscapes greater emphasis has been placed on the way the physical landscape might contribute to immaterial dimensions (for example how land cover patterns can contribute to perceptual dimensions such as colours and smells).

In German landscape planning the user-independent methods for assessing the aesthetic quality of landscapes are well-developed and widespread in practice. Basically, the criteria diversity, uniqueness and naturalness are used to assess the visual landscape in user-independent methods. User-independent and user-related methods are applied for different planning purposes. In German practice, the application of methods for assessing actual recreational use and economic valuations of CES are less common, but when they are used create a considerable added value for informing landscape planning responses.

Regardless of the different approaches in European countries, Table 15.2 shows that there is much common ground. There is considerable agreement on the set of natural, cultural and perception indicators that satisfy people's aesthetic needs and

Table 15.2 Natural, cultural and perceptual indicators for assessing landscape aesthetics capacity

Case study/ criteria	Indicators		
	Natural	Cultural and social	Perceptual and experiential
LCA in England, Scotland and Wales	Geology, landform, air and climate, hydrology, soils, land cover, flora and fauna	Land use, settlement, enclosure, land ownership, time depth	Tranquility, light pollution, intrusion, perceived character, wilderness
LCA in Hungary	Relief, geomorphology Land cover dominance	Human impact, heterogeneity and intensity	Smell, sound memories, sense of place
Formal delivery assessment and use oriented assessment in Germany	(Perceived) naturalness/ hemeroby, geomorphologic characteristics, characteristic fauna and flora, natural preconditions for specific activities	Heterogeneity of land cover; uniqueness/ historicity (landscape character) Rare and characteristic landscape elements e.g. single trees, alleys etc. – historic cultural landscape elements such as castles.	Individual preferences, actual recreational use; perceived disturbances (visual e.g. settlement, buildings and other artificial structures), noise, smells etc.)
Landscape Preferences Spatial Framework (LPSF) in Portugal	Relief, low input land covers e.g. natural forests	Cultural heritage castles, bridges Montado agro forestry churches	Colours, smells and sounds, landscape identity, stewardship
(EU scale) Societal awareness of landscape	Protected areas	High quality traditional products (e.g. Denominazione di origine protetta – DOP), farmland tourism	
Other possible indicators cited in European literature			
(Low) disturbance	Hemeroby		
Historicity		Heritage buildings	
Visual scale	Size of viewsheds		Viewpoints
Imageability			Stones with undefined formats
Diversity		High number of land cover classes per view shed	
Ephemera	Specific land covers, flowering plants		
Coherence			The whole landscape mosaic is greater than the sum of the individual parts
Cared for landscapes		Hedges cared for	
Stewardship			

desires. Additionally, the indicators used in the different case studies are often cited in the wider literature on landscape preferences in Europe (see the lower part of Table 15.2).

15.6 Conclusions

Although the role of ES is recognized in policy and management, CES are not commonly integrated into spatial planning and related decision making processes across Europe. This is arguably due to their supposed 'subjectivity and immaterialness' (Carpenter et al. 2009; Kushner et al. 2012).

This review highlights the different ways in which several European countries have dealt with the assessment of LAC. Despite the differences in methods, the case studies presented here reveal that there are similarities in the indicators used to capture aesthetic capacities. Although perceptual, cultural and social factors are often easier to collect in field surveys, some relevant datasets are being developed that represent such factors, without the necessity of mapping them. At the European scale, for example, the Eurostat LUCAS survey (http://ec.europa.eu/eurostat/statistical-atlas/gis/viewer/?myConfig=LUCAS-2012.xml) includes photographs of each data point that contributed to the classification of European landscapes according to perceptual dimensions. It is still not known whether this dataset is adequate to create indicators of LAC in a systematic manner. However, it is worthwhile to explore the possibilities of such an extensive dataset.

This chapter has aimed to summarise both theories and empirical work conducted in four different European countries regarding the assessment of LAC. The indicators listed in Table 15.2 provide a basis for enhancing European-wide frameworks. We acknowledge, however, that there is still a long way to go to comprehensively include Landscape aesthetics capacity into planning all over Europe.

References

Abraham, A., Sommerhalder, K., & Abel, T. (2009). Landscape and well-being – A scoping study on the health-promoting impact of outdoor environments. *International Journal of Public Health, 55*, 59–69.

Abreu, C., Pinto Correia, T., & Oliveira, R. (2004). *Contributos para a Identificação e caracterização da paisagem em Portugal Continental.* Coordenação/DGOTDU. http://www.dgterritorio.pt/static/repository/2013-12/2013-12-02112022_f7664ca7-3a1a-4b25-9f46-2056eef44c33$$822A9394-2740-4D34-A6C1-8ED04570B3B5$$AB974510-DCBF-4B12-9BD8-9459CB6B545F$$storage_image$$pt$$1.pdf. Accessed 25 June 2018.

Adamowicz, W., Louviere, J., & Williams, M. (1994). Combining revealed and states preference methods for valuing environmental amenities. *Journal of Environmental Economics and Management, 26*(3), 271–292.

Antrop, M. (2000). Background concepts for integrated landscape analysis. *Agriculture, Ecosystems and Environment, 77*(1–2), 17–28.

Antrop, M. (2005). Why landscapes of the past are important for the future. *Landscape and Urban Planning, 70*(1–2), 21–34.

Appleton, J. (1975). *The experience of landscape*. London: Wiley.
Bieling, C., & Plieninger, T. (2013). Recording manifestations of cultural ecosystem services in the landscape. *Landscape Research, 38*(5), 649–667.
Blume, H. P., & Sukopp, H. (1976). Ökologische Bedeutung anthropogener Bodenveränderungen. *Schrreihe Vegkd, 10*, 74–89.
Boll, T., von Haaren, C., & von Ruschkowski, E. (2014). The preference and actual use of different types of rural recreation areas by urban dwellers – The Hamburg case study. *PLoS One, 9*(10), e108638.
Bourassa, S. C. (1999). *The aesthetics of landscape*. London: Belhaven Press.
Butler, A., & Åkerskog, A. (2014). Awareness-raising of landscape in practice. An analysis of landscape character assessments in England. *Land Use Policy, 36*, 441–449. https://doi.org/10.1016/j.landusepol.2013.09.020.
Butler, A., & Berglund, U. (2012). Landscape character assessment as an approach to understanding public interests within the European landscape convention. *Landscape Research, 39*(3), 219–236. https://doi.org/10.1080/01426397.2012.716404.
Carpenter, S. R., Mooney, H. A., Agard, J., et al. (2009). Science for managing ecosystem services: Beyond the millennium ecosystem assessment. *Proceedings of the National Academy of Sciences of the United States of America, 106*, 1305–1312.
Carvalho Ribeiro, S. M., & Lovett, A. (2011). Is an attractive forest also considered well managed? Public preferences for forest cover and stand structure across a rural/urban gradient in northern Portugal. *Forest Policy and Economics, 13*(1), 46–54.
Carvalho Ribeiro, S. M., Migliozzi, A., Incerti, G., et al. (2013). Placing land cover pattern preferences on the map: Bridging methodological approaches of landscape preference surveys and spatial pattern analysis. *Landscape and Urban Planning, 114*, 53–68.
Casalegno, S., Inger, R., Desilvey, C., et al. (2013). Spatial covariance between aesthetic value & other ecosystem services. *PLoS One, 8*(6), e68437. https://doi.org/10.1371/journal.pone.0068437.
Cash, D. W., Adger, W. N., Berkes, F., et al. (2006). Scale and cross-scale dynamics: Governance and information in a multilevel world. *Ecology and Society, 11*(2), 8.
Cassatela, C., & Peano, A. (2011). *Landscape indicators: Assessing and monitoring landscape quality*. Dordrecht: Springer.
Council of Europe. (2000). *European landscape convention*. http://conventions.coe.int/Treaty/en/Treaties/Html/176.htm. Accessed 8 Apr 2015.
CPRE – Campaign to Protect Rural England. (2007). *A map showing the range of tranquillity in England*. http://www.cpre.org.uk/resources/countryside/tranquil-places/item/1839. Accessed 24 Aug 2018.
Croucher, K., Myers, L., & Bretherton, J. (2007). *The links between greenspace and health: A critical literature review*. Stirling: University of York.
Daniel, T. C. (2001). Whither scenic beauty? Visual landscape quality assessment in the 21st century. *Landscape and Urban Planning, 54*, 267–281.
Dick, J., Maes, J., Smith, R. I., et al. (2014). Cross-scale analysis of ecosystem services identified and assessed at local and European level. *Ecological Indicators, 38*, 20–30.
Dramstad, W. E., Tveit, M. S., Fjellstad, W. J., et al. (2006). Relationships between visual landscape preferences and map-based indicators of landscape structure. *Landscape and Urban Planning, 78*, 465–474.
Fairclough, G. (2004). History and time: Managing landscape and perceptions. In M. Berlan-Darqué, Y. Luginbühl, & D. Terrasson (Eds.), *Landscape, from knowledge to action* (pp. 147–160). Versailles Cedex: Éditions Quæ.
Frank, S., Fürst, C., Koschke, L., et al. (2013). Assessment of landscape aesthetics – Validation of a landscape metric-based assessment by visual estimation of the scenic beauty. *Ecological Indicators, 32*, 222–231.
Gobster, P. H., Nassauer, J. I., Daniel, T. C., et al. (2007). The shared landscape: What does aesthetics have to do woth ecology? *Landscape Ecology, 22*(7), 959–972.

Hanstein, U. (1967). Über die Gewohnheiten, Ansichten und Wünsche der Waldbesucher. *Allgemeine Forstzeitschrift, 22*, 465–467.

Herbst, H., Förster, M., & Kleinschmidt, B. (2009). Contribution of landscape metrics to the assessment of scenic quality – The example of the landscape structure plan Havelland/Germany. *Landscape Online, 10*, 1–17.

Hermes, J., van Berkel, D., Burkhard, B., et al. (2018a). Assessment and valuation of recreational ecosystem services of landscapes. *Ecosystem Services, 31*, 289–295. https://doi.org/10.1016/j.ecoser.2018.04.011.

Hermes, J., Albert, C., & von Haaren, C. (2018b). Assessing the aesthetic quality of landscapes in Germany. *Ecosystem Services, 31*, 296–307. https://doi.org/10.1016/j.ecoser.2018.02.015.

Jacobs M (2011) Psychology of the visual landscape. In: Nijhuis S, von Lammeren R, van der Hoeven F (eds) Exploring the visual landscape – Advances in physiognomic landscape research in the Netherlands. IOS Press, Wageningen, p 41–55.

Jones, P. J., Andersen, E., Capitani, C., et al. (2016). The EU societal awareness of landscape indicator: A review of its meaning, utility and performance across different scales. *Land Use Policy, 53*, 112–122.

Kahn, P. H., Friedman, B., Gill, B., et al. (2008). A plasma display window? The shifting baseline problem in a technologically mediated natural world. *Journal of Environmental Psychology, 28*, 192–199.

Kaplan, R. (2001). The nature of the view from home: Psychological benefits. *Environment and Behavior, 33*, 507–542.

Kiemstedt, H. (1967). *Zur Bewertung natürlicher Landschaftselemente für die Planung von Erholungsgebieten*. Sehnde: Jänecke, Hannover.

Kowarik, I. (2006). Natürlichkeit, Naturnähe und Hemerobie als Bewertungskriterien, [Naturalness, closeness to nature and hemeroby as evaluation criteria]. In O. Fränzle, F. Müller, & W. Schröder (Eds.), *Handbuch der Umweltwissenschaften: Grundlagen und Anwendungen der Ökosystemforschung* (Vol. 16, pp. 1–18). Weinheim: Wiley-VCH.

Kushner, B., Waite, R., Jungwiwattanaporn, M., et al. (2012). *Influence of coastal economic valuations in the Caribbean: Enabling conditions and lessons learned* (Working Paper). Washington DC: World Resources Institute.

Land Use Consultants. (2011). *Landscape character assessment guidance for England, Scotland & Wales (Unpublished Consultation Draft)*. London: Land Use Consultants.

Lefebvre, M., Espinosa, M., Paloma Gomez, S., et al. (2015). Agricultural landscapes as multi-scale public good and the role of the Common Agricultural Policy. *Journal of Environmental Planning and Management, 58*(12), 2088–2112.

Martínez Pastur, G., Peri, P. L., Lencinas, M. V., et al. (2015). Spatial patterns of cultural ecosystem services provision in Southern Patagonia. *Landscape Ecology, 31*(2), 383–399.

Martín-López, B., Gómez-Baggethun, E., Lomas, P. L., et al. (2009). Effects of spatial and temporal scales on cultural services valuation. *Journal of Environmental Management, 90*(2), 1050–1059.

Milcu, A. I., Hanspach, J., Abson, D., et al. (2013). Cultural ecosystem services: A literature review and prospects for future research. *Ecology and Society, 18*(3), 44.

Mücher, C. A., Klijn, J. A., Wascher, D. M., et al. (2010). A new European Landscape Classification (LANMAP): A transparent, flexible and user-oriented methodology to distinguish landscapes. *Ecological Indicators, 10*, 87–103.

Nohl, W. (2001). *Landschaftsplanung: Ästhetische und rekreative Aspekte*. Berlin/Hannover: Patzer.

Ode, A., & Miller, D. (2011). Analysing the relationship between indicators of landscape complexity and preference. *Environment and Planning B: Urban Analytics and City Science, 38*(1), 24–40.

Ode Sang, Å., & Tveit, M. S. (2013). Perceptions of stewardship in Norwegian agricultural landscapes. *Land Use Policy, 31*, 557–564.

Ode, Å., Fry, G., Tveit, M. S., et al. (2009). Indicators of perceived naturalness as drivers of landscape preference. *Journal of Environmental Management, 90*(1), 375–383. https://doi.org/10.1016/j.jenvman.2007.10.013.

Opdam, P., Foppen, R., & Vos, C. (2001). Bridging the gap between ecology and spatial planning in landscape ecology. *Landscape Ecology, 16*(8), 767–779.

Paracchini, M. L., & Capitani, C. (2011). *Implementation of a EU wide indicator for the rural-agrarian landscape* (JRC Scientific and Technical Reports EUR25114EN-2011). Brussels: European Commission, Joint Research Centre.

Paracchini, M. L., Capitani, C., Schmidt, A. M., et al. (2012). *Measuring societal awareness of the rural agrarian landscape: Indicators and scale issues. EUR 25192 EN – 2012*. Luxembourg: Joint Research Centre. Publications Office of the European Union.

Paracchini, M. L., Zulian, G., Kopperoinen, L., et al. (2014). Mapping cultural ecosystem services: A framework to assess the potential for outdoor recreation across the EU. *Ecological Indicators, 45*, 371–385.

Plieninger, T., Dijks, S., Oteros-Rozas, E., et al. (2013). Assessing, mapping, and quantifying cultural ecosystem services at community level. *Land Use Policy, 33*, 118–129.

Rockström, J., Steffen, W., Noone, K., et al. (2009). A safe operating space for humanity. *Nature, 461*, 472–475.

Roser, F. (2011). *Entwicklung einer Methode zur großflächigen rechnergestützten Analyse des landschaftsästhetischen Potenzials*. Berlin: Weißensee Verlag.

Roth, M., & Gruehn, D. (2006). Die Bedeutung von Landschaftselementen für das Landschaftserleben: Vorstellung eines empirisch basierten Ansatzes zur validen Landschaftsbildbewertung auf der Ebene des Landschaftsprogramms. In B. Kleinschmit & U. Walz (Eds.), *Landschaftsstrukturmaße in der Umweltplanung. Beiträge zum Workshop der IALE-AG Landschaftsstruktur – Berlin 2006 (Landschaftsentwicklung und Umweltforschung)* (pp. 154–168). Berlin: TU Berlin.

Rüdisser, J., Tasser, E., & Tappeiner, U. (2012). Distance to nature – A new biodiversity relevant environmental indicator set at the landscape level. *Ecological Indicators, 15*, 208–216.

Russel, R., Guerry, A. D., Balvanera, P., et al. (2013). Humans and nature: How knowing and experiencing nature affect Well-being. *Annual Review of Environment and Resources, 38*, 473–450.

Schipperijn, J., Stigsdotter, U. K., Randrup, T. B., et al. (2010). Influences on the use of urban green space – A case study in Odense, Denmark. *Urban Forestry and Urban Greening, 9*(1), 25–32. https://doi.org/10.1016/j.ufug.2009.09.002.

Sen, A., Harwood, A. R., Bateman, I. J., et al. (2014). Economic assessment of the recreational value of ecosystems: Methodological development and national and local application. *Environmental and Resource Economics, 57*(2), 233–249.

Steinitz, C. (2010). An assessment of the visual landscape of the autonomous region of Valencia, Spain: A case study in linking research, teaching and landscape planning. *Landscape, 21*(2010), 14–33.

Sukopp, H. (1976). Dynamik und Konstanz in der Flora der Bundesrepublik Deutschland, [Dynamics and stability in Flora of the Federal Republic of Germany]. *Schrreihe Vegkd, 10*, 9–26.

Surova, D., & Pinto-Correia, T. (2008). Landscape preferences in the cork oak Montado region of Alentejo, southern Portugal: Searching for valuable landscape characteristics for different user groups. *Landscape Research, 33*, 311–330.

Swanwick, C. (2002). *Landscape character assessment. Guidance for England and Scotland.* Cheltenham: Countryside Agency.

Swanwick, C. (2009). Society's attitudes to and preferences for land and landscape. *Land Use Policy, 26*, 62–75.

Szerdahelyi, I. (2003). *Bevezetés az esztétikába (Introduction to the Aesthetics) Student Manual.* Budapest: Highschool Zsigmond Király.

Tahvanainen, L., Tyrvainen, L., Ihalainen, M., et al. (2001). Forest management and public perceptions – Visual versus verbal information. *Landscape and Urban Planning, 53*, 53–70.

TEEB. (2010). *The economics of ecosystems and biodiversity: Mainstreaming the economics of nature: A synthesis of the approach, conclusions and recommendations of TEEB*. http://www.teebweb.org. Accessed June 2018.

Tudor, C. (2014). *An approach to landscape character assessment*. London: Natural England.

Tveit, M., Ode, Å., & Fry, G. (2006). Key concepts in a framework for analysing visual landscape character. *Landscape Research, 31*(3), 229–255. https://doi.org/10.1080/01426390600783269.

van Zanten, B. T., Verburg, P. H., Koetse, M. J., et al. (2014). Preferences for European agrarian landscapes: A meta-analysis of case studies. *Landscape and Urban Planning, 132*, 89–101.

von Haaren, C., Galler, C., & Ott, S. (2008). *Landscape planning. The basis of sustainable landscape development*. Bonn: Bundesamt für Naturschutz.

Waltert, F., Schulz, T., & Schläpfer, F. (2011). The role of landscape amenities in regional development: Evidence from Swiss municipality data. *Land Use Policy, 28*(4), 748–761. https://doi.org/10.1016/j.landusepol.2011.01.002.

Walz, U., & Stein, C. (2014). Indicators of hemeroby for the monitoring of landscapes in Germany. *Journal for Nature Conservation, 22*, 279–289.

Warnock, S., & Griffiths, G. (2015). Landscape characterisation: The living landscapes approach in the UK. *Landscape Research, 40*(3), 261–278. https://doi.org/10.1080/01426397.2013.870541.

Whitten, S. M., & Bennett, J. W. (2002). A travel cost study of duck hunting in the upper south east of South Australia. *Australian Geographer, 33*(2), 207–221.

Wilson, E. O. (1986). *Biophilia – The human bond with other species*. Cambridge, MA: Harvard University Press.

Wood, S. A., Guerry, A. D., Silver, J. M., et al. (2013). Using social media to quantify nature-based tourism and recreation. *Scientific Reports, 3*, 2976. https://doi.org/10.1038/srep02976.

Geodiversity: The Natural Support System of Ecosystems

16

Jenni A. Turner

Abstract

Geodiversity, the diverse range of properties and processes of the abiotic natural world, provides services which support biodiversity, thus geodiversity is intrinsically linked with ecosystem services (ES). Many countries have adopted strategies and frameworks for identifying sites that exhibit valuable geodiversity and have implemented measures for conserving them, including recognition of their distinctive nature within the planning process. However, there is no international legislation enforcing protection of geodiversity, and since there is much spatial variability in the services provided the responsibility for identifying sites and informing planning policies is often devolved to local bodies. This has resulted in a variety of approaches, so this chapter offers a broad framework for formulating geodiversity action plans, carrying out site audits and assessing the value of sites in terms of their geosystem services. Useful resources for both theoretical and applied geodiversity practice are identified and the examples of geosite assessments in this chapter can be expanded and adapted to suit user requirements in order to demonstrate links to ES and planning.

Keywords

Geodiversity · Geoparks · Action plans · Site audits

16.1 Introduction to Geodiversity

Geodiversity refers to the variety of abiotic nature that is the non-biological part of the natural world. It has been defined by Murray Gray (2004: 8) as:

J. A. Turner (✉)
School of Environmental Sciences, University of East Anglia, Norwich, UK
e-mail: jenni.turner@uea.ac.uk

C. von Haaren et al. (eds.), *Landscape Planning with Ecosystem Services*,
Landscape Series 24, https://doi.org/10.1007/978-94-024-1681-7_16

> the natural range (diversity) of geological (rocks, minerals, fossils), geomorphological (landforms, processes) and soil features. It includes their assemblages, relationships, properties, interpretations and systems.

To better appreciate the value of geodiversity Gray asks the reader to imagine a landscape lacking in such variations and conjures an image of a hypothetical smoothly spherical earth where the landscape is monotonously flat and so without mountains, valleys and deep basins occupied by seas; the rock type is homogenous so lacks a variety of goods such as minerals, building materials or hydrocarbons. Capacities such as the ability to store water, weathering of diverse rock types to produce a variety of soils for growing different crops or supporting different ecosystems also lack variety.

Diversity in the abiotic (geo) world is therefore intrinsically linked with ecosystem services and landscape planning. As illustrated in Fig. 16.1 it underpins many types of provisioning or regulating services (e.g. those relating to food production and supply of materials, see Chap. 10), as well as providing the medium which hosts biotic nature (Chap. 18). In addition to these indirect associations, geodiversity directly supplies other services and benefits, including those associated with scientific knowledge, human experience of landscapes and use for recreational activities.

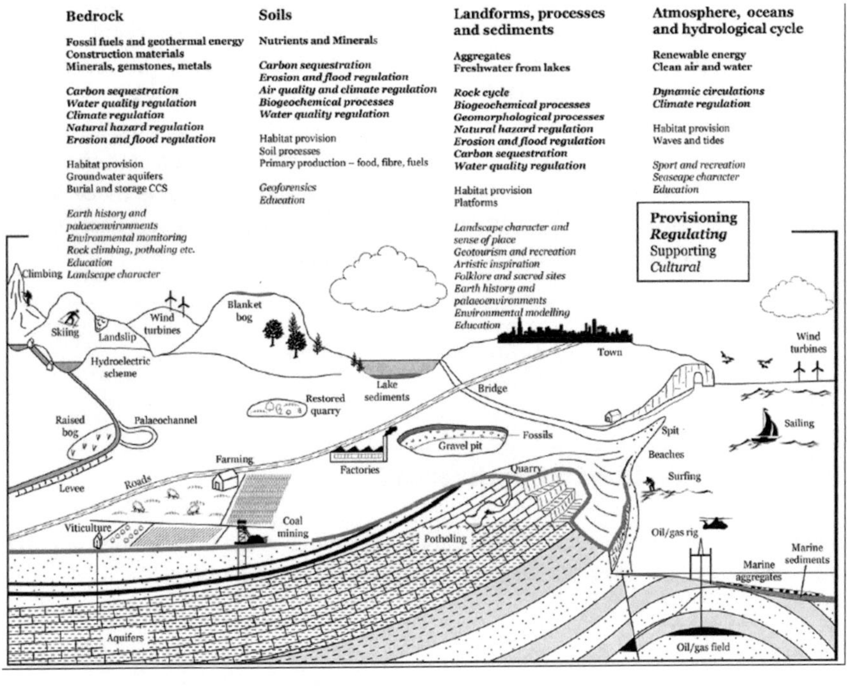

Fig. 16.1 Examples of goods and services derived from geodiversity (Gray et al. 2013, permission to reproduce kindly granted by Elsevier)

The recognition of the importance of geodiversity in landscape planning has increased rapidly during the past decade (Gray 2009). However, awareness of the services associated with geodiversity lags behind those for biodiversity, possibly because the geosphere is perceived as more robust and durable so not requiring conservation. Nevertheless, the importance of the connection between the geosphere and biosphere has received more attention in landscape planning and criteria such as endangerment or rareness are being used to prioritize protective or management actions with respect to geodiversity.

16.2 Planning and Implementation Framework

Examples of individual areas being designated because of their geological or geomorphological features can be traced back to the nineteenth century. For instance, one of the earliest nature conservation initiatives in Germany (1836) sought to protect the silhouette of the Drachenfels mountain. However the first global initiative for geoconservation did not occur until 1972 at the United Nations, Educational, Cultural and Scientific Organisation (UNESCO) World Heritage Convention when 189 UN member states signed a treaty to conserve sites of:

> cultural and/or natural significance which is so exceptional as to transcend national boundaries and to be of common importance for present and future generations (UNESCO 2013: Para 49).

Specific reference to geological and physio-graphical formations ensured geoconservation of World Heritage Sites such as the Veneto mountains in the Dolomites (Italy), Surtsey volcano in Iceland and the Great Copper Mountain in Falun (Sweden).

The value of geodiversity at regional and local scales was given more recognition when the European Geoparks Network was established in 2000 (subsequently extended as a global network). The adoption of Recommendation Rec(2004)3 by the Council of Europe Committee of Ministers placed further responsibility on member states to identify and manage areas of special geological interest. European Geoparks such as the Vulkaneifel European Geopark (Germany) and Parco di Madonie (Italy) have since flourished. These policy developments at the EU level have been driven by a combination of bottom-up efforts by local and regional bodies and pan-national organizations such as ProGeo (the European Association for the Conservation of Geological Heritage, http://www.progeo.se/). ProGeo has been particularly effective in translating recommendations into actions and as a driving force working with the International Union for Conservation of Nature (IUCN) and the United Nations Educational, Scientific and Cultural Organization (UNESCO, which manages important World Heritage Sites and Global Geoparks, http://whc.unesco.org/en/list) to promote geoconservation and integration with nature conservation. However, the degree to which individual European countries have adopted geodiversity in landscape planning, embraced geoconservation and declared new geoheritage sites varies considerably.

An international milestone was reached in 2008 when for the first time in its 60 year history the IUCN adopted a resolution relating to the Conservation of Geodiversity and Geological Heritage. It is probably no coincidence that the United Nations General Assembly proclaimed 2008 to be the International Year of Planet Earth, initiated jointly by the International Union for Geological Science (IUGS) and UNESCO in order to (World Conservation Congress 2008: 102):

> increase awareness of the importance of earth sciences in achieving sustainable development and promoting local, national, regional and international action

[and emphasizing]

> that geodiversity, understood to include geological and geomorphological diversity, is an important natural factor underpinning biological, cultural and landscape diversity, as well as an important parameter to be considered in the assessment and management of natural areas.

Contrary to biodiversity protection, there is no international legislation covering geodiversity and consequently at this level there are no frameworks, standards or guidelines on procedures for measuring, recording and valuing geodiversity. However, several European Directives and Conventions require an understanding of the functional support and underpinning that geodiversity provides for biodiversity and landscape, for example the Habitats and Species Directives, the Water Framework Directive, the Floods Directive and the European Landscape Convention. In addition to IUCN, UNESCO and ProGeo there are several international organizations supporting geodiversity, for example the Committee of Ministers of the Council of Europe, the Nordic Council of Ministers, the European Federation of Geologists (EFG, www.eurogeologists.de), International Association of Geomorphologists (www.geomorph.org) and Coastal and Marine Union (EUCC, www.eucc.nl/en/).

An increasing number of EU countries have included geoconservation in policy frameworks, two examples being the revision of the Spanish National Law on Protection of Natural Areas and Wildlife (Carcavilla et al. 2009) and guidance published for England (http://publications.naturalengland.org.uk/category/30050). Such legislation and official documents provide a basis for incorporating considerations regarding geodiversity and associated ecosystem services into landscape planning, spatial planning and environmental impact assessments. This can be achieved through a variety of mechanisms including designating areas for protection, considering geodiversity in land use zoning strategies and investing in infrastructure to support the development of geosites for recreation and tourism purposes. Implementing such strategies typically requires the application of assessment criteria to identify priorities on a regional scale, but also with regard to how unique specific features may be on a national, European or even global scale. This is particularly true if the nature of the geodiversity is such that it has implications for recreation and tourism, in which case methods to estimate monetary use and non-use values may also help in the assessment of options.

16.3 Assessment Framework

There are no EU frameworks for geodiversity assessments, thus countries may implement their own guidelines for auditing and assessing sites for geodiversity value. In practice the available guidance varies greatly in scope and content. Ideally, each country will have a National Geodiversity Action Plan (GAP), which both feeds into and is fed by local plans – people tend to engage more readily in initiatives that have a local (and therefore directly relevant) impact rather than those at a less tangible national scale. The importance of localness was emphasised in Agenda 21 which was agreed at the United Nations Conference on Environment and Development (UNCED) Summit in 1992 and has been very influential in environmental planning. The aim of local GAPs is to formulate a management framework for observing, conserving and enhancing the valued geology, landforms, soils and associated earth heritage features within a defined area (e.g. as agreed by local government or other statutory organizations).

Figure 16.2 sets out a typical sequence of steps for the development of a Local Geodiversity Action Plan (LGAP) which meets policy requirements, but is tailored

Fig. 16.2 A framework for developing and implementing a Local Geodiversity Action Plan

to the needs of identified stakeholders and local geo-diversity processes and properties. Whilst most LGAPs are written by local interest groups, increasingly organizations whose industries utilize geo-goods are adopting GAPs. For example the UK's Department for Environment, Food and Rural Affairs (DEFRA) was funded by the aggregates sector to draw up a GAP for that industry (Thompson et al. 2006). There are a range of useful documents in the English language readily available online which provide guidance on setting up GAPs. At the strategic level these include the UK Geodiversity Action Plan (Burek and Potter 2006, http://www.ukgap.org.uk/) and a good practice guide by Prosser et al. (2006) which is available from the Natural England web-site listed in Sect. 16.2 above).

The key deliverable of LGAPs is to identify sites that have properties and processes providing services of such value that conservation is required. The aims and objectives of geoconservation should be clearly defined by the partnership, with key criteria likely to include quality, rareness and value of the site properties and landscape elements. The specific criteria are usually decided by the partnership in the context of abundance and quality of similar properties in the area covered by the LGAP. Where properties are also considered of value on a broader scale, such as national or international, the site may be considered as a candidate for higher level conservation such as SSSI, or World Heritage Site.

A LGAP may include information on, and examples of, sites which typify the geodiversity which characterizes the area, examples of existing geo-conservation sites and future priorities such as developing partnerships and education opportunities. An example for one county in the UK is provided by Holt-Wilson (2011) and a review of UK experience is given by Dunlop et al. (2018).

Geoconservation sites are typically identified through a clearly defined and sequential process. As illustrated in Fig. 16.3 this commences with identification of potential sites, progressing through site audit and selection for conservation, to geosite designation through a statutory or non-statutory framework and site conservation, positive management and promotion.

The next section provides an example of the audit and evaluation processes. This procedure is also applicable to audits of existing geoconservation sites as part of management plans to monitor and report on changes in their state.

16.3.1 Identifying and Characterising Geodiversity

A general framework by which a geosite is evaluated will include diverse criteria for site assessment, ranging from uniqueness and quality of the site geodiversity through social and aesthetic values, to importance in relation to capacity for supporting ecosystems services (Table 16.1).

Where a location is a potential candidate as a geodiversity site the evaluation will follow an agreed audit of properties and processes that is typically developed at regional or local level. The methods involved will vary widely depending on the criteria employed and extent of existing databases, but site selection typically involves both desk and field work, a typical audit form will include recognition of how the geosphere supports ecosystem services, for example:

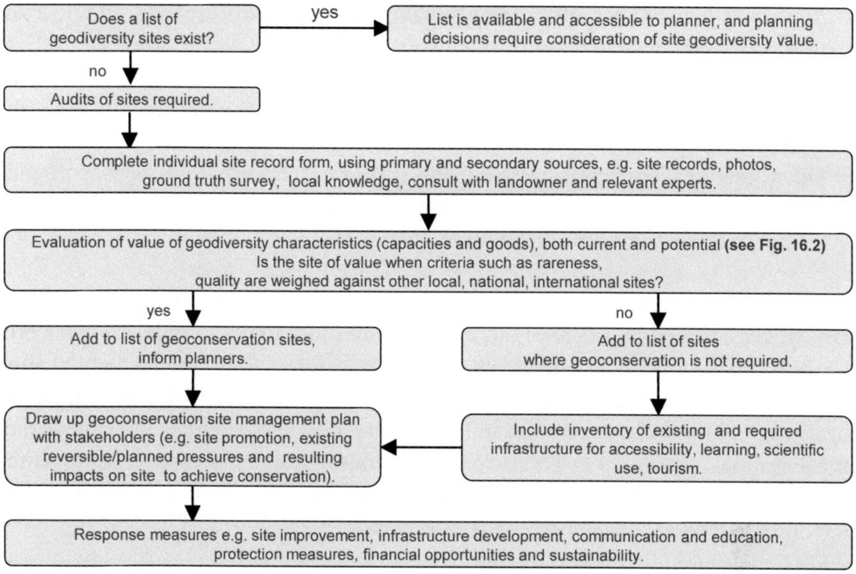

Fig. 16.3 Typical workflow to identify priority geoconservation sites

Table 16.1 Examples of criteria used to assess site geodiversity with details of methods and data sources that can be employed

Criteria	Parameters	Evaluation standards	Sources and methods
Rareness, endangerment, quality	Frequency and proportion of geosite properties and/or processes, quality relative to similar occurrences	Occurrence, existing recognition and protection (e.g. SSSI), statistical analysis	GIS analysis, existing data bases, historical and geological maps
Cultural, heritage value	Age, continuity of land use, historical connections	Existing recognition and protection, reputation	Published records and reports, measurable assets
Economic, social, aesthetic value	Willingness to pay, measurable revenue, accessibility	Frequency of use, footfall, monetary revenue, popularity	Surveys, published records and reports, publicity material
Benefit to education, science	Derived learning, contribution to scientific understanding and research	Derived learning, value to science	Educational bodies, scientific publications and citations
Ecosystem service provider	Dependence of a valued ecosystem service on a geosite	Extent of ecosystem service dependence, and the value of dependent ecosystem services	Ecosystems service records and reports

- Site status e.g. site conditions and threats, current and historic use, site description, any existing designations or conservation (e.g. of wildlife status).
- Geofeatures e.g. quantitative and qualitative record of site properties and processes, such as the geology, paleontology, soil characteristics, water, and geomorphology.
- Other features e.g. educational value, historical information, archaeological records, any particular cultural, wildlife, aesthetic and recreational attributes, references for sources of information.

The geosite services may well not be identified as 'capacities' and 'goods' on the sorts of site audit forms currently in use; they are more likely to follow a version of a popular classification of geosystem services (e.g. Gray 2011). As part of the audit process the current *and* potential beneficiaries should be identified, including organizations, public bodies, societies, industries and local communities who already, or in future, may benefit from geological and geomorphological conservation. When identifying possible future beneficiaries of the geosystem services at a site Prosser et al. (2006) suggest the following as examples:

- Those involved in geological and geomorphological research, seeking to understand the earth and the environmental change impacting on it.
- Geologists working in those industries seeking to find, utilise and manage mineral and water resources, or manage the natural environment.
- Land owners, land managers, public utilities, planning authorities and others, who require some understanding of geology, geomorphology and landscape planning, to better inform their decisions and actions.
- Ecologists, and those involved in nature conservation more generally, who need some understanding of geology, to help, for example, in planning habitat re-creation projects.

Following the site audit there should be an evaluation of the value of the identified capacities and services which, when summed and compared to other local sites, form the basis of a decision as to whether the site should be recommended for geoconservation. Whilst the method for measuring and expressing the overall value of a site varies in practice, judgments usually include quantitative and qualitative indicators of the overall state, including the uniqueness, quality, condition and potential worth of each property and process in terms of its services. The value assessment process is likely to, and should, vary to reflect relevant local values, but where possible the degree of subjectivity should be minimized and the steps involved in arriving at a final assessment should be transparent.

16.3.2 Valuation of Geosite States and Impacts

The priorities assigned to different properties and services are likely to vary between GAPs, but should be transparent in the workflow for identifying geoconservation

	1 Present	2 Fair	3 Good	4 V Good	5 Outst	n/a	Don't know
SCIENTIFIC VALUE							
▪ Stratigraphy							
Time zones represented							
Stratotype present							
Major unconformity							
▪ Lithology / rock types							
▪ Sedimentary structs							
▪ Palaeontology							
▪ Landforms							
▪ Geomorpho processes							
▪ Minerals							
▪ Archaeology							
▪ Soil features							
▪ Tectonic and other							
▪ Published refs							
CULTURAL VALUE							
Earth Science assoc							
Economic / built envt							
History & folklore							
Landscape & aesthetic							
GEODIVERSITY VALUE							
Uniqueness at county level							
USABILITY							
Physical accessibility							
Free public access							
Condition e.g. overgrown							
Known hazards							
Amenity / leisure use							
EDUCATION VALUE							
Primary Schools Secondary Schools Tertiary Level General Interest Groups							
SUBTOTALS							
SCORE							

EXISTING DESIGNATIONS	SSSI geo	SSSI bio	GCR	RIGS	CWS	Other

RECOMMENDATIONS	
National designation SSSI geo	
County designation CGS	
Local geosite only	
Get more information	

Reasons / comments	

Fig. 16.4 An example site evaluation form. (Source: Norfolk Geodiversity Partnership)

sites (Fig. 16.4). Assigning a relative worth (e.g. on a scale of 1–5) to the various capacities and goods (e.g. scientific value, cultural value) identified in each site audit should be a stage that is distinct from the audit itself in order to provide clarity.

Using such ratings also helps to facilitate future site surveys where site characteristics may not change but their value does; a separate assessment stage will allow changes in state over time to be measured. In addition, the evaluation methods and assessment criteria should be standardized to facilitate intra- and inter-geosite comparison at least within a LGAP area. These assessment criteria are likely to include degree of rareness and quality of the site, noting both the current and potential values. An example assessment form is shown in Fig. 16.4 where a score of 4 or more for any characteristic qualified a site for recommendation as a Local Geoconservation Site.

Whilst ratings such as the 1–5 scale in Fig. 16.4 are easy to use they also embed the user's opinions about rareness, quality and other attributes, a part of the assessment that ideally should be transparent. For some services (e.g. amenity and leisure use) alternative metrics such as access payments may be possible. Bruschi et al. (2011) reviewed the range of values given by experts to various geodiversity sites and found a broad commonality in the weightings employed, but also noted there was difficulty creating a completely objective methodology that would be widely applicable. As a general rule, consulting experts where there is an agreed lack of expertise amongst the assessors and formulating a reference set of weightings based on the LGAP priorities are likely to help improve transparency and reduce user subjectivity.

When all site attributes are weighted and summed the total value is a good indication of the site suitability for geoconservation and the contributing values help to identify the important features for site management plans. For instance, recreational value could be quantified by monitoring payments at a designated car park, visitor numbers, or questionnaires which determined a willingness to pay to visit the site. In assessing the overall value of recreation a low weighting might be applicable if there were numerous sites of similar tourist attraction in the area. However, if a geosite is an exemplar for a particular geological formation then a high value weighting for scientific interest would be appropriate and could mean that it was identified for conservation even if it scored lowly on all other categories.

To increase the rigour of the assessment process, statistical methods may be applied, either in addition to or instead of weighting of values. Benito-Cavlo et al. (2009) used GIS to classify climatic, geological and land surface characteristics in the Iberian peninsula and compared several statistical techniques (Shannon's Index, Simpson's Index, Patch Richness Density) to assess regional geodiversity. They found that while the Patch Richness Density undervalued the geodiversity, the Shannon and Simpson indices were reliable as an objective evaluation of the *relative* regional geodiversity and so allowed a large area to be rapidly assessed. Experience with audits suggests that many sites have features of local interest so such techniques can be of particular value in identifying the top geoconservation priorities in regional or national planning frameworks.

16.4 Examples of Geosite Audits and Management

As the implementation of geosite audits and assessments depends appreciably on the sources of available information and the nature of any survey, three case studies are provided here to illustrate different approaches.

16.4.1 Case Study 1: GIS Audit

Where an audit is a review of site characteristics as part of an established monitoring programme there are likely to be many sources of readily available information, for example in a Geopark there will be data on footfall and opportunities for surveys to obtain quantitative values on the site as a tourist attraction (see Webber et al. 2006 for an assessment of the social and economic value of UK geodiversity). At the other extreme an audit may be a primary evaluation of a large, remote and inaccessible area with little existing data so a more suitable approach could be remotely sensed data collected by satellites with analyses using GIS software.

An example is provided by Hjort and Luoto (2012) who conducted a geodiversity audit of a remote and large area in Finland (Fig. 16.5). As ground truthing was impractical and primary data were lacking the authors used GIS processing and statistical techniques to explore relationships between numbers of geodiversity features and landscape variables (e.g. elevation, slope, solar radiation). Capacities and goods were not explicitly identified, but secondary data sets regarding services such

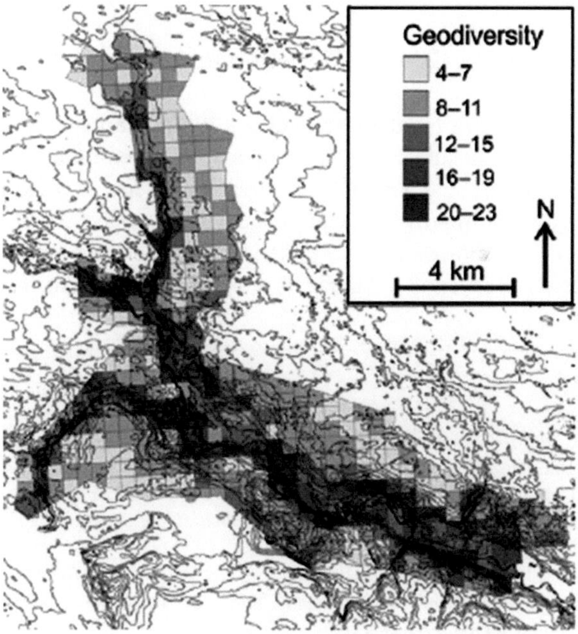

Fig. 16.5 Case Study 1: Can geodiversity be evaluated from space using GIS? The figure shows spatial patterns of geodiversity features in northern boreal Oulanka, Finland, with the legend illustrating the number of different features per cell. (Hjort and Luoto 2012, Fig 2, permission to reproduce kindly granted by Elsevier). The authors draped remote sensed data on bedrock geology, geomorphology and hydrology over a DEM to measure topographic heterogeneity. Multivariate statistical techniques were applied to evaluate relationships with geodiversity and mapped a range of features including springs, tors, sandbars and deflation surfaces

as agricultural cropping and water resources were generated. This example suggests that geodiversity may underpin connections between capacities and goods.

16.4.2 Case Study 2: Site Assessment in a Landscape Context

The assessment of state or impacts at a site scale may require consideration of processes operating across a larger geographical area. For instance, Stace and Larwood (2006) discuss the impact of agricultural drainage and activity on the distinctive peat soils and landscape of the Fenlands in eastern England. Inappropriate land management has resulted in breakdown of the organic matter that makes up the bulk of these soils, resulting in loss of mass and wind erosion that potentially limits their agricultural capacity. The impacts extend over a wide area as the blown soils are carried in the atmosphere and can discolour water bodies, requiring additional treatment of abstracted water used for human consumption. Resolving these issues requires changes in farming practice to conserve the peat soils and protect water resources. Responses to impacts and the management of capacities and goods in the context of a broader landscape planning framework is further discussed in Chap. 19.

16.4.3 Case Study 3: Geosite Management

The Giant's Causeway in Northern Ireland is a World Heritage Site valued for the striking basalt rock exposures. The site supports protected flora and fauna, as well as having many cultural associations. As a result of a fire at the site management plans were revised to improve education and visitor awareness of the geological features and this has improved visitor understanding, demonstrating the benefits of including review and revision of strategies within geosite management plans.

References

Benito-Calvo, A., Pérez-González, A., Magriand, O., et al. (2009). Assessing regional geodiversity: The Iberian Peninsula. *Earth Surface Processes and Landforms, 34*, 1433–1445.

Bruschi, V. M., Cendrero, A., & Alertose, J. A. C. (2011). A statistical approach to the validation and optimisation of geoheritage assessment procedures. *Geoheritage, 3*, 131–149. https://doi.org/10.1007/s12371-011-0038-9.

Burek, C., & Potter, J. (2006). *Local geodiversity action plans – Setting the context for geological conservation. English nature research report 560.* http://publications.naturalengland.org.uk/category/47017. Accessed 21 May 2018.

Carcavilla, L., Durán, J. J., García-Cortés, A., et al. (2009). Geological heritage and geoconservation in Spain: Past, present, and future. *Geoheritage, 1*, 75–91. https://doi.org/10.1007/s12371-009-0006-9.

Dunlop, L., Larwood, J. G., & Burek, C. V. (2018). Geodiversity action plans – A method to facilitate, structure, inform and record action for geodiversity. In E. Reynard & J. Brilha (Eds.), *Geoheritage – Assessment, protection and management* (pp. 53–65). Amsterdam: Elsevier.

Gray, M., Gordon, J. E., & Brown, E. J. (2013). Geodiversity and the ecosystem approach: The contribution of geoscience in delivering integrated environmental management. *Proceedings of the Geologists Association, 124*(4), 659–673.

Gray, M. (2011). Comment other nature: Geodiversity and geosystems services. *Environmental Conservation, 38*(3), 271–274. https://doi.org/10.1017/S0376892911000117.

Gray, M. (2009). Saving the stones. *Geographical Magazine, 81*(3), 34.

Gray, M. (2004). *Geodiversity: Valuing and conserving abiotic nature*. Chichester: Wiley. http://geoduma.files.wordpress.com/2010/02/geodiversity.pdf. Accessed 18 Oct 2017.

Holt-Wilson, T. (2011). *Norfolk's earth heritage – Valuing our geodiversity*. Norwich: Interprint. http://www.nbis.org.uk/sites/default/files/documents/Norfolk's%20Earth%20Heritage_Screen%201-17.pdf. Accessed 18 Oct 2017.

Hjort, J., & Luoto, M. (2012). Can geodiversity be predicted from space? *Geomorphology, 153*(154), 74–80.

Norfolk Geodiversity Partnership https://sites.google.com/site/norfolkgeodiversity. Accessed 16 June 2018.

Prosser C, Murphy M, Larwood J (2006) *Geological conservation: a guide to good practice. English nature report 145.* http://publications.naturalengland.org.uk/publication/83048. Accessed 18 Oct 2017.

Stace, H., & Larwood, J. G. (2006). *Natural foundations: Geodiversity for people, places and nature*. Peterborough: English Nature.

Thompson, A., Poole, J., & Carroll, L. (2006). Geodiversity action plans for aggregate companies. In: G. Walton (ed) *Proceedings of the 14th extractive industry geology conference*. EIG Conferences.. http://www.eigconferences.com/2006-proceedings/. Accessed 21 May 2018.

United Nations Educational, Scientific and Cultural Organisation. (2013). *Operational guidelines for the implementation of the world heritage convention. Source paragraph 49.* http://whc.unesco.org/en/guidelines. Accessed 13 June 2018.

Webber, M., Christie, M., & Glasser, N. (2006). *The social and economic value of the UK's geodiversity* (English nature research report 709). http://publications.naturalengland.org.uk/publication/62015. Accessed 18 Oct 2017.

World Conservation Congress, Barcelona. (2008). *CGR4.MOT055 conservation of geodiversity and geological heritage*. https://cmsdata.iucn.org/downloads/motions_english_collated.pdf. Accessed 13 June 2018.

Identification and Evaluation of Habitat Development Potentials

17

Christina von Haaren, Jan Bug, and Jan Barkmann

Abstract

The habitat development potential (HDP), as presented here, is the capability of the abiotic components of the ecosystem to produce biodiversity. We present a method for classifying sites with reference to the value of their possible habitats, including their phytocenosis. The method is based on the observation that the greatest chance for developing biocenoses, with conservation values, exists when site conditions are extreme and vary the most from the standard, 'normal', conditions on agricultural land. The input parameters for the model include: soil moisture level, nutrient content, and pH value. If these parameters are not available they may be deduced with the help of pedotransfer rules from primary soil data. The results offer added value for choosing sites on which; (i) habitats should be mapped in more detail, in order to check how well the state represents the potential; (ii) agri-environmental measures can be allocated with best chances of successfully developing valuable biodiversity; (iii) compensatory mitigation generates the most valuable habitats; (iv) landscapes with small shares of habitat structures can be augmented with best value for money; and (v) a market for soils/sites with a high HDP already exists in the context of habitat banking and

C. von Haaren (✉)
Institute of Environmental Planning, Leibniz University Hannover, Hannover, Germany
e-mail: haaren@umwelt.uni-hannover.de

J. Bug
Landesamt für Bergbau, Energie und Geologie, Hannover, Germany

Institute for Physical Geography and Landscape Ecology, Leibniz University Hannover, Hannover, Germany
e-mail: Jan.Bug@lbeg.niedersachsen.de

J. Barkmann
Fachbereich Gesellschaftswissenschaften, University of Applied Sciences Darmstadt, Darmstadt, Germany
e-mail: jan.barkmann@h-da.de

C. von Haaren et al. (eds.), *Landscape Planning with Ecosystem Services*,
Landscape Series 24, https://doi.org/10.1007/978-94-024-1681-7_17

result-oriented agri-environmental measures. In these latter cases a monetary HDP value can be assigned to soils.

Keywords
Habitat potential · Abiotic components · Soils · Biodiversity · Impact assessment

17.1 Introduction

The entirely new development or substantial rehabilitation of habitats is nowadays a common task in landscape planning. In order to determine where to undertake such activities most effectively, it is important to know which sites offer the best chances for successful development of especially valuable habitats and species. More specifically, it is important to assess the Habitat Development Potential (hereafter HDP) of a site. In knowing the HDP of a site, there is a greater possibility to protect the area from land uses that would destroy its potential. The ecosystem service (ES) values that these sites hold are option values 'borrowed' from the value of the entire habitat, including any vegetation potentially occurring within the habitat area. Therefore, the HDP incorporates elements of both geodiversity and biodiversity (Chaps. 16 and 18). The ES of the HDP values can be evaluated predominantly on the basis of shared societal values, which may be existence values or bequest values expressed in legislation, legitimised agreements etc. (e.g. about the handing over of natural capital unimpaired to the next generation, see Chap. 4). Particularly relevant for the evaluation of the HDP are the goals of EU directives about biodiversity or national soil protection laws.

17.2 Definition and Concept (Box 17.1)

Box 17.1: Definition of Habitat Development Potential
Habitat development potential is the capability of the abiotic components of the ecosystem to produce biodiversity. The corresponding method describes and classifies this potential, with regard to the conservation value of the possible habitats, including their phytocenosis. It is based on the assumption that the greatest chance for developing biocenoses, with conservation values, exists when the site conditions vary most from the standard, 'normal', conditions. In other words, extreme sites have a higher habitat development potential.

The goal of habitat potential assessment is to evaluate whether and how habitat development activities (e.g. types of management) can contribute to the value of a specific site for habitat and species conservation. Information about this scope is especially applicable for improving intensively used areas that presently have little habitat value. The abiotic environment is selectively assessed in light of this goal. Other services such as food or water provision, which are also based on soil, geology, and hydrologic properties, are assessed by methods regarding provisioning and regulating services (Chaps. 10, 11, 12, 13, and 14). The HDP may complement information about geodiversity (Chap. 16), but is less concerned about distinctiveness and possible uses by people. Nevertheless, there may be overlaps between the HDP and geodiversity as distinctive sites are also more likely to be extreme in HDP terms. As a rule, extreme sites offer suitable living conditions for the more sensitive, vulnerable and rare specialists rather than for generalist species (cf. Ellenberg 1996; Preising 1954; Brahms et al. 1989; von Haaren and Bathke 2007; Bredemeier et al. 2015a (see Fig. 17.1)). In the 20th century, predominant land management practices, for instance, draining, fertilising, and liming, have levelled the differences between sites. Due to this widespread impact, the 'normal site' can be characterised as having a well-balanced fresh water regime, good nutrient supply, medium pH value, and agricultural or forestry use (generally intensive).

In order to define and classify the habitat development potential, the abiotic factors (water, climate, and especially soil) are considered. Additional parameters that influence the ability of a site to produce a particularly valuable habitat, and respectively valuable biocenosis, include the seed bank of the soils (e.g. Franke et al. 2009) and the landscape matrix that defines the chances for recolonization (e.g.,

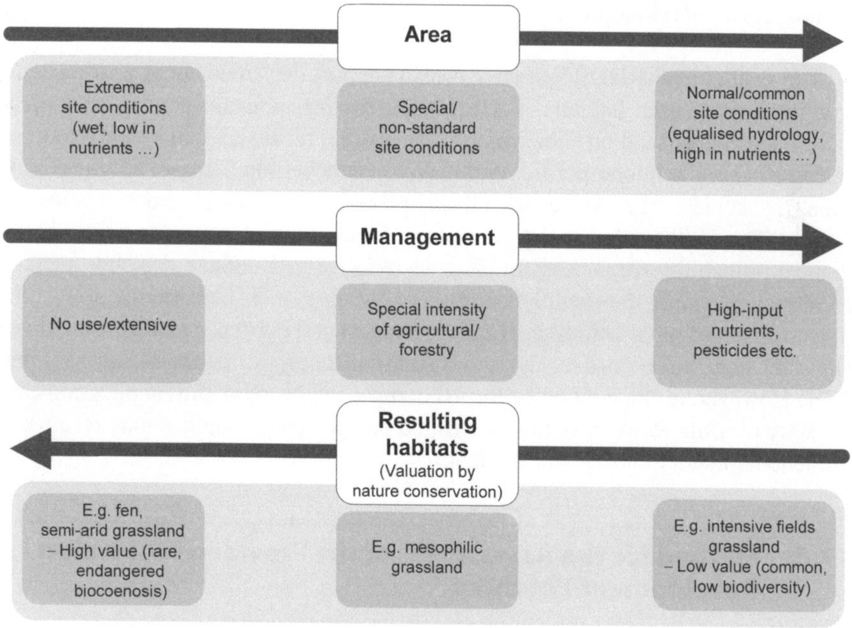

Fig. 17.1 Which combination of site and land use produces rare and endangered biocenosis? The conservation value generally increases with extreme site conditions and with a decrease in intensity (of nutrient and pesticide input, tillage etc.)

Bredemeier et al. 2015b; Rusch et al. 2012; Roschewitz et al. 2005). Irreversible anthropogenic changes of soil and hydrology are considered as part of the HDP. However, the present land use of a site is not included in the evaluation. Instead it is understood as a variable which could be changed in order to improve the habitat qualities of a site.

17.3 Practical Relevance of the Habitat Development Potential and Resulting Demands for Its Representation in Planning

The representation of habitat development potentials is particularly important for defining restoration opportunities and priorities on intensively-used areas. The HDP supports the following planning tasks:

- Allocation of agri-environmental measures to sites with good chances of success (important in the context of result-oriented remuneration).
- Selection of sites for compensatory mitigation in the context of impact mitigation regulations.
- Augmentation and development of habitats in landscapes with small shares of habitat structures.
- Protection of sites against destruction, in order to safeguard their capacity for future habitat development.

The results of the HDP assessment also support decisions about restoration of less intensively used habitats. For example, the reestablishment of low nutrient, semi-natural grassland on sites which are at present fertilised, will be only successful when the conditions permit such development within a reasonable period of time. To achieve the above objectives, place-based, ordinal-scaled results are needed. They will be used to direct response measures (and money) to sites with the highest probability for successful changes (relative to a defined decision area). In local scale planning the spatial resolution of the maps must be sufficient to identify plots or parts of plots with high HDP. As bequest and existence values of biodiversity are hard to monetise, it is similarly difficult to economically validate HDP. However, in the context of formal compensatory mitigation applications, the capacity of a site for developing the desired compensatory habitats may be a factor influencing the market price of the land.

17.4 Method for the Assessment of the Provided ES: Habitat Development Potential

The parameters, criteria, and the methodological workflow that guide the assessment of the HDP are shown in Fig. 17.2. The specific parameters used are "soil moisture levels", "nutrient content/supply expressed as cation exchange capacity (CEC) effective in the root penetration zone", and "soil reaction/carbonate level"

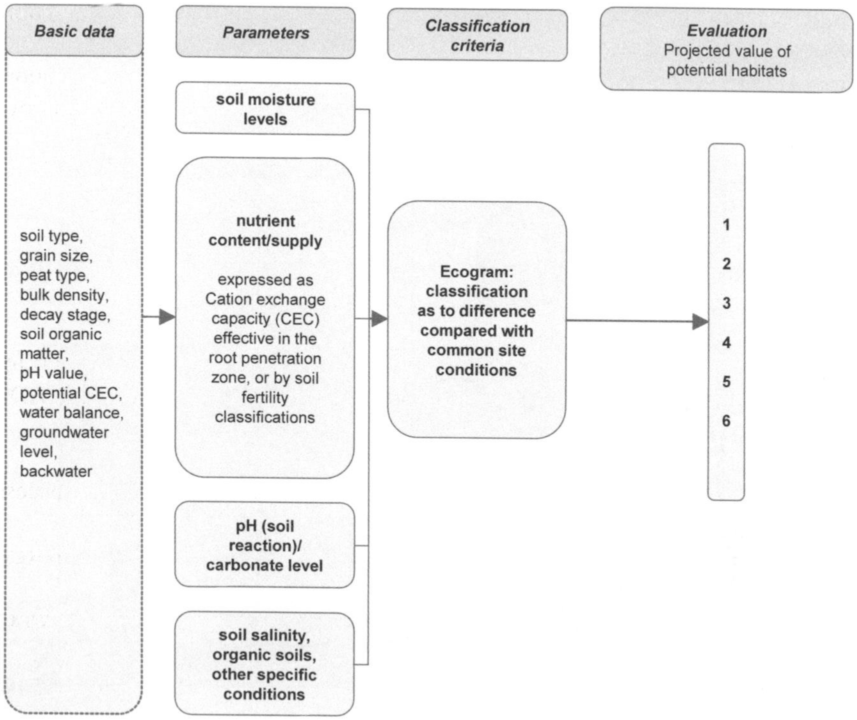

Fig. 17.2 Workflow, parameter and criteria for the evaluation of habitat development potential

(DIN 4220 (2008); Ad-hoc AG Boden 2005). These parameters may be available from regional or local soil maps with scales from 1:5,000 to 1:50,000 in many European states.

If these characteristics are not available they may be deduced with the help of pedotransfer rules from primary soil data, for example from soil type and grain size, peat type, bulk density, decomposition stage, soil organic matter content, pH value, potential CEC, water balance and groundwater level (Müller and Waldeck 2011; Mueller et al. 2007). In general, the scale and detail of the input data determines the degree of differentiation of the results and their applicability.

With some soil parameters, it is important to differentiate between the **actual potential** (average pH for the complete profile) and the **basic potential.** The basic potential can greatly differ from the actual potential if, for example, the subsoil has extremely low pH values and the pH of the top soil has been changed by liming. This may be very relevant for choosing habitat development measures. If precise information about the pH value of the soil profile is not available, then information from the soil type map must be interpreted in order to identify areas with extreme soil conditions (areas with very alkaline or acidic soils) (Horn et al. 1982).

Based on a literature analysis of site conditions of vegetation types (Brahms et al. 1989; also Bredemeier et al. 2015a), the soil parameters are combined in an **ecogram** (see Fig. 17.3) in order to demonstrate the link between different combinations of site conditions and habitats (characterised by vegetation).

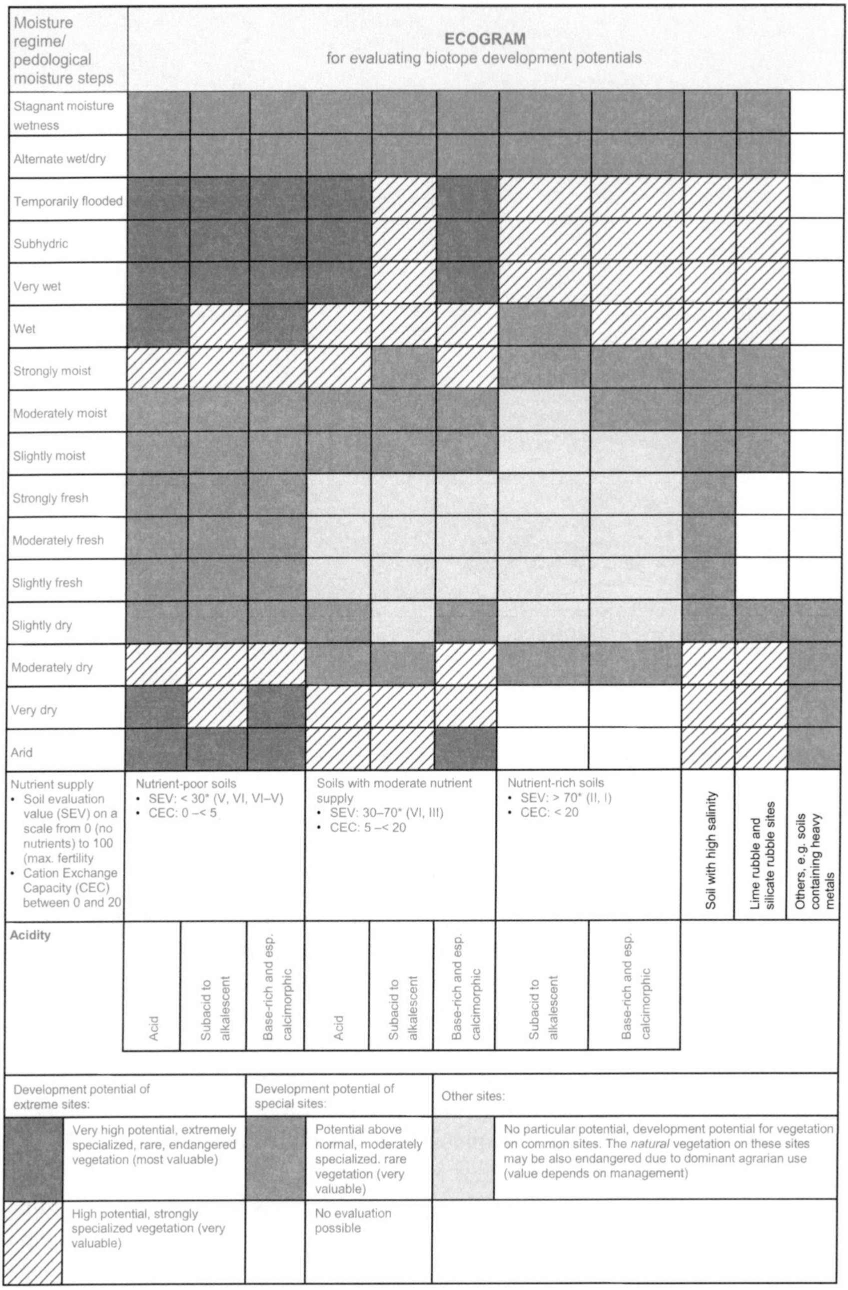

Fig. 17.3 Ecogram for the evaluation of biotope development potentials. (Adapted from Brahms et al. 1989)

On this basis the HDP can be evaluated on an ordinal scale by roughly projecting the rareness of theoretically possible habitat types on the site conditions (Fig. 17.3, for need to adapt the values see Engel 2013). The values assigned in the ecogram are based on rareness and endangerment of EU and German habitats (see Chap. 18) as well as the finding that extreme conditions produce rare habitats if managed accordingly or left to succession. Usually if two of the parameters have extreme values, then a development potential for very valuable, highly specialised vegetation is assumed. Similarly, if only one parameter value is extreme, then it is assumed that development possibilities exist for strongly specialised vegetation. Soils with properties other than those already mentioned, for example, high soil salinity, lime, a high content of coarse particles, or soils containing heavy metal, also represent a high HDP, but should additionally be considered on an individual basis. In general, the site conditions are not evaluated as a (potential) habitat for animals.

The evaluation scale ranges from 'no particular HDP' to 'very high HDP'. If recent soil data are evaluated then the present site conditions in the upper soil layer are evaluated. More often, 'historic' soil conditions represented in the available soil maps will be the basis. This is not necessarily a disadvantage because the soil map data show the potential without recent changes by land use (drainage, nutrient input…). In the ideal case both recent and basic soil data are available. The comparison between actual and basic potential specifies the management efforts which would be necessary for rehabilitation of the site conditions (Engel 2013). If the site is too greatly influenced it may take too long or be impossible to activate the HDP (cf. Wellstein et al. 2007; Bekker et al. 1997).

The results can and should be further differentiated by using available site-specific information as well as additional field data. For example, topography may not have been sufficiently covered by the input data for the ecogram. The effect of this factor is indirectly included in data regarding the soil type, as well as in the water and nutrient balance data. However, the results of the ecogram could alter in cases where topographic conditions are extreme, like steep slopes. This is relevant particularly when including the intensity of the solar irradiation and its effect on vegetation. For this purpose, solar energy potential maps (Dubayah and Rich 1995), as discussed in Chap. 12, could be used. In general, the results should be validated in the field, for example parallel to habitat mapping (Chap. 18), by looking for visible indicators of soil moisture, ground water table or indicator plants (Ellenberg 1996).

17.5 Economic Valuation of a Market Value or Potential Use Value

Economic valuation of the HDP can take place when the different HDP values become part of the property value of a site (see Sections 20.3 and 20.4). This may be the case when the nature protection authority is interested in buying sites with a high HDP for habitat banking (Chap. 25) or if a farmer has to consider the HDP in choosing sites for result-oriented agri-environmental measures (e.g. Klimek et al. 2008; Gerowitt et al. 2003). In these cases a monetary value is assigned to the sites either on a commercial market as a result of demand and available supply or calculated by the land user from result-oriented payment minus necessary 'production cost' of the desired habitat. The production and management cost will be lower in case of a high HDP. It is also possible to assign 'virtual' habitat credits to the sites, which represent the conservation value of the habitat to be developed. When the habitat is developed, these credits can be traded and used for habitat banking (Chap. 25).

17.6 Impact Assessment

To assess the impact of land use change on the HDP it is necessary to judge the influence of different pressures on the soil types (Table 17.1).

Of course all soils will be lost in the case of soil excavation or soil sealing. In the instances where the hydrology of a site or the soil chemistry is changed, the reaction of the different soils will vary and the level of pressure will define to what degree the soil is damaged. In order to judge the relevance of such impacts the value of the HDP also has to be considered.

Table 17.1 Sensitivity of different types of soils with high habitat potential against the most relevant pressure factors

	Pressure factor			
Sensitivity of ES: soil types	Complete destruction of soil, morphology e.g. by development, excavation …	Changing hydrology e.g. by draining	Changing soil structure e.g. by ploughing up grassland	Permanent change of soil chemistry by adding mineral substrates, lime …
Soils with high moisture level	**5**	**5**	**2**	**3**
Organic soils	**5**	**4**	**5**	**4**
Soils with low nutrient content and supply	**5**	**1**	**1**	**5**
Low pH or high carbonate level	**5**	**1**	**1**	**5**

Scale of sensitivity: 5 – very high; 4 – high; 3 – medium; 2 – low; 1 – very low

17.7 Conclusion

The HDP is a very useful analytical component of landscape planning. The practical value of the HDP assessment is primarily in supporting the definition of priorities, for instance where development measures should be located under conditions of limited resources. Furthermore it may be used as a basis for the economic valuation of the sites in the context of a market for nature conservation services.

References

Ad-hoc AG Boden. (2005). *Bodenkundliche Kartieranleitung*. KA5.

Bekker, R. M., Verweij, G. L., Smith, R. E. N., et al. (1997). Soil seed banks in European grasslands: does land use affect regeneration perspectives? *Journal of Applied Ecology, 34*, 1293–1310.

Brahms, M., von Haaren, C., & Jansen, J. (1989). Ansatz zur Entwicklung der Schutzwürdigkeit der Böden im Hinblick auf das Biotopentwicklungspotential. *Landschafts- und Stadtplanung, 21*(3), 110–114.

Bredemeier, B., von Haaren, C., Rüter, S., et al. (2015a). Evaluating the nature conservation value of field habitats: A model approach for targeting agri-environmental measures and projecting their effects. *Ecological Modelling, 295*, 113–122.

Bredemeier, B., Rüter, S., von Haaren, C., et al. (2015b). Spatial congruence between organic farming and biodiversity related landscape features in Germany. *International Journal of Biodiversity Science, Ecosystem Services & Management, 11*(4), 330–340.

DIN 4220. (2008). Bodenkundliche Standortbeurteilung – Kennzeichnung, Klassifizierung und Ableitung von Bodenkennwerten.

Dubayah, R., & Rich, P. (1995). Topographic solar radiation models for GIS. *International Journal of Geographical Information Systems, 9*(4), 405–419.

Ellenberg, H. (1996). *Vegetation Mitteleuropas mit den Alpen in ökologischer, dynamischer und historischer Sicht*. Stuttgart: Ulmer.

Engel, N. (2013). Bodenfunktionsbewertung auf regionaler und kommunaler Ebene. Ein niedersächsischer Leitfaden für die Berücksichtigung der Belange des vorsorgenden Bodenschutzes in der räumlichen Planung. *GeoBerichte* 26.

Franke, A. C., Plotz, L. A., van der Burg, W. J., et al. (2009). The role of arable weed seeds for agroecosystem functioning. *Weed Research, 49*, 131–141.

Gerowitt, B., Isselstein, J., & Marggraf, R. (2003). Rewards for ecological goods – Requirements and perspectives for agricultural land use. *Agriculture, Ecosystems and Environment, 98*, 541–547.

Horn, D. P., Ally, M. M., & Bertsch, P. (1982). Cation exchange capacity measurements. *Communications in Soil Science and Plant Analysis, 13*(10), 851–862.

Klimek, S., Richter gen Kemmermann, A., Steinmann, H. H., et al. (2008). Rewarding farmers for delivering vascular plant diversity in managed grasslands: A transdisciplinary case-study approach. *Biological Conservation, 141*, 2888–2897.

Mueller, L., Schindler, U., Behrendt, A., et al. (2007). *The Muencheberg soil quality rating (SQR). Field manual for detecting and assessing properties and limitations of soils for cropping and grazing. Draft Nov 2007*. Muencheberg: Leibniz-Centre for Agricultural Landscape Research (ZALF) e. V..

Müller, U., & Waldeck, A. (2011). Auswertungsmethoden im Bodenschutz – Dokumentation zur Methodenbank des Niedersächsischen Bodeninformationssystems (NIBIS®). *GeoBerichte* 19.

Preising, E. (1954). Übersicht über die wichtigsten Acker- und Grünlandgesellschaften. *Angew Pflanzensoziologie, 8*, 19–30.

Roschewitz, I., Gabriel, D., Tscharntke, T., et al. (2005). The effects of landscape complexity on arable weed species diversity in organic and conventional farming. *Journal of Applied Ecology, 42*, 873–882.

Rusch, A., Valantin-Morison, M., Roger-Estrade, J., et al. (2012). Using landscape indicators to predict high pest infestations and successful natural pest control at the regional scale. *Landscape and Urban Planning, 105*, 62–73.

von Haaren, C., & Bathke, M. (2007). Integrated landscape planning and remuneration of agri-environmental services – Results of a case study in the Fuhrberg region of Germany. *Journal of Environmental Management, 89*(3), 209–221.

Wellstein, C., Otte, A., & Waldhardt, R. (2007). Impact of site and management on the diversity of central European mesic grassland. *Agriculture, Ecosystems and Environment, 122*, 203–210.

Habitat Capacity

18

Stefan Rüter and Paul Opdam

Abstract

We present two complementary approaches for the assessment and evaluation of habitat capacity that are based on (i) habitats and (ii) habitat networks. The first approach allows for an area-wide evaluation of the habitat qualities of a landscape to be made on the basis of habitat types. This can be underpinned through selective registration of, for example, endangered species. The assessment of habitat networks is an interpretation of the amount of habitat and its spatial distribution within the landscape. It considers man-made structures that interfere with horizontal relations, such as roads that act as barriers to flows of individuals. This interpretation is done with reference to life-history traits and can either be made for particular target species or for species ecoprofiles. Altogether both approaches enable an adequate consideration of habitat issues in landscape planning by applying relatively simple but robust models that are based on available legislation and expert standards.

Keywords

Habitat evaluation · Habitat types · Habitat networks · Species requirements · Metapopulations

S. Rüter (✉)
Leibniz University Hannover, Institute of Environmental Planning, Hannover, Germany
e-mail: rueter@umwelt.uni-hannover.de

P. Opdam
Land Use Planning Group and Alterra Wageningen UR Nature and Society Group, Wageningen University, Wageningen, The Netherlands
e-mail: paul.opdam@wur.nl

C. von Haaren et al. (eds.), *Landscape Planning with Ecosystem Services*, Landscape Series 24, https://doi.org/10.1007/978-94-024-1681-7_18

18.1 Introduction

The protection of habitats and threatened species of wild fauna and flora are important objectives of the EU nature legislation (Birds and Habitats Directives, 79/409/ECC, 92/43/ECC) and international agreements such as the Convention on Biological Diversity. In order to meet objectives of maintaining the biological diversity of the environment, conservation activities have often focused on the designation of protected sites within the Natura 2000 network (Trochet and Schmeller 2013). More recently, biodiversity conservation is being viewed more comprehensively, to include the potential to maintain or improve the supply of ecosystem services (ES) (Egoh et al. 2014; Haslett et al. 2010; European Commission 2011). Consequently, besides the intrinsic values of nature for humans, efforts to conserve habitats and species should consider the key roles of biodiversity at different levels of the ES hierarchy: as a regulator underpinning ecosystem processes (Mace et al. 2012) and the precondition of provisioning and regulation services (Maes et al. 2012). Transforming these expanding demands into pragmatic guidelines is a challenge for landscape planning. In this chapter, we propose a practical approach for the assessment and evaluation of habitat capacity that is specifically adapted to the requirements of landscape planning at the regional level. With respect to the ES approach described in Chap. 3, we concentrate on 'offered ES' (von Haaren et al. 2014; in this book termed 'delivered') and habitat non-use values such as existence values or bequest values of handing over the natural capital to the next generation (cf. de Groot et al. 2010).

18.2 Definition and Concept

What is understood by the term habitat may vary between disciplines and in the different expert languages, so it is important to establish a clear definition (Hall et al. 1997). The term habitat was traditionally used by biologists to describe the place where a species lives and can survive, as an individual or as a population. Habitat is sometimes used to describe the type of environment a species is usually found in. In some instances, the term refers to a concrete spatial unit providing the living conditions necessary for survival. In the context of landscape planning a third definition of habitat is useful: the totality of biotic and abiotic living conditions for a population system of organisms of different animal and plant species (= biocenosis), characterising a defined area (Evert 2010). This definition of habitat is in line with the meaning of biotope as exists in German or French (cf. von Haaren et al. 2012). For the purpose of this chapter such a definition is used.

Habitat types are idealised types, defined by characteristic environmental conditions (e.g. soil, climate, and topography) and biota, which are similar between the areas but that differ from areas of other habitat types (e.g. the composition of typical species and their relative abundances). An example is the habitat type classification according to Annex I of the European Habitats Directive. Habitat type examples include dry heathland, regularly flooded riparian woodland, or oligotrophic standing water bodies.

Within landscapes, combinations of habitat types can be found in a particular configuration, called habitat mosaics. The most commonly occurring and characteristic, functional mosaics can be defined as a habitat complex. This includes patterns of habitats along an ecological gradient (e.g. gradient of humidity in a floodplain) as well as anthropogenic distributions caused by the historical and cultural development of a landscape (Blab et al. 1995; Noss 1987).

We define habitat capacity as the potential ability of landscapes to provide suitable habitat for populations of plant and animal species. The potential spectrum of habitat types and the species they support in a landscape depends upon the physical environment, land-use, and biogeographic regions (Haines-Young 2009; Kreft and Jetz 2010). The resulting variety of species populations furthermore depends on the amount and spatial arrangement of specific habitats, so called habitat networks, in the landscape and in its wider surroundings (Opdam et al. 2003). However, the relations between habitat and species assemblages are not predetermined. For example, fluctuations in weather conditions may cause perturbations in the size of populations and trends in the climate may cause geographical shifts in species distributions on a large spatial scale (Parmesan et al. 1999). Therefore, species are frequently temporarily absent or underrepresented in suitable habitats.

An adequate consideration of habitat capacity issues in planning requires that the range of interactions among ecological components be considered and that the unique characteristics of each ecosystem be evaluated (e.g. small-patch habitat at the local scale, inherent dynamics of metapopulations at the regional scale) (cf. Pickett et al. 1992; Poiani et al. 2000). At the same time, methods for the assessment and evaluation of habitat capacity should address the application context of landscape planning by applying relatively simple but robust models that are based on available legislation and expert standards. Therefore, we suggest two complementary approaches of assessment and evaluation of habitat capacity based on:

1. Habitats
2. Habitat networks.

The first approach focuses on the vertical dimension in landscapes: the interaction between abiotic conditions, vegetation type, and spatial and temporal variability in habitat quality (partly due to human interventions). The second approach focusses on the horizontal dimension in landscapes: the spatial distribution of habitat type cover and man-made structures that interfere with horizontal relations, such as roads that act as barriers to flows of individuals (Fig. 18.1).

Habitat-based assessment (the vertical dimension) is very common in landscape planning as it allows for an area-wide evaluation of the importance of a given area for overall biodiversity. Because of the link between habitats, communities (defined as a group of populations living and interacting with each other in an area), and species in ecosystems, there is often likely to be some overlap between their assessment and evaluation (Bunce et al. 2013; Southerland 1993; Tucker 2005). For example, species are often essential attributes that define a habitat (e.g. particular tree species in woodlands and forests). At the same time habitats may provide

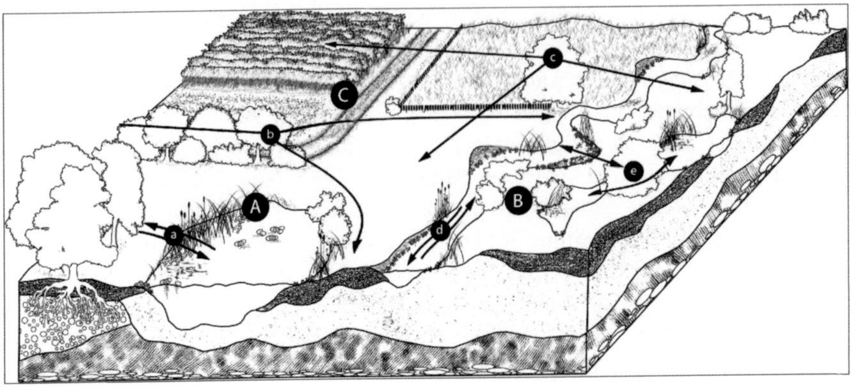

Fig. 18.1 Habitats reflect biotic and abiotic living conditions for species or communities in the vertical dimension of landscapes (for example, A: Mesotrophic standing water body including littoral zone, B: River floodplain wetland complex, C: Farmland interspersed with semi-natural vegetation). The variety of species populations also depends on the spatial distribution and configuration of habitats in the horizontal dimension in landscapes (for example, a: Habitat use along an environmental gradient (e.g. amphibians, dragonflies), b: Bats follow flight paths to hunting grounds, c: Birds of prey perch on exposed trees, d: Streams serve as corridor for migratory species, e: Habitat patches contribute to the persistence of metapopulations)

specific functions for species or communities of conservation concern (e.g. dead or old-growth trees provide suitable roosts for bats). Therefore, evaluations based on habitats are useful for formulating management options in landscape planning.

However, limiting the assessment to the vertical dimension in landscapes results in an incomplete view of the habitat capacity. Due to processes in the horizontal dimension, species can be absent from a landscape even if the available habitat sites are of good quality. The spatial distribution and configuration of habitats and the landscape pattern in between habitat sites are inherent components of the habitat capacity of the landscape (Opdam et al. 2003). For example, a breeding pair of a bird species requires a minimum amount of habitat to collect enough food. Hence smaller and isolated fragments do not contribute to the species habitat in a landscape. Furthermore, local populations living in patches of habitat are more likely to fluctuate in numbers due to variability in weather conditions and food resources – and as a result may go extinct. It is not unusual that a percentage of habitat sites in an area are unoccupied by the species. Depending on their place in the habitat network and the permeability of the intermediate landscape for individuals searching for a place to live, deserted patches may be reoccupied at a later time. The spatial distribution of a habitat determines whether such site populations can function as a metapopulation (Hanski 1999). A network of habitats can thus support persistent populations while individual sites cannot. When the configuration of a habitat does not allow site populations to interact as a population network, species can be absent in the area despite the availability of good habitats. The degree to which the habitat sites of a landscape support such a population network is also dependent on the life-history characteristics of species, for instance their body size and dispersal capacity. Thus, habitat capacity should also be considered in terms of the horizontal dimension of landscapes and in light of the wide range of ecological profiles and populations of the focal biodiversity.

Nature conservation strategies should also consider genetic variation within species. For example, low levels of genetic diversity can result in a reduction of population fitness and an increase in the probability of population extinction (Saccheri et al. 1998). However, specific assessment and evaluation of genetic diversity goes beyond the scope of current general landscape planning applications at the regional level. Therefore, we do not consider genetic diversity in this chapter. Nevertheless, it could be argued that genetic diversity can be conserved by ensuring that separate populations or groups of species are conserved.

18.3 Practical Relevance in Planning

The assessment and evaluation of habitat capacity are common tasks in landscape planning (von Haaren et al. 2008). Important issues addressed include, for example, the protection and maintenance of endangered habitats and species (Noss et al. 1997) and the development of ecological networks (Jongman and Pungetti 2004). Evaluations are often based on legal standards, principles of nature conservation and landscape management. Examples include the development of the European ecological network Natura 2000. Another example is the contribution to environmental assessment studies (EIA, SEA). A further example is the assessment of compliance with international conventions (e.g. Ramsar Convention on Wetlands 1971; Rio Convention on Biological Diversity 1992).

Apart from assessments that aim to determine the performance of governmental nature and landscape conservation policies, evaluations may also consider the planning of multifunctional landscapes (see Chap. 19). Here, a range of aims and values from a variety of stakeholders and disciplines have to be merged into a single view of a future landscape. An example of this is green infrastructure planning. Here, a network of semi-natural elements is considered as the provider of different landscape benefits (cf. ecosystem services) (Steingröver et al. 2010). In this context landscape planning supports the task of providing members of the public with environmental information about habitats and species. This information may be tempered by the particular interest groups or stakeholders concerned – whether they be citizens, organisations or private companies. Because governments at national and regional scale often decentralise their responsibility for conserving the common values of the environment, community-based landscape planning has become more commonly practiced (Opdam 2013).

18.4 Assessment and Evaluation of the State

18.4.1 Habitats

18.4.1.1 General Approach

The habitat-based assessment provides the basis to improve the overall quality of habitat capacity in the study area. The major steps involved in planning and executing this approach are illustrated in Fig. 18.2.

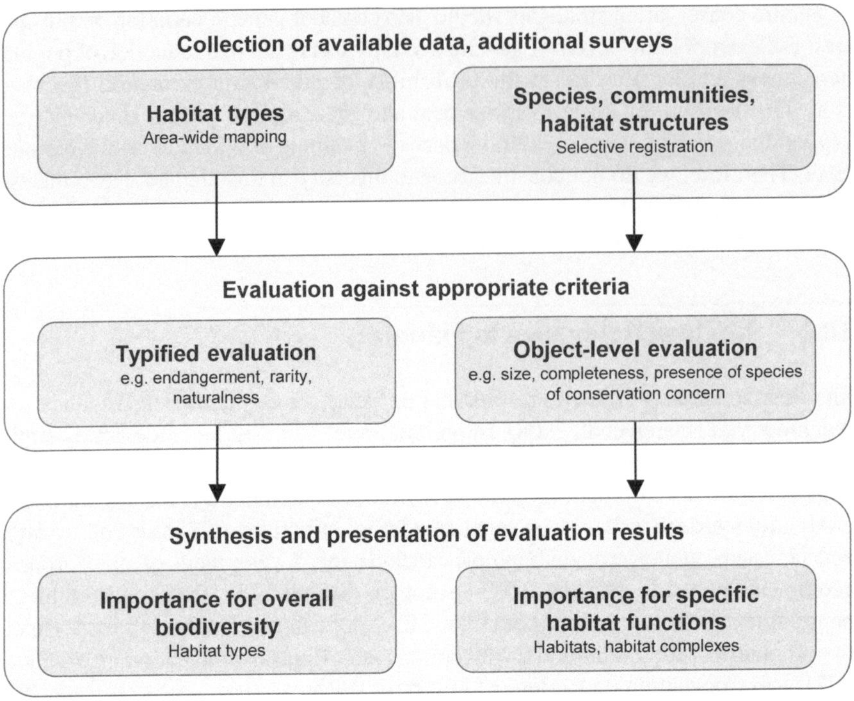

Fig. 18.2 A generic framework for an area-wide assessment of the habitat qualities of a landscape

Extensive mapping of habitat types is an essential step in the assessment. The evaluations made on the basis of habitat types (based on criteria such as endangerment or rarity of certain habitat types which stem from the Habitats Directive and national legislation) can be underpinned through the selective recording of species (especially with respect to the presence of rare or endangered species) and the analysis of specific functional relationships (e.g. description of habitat complexes).

The following key considerations should be addressed when designing the assessment program: the habitats and species of concern in the region should be identified (e.g. review the status and trends, identify potential pressures and impacts on these habitats) and the conservation objectives defined (e.g. consider objectives of higher-level landscape planning, legal objectives of governmental nature and landscape conservation policies). The assessment and evaluation of habitat capacity may not meet its full potential unless the objectives are properly defined.

18.4.1.2 Assessment

Once the objectives for the study area have been defined the relevant features for the assessment need to be selected. These features should include:

- Habitat types and
- Selected species, communities, and habitat structures.

The available data and information are collected and, where necessary, supplemented with additional surveys. An area-wide map of habitat types usually gives baseline information on the current state of nature and the landscape. At present, many EU countries already have habitat classifications, and others are working toward regional classifications (European Environment Agency 2014). More recently, habitat type maps have been increasingly produced to address policy-related issues. Up-to-date habitat maps (not older than 5–10 years) can be used or the underlying classification systems be applied in the assessment. However, if these classifications are not possible or available, the pan-European EUNIS habitat type classification (Davies et al. 2004) is recommended as a basis for an area-wide habitat type mapping (Fig. 18.3).

The EUNIS classification affords the opportunity of a sound scientific cross-reference between widely accepted European habitats, including those listed in Annex I of the Habitats Directive, and phytosociological definitions of vegetation types (Rodwell et al. 1998). Thus, the use of EUNIS also helps to link regional and national to European habitat classifications. However, often EUNIS is not detailed enough for representing the differences in habitat characteristics which may be important for either setting priorities or for making changes in habitat qualities visible as a consequence of restoration measures. Therefore, subclassification may be necessary.

Although a regional scale is applied in this book, it is important to remember that habitat structures of concern that go beyond a regional classification should also be considered in habitat mapping (e.g. unvegetated areas within habitats; Fig. 18.3). In view of local issues or problems it is often necessary to collect such additional data which go beyond the habitat type description and describe the individual habitat (object), as well as data about species and communities. For instance, a survey should be carried out if there is a reasonable likelihood of protected species or communities being present in a site, or if a site is likely to be affected by future developments (e.g. those compiled by national red data books). Usually there is little need for further investigation in intensively-used landscapes, but a high need in extensively-used landscapes with high biodiversity. However, some species have general characteristics which can be used to better assess habitat capacity (e.g. those that require specific management, or those that integrate functional relationships in the landscape). For instance, amphibians are regarded as good ecological indicators in wetlands for habitat loss and fragmentation. Additional surveys should address such species or taxonomic groups that can be used as indicators for specific habitat qualities and sensitivities to impacts of concern in the study area (Table 18.1).

When the habitat structures, species and communities of concern have been defined, appropriate methods for the surveys that best reflect their condition need to be decided (e.g. presence-absence, number of individuals, reproductive success; cf. Tucker 2005). For specific survey methods for habitats and species the reader is referred to, e.g. Hill et al. (2005) and Plachter et al. (2002).

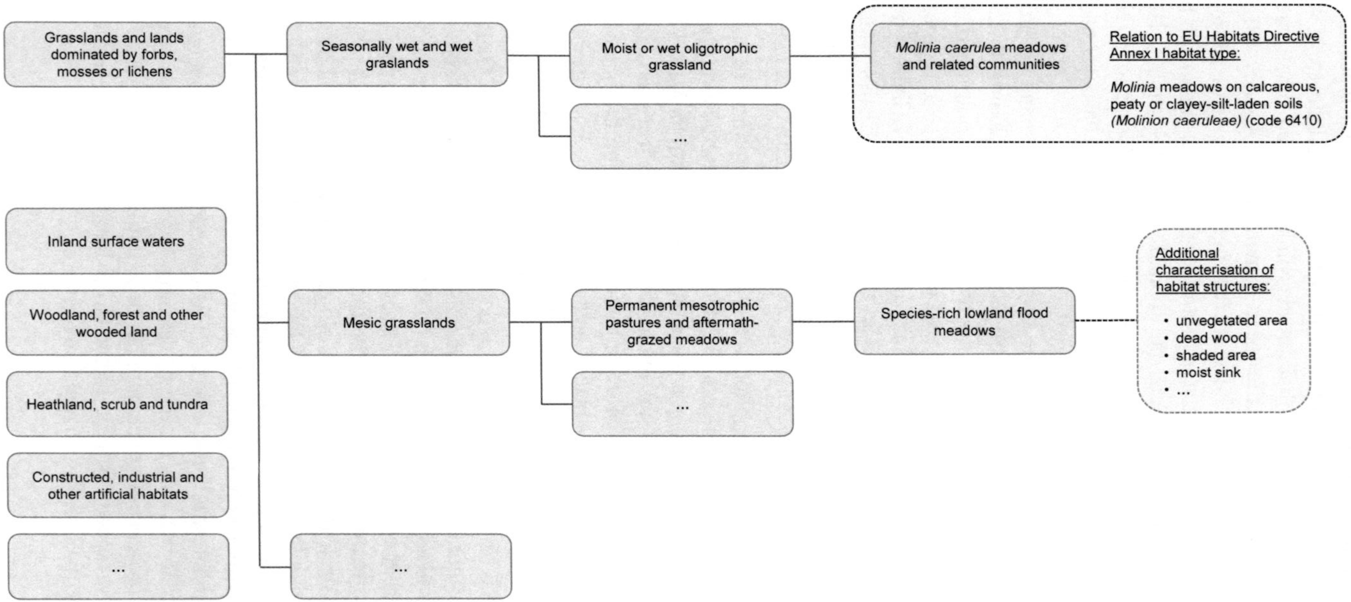

Fig. 18.3 Illustration of the hierarchical EUNIS habitat classification showing linkages to EU Habitats Directive Annex I habitat types and potentials for additional characterisation of habitat structures. (Based on Davies et al. 2004)

Table 18.1 Taxonomic groups that can be used as indicators for specific habitat qualities and sensitivities to impacts

	Bats	Other mammals	Birds	Reptiles	Amphibians	Fish	Macrozoobenthos	Dragonflies	Grasshoppers	Butterflies	Spiders	Carabid beetles
Demands for specific habitat qualities												
Functional connectivity	++	++	++	++	++	++	+	+	+	++	–	+
Long-distance migration patterns	++	++	++	–	–	++		–				
Abiotic factors				–	+	+	++	+			++	++
Microhabitats, structural qualities	++	++	++	++		+	+	++	++	++	++	+
Soil substrate						++	++					+
Dynamic processes			+		+	++	++	+	+	+		+
Habitat tradition	++	+	+	++	+	+	–		–	–	+	+
Sources of food	–	–	+	–		+	+	–	+	++		
Interactions with other species						–				+		
Specific sensitivities to impacts												
Fragmentation, barriers to migration	+	++	+	++	++	++	++		+	+		+
Intensive land-use			++		+	+		++	+	+	+	+

(continued)

Table 18.1 (continued)

	Bats	Other mammals	Birds	Reptiles	Amphibians	Fish	Macrozoobenthos	Dragonflies	Grasshoppers	Butterflies	Spiders	Carabid beetles
Human disturbance	++	++	++									
Noise/ light emission	+	++	++									
Pollution (e.g. pesticides)	+		+		–	–	+	–				–
Eutrophication, saproby			–				+		+	+		

++ very important, + important, – less important; Bernotat et al. (2002)

18.4.1.3 Evaluation

The next step is the linking of the collected data to relevant evaluation criteria. However, the list of potential criteria that can be examined to evaluate habitats and species is extensive (e.g. Noss 1990). Landscape planners do not have the ability and resources to examine all of them, nor do all existing criteria encompass the critical patterns and processes at the regional level. We therefore suggest the use of criteria that are already widely accepted for both typified and object-level habitat evaluation. Evaluating these criteria should be relatively straightforward, while remaining objective, repeatable and based on scientific ecological principles, to achieve best practices in evaluation (for more details see, e.g. Spellerberg 1991).

Habitat types are easy to evaluate by typified evaluation criteria and allow for an area-wide and objective, rather than subjective, evaluation. We suggest evaluating a set of well-known and predefined criteria such as endangerment or rarity of certain habitat types (Table 18.2).

Table 18.2 Criteria for typified and object-level evaluation of habitat capacity

Evaluation criteria	Characterisation and application
Typified evaluation	
Endangerment	Applied either to habitats or to species. The categories of endangerment are defined in Red Lists of threatened habitats and species that have been developed in many European countries on national and regional level
Rarity	Rare habitats and species are generally more highly regarded than common ones. Rarity can be evaluated by comparisons with national or regional data (e.g. population size, abundance). Evaluation of rarity can be useful, especially when no data about endangerment is available
Naturalness/ hemeroby	Natural habitats that are least intensively modified by humans are of special importance for nature conservation (e.g. peat bogs, forests). In landscape planning, the criterion is also used to evaluate human influence on habitats in terms of hemeroby. The level of hemeroby depends on the degree of human impacts that prevents a system from developing towards a natural endpoint situation. So it is possible to determine the deviation from naturalness as a result of specific land use types
Object-level evaluation	
Presence of species of conservation concern	Applied to habitats and habitat complexes. The presence of important animal or plant species (e.g. endangered species, indicator species) is evaluated by comparisons with regional data. Evaluation generally takes account of species abundance rather than mere presence-absence data
Completeness	A measure of how well a habitat or a habitat complex includes typical combinations or mosaics of habitat features and species (e.g. occurrence of animal and plant species that represent typical functional relationships). Evaluation requires target setting, e.g. list of target species
Habitat continuity	Old-growth, continuous habitats support a high species richness including typical functional relationships. Therefore, habitats with higher continuity (including non-intensive land use) are generally held to be more important for nature conservation than those with low continuity
Size	Including both area of habitats and population sizes for individual species. Large habitats are generally more highly regarded, due to the reduction in the negative impacts of patch edge effects, increasing species richness, and the ability of larger habitats to support greater populations with lower extinction probabilities

Based on Tucker (2005)

These criteria are based on the European Birds Directive and Habitat Directive as well as on the standards for protected species or habitat types in many examples of national nature conservation legislation. On an EU level the habitat types of the Habitat Directive represent the highest level of rarity and endangerment. However, there are a variety of habitat attributes that may also be used as evaluation criteria (e.g. naturalness).

Specific functions of individual habitats or habitat complexes, although of great importance for habitat capacity, are more difficult to define in terms that allow objective evaluations. Nevertheless, some criteria exist that allow for object-level evaluation. These criteria commonly focus on attributes such as habitat continuity, the presence of species of conservation concern, and the completeness of habitat features (Table 18.2). For example, access to erosion banks along rivers is critical for the Sand Martin (*Riparia riparia*) to excavate tunnels for breeding. Therefore, the existence of both, the habitat feature (erosion bank) and the birds, may give important additional information about the conservation value of river habitat complexes. Evaluation of such functional relationships depends on the needs of the focal species and thus requires target setting (e.g. list of target species).

Important steps to be considered in the evaluation are the following: As a first step, values are given to the habitat types (typified evaluation). These values are graded using an ordinal scale. Evaluation should be based on existing national standards (e.g. JNCC 1995; Riecken et al. 1994), or if available, regional standards (e.g. von Drachenfels 2010). The grades reflect the significance of each habitat type for the selected evaluation criteria. The results of the typified evaluation can be synthesised as importance for overall biodiversity (*Synthesis I*, Table 18.3).

The second step produces values for the individual habitats and habitat complexes (object-level evaluation). Note that this step may lead to different values for habitats that were initially classified as the same type. For example, large sites of traditional orchard are considered to support higher species richness than small sites and, thus, are valued as having higher importance for habitat capacity. Thresholds

Table 18.3 Example for the synthesis of typified and object-level evaluation of habitats

Habitat type	Traditional orchard			Mesotrophic meadow	
Habitat No.	1	2	3	4	5
Results of typified evaluation					
Endangerment	2	2	2	3	3
Rarity	2	2	2	4	4
Naturalness/hemeroby	3	3	3	3	3
Synthesis I: Typified evaluation	***2***	***2***	***2***	***3***	***3***
Results of object-level evaluation					
Presence of species of conservation concern	3	2	1	2	4
Completeness	x	x	x	2	3
Habitat continuity	4	3	1	2	3
Size	4	2	2	3	3
Synthesis II: Overall evaluation	***3***	***2***	***1***	***2***	***3***

Values for importance: 1 very high, 2 high, 3 medium, 4 low, 5 very low, x no data

of significance (e.g. size) may differ substantially between habitat types or biogeographical regions. It is therefore advisable to group sites for evaluation into similar types before comparisons are carried out (Tucker 2005).

Finally, the results of the typified evaluation and the object-level evaluation of habitats can be synthesised and presented as overall evaluation results (*Synthesis II*, Table 18.3). These results can be cartographically depicted, e.g. in an evaluation map for the 'importance for specific habitat functions', with a five-part colour scale ranging from red (very high importance) to a pale yellow (currently low importance). See also Chap. 6 for the use of GIS in analysing and presenting habitat and species data into formats that are useful to planners.

Recent approaches to habitat evaluation are modelling species diversity as an evaluation criterion for differentiating the basic value assigned to habitat types (examples for field habitats: Sybertz et al. 2017, Bredemeier et al. 2015; for urban habitats: Rüter et al. 2017). These approaches may suffice for many tasks in landscape planning if a general improvement of the habitats for a wide variety of species is to be achieved, e.g. by changing managing practices or landscape context. The advantage of these approaches is that they improve the efficiency of habitat assessment because species information is provided at lower cost (time) and do not require extensive mapping of species. However, complete field surveys are more precise and, therefore, enable an evaluation of the true species diversity as well as helping to optimise the habitat for specific endangered species.

18.4.2 Habitat Networks

18.4.2.1 General Approach

The assessment of habitat networks builds on the vertical dimension in landscapes, because available habitat types are interpreted as suitable places to live but the assessment explicitly adds an interpretation of the amount of habitat and its spatial distribution within the landscape area. This interpretation is done with reference to the life-history traits of a species, and hence is species specific. It can either be done for particular target species or for species ecoprofiles. A species ecoprofile represents a particular set of species traits, such as suitable habitat, body size (which correlates with area requirements of a reproductive unit) and dispersal mobility (the capacity to explore heterogeneous landscape surrounding the place of birth to find a place to live and reproduce) (Opdam et al. 2008).

The concept of habitat networks is based on metapopulation theory (Hanski 1999). The fundamental assumption is that the summed habitat capacity of all individual pieces of habitat increases if the pieces are able to function as a network. In such a network the population has better chances to survive as connecting habitat sites into networks can solve fragmentation problems.

The metapopulation is a network of populations living in the habitat network. These populations show partly independent dynamics and some of them disappear from one year to another. In many habitat networks, in a particular year, some patches are not occupied even though they are suitable. Over the following years, some of these unoccupied patches become inhabited due to successful colonisation by individuals from elsewhere in the network. Hence, the pattern of occupied patches changes over time. This implies that all patches, including the ones not occupied, contribute to the persistence of the metapopulation.

The habitat capacity of habitat networks equals the average number of reproductive units (territories, couples, etc.) that inhabit the network. This number depends on the spatial cohesion of the habitat network. Spatial cohesion has multiple dimensions: it includes the total amount of habitat weighted for habitat quality and the average distance between patches weighted for landscape permeability (Opdam et al. 2013). Spatial cohesion is a landscape index which is scaled to the dispersal capacities and area requirements of a species (Vos et al. 2001; Verboom et al. 2001). It means that two pieces of habitat may be functionally connected if individuals of the species can move between them, even though there is no physical connection (e.g. a corridor). Hence, spatial cohesion is a functional interpretation of the spatial pattern of habitat patches and other elements that influence movement of individuals (or seeds) of a species.

Habitat patches within a network vary in size. Relatively large patches (so called key patches), with a small probability that the population there goes extinct, have a relatively large contribution to metapopulation viability. A fundamental premise of habitat networks is that individual patches are not able to support a viable population. Hence, population viability is realised through the interaction of patch populations, which results from a stream of individuals or seeds dispersing across the landscape. Only through this cooperation can a species survive as a metapopulation (Opdam et al. 2006; Rowland and Wisdom 2009).

The assessment of habitat networks can be done by detailed studies regarding the occurrence of species in suitable habitat patches over time. Distribution maps can be fed into metapopulation models to determine the viability of the metapopulation in the habitat network (cf. Vos et al. 2001; Rüter 2009). However, in planning cases usually neither time nor budget are available for such detailed studies.

The best alternative and more practical approach is to analyse the habitat network pattern and the landscape matrix in which it is embedded by using rules that are inferred from detailed ecological and modelling studies (for example rules for key patches see Verboom et al. 2001). An example of such a rule is the minimum area of interconnected habitat required for a viable population. Such rules are incorporated in GIS models (Verboom and Pouwels 2004). A full conceptual framework is given in Opdam and Steingröver (2008).

18.4.2.2 Collection of Data and Selection of Criteria

Because spatial cohesion is a species-specific index, habitat network assessment starts with a focus on the specific requirements of target species or a balanced set of ecoprofiles. If no proper data for a target species are available, data taken from a species that has similar habitat choice, body size, and mobility can be used. In case of an assessment for a broader range of species, indicator species may be preferred. For landscape planning purposes, we recommend the ecoprofile species approach (Opdam et al. 2008). In this approach, a matrix is constructed in which virtual species classes are distinguished based on area requirements and dispersal range (Fig. 18.4).

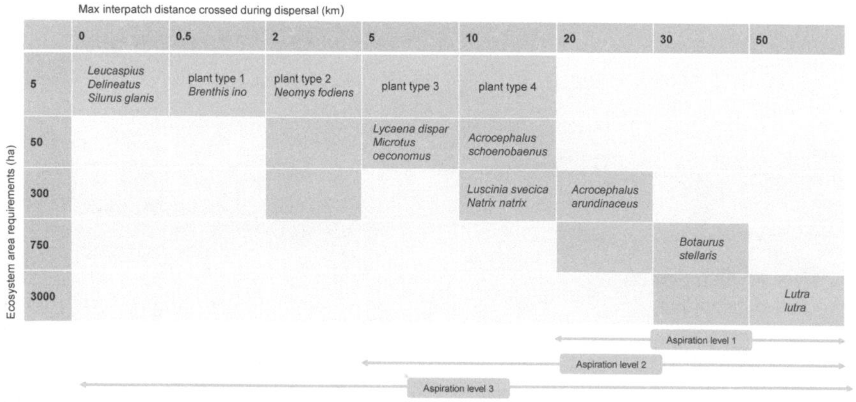

Fig. 18.4 Ecoprofile matrix for ecosystem type 'marshland' as applied in planning a Robust Corridor zone in the Dutch National Nature Network. All 398 target species of the Dutch Nature Conservation Policy were assigned to the cells of the matrices for 7 ecosystem types. Only 15 cells were used (as indicated by grey shade). Possible aspiration levels are indicated by the arrows, showing different combinations of ecoprofiles which require different efforts: level 3 requires a high degree of connectivity

Each cell in the matrix represents a cluster of species for a specific ecosystem type with area requirements and dispersal distance in the same order of magnitude. For communication purposes, a species representative of such a class can be used ('the grass frog profile').

When choosing a target species, *step 1* (Fig. 18.5) is to collect data about the vegetation it inhabits; if necessary, also gather data about the seasonal changes in the chosen habitat.

Then the area required for a reproductive unit (i.e. a pair of birds, a badger social group, a bee colony) is determined. Such a minimum area may vary throughout the geographic range of a species, so ideally figures are used that are available for the planning area or a nearby region. Also required are data about the minimum number of reproductive units required for a viable metapopulation. This can be collected from literature or from experts (see Table 18.4 for an example).

By combining these two figures the minimum required total area of a habitat network is obtained (Fig. 18.5). See Verboom and Pouwels (2004) for further information.

Also required in *step 1* are approximations of the distances that individuals of the species can move through the landscape matrix. Two points need to be considered here. First, the usual pattern is that many individuals move short distances, a minority move intermediate distances, and only a very small number cover long distances. Ideally, dispersal range should be used, which includes approximately 95% of all individual movements. The remaining 5% will probably not contribute to the metapopulation in the planning area.

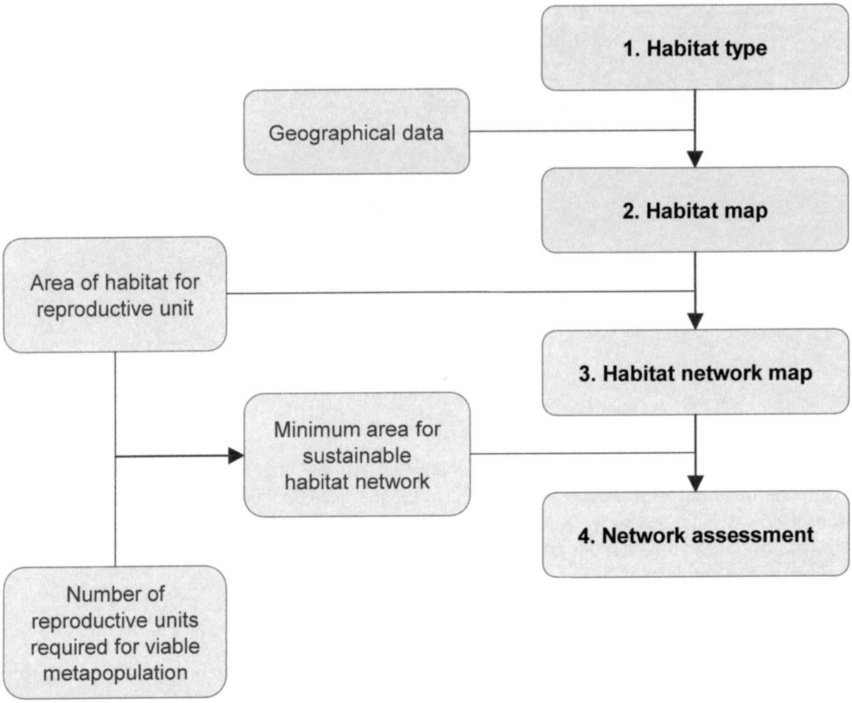

Fig. 18.5 Overview of habitat network assessment method. The four steps are indicated in the boxes in the vertical column. The other boxes represent basic data collected in step 1

Table 18.4 Minimum carrying capacity, for habitat networks with and without key patches (Verboom et al. 2001). A key patch is a relatively large patch in the network that contributes most to the stability of a metapopulation. The numbers can be translated into area requirements

	Number of pairs or family groups for minimum network carrying capacity		
Species groups	*Key patch*	*Network with key patch population*	*Network with no key patch population*
Long-lived/large vertebrates	20	80	120
Middle long-lived/medium sized vertebrates	40	120	200
Short-lived/small vertebrates	100	150	200

In *step 2* a map of the planning area is made 'through the eyes of the target species'. This incorporates the land cover types or vegetation types that represent its habitat, and the landscape elements that would be likely to influence its dispersal movements. In case of a woodland species, such a map could show deciduous woodland patches, hedgerows and tree lines, and clumps of trees near farmsteads. If relevant, barriers inhibiting dispersal, for example roads and canals, should be included in the map.

In *step 3* this species-specific habitat map is transformed into a map of habitat networks. The basic way to do this is i) to select all patches large enough for a reproductive unit (which provides the total area of potential habitat), and ii) to link all patches that are closer than the maximum dispersal distance (which provides a habitat network map). Note that landscape elements that enhance movements (such as hedgerows, riparian forest) or inhibit them influence the linking distance. As a result, it is possible to obtain a map in which all patches in the area constitute one single network. More often however, several networks appear that are isolated from each other by a highway or by tracts of unsuitable land too wide to be overcome by the species. Various more sophisticated (but complicated) methods to calculate connectivity are available (e.g. Saura and Pascual-Horta 2007). Some of these methods result in maps showing zones with increasing levels of connectivity.

In the *fourth and final* step, the map with the habitat networks is interpreted against the minimum required network size obtained in *step 1*. This approach results in a map showing (for a single species) which habitat patches are part of a sustainable network, those that are part of an unsustainable network (i.e. not large enough), and the patches which are completely isolated. If a series of maps for a range of target species (or ecoprofiles) is available, the assessment can be framed as (for example) the percentage of the species having sustainable networks, or as the percentage of available habitat patches that contributes to sustainable networks (calculated over all species). Also, maps of connectivity levels within the network can be made, for example by using the LARCH model (van der Grift and Pouwels 2006; Rüter et al. 2014) (Fig. 18.6).

More information can be found in van der Grift and Pouwels (2006) and Verboom and Pouwels (2004).

18.4.2.3 Comments on the Basic Method

The method described in the previous section can essentially be done by hand. More sophisticated methods using GIS-based models allow a better approximation of dispersal through the landscape (see Chap. 6). Dispersal is more complex than individuals leaving patch A and arriving in the nearest patch (B) by following a straight line. Instead, dispersal is a more or less a random searching process, with individuals dispersing in all directions, sometimes using dispersal-aiding landscape elements. Individuals may roam through unsuitable landscape for weeks, risking starvation and predation. Thus, probabilities play a large role. Models based on probabilities of reaching a certain distance are an important improvement, but are much more complex. The use of models also makes it possible to handle many species and integrate the assessment results.

Rarely is a planning area an island. Habitat networks often extend across planning borders; a proper assessment therefore includes a zone of adjacent area. If this beyond-border landscape is not included, wrong conclusions are likely to be made. For example, a network considered within the limits of the planning area may seem unsustainable because it does not match the minimum size criterion, but in fact it may be part of a larger sustainable network that extends beyond the limits of the planning area.

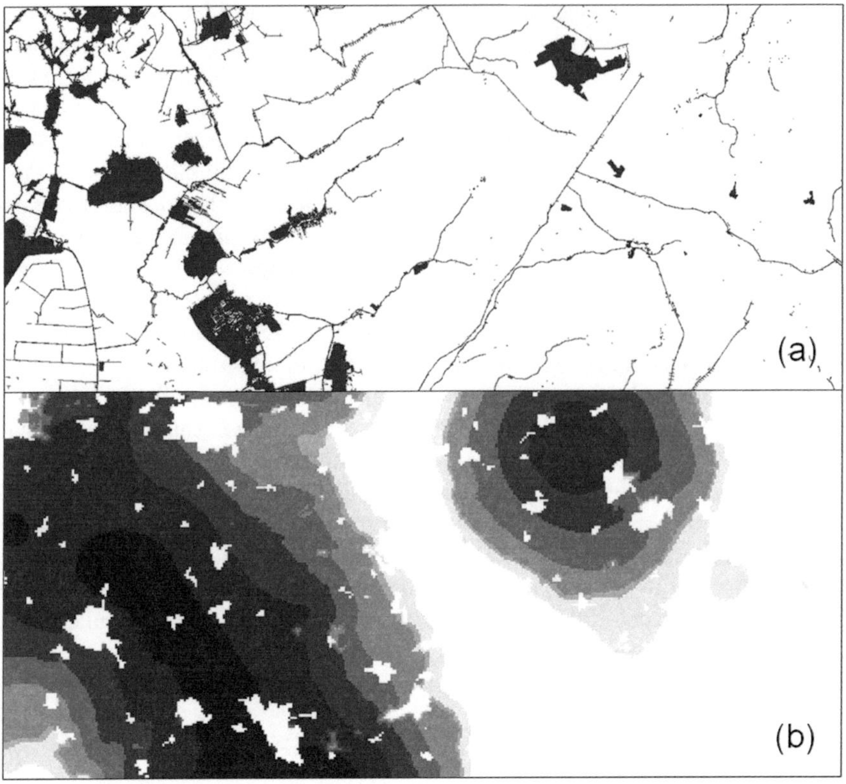

Fig. 18.6 Habitat map for the European otter (*Lutra lutra*) (**a**) and zones with increasing levels of connectivity within the habitat network (**b**). The habitat map shows wetland and aquatic habitats, including streams and lakes with bank side vegetation, as well as marshes and swamp forests. The network analysis with the LARCH model uses all habitat patches large enough for a reproductive unit, the maximum dispersal distance, as well as the permeability of the landscape matrix between habitat patches. The connectivity map indicates which habitat patches are part of a sustainable network (dark grey zones), and which patches are isolated (white zone)

Habitat for a species is often not a single land cover class or vegetation type. For example, a species with high densities in deciduous woodland can also occur in lower densities in coniferous woodland. This means that for viability more area of coniferous woodland is necessary than of deciduous woodland. Such habitat variability can be shown on the species-specific habitat map in *step 2*, and incorporated in the index for minimum required network size.

Note that the assessment map of *step 4* can be used to identify where the network can be improved, for example where stepping stone patches can best be developed to increase connectivity or to connect two unsustainable networks. See van der Grift and Pouwels (2006) and van Rooij et al. (2003) for examples.

As previously described for habitats, very simplified assessment and evaluation methods also exist for habitat networks. These approaches model the structural connectivity between habitats of a type (e.g. hedgerows, grassland) according to distances that can be overcome by certain species groups (for an example see von Haaren et al. 2012). It is assumed that the more species groups can reach the next suitable habitat the better the connectivity. Again, such approaches can be useful as a basis for improving the overall connectivity of the landscape but not as a basis for managing the landscape for certain endangered species or metapopulations (Crooks and Sanjayan 2006).

18.5 Impact Assessment

In addition to the assessment of habitat capacity (and also habitat development potentials, see Chap. 17), policy statements for specific areas should consider the sensitivity of habitats and species to impacts, as well as on their ability to restore their performance and functional capability. The assessment should focus on pressures from human activities that can be classified into structural, material, mechanical, acoustic/visual and biological effects (Table 18.5).

Following the DPSIR framework (European Environment Agency 1999), the differences between the state before and after pressures have altered the state, can be described as impacts. The interactions and cumulative effects can be estimated (e.g. changes in agricultural structure, production of renewable energy). The impact analysis will mainly concentrate on existing, but variable pressures and reversible

Table 18.5 Pressures by human activities and potential impacts on habitats and species

Pressures		Potential impacts
Structural	E.g. intensification of agricultural land, soil sealing, commercial development, roads, dams	Habitat loss, fragmentation, disruption of critical processes (e.g. river dynamics, fire)
Material	E.g. use of fertilisers and pesticides, untreated sewage, waste, acid rain, traffic, industrial emissions	Habitat degradation (e.g. water quality), pollution, eutrophication, siltation
Mechanical	E.g. operation of wind turbines, traffic, transmission lines, tillage	Mortality and injury from collision, soil degradation
Acoustic/visual	E.g. traffic, recreational activities, construction, agricultural land use	Physiological stress, spatial deterrence, behavioural interruption, reduced reproductive success
Biological	Invasive species	Crowding out of native species, by more competitive imported species. Impact is considered negative if this endangers the native species (loss of biodiversity) or if the invasive species do not provide as well to ecosystem services or even impair those in comparison to indigenous species

changes of the habitats. Under such conditions, there is a good chance, that the habitat may be restored. Whether there has been, or will be, a negative impact on the habitats and their value can be deduced by assessing the harmful strength of pressure in combination with the sensitivity (degree and reversibility of impact) of the habitat. The value of the habitat in terms of endangerment may be used to judge the relevance of this impact for nature conservation. With regard to planning pressures, the relevance of the impact as to its effects on overall biodiversity can be estimated by comparing the values of habitat capacity before and after impacts occur. These effects can be shown in scenario form (e.g. Kowalski et al. 2009). Decision makers require such visualisation of longer-term developments in order to make strategic decisions for the communities and regions, while taking risks into consideration. If significant negative effects are determined, this in turn may initiate response measures to prevent, reduce or compensate these effects (see Chap. 25).

18.6 Habitat and Species Option Values

The assessment of habitat capacity should increasingly be coordinated with the value of habitats and species to benefit other ES in order to help optimise environmental measures for landscape multifunctionality (see Chap. 19). For example, target species can be chosen that match the scale level of the habitat networks in the planning area (e.g. on a local scale immobile butterflies can be chosen as species capable of flying, next to amphibians as barrier-sensitive species). In multifunctional planning, such a choice may also be based on preferences by the local community, for example for a species they associate with the landscape identity of their area. Alternatively, landscape benefits (cf. ecosystem services) may be chosen as planning targets. In particular, benefits from regulatory landscape services may depend on sufficient species diversity. For example, water purification, pollination, and natural pest control, have been shown to increase in efficiency and reliability with increasing species diversity (Vos et al. 2014). A good example of how the rapidly growing body of knowledge about this role of species diversity can be utilised in collaborative landscape planning is given in Steingröver et al. (2010).

Implementation strategies should use the communicative power of linking habitat and species issues with landscape benefits, as humans are mostly interested in positive changes for their well-being (Liu and Opdam 2014). However, knowledge about the relation between habitat capacity and other ES can only facilitate target setting and the design of spatial solutions if it supports a social learning process, for example a process of negotiating about values (Opdam et al. 2008).

References

Bernotat, D., Schlumprecht, H., Brauns, C., et al. (2002). Gelbdruck "Verwendung tierökologischer Daten". In H. Plachter, D. Bernotat, R. Müssner, et al. (Eds.), *Entwicklung und Festlegung von Methodenstandards im Naturschutz* (SchrR Landschaftspfl Natursch) (Vol. 70, pp. 109–217).

Blab, J., Riecken, U., & Ssymank, A. (1995). Proposal on a criteria system for a national red data book of biotopes. *Landscape Ecology, 10*, 41–50.

Bredemeier, B., von Haaren, C., Rüter, S., et al. (2015). Evaluating the nature conservation value of field habitats: A model approach for targeting agri-environmental measures and projecting their effects. *Ecological Modelling, 295*, 113–122.

Bunce, R. G. H., Bogers, M. M. B., Evans, D., et al. (2013). The significance of habitats as indicators of biodiversity and their links to species. *Ecological Indicators, 33*, 19–25.

Crooks, K., & Sanjayan, M. (Eds.). (2006). *Connectivity conservation*. Cambridge: Cambridge University Press.

Davies, C. E., Moss, D., & Hill, M. O. (2004). *EUNIS habitat classification revised 2004*. Copenhagen: European Environment Agency.

de Groot, R. S., Alkemade, R., Braat, L., et al. (2010). Challenges in integrating the concept of ecosystem services and values in landscape planning, management and decision making. *Ecological Complexity, 7*, 260–272.

Egoh, B. N., Paracchini, M. L., Zulian, G., et al. (2014). Exploring restoration options for habitats, species and ecosystem services in the European Union. *Journal of Applied Ecology, 51*, 899–908.

European Commission. (2011). Our life insurance, our natural capital: An EU biodiversity strategy to 2020. Communication from the commission to the European parliament, the council, the economic and social committee and the committee of the regions. COM(2011) 0244 final. Brussels.

European Environment Agency. (1999). *Environmental indicators: Typology and overview* (EEA Technical report, no 25). Copenhagen: EEA.

European Environment Agency. (2014). *Terrestrial habitat mapping in Europe: an overview* (Joint MNHN-EEA report. EEA Technical report, no 1/2014). Luxembourg: Publications Office of the European Union.

Evert, K.-J. (Ed.). (2010). *Encyclopedic dictionary of landscape and urban planning*. Berlin: Springer.

Haines-Young, R. (2009). Land use and biodiversity relationships. *Land Use Policy, 26*, 178–186.

Hall, L. S., Krausman, P. R., & Morrison, M. L. (1997). The habitat concept and a plea for standard terminology. *Wildlife Society Bulletin, 25*, 173–182.

Hanski, I. (1999). *Metapopulation ecology*. New York: Oxford University Press.

Haslett, J. R., Berry, P. M., Bela, G., et al. (2010). Changing conservation strategies in Europe: A framework integrating ecosystem services and dynamics. *Biodiversity and Conservation, 19*, 2963–2977.

Hill, D., Fasham, M., Tucker, G., et al. (Eds.). (2005). *Handbook of biodiversity methods*. Cambridge: Cambridge University Press.

JNCC. (1995). Biodiversity: The UK steering group report. Volume 2: Action plans (Annex F and Annex G). Joint Nature Conservation Committee, Londone.

Jongman, R. H. G., & Pungetti, G. (Eds.). (2004). *Ecological networks and greenways: Concept, design, implementation*. Cambridge: Cambridge University Press.

Kowalski, K., Stagl, S., Madlener, R., et al. (2009). Sustainable energy futures: Methodological challenges in combining scenarios and participatory multi-criteria analysis. *European Journal of Operational Research, 197*, 1063–1074.

Kreft, H., & Jetz, W. (2010). A framework for delineating biogeographical regions based on species distributions. *Journal of Biogeography, 37*, 2029–2053.

Liu, J., & Opdam, P. (2014). Valuing ecosystem services in community-based landscape planning: introducing a wellbeing-based approach. *Landscape Ecology, 29*, 1347–1360.

Mace, G. M., Norris, K., & Fitter, A. H. (2012). Biodiversity and ecosystem services: A multilayered relationship. *Trends in Ecology & Evolution, 27*, 19–26.

Maes, J., Paracchini, M. L., Zulian, G., et al. (2012). Synergies and trade-offs between ecosystem service supply, biodiversity, and habitat conservation status in Europe. *Biological Conservation, 155*, 1–12.

Noss, R. F. (1987). Protecting natural areas in fragmented landscapes. *Natural Areas Journal, 7*, 2–13.

Noss, R. F. (1990). Indicators for monitoring biodiversity: A hierarchical approach. *Conservation Biology, 4*, 355–364.

Noss, R. F., O'Connell, M. A., & Murphy, D. D. (1997). *The science of conservation planning: Habitat conservation under the endangered species act*. Washington, DC: Island Press.

Opdam, P. (2013). Using ecosystem services in community based planning: Science is not ready to deliver. In B. Fu & K. B. Jones (Eds.), *Landscape ecology for sustainable environment and culture* (pp. 77–101). Dordrecht: Springer.

Opdam, P., & Steingröver, E. (2008). Designing metropolitan landscapes for biodiversity: Deriving guidelines from metapopulation ecology. *Landscape Journal, 27*, 69–80.

Opdam, P., Verboom, J., & Pouwels, R. (2003). Landscape cohesion: An index for the conservation potential of landscapes for biodiversity. *Landscape Ecology, 18*, 113–126.

Opdam, P., Steingröver, E., & van Rooij, S. (2006). Ecological networks: A spatial concept for multi-actor planning of sustainable landscapes. *Landscape and Urban Planning, 75*, 322–332.

Opdam, P., Pouwels, R., van Rooij, S., et al. (2008). Setting biodiversity targets in participatory regional planning: Introducing ecoprofiles. *Ecology and Society, 13*, 20.

Opdam, P., Nassauer, J., Wang, Z., et al. (2013). Science for action at the local landscape scale. *Landscape Ecology, 28*, 1439–1445.

Parmesan, C., Ryrholm, N., Stefanescu, C., et al. (1999). Poleward shifts in geographical ranges of butterfly species associated with regional warming. *Nature, 399*, 579–583.

Pickett, S. T. A., Parker, V. T., & Fiedler, P. L. (1992). The new paradigm in ecology: Implications for conservation biology above the species level. In P. L. Fielder & S. K. Jain (Eds.), *Conservation biology: The theory and practice of nature conservation, preservation, and management* (pp. 66–88). New York: Chapman & Hall.

Plachter, H., Bernotat, D., Müssner, R., et al. (Eds.). (2002). *Entwicklung und Festlegung von Methodenstandards im Naturschutz* (SchrR Landschaftspfl Natursch) (Vol. 70, p. 566).

Poiani, K. A., Richter, B. D., Anderson, M. G., et al. (2000). Biodiversity conservation at multiple scales: Functional sites, landscapes, and networks. *Bioscience, 50*, 133–146.

Riecken, U., Ries, U., & Ssymank, A. (1994). *Rote Liste der gefährdeten Biotoptypen der Bundesrepublik Deutschland* (SchrR Landschaftspfl Natursch) (Vol. 41, p. 184). Greven: Kilda Verlag.

Rodwell, J.S., Schaminée, J. H. J., Mucina, L. et al. (1998). *The scientific basis of the EUNIS Habitat Classification* (Report to the European Topic Centre on Nature Conservation, Unit of Vegetation Science). Lancaster University.

Rowland, M. M., & Wisdom, M. J. (2009). Habitat networks for terrestrial wildlife: Concepts and case studies. In J. J. Millspaugh & F. R. Thompson (Eds.), *Models for planning wildlife conservation in large landscapes* (pp. 501–531). San Diego: Academic.

Rüter, S. (2009). Multifunctional grass strips – scenario-based modeling of habitat connectivity and water retention. In J. Breuste, M. Kozová, & M. Finka (Eds.), *European landscapes in transformation: Challenges for landscape ecology and management* (pp. 223–227). Salzburg: University of Salzburg.

Rüter, S., Vos, C. C., van Eupen, M., et al. (2014). Transboundary ecological networks as an adaptation strategy to climate change: The example of the Dutch-German border. *Basic and Applied Ecology, 15*, 639–650.

Rüter, S., Matthies, S., & Zoch, L. (2017). Applicability of Modified Whittaker plots for habitat assessment in urban forests: Examples from Hannover, Germany. *Urban Forestry and Urban Greening, 21*, 116–128.

Saccheri, I., Kuussaari, M., Kankare, M., et al. (1998). Inbreeding and extinction in a butterfly metapopulation. *Nature, 392*, 491–493.

Saura, S., & Pascual-Hortal, L. (2007). A new habitat availability index to integrate connectivity in landscape conservation planning: Comparison with existing indices and application to a case study. *Landscape and Urban Planning, 83*, 91–103.

Southerland, M. (1993). *Habitat evaluation: Guidance for the review of environmental impact assessment documents*. Washington, DC: U.S. EPA office of Federal Activities.

Spellerberg, I. F. (1991). *Monitoring ecological change*. Cambridge: Cambridge University Press.

Steingröver, E. G., Geertsema, W., & van Wingerden, W. K. R. E. (2010). Designing agricultural landscapes for natural pest control: A transdisciplinary approach in the Hoeckse Waard (the Netherlands). *Landscape Ecology, 25*, 825–838.

Sybertz, J., Matthies, S., Schaarschmidt, F., et al. (2017). Assessing the value of field margins for butterflies and plants: How to document and enhance biodiversity at the farm scale. *Agriculture, Ecosystems & Environment, 249*, 165–176.

Trochet, A., & Schmeller, D. (2013). Effectiveness of the Natura 2000 network to cover threatened species. *Nature Conservation, 4*, 35–53.

Tucker, G. (2005). Biodiversity evaluation methods. In D. Hill, M. Fasham, G. Tucker, et al. (Eds.), *Handbook of biodiversity methods* (pp. 65–104). Cambridge: Cambridge University Press.

van der Grift, E., & Pouwels, R. (2006). Restoring habitat connectivity across transport corridors: Identifying high-priority locations for defragmentation with the use of an expert-based model. In J. Davenport & J. L. Davenport (Eds.), *The ecology of transportation: Managing mobility for the environment* (pp. 205–231). Dordrecht: Springer.

van Rooij, S. A. M., Steingröver, E. G., & Opdam, P. F. M. (2003). *Corridors for life. Scenario development of an ecological network in Cheshire County* (Alterra report 699). Wageningen.

Verboom, J., & Pouwels, R. (2004). Ecological functioning of ecological networks: A species perspective. In R. H. G. Jongman & G. Pungetti (Eds.), *Ecological networks and greenways: Concept, design, implementation* (pp. 65–72). Cambridge: Cambridge University Press.

Verboom, J., Foppen, R., Chardon, P., et al. (2001). Introducing the key patch approach for habitat networks with persistent populations: An example for marshland birds. *Biological Conservation, 100*, 89–101.

von Drachenfels, O. (2010). *Klassifikation und Typisierung von Biotopen für Naturschutz und Landschaftsplanung. Ein Beitrag zur Entwicklung von Standards für Biotopkartierungen, dargestellt am Beispiel von Niedersachsen*. (Natursch Landschaftspl Niedersachsen) (Vol. 47, p. 322).

von Haaren, C., Galler, C., & Ott, S. (2008). *Landscape planning. The basis of sustainable landscape development*. Bonn: Federal Agency for Nature Conservation.

von Haaren, C., Kempa, D., Vogel, K., et al. (2012). Assessing biodiversity on the farm scale as basis for ecosystem service payments. *Journal of Environmental Management, 113*, 40–50.

von Haaren, C., Albert, C., Barkmann, J., et al. (2014). From explanation to application: Introducing a practice-oriented ecosystem services evaluation (PRESET) model adapted to the context of landscape planning and management. *Landscape Ecology, 29*, 1335–1346.

Vos, C. C., Verboom, J., Opdam, P. F. M., et al. (2001). Towards ecologically scaled landscape indices. *The American Naturalist, 157*, 24–51.

Vos, C. C., Grashof-Bokdam, C. J., & Opdam, P. F. M. (2014). *Biodiversity and ecosystem services: Does species diversity enhance effectiveness and reliability?* (WOt-technical report 25). Wageningen.

Evaluation of Multifunctionality and Aggregated Benefits

19

Peter Stubkjær Andersen and Carolin Galler

Abstract

This chapter discusses the concept of multifunctionality and how it can be interpreted and evaluated in the context of landscape planning. The importance of scale is emphasised and examples of assessment frameworks from five studies are reviewed. The concept of aggregated benefits and additionally, the potential complementarity of multifunctionality and ES frameworks are also discussed.

Keywords

Multifunctionality · Land sharing · Land sparing · Scale · Aggregated benefits · Evaluation methods

19.1 Introduction: Multifunctionality in the Context of Landscape Planning

Landscape multifunctionality refers to the multiple functions or services that landscapes provide to humans (Willemen et al. 2010). The basic – and rather simple – idea of addressing landscape from a multifunctionality perspective is justified by the need to describe landscapes, not only by a structural land use/land cover approach, but also from a functional dynamics approach (Andersen 2013). As a starting point, the main question to address in an evaluation of landscape multifunctionality is

P. S. Andersen (✉)
Department of Geosciences and Natural Resource Management, Landscape Architecture and Planning, University of Copenhagen, Frederiksberg, Denmark
e-mail: stub@ign.ku.dk

C. Galler
Niedersächsisches Ministerium für Umwelt, Energie und Klimaschutz, Hannover, Germany
e-mail: carolin.galler@mu.niedersachsen.de

C. von Haaren et al. (eds.), *Landscape Planning with Ecosystem Services*,
Landscape Series 24, https://doi.org/10.1007/978-94-024-1681-7_19

"which functions are present in this well-delineated landscape at this specific scale?" Based on this initial identification of functions a further assessment of whether these functions are synergetic, indifferent, or conflicting is needed (Galler et al. 2013) in order to evaluate the multifunctionality of the landscape.

When multifunctionality is considered in the context of landscape planning the broad and multifaceted discussion about multifunctionality can be restricted to a narrower discussion of what is relevant for landscape planning. In the framework of landscape planning, the definition of the term *function* is understood – not in the descriptive sense of fluxes and processes found in landscape ecology or biology (Jax 2005) – but rather in the narrative sense of function for a human demand (which is almost identical with *delivered ecosystem services,* cf. Chap. 3). Thus, the main challenge is to understand the link between the multiple functions (or services) of ecological systems and human well-being (Potschin and Haines-Young 2011).

Accordingly, landscape multifunctionality can be used as an analytical concept to reveal the overlaps of different landscape functions and the connections between them (synergies, indifferences, conflicts). This outcome is relevant for landscape planning as synergies are generally supported while conflicts have to be solved. Furthermore, multifunctional solutions can be much more efficient – from a cost or land use perspective – than monofunctional solutions (Selman 2008).

Several reasons for evaluating multifunctionality in landscapes can be identified. Selman (2009) states that "multifunctionality is generally desirable, as it encourages efficient use of land, delivers wider public benefit and builds partnerships of user groups, leading to better stewardship". More specifically, reasons for an evaluation of multifunctionality pertain to:

- Understanding the basic functionality of the landscape
- Understanding the interactions (conflicts and synergies) between landscape functions
- Planning for landscape functions in terms of prioritizing different interests e.g., nature, environment, recreation, production (e.g., food, energy)
- Giving quantitative and qualitative input in relation to subsidy allocation (e.g., from the EU) that focuses on supporting the development of one or more functions

A central discussion in relation to landscape multifunctionality and the need for evaluation concerns two overall approaches to the concept: *neutral* and *normative*. In brief, the first approach deals with multifunctionality as a characteristic of the landscape, the simple and neutral fact that multiple functions exist in any given landscape (with very few exceptions) (OECD 2003); the second approach recognizes multifunctionality in certain situations as a goal and suggests that the functional status of the landscape is deliberately changed into something more multifunctional through necessary landscape management and planning (OECD 2003). This further implies that a value judgement is included where increased multifunctionality – or at least a change in current landscape functionality – is considered desirable. However, neither a neutral approach nor a strictly normative approach that favors multifunctionality as a goal is directly useful for practical landscape planning. Multifunctionality can only be judged as desirable or conflicting in the context of a specific set of functions in a certain spatial and temporal configuration – and in comparison with a monofunctional alternative.

A further perspective on landscape multifunctionality in relation to planning concerns the notions *land sharing* and *land sparing*. These have recently received increased interest in the context of landscape sustainability. The concepts originate from discussions on biodiversity conservation in relation to agricultural production landscapes (Herzog and Schüepp 2013). In this context, land sharing and sparing represents two contrasting multifunctionality strategies in order to reach improvements in both biodiversity and in production (Law and Wilson 2015). The improvements can be measured e.g. in terms of species numbers and habitat quality (for biodiversity) and in terms of increased productivity (for agriculture). The terms land sharing and land sparing may be seen as a very concrete land-based manifestation of functional integration and functional segregation. In a situation of land sharing, the land is shared by two or more functions and it is assumed that the functions are fully integrated in space and time. In the situation of land sparing, the land is subdivided into functional zones that are reserved for focus on individual functions, thus the functions are segregated in space and time. However, the conceptual framework of sharing and sparing mainly focuses on two of the primary landscape functions (habitat protection and agricultural production). For a comprehensive assessment of landscape functionality, additional functions (such as recreation and hazard regulation) should be included. Furthermore, the scale of analysis and implementation of such strategies determines whether improvements in both biodiversity and production are possible.

In relation to operational landscape planning the main question to address is where strategies of land sharing or land sparing should be implemented and furthermore, at which spatial scale the different strategies should be applied. The two strategies are not conceptually tied to a specific scale (Herzog and Schüepp 2013). A possible consequence is that the outcome may depend on perspective e.g. what is considered a land sparing example by some may be considered land sharing by others (Fischer et al. 2014). This observation is also relevant in more general considerations regarding multifunctionality and scale (see Sect. 19.3). Segregated land uses are often the best solution at site level. At the landscape level however, multiple ecosystem services can be produced simultaneously following the concept of multifunctionality. Regardless of the level in focus, landscape planning needs to assess possible synergies, conflicts and trade-offs between land management actions targeting different ecosystem services. Furthermore, landscape planning needs to reflect upon which ecosystem services to prioritize in the case of conflicts.

19.2 The Concept of Multifunctionality in Multiple Contexts

There is nothing new about multifunctionality in relation to land use and landscape management or the idea that it should lead to or support sustainable development in terms of resource use. The first historical examples are found in forestry practices where it has been pointed out for more than a century that multiple benefits of forests exist and furthermore, a focus on these multiple benefits increases the sustainability of the forest resource (Larsen 2005).

In recent decades the importance of the concept of multifunctionality has been recognized in politics and research, primarily as a way of describing agriculture and rural landscapes (Renting et al. 2009) and secondly, as an analytical framework. From the perspective of politics, multifunctionality found its way into agricultural trade discussions within the WTO in the 1990s. The concept of multifunctionality was especially used by EU trade negotiators to describe European agriculture and thus as a strategic argument for continued agricultural support for European farmers (Andersen 2013).

In research the concept was simultaneously adopted and used in academic models so that agriculture could be described with a specific focus on functionality – in contrast to a structural focus. In short, the recent history of the concept can be described as being first a political introduction of the term and a subsequent and ongoing conceptual development within academia. This ongoing, and to some extent sequential development, consists of work on the definitions of multifunctionality (and functions), work on the classification of functions that constitute multifunctionality, and current work on the operationalization of multifunctionality (Andersen 2013) that is relevant for the evaluation of multifunctionality.

In evaluations of multifunctionality it is useful to make a distinction between the supply side – which functions are supplied in a specific landscape – and the demand side – which functions are demanded (by society) from a particular landscape (van Huylenbroeck et al. 2007; Vejre et al. 2007). This distinction allows insight into the current status, the supply-demand gaps, and can facilitate discussions on normative aspects regarding current and future landscape multifunctionality. In addition, an analytical framework that focuses on supply and demand will also need to address aspects of scale in relation to supply and demand, which is considered to be of key importance in landscape research (Herod 2011).

19.3 Scale Matters

A basic characteristic of landscape evaluation methods is that they operate at a specific geographical scale. This means that they relate to a specific area and a functional unit – which serves as the unit of analysis that is combined with a specific level of detail. This is, for example, the plot level, field level, the farm scale, the landscape or supra landscape scale (e.g., regional, national). The choice of operational scale is mainly linked to the purpose of evaluation but may also be defined by data availability in terms of which data are used for the evaluation and the scale at which they are collected.

The farm level is often considered to be an appropriate scale for evaluating multifunctionality (Wilson 2009) and has also been used for evaluating ecosystem services (Lerouge et al. 2016). This has to do with the fact that the farm is often the main operational unit of cultural landscape from a management perspective. Furthermore, a farm is the smallest legal unit in administrative terms and it is the unit in which the owner makes land use decisions, which are important for defining functionality (Andersen 2013). In Denmark this means that farm level data are available in a well-defined way and obtainable either through the national farm registry or through direct contact with the farm owner. However, access to information from farmers very much depends on country/culture specific characteristics, and whether the farm owners want to share information on their practices and management decisions between individuals. A multifunctionality analysis with the whole farm as a reference unit is not very useful for landscape planning, but farm scale data about the multifunctionality of functional spatial units or cells on the farm can be used in landscape planning for various purposes. For example pointing out areas where farmers can combine their production services most effectively with habitat or water retention services is information that can be communicated very well by local landscape planning.

Based on a method for whole farm multifunctionality analysis (Andersen et al. 2013) an approach to measuring multifunctionality which is fit for use in landscape planning has been developed. In this method the functional unit is the farm enterprise area. The study provides an overview of the functional strength at the farm scale but also offers the possibility of aggregating the farm level data to a higher hierarchical level, the landscape scale. This demands map-based approaches and interpretations of the functionality patterns found at the farm level. Even if the method ideally provides precise and detailed information on functionality, this information may be fragmented if data are not obtained from all farm units in the landscape. System theory also suggests that information may be lost by transferring data from one hierarchical level to another (von Bertalanffy 1972).

This is also very relevant for one issue related to the management of landscapes: functional demand and supply. From a societal perspective the farm level is the basic administrative scale at which many functions are supplied. However, societal demands may be at different scales, e.g. soil or field units for the provision of food, landscape units for aesthetic experience and river catchment areas in relation to flood prevention. These are often different from the administrative level. For example, the demands for the provision of bird habitats (for rare species) or for recreational tracks through a farmed landscape (to connect existing recreational hotspots). In these examples it is unlikely that the ideal supply unit is limited to one farm, thus leading to a mismatch between functional demand and supply (van Huylenbroeck et al. 2007).

From a planning perspective, the identification of the areas where functional supply does not meet the needs of society should be useful in terms of formulating concrete initiatives to rectify the situation. Generally, what is needed is cross-scale coordination and planning (Chap. 3). Specifically, planners need to involve interest groups and stakeholders concerned with, and therefore representing, the different functions at different spatial scales. This should, theoretically, enable decision-making that leads to actual change in landscapes (Selman 2004). However, it is important to realize that multifunctionality at the landscape scale may encompass both whole farms that are largely monofunctional (e.g. mono crop production), and multifunctional farms that include monofunctional plots. It is stressed that in relation to landscape planning the benefits of multifunctionality have to be measured at different scales from plot level upwards in order to develop the most efficient strategy of landscape development.

19.4 Landscape Multifunctionality and Aggregated Benefits

In discourses regarding the concept of multifunctionality, the concept of aggregated benefits is often mentioned (e.g. Galler et al. 2015). As previously discussed, evaluations of landscape multifunctionality are either neutral or normative. Aggregated benefits directly implies normative (positive) aspects of the landscape in relation to human well-being. Hence, within the multifunctionality framework normative aspects are already incorporated by statements that link the level of landscape multifunctionality and the sustainability of the landscape, which implies positive characteristics of multifunctionality (Selman 2009; Willemen et al. 2010). This emphasizes the policy relevance of working with both multifunctionality and aggregated benefits, in a planning context. As such the two concepts should be considered in combination rather than in opposition.

The concept of aggregated benefits directly stresses that benefits are associated with landscapes. In order to assess the aggregated benefits, the name itself further implies that the assessment could be a relatively simple task of calculating the functional synergies (positive) and trade-offs (negative) of selected functions. Even if some differences in the concepts of multifunctionality and aggregated benefits are obvious, what unites the two concepts is their assessment of landscapes from a holistic perspective: namely acknowledging that landscapes have multiple aspects (functions) that are all potentially important and therefore cannot be looked upon in isolation for long. This is a key basis of landscape planning because sound decision-making implies taking all important factors into account.

19.5 Methods for Evaluating Multifunctionality and Their Applicability to Landscape Planning

In the following, examples for multifunctionality assessment schemes are given and commented upon in terms of their usability and their suitability with regard to landscape planning. The methods have been selected on the basis of the following criteria: the spatial scale and the spatial units to which the evaluation of multifunctionality refers, how the methods contribute to defining priorities, whether they measure efficiency in relation to allocation of functions, and their possible application in landscape planning.

In relation to application in landscape planning, a general guideline is that methods should assess all relevant functions/ecosystem services. The services covered in this book can be used as a preliminary list, complemented by further services which may be relevant in the specific planning case. If certain functions are not considered the results of landscape assessment or the responses may be jeopardized because synergies or conflicts do not become obvious. A criterion for choosing functions should be their relevance to the overall purpose of the planning initiative. Consequently, if functional conflicts and solving these conflicts are in focus then the analysis should include overlaying potentially conflicting functions (e.g. habitat functions vs. provisioning functions) in order to identify conflict zones where interventions could happen – either driven from a bottom-up or top-down perspective. Conversely, if identifying synergies is the main focus then the generally synergistic functions have to be included.

Five examples from the academic literature have been chosen in order to illustrate the variety of approaches (see also Table 19.1). These could potentially serve as a foundation and inspiration for developing integrated planning tools for evaluating multifunctionality that also have a high degree of usability.

Example 1: Gimona and van der Horst (2007) developed a method of mapping multifunctionality hotspots based on selected landscape functions: biodiversity, visual amenity, and woodland recreation. The evaluation of functions was based on a grid of cells (500 m x 500 m) and within this framework existing functional hotspots were identified using four different weighting schemes for the three functions; the first being an equal weighting and each of the other three giving priority to one of the functions. This allows the possible incorporation of priorities into the mapping process, which may serve as valuable background information for evaluating consequences of landscape planning decisions e.g. as focused on by Gimona and van der Horst (2007) in the case of a landscape zonation.

The method is limited to cultural ES. However, the application in landscape planning is that inconsistencies between visual landscape amenities and actual recreation can be identified. For example high recreation use due to proximity to settlement and low visual landscape quality would be a trigger for improving landscape amenities. Also the overlap of high biodiversity values and high recreational use could be taken as an indicator for conflicting services with a need for problem solving measures.

Example 2: Willemen et al. (2008, 2010) provide an evaluation of multifunctionality that also focuses on identifying multifunctional hotspots in landscapes. The method is centered on the assumption that the spatial pattern of landscape functions is a result of the interactions (synergies and conflicts) between the functions. The authors identify seven landscape scale functions: residence, livestock production, cultural heritage, tourism, plant habitat, arable production, and leisure cycling. For each function a number of indicators are chosen and weighted, in some cases indicators are also included which have an assumed and experienced negative influence on the functional strength.

An application of the method results in several maps illustrating both basic expression of the potential number of landscape functions in grid cells (100 m × 100 m) and the capacity of the landscape to provide multiple functions. What is

Table 19.1 Overview of selected methods for evaluating multifunctionality in relation to landscape planning

Author	Gimona & v. der Horst	Willemen et al.	Lovell et al.	Waldhardt et al.	Galler et al.
Years	(2007)	(2008)/(2010)	(2010)	(2010)	(2015)
Evaluation unit	Benefit score	Presence/ non-presence	Normative score	Means per area unit	Multiple
Scale (reference area for multifunctionality evaluation)	Landscape (grid cells 500 × 500 m)	Landscape (grid cells 100 × 100 m)	Farm to landscape (area 187–513 ha)	Multiple grid sizes (20 × 20 m to 500 × 500 m) and land parcels	Large landscape (intersection of functional areas such as plot, habitat, soil unit)
Functions evaluated	Biodiversity	Residential	Production	Metal in soils	Water quality
	Visual amenity	Intensive livestock	Ecological	Nitrate loads	Soil erosion
	Recreation	Cultural heritage	Cultural	Plant species richness	Climate change mitigation
		Tourism		Bird population	Biodiversity
		Plant habitat		Land rent	
		Arable production		Landscape scenery	
		Leisure cycling			
Functional evaluation approach	Calculation of potential functional scores based on key variables	Indicators of different functions and focus on interactions between functions	Indicators related to three main functions that are rated in a scale from −2 to +2	Spatially explicit indicators	Focus on multifunctionality of environmental measures
Strengths – in relation to landscape planning	Straight-forward in terms of calculations of the benefit scores and mapped overview of functional hotspots	Identifies hotspots using different approaches and provides an overview of the functional interactions	Simple to use and gives detailed information at farm scale	The framework integrates numerous aspects of landscape multifunctionality and proposes a future scenario	Evidence-based approach. The spatial and cost effectiveness of measures in regard to multiple ES can be quantified

Weaknesses – in relation to landscape planning	Evaluates very few functions meaning that important aspects of landscape functionality may be overlooked	Some of the chosen functions in the methods could also be understood as simply activities rather than functions	The score results are based on a rating approach that can only be understood in comparison with other farms/ landscapes	Comprehensive and complex to use for planners and other user groups	Rather complex framework that may only be used by trained planners and professionals
Applicability in landscape planning	The GIS interface gives spatial information of functions and the weighting of functions allows priorities from a functional perspective to be applied	The focus on interactions of functions enables planners (users) to focus on trade-offs and synergies which is needed for identifying planning initiatives	The simplicity of use of the method combined with the detailed information is attractive but may be too detailed for landscape level planning	The high level of information and the combination of numerous methods suggest dependency on experts	Landscape facts are in focus allowing unbiased application of information into a planning context; measures can be rated according to their cost-effectiveness

particularly interesting about the method is the focus on the interactions between functions. For instance, which functions are found to be in conflict and which functions experience synergies. The interactions are used as a basis for identifying the multifunctionality hotspot and coldspots spatially which can be used to discuss the advantages and disadvantages of monofunctional vs. multifunctional land strategies at local and regional level. In relation to landscape planning the spatial information relating to hotspots and coldspots may serve as a foundation for targeting spaces for response actions to enhance the current functional composition of the landscape.

Example 3: Lovell et al. (2010) provide a framework for evaluating multifunctionality with the aim of designing sustainable agricultural landscapes based on an integration of knowledge from agroecology and multifunctionality theory. The framework is supported by information collected from interviews with farmers. In a landscape planning context input through interviews should be considered valuable, however it demands that the interview questions have a generic format to assure comparability across different landscapes. Here the interview data inform in-depth descriptions of the key functions of the landscape categorized as production functions, ecological functions, and cultural functions. This results in a multifunctional landscape assessment based on normative ratings of different indicators assigned to the key functions. The assessment is subdivided into land use categories in order to achieve a so-called "performance sum" for each land use category.

The resulting framework has an ordinal scale. Furthermore, the assessments are at different spatial levels, what could be considered as a farm scale and a landscape levels. The authors provide two case studies in landscapes of only 187 ha and 513 ha, respectively.

Example 4: Waldhardt et al. (2010) present a comprehensive framework for evaluating landscape multifunctionality. The framework consists of a number of tools to assess different aspects of the landscape. The tools include five GIS-based ITE^{2}M models: ATOMIS, SWAT, ProF, GEPARD, and ProLand, for assessing metals in soils, water quality and quantity, plant species richness, bird species abundance, and land rent respectively.

The method is described as normative in the sense that once the evaluation of the current status of multifunctionality is performed, this is supplemented with an assessment of the so-called functional deficits of the current landscape considering the selected single functions. The assessment includes input on where the landscape is dysfunctional. This is used as a point of departure for suggesting land use-based scenarios to mitigate the functional deficit. Consequently, actions associated with the scenarios should lead to desirable futures in terms of more multifunctional landscapes to meet demands (especially for the local inhabitants), according to Waldhardt et al. (2010). The method uses comprehensive modelling and thus has a demanding data input for the planning region. In the specific landscape planning context the planner will have to judge whether detailed results will be needed (e.g. because problems with soil contamination are suspected and limited information on this aspect will impair discussions with local land owners).

Example 5: Galler et al. (2015) recently developed a method which focusses on measuring the multifunctionality of response measures. The first step is to evaluate for different ES the fulfilment of environmental quality objectives against official reference levels. The method assesses the spatial and cost efficiency of multifunctional optimization versus optimizing the landscape functions (in sense of delivered ES) separately in the planning area until an official objective is reached. This includes processes of scenario-building consisting of a variety of restoration and improvement measures that already exist in the agricultural policy of the European Union. The four presented scenarios aim at optimizing different environmental objectives concerning: erosion, water quality, climate change, and biodiversity.

The method incorporates economic aspects in terms of assessments of production costs as well as implementation costs for restoration and improvement measures. In this way the framework supports the optimization of a landscape in terms of ES provision and optionally the most area-effective or cost-effective mix of measures. This takes landscape planning a step further in making it a basis for decisions about where and on which measures public spending is best invested (e.g. minimize the land you need to achieve a given set of objectives or obtain best multiple achievements for money invested).

19.6 Linking Multifunctionality and Ecosystem Services

The concept of ecosystem services (ES) is strongly interlinked with the concept of aggregated benefits as ES are defined as "the *benefits* people obtain from ecosystems". However, there are also similarities between the frameworks of multifunctionality and ES.

One of the links between the two frameworks relates to the terms "function" and "service". These are very similar in their meaning and in some cases used interchangeably. In attempts to separate the meaning of the two, it has been noted that landscape function should be understood as the core capacity of a landscape to provide goods and services to society (de Groot et al. 2010). Goods and services are the benefits people obtain (Millennium Ecosystem Assessment 2005).

Further integration of ES into landscape planning frameworks is needed (Albert et al. 2016) in order to attain win-win situations for the environment and for economic development, with the overarching goal of achieving sustainable development (de Groot et al. 2010). In order for this to occur, a number of challenges need to be overcome. This could possibly be done by drawing on the landscape-oriented approaches found in frameworks concerning landscape multifunctionality.

One way of analyzing ES is in the framework of 'ecosystem service bundles' (i.e. sets of ecosystem services that repeatedly appear together across space or time) which has been suggested as an expansion of the ES framework in relation to the evaluation of ES trade-offs (Raudsepp-Hearne et al. 2010). The framework has been used to identify potential synergies between services with the prospect of increasing landscape multifunctionality (Turner et al. 2014). Thus the ES framework, though not specifically focused on increasing multifunctionality, through its analysis of services in an integrated manner may highlight opportunities for functional integration (or segregation).

Several examples of integration of ES into a planning context can be found (Albert et al. 2014). One suggestion from Liu and Opdam (2014) is to incorporate a human well-being dimensions with ES. The idea is that it is emphasized within the concept of ES that ecosystems contribute to human well-being. Consequently, if the overall objective of planning is to improve the life of humans then the link between well-being dimensions (such as health, education, living standards and happiness) and different ES (e.g. food production, habitat provision, recreation) should be clarified in order to assure specific landscape planning actions for the benefit of humans. García-Llorente et al. (2012) analyze the relationship between multifunctionality and social preferences through studies of several provisioning, regulating and cultural ES. Their results show a clear connection between a diverse provision of ES and people's favorites views. The approach seeks to link people's landscape view preferences and the delivered ES from the landscapes. In that sense the landscape "as perceived by people" (as defined in the European Landscape Convention, Council of Europe 2018) is directly associated with the biotic and abiotic condition of the landscape, expressed through an evaluation of ES. The knowledge of the delivered ES and the link to people's preferences can be applied in planning initiatives in order to guide development in a direction where ES provision is improved without compromising social preferences. Fürst et al. (2014) give suggestions on how the ES framework can create an increased awareness amongst landscape planning stakeholders regarding the benefits of landscapes. They propose a so-called balanced score card as a tool to identify where imbalances regarding groups of ES may exist and could be rectified through appropriate planning initiatives. These examples suggest that – from a planning perspective – the frameworks of multifunctionality and ES may actually supplement each other quite well in an attempt to reach an integrated framework for evaluating the aggregated benefits of landscapes.

19.7 Conclusion

In recent decades landscape practitioners and academics have struggled to truly operationalize the concept of multifunctionality. Operationalizing means taking the concept from being more than just a vague – but in theory reasonable – guiding principle for land management to actually achieving landscape sustainability. Landscapes should be assessed with a strong focus on quantifying the functions in a generally accepted way and by further focusing on evaluating the strength of the correlations (synergies and trade-offs) between functions. In this chapter, selected examples of promising methods for quantifying functions have been presented.

Further development of these methods and possible combination of them to reach a unifying quantification methodology could be the next step forward. This may also include integrating key aspects of other conceptual frameworks such as those regarding land sharing and land sparing. Making tools for assessment of landscape multifunctionality and the aggregated benefits available for planners and the public should be a key task for researchers.

References

Albert, C., Aronson, J., Fürst, C., et al. (2014). Integrating ecosystem services in landscape planning: Requirements, approaches, and impacts. *Landscape Ecology, 29*, 1277–1285.

Albert, C., Galler, C., Hermes, et al. (2016). Applying ecosystem services indicators in landscape planning and management: The ES-in-Planning framework. *Ecological Indicators, 61*, 100–113.

Andersen, P. S. (2013). *Temporal and spatial quantification of farm and landscape functions*. PhD thesis, University of Copenhagen.

Andersen, P. S., Vejre, H., Dalgaard, T., et al. (2013). An indicator-based method for quantifying farm multifunctionality. *Ecological Indicators, 25*, 166–179.

Council of Europe. (2018). *European landscape convention*. https://www.coe.int/en/web/landscape/reference-texts. Accessed 27 July 2018.

de Groot, R. S., Alkemade, R., Braat, L., et al. (2010). Challenges in integrating the concept of ecosystem services and values in landscape planning, management and decision making. *Ecological Complexity, 7*(3), 260–272.

Fischer, J., Abson, D. J., Butsic, V., et al. (2014). Land sparing versus land sharing: Moving forward. *Conservation Letters, 7*(3), 149–157.

Fürst, C., Opdam, P., Inostroza, L., et al. (2014). Evaluating the role of ecosystem services in participatory land use planning: Proposing a balanced score card. *Landscape Ecology, 29*(8), 1435–1446.

Galler, C., von Haaren, C., & Albert, C. (2013). Planning multifunctional measures for efficient landscape management: Quantifying and comparing the added value of integrated and segregated management concepts. In *Landscape ecology for sustainable environment and culture* (pp. 249–284). Dordrecht: Springer.

Galler, C., von Haaren, C., & Albert, C. (2015). Optimizing environmental measures for landscape multifunctionality: Effectiveness, efficiency and recommendations for agri-environmental programs. *Journal of Environmental Management, 151*, 243–257.

García-Llorente, M., Martín-López, B., Iniesta-Arandia, I., et al. (2012). The role of multifunctionality in social preferences toward semi-arid rural landscapes: An ecosystem service approach. *Environmental Science & Policy, 19*, 136–146.

Gimona, A., & van der Horst, D. (2007). Mapping hotspots of multiple landscape functions: A case study on farmland afforestation in Scotland. *Landscape Ecology, 22*(8), 1255–1264.

Herod, A. (2011). *Scale (key ideas in geography)*. Oxon: Routledge.

Herzog, F., & Schüepp, C. (2013). Are land sparing and land sharing real alternatives for European agricultural landscapes. *Aspects of Applied Biology, 121*, 109–116.

Jax, K. (2005). Function and "functioning" in ecology: What does it mean? *Oikos, 111*(3), 641–648.

Larsen, J. B. (2005). Functional forests in multifunctional landscapes: Restoring the adaptive capacity of landscapes with forests and trees. In V. Veltheim & B. Pajari (Eds.), *Forest landscape restoration in Central and Northern Europe* (Vol. 53, pp. 97–102). Joensuu: Proceedings of the European Forest Institute.

Law, E. A., & Wilson, K. A. (2015). Providing context for the land-sharing and land-sparing debate. *Conservation Letters, 8*(6), 404–413.

Lerouge, F., Sannen, K., Gulinck, H., et al. (2016). Revisiting production and ecosystem services on the farm scale for evaluating land use alternatives. *Environmental Science & Policy, 57*, 50–59.
Liu, J., & Opdam, P. (2014). Valuing ecosystem services in community-based landscape planning: Introducing a wellbeing-based approach. *Landscape Ecology, 29*(8), 1347–1360.
Lovell, S. T., Nathan, C. A., Olson, M. B., et al. (2010). Integrating agroecology and landscape multifunctionality in Vermont: An evolving framework to evaluate the design of agroecosystems. *Agricultural Systems, 103*(5), 327–341.
Millennium Ecosystem Assessment. (2005). *Ecosystems and Human Well-being: Synthesis*. Washington, DC: Island Press.
OECD. (2003). *Multifunctionality. The policy implications*. Paris: OECD Publications.
Potschin, M. B., & Haines-Young, R. H. (2011). Ecosystem services: Exploring a geographical perspective. *Progress in Physical Geography, 35*(5), 575–594.
Raudsepp-Hearne, C., Peterson, G. D., & Bennett, E. M. (2010). Ecosystem service bundles for analyzing tradeoffs in diverse landscapes. *Proceedings of the National Academy of Sciences of the United States of America, 107*(11), 5242–5247.
Renting, H., Rossing, W. A. H., Groot, J. C. J., et al. (2009). Exploring multifunctional agriculture. A review of conceptual approaches and prospects for an integrative transitional framework. *Journal of Environmental Management, 90*, S112–S123.
Selman, P. (2004). Community participation in the planning and management of cultural landscapes. *Journal of Environmental Planning and Management, 47*(3), 365–392.
Selman, P. (2008). What do we mean by sustainable landscape? *Sustainability: Science, Practice, & Policy, 4*(2), 23–28.
Selman, P. (2009). Planning for landscape multifunctionality. *Sustainability: Science, Practice, & Policy, 5*(2), 45–52.
Turner, K. G., Odgaard, M. V., Bøcher, P. K., Dalgaard, T., & Svenning, J. C. (2014). Bundling ecosystem services in Denmark: Trade-offs and synergies in a cultural landscape. *Landscape and Urban Planning, 125*, 89–104.
van Huylenbroeck, G., Vandermeulen, V., Mettepenningen, E., et al. (2007). Multifunctionality of Agriculture: A Review of Definitions, Evidence and Instruments. *Living Reviews in Landscape Research, 1*(3), 43.
Vejre, H., Abildtrup, J., Andersen, E., et al. (2007). Multifunctional agriculture and multifunctional landscapes – Land use as an interface. In Ü. Mander, K. Helming, & H. Wiggering (Eds.), *Multifunctional land use: Meeting future demands for landscape goods and services* (pp. 93–104). Berlin: Springer.
von Bertalanffy, L. (1972). *General system theory: Foundations, development, applications* (pp. 30–53). London: Allen Lane.
Waldhardt, R., Bach, M., Borresch, R., et al. (2010). Evaluating today's landscape multifunctionality and providing an alternative future: A normative scenario approach. *Ecology and Society, 15*(3), 30.
Willemen, L., Verburg, P. H., Hein, L., et al. (2008). Spatial characterization of landscape functions. *Landscape and Urban Planning, 88*(1), 34–43.
Willemen, L., Hein, L., van Mensvoort, M. E., et al. (2010). Space for people, plants, and livestock? Quantifying interactions among multiple landscape functions in a Dutch rural region. *Ecological Indicators, 10*(1), 62–73.
Wilson, G. A. (2009). The spatiality of multifunctional agriculture: A human geography perspective. *Geoforum, 40*, 269–280.

Economic Valuation of Services

20

Andrew A. Lovett

Abstract

This chapter discusses the economic methods that can be used to derive monetary values for the different benefits provided by ecosystem services. It begins by considering the basis of valuation, particularly the importance of identifying the benefits involved. This leads on to a review of three main classes of techniques (i) market valuation, (ii) revealed preference methods and (iii) stated preference methods. Subsequent sections examine the challenges of transferring economic values from one setting to another and the current provision of databases and decision support tools to assist in valuation exercises.

Keywords

Monetary valuation · Use and non-use values · Revealed preference methods · Stated preference methods · Benefit transfer

20.1 Introduction

Chapter 4 reviewed the core frameworks that are widely employed to assess different planning options regarding the provision of ecosystem services (ES). Typically such assessments need to consider a variety of *use* and *non-use* values (Pascual et al. 2010) in combination with economic and social methods for evaluating individual preferences. This provides particularly powerful indicators of use values (e.g. market prices), as well as measures of non-use (e.g. existence values). Legislative frameworks provide a societal expression of attitudes regarding collective use and non-use values, often setting the bounds within which preference-based measures

A. A. Lovett (✉)
School of Environmental Sciences, University of East Anglia, Norwich, UK
e-mail: a.lovett@uea.ac.uk

C. von Haaren et al. (eds.), *Landscape Planning with Ecosystem Services*, Landscape Series 24, https://doi.org/10.1007/978-94-024-1681-7_20

shape decisions (Chap. 4). However, it is important to recognize that the balance of power between the two frameworks can vary appreciably both between and within countries, the role of the state being very prominent in some instances and a much more free-market situation existing in others (see Part F of this book, 'Global context and conclusion', for examples).

This chapter considers the economic techniques that can be used to place monetary values on ecosystem services and should be regarded as a methodological complement to the overview presented in Chap. 4. The literature on such techniques has expanded enormously since the 1990s, with extensive reviews presented in Bateman and Willis (1999), Bennett (2013) and Freeman et al. (2014). One reason for this interest is undoubtedly the attraction of measuring environmental costs and benefits in monetary terms because of the comparability with other investment options it enables, giving a voice to the natural world in decision making that would otherwise be weaker (Fisher et al. 2015; Badura et al. 2016). On the other hand, there is no doubt that such monetary valuations can be contentious, with arguments that they encourage a commodification of nature (Norgaard et al. 1998; Gómez-Baggethun and Ruiz-Pérez 2011) and undermine other important reasons for conservation. The user of such techniques therefore needs to recognize that they are likely to be more robust, and appropriate, in some circumstances than others.

The remainder of this chapter is organized into six main sections. Next, the basis of monetary valuation and the techniques that can be applied to different ecosystem services and benefits are outlined. Three subsequent sections discuss the main categories of techniques, namely market valuations and revealed or stated preference methods for non-market benefits. Means of transferring monetary values from one study site to another are then considered, leading to a final section on databases and tools to support valuation exercises.

20.2 The Basis of Monetary Valuation

Ecosystem services are fundamentally anthropocentric, representing the benefits that nature provides for human well-being. Any robust monetary values therefore need to relate to those *benefits* and separate out underpinning natural processes from the key contributions to welfare-bearing goods and services (Fisher et al. 2009). Typically, it is also increases or decreases in these benefits that are of policy concern (rather than absolute presence/absence), so most economic assessments focus on *marginal* changes in provision (i.e. a proportionally small increase or decrease in the total quantity). This emphasis on marginal change contrasts with the permit or prevent focus of many legislation-based frameworks for the assessment of options and further highlights the complementary nature of the two approaches.

Monetary estimates of value are usually based on what individuals are *willing to pay* (WTP) to secure a benefit of a positive change in the natural environment, or what they are *willing to accept* (WTA) as compensation to forgo it. If a change is negative the economic value is measured by what individuals are WTP to avoid such a cost or what they are WTA as compensation to tolerate it (Ozdemiroglu and Hails

2016). These changes can be in the quality and/or quantity of the environment or in individual access to it. The important point for valuation is that the change is demonstrably linked to the benefit received by individuals. Rolls and Sunderland (2014) provide some excellent examples of such 'theories of change' for a range of benefits. One for measures related to freshwater flood risk management is shown in Fig. 20.1, where woodland planting and channel restoration lead to changes in infiltration and runoff which, in turn, generate benefits linked to reduced probability of downstream flooding.

It follows from the above that the economic valuation of a proposed policy or measure often involves the following three stages (Ozdemiroglu and Hails 2016).

- Understanding how a decision will influence the environment (qualitative assessment)
- Measuring the change in the environment and related benefits (quantitative assessment)
- Valuation of the change in benefits in monetary terms

The information and methodological requirements of the first two stages often mean that input is required from the natural sciences or other social sciences as part of the economic valuation of ecosystem services (Badura et al. 2016). For the economic valuations themselves three main types of data (market prices, observed human behavior and individual statements of value) are commonly used, but the applicability of these sources varies according to the types of ecosystem service and benefit involved (Goulder and Kennedy 2011). As a broad generalization, provisioning services are most strongly associated with direct use values, regulating services with indirect use values and cultural services with aspects of non-use (e.g. bequest or existence values) (Pascual et al. 2010). However, some types of natural asset (e.g. an urban park) offer a number of ecosystem services (e.g. aesthetic or recreational) and are linked to both market (e.g. property price premiums) and non-market benefits (e.g. WTP to maintain the park) (Ozdemiroglu and Hails 2016). This means that several different valuation methods are potentially applicable and just highlights the importance of clarifying what the relevant benefits are (and to whom) at an early stage in such exercises.

Where the benefits involved are directly linked to exchange of goods or services within markets then the derivation of monetary values is relatively straightforward. If the values involved are more related to aspects of indirect use or non-use then it is less likely that the goods or services pass through established markets and

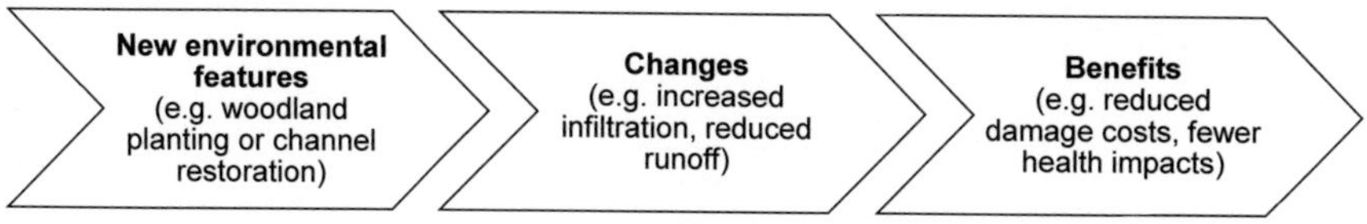

Fig. 20.1 A theory of change for measures related to freshwater flood risk management. (Source: Rolls and Sunderland 2014 p. 37)

techniques of *non-market valuation* are required (Fisher et al. 2015). Often a legislation-based evaluation framework (Chap. 4) may be sufficient to represent the importance of such ES, but sometimes an additional monetary evaluation can help to reinforce the individual or group benefits involved. As discussed further below the economic methods are potentially very flexible in the benefits they can value, though their robustness and precision have been debated as the preferences involved become more hypothetical (i.e. regarding non-use values) (Badura et al. 2016).

20.3 Market-Based Valuations

When goods such as food, water, timber, fuel and minerals are bought or sold in markets then the prices paid in response to changes in the quality and quantity available provide an indicator of the benefit received. However, it is important to recognize that economic value is not the same as market price, since the latter only represents the marginal WTP (Fisher et al. 2015). As illustrated in Fig. 20.2 the total economic value is larger because it includes all of the shaded area under the demand curve up to quantity q rather than just price p multiplied by quantity q.

One challenge for market-based valuation methods is therefore to define the form of the supply and demand curves. Where this is not feasible, prices can be used as a measure of marginal value, but there may still be a need to adjust them for market distortions such as government subsidies or taxes. Another issue is that ecosystem services may represent only part of the inputs underpinning the production of a given market good, with other aspects of manmade capital also required (Bateman et al. 2011a; Diaz et al. 2015). For instance, this is true of most types of crop production which are underpinned by a combination of natural and anthropogenic assets. In such cases it is necessary to use a *production function* approach where the marketed good or service is decomposed into the separate characteristics which define

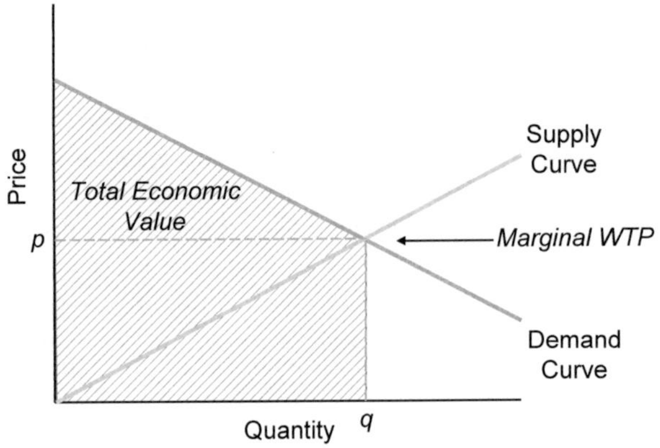

Fig. 20.2 An example of the difference between economic value and market price. (Source: adapted from Fisher et al. 2015 p. 45)

value (Barbier 2007). This usually involves the application of statistical techniques to estimate how much a given environmental good or service contributes to the delivery of market goods or services. For example, the value of pollination services can be assessed in terms of their contribution to the crop yield and valued at the market price of the crop (e.g. Ricketts et al. 2004). A similar issue would arise in distinguishing the habitat development potential of a parcel of land site from other attributes that influence its purchase price (see Sect. 17.4). The requirements in terms of data and understanding for such analyses can be considerable (Pascual et al. 2010), but they are potentially applicable to a wide variety of situations (e.g. see Fezzi et al. 2014 for a study of climate change implications).

As an alternative to prices some researchers have used cost-based approaches which focus on the expenditure that would be incurred if ecosystem service benefits needed to be recreated through artificial means (Garrod and Willis 1999; Pascual et al. 2010). A number of different techniques exist including:

- *Avoided costs* which estimate the damage costs that would have been incurred in the absence of ecosystem services
- *Replacement costs* which calculate the costs incurred by replacing ecosystem services with artificial technologies
- *Mitigation or restoration costs* which refer to the cost of mitigating the effects caused by to the loss of ecosystem services or the cost of getting those services restored.

The avoided damage costs are often used to assess carbon storage services (e.g. Turner et al. 2007a). Replacement costs can be applicable where artificial defenses may need to be constructed because coastal wetlands have been degraded and no longer protect nearby settlements from coastal flooding. In such a case the cost of the defenses represents a minimum estimate of the value of protection service provided by the wetlands (Badola and Hussain 2005). The costs of wetland restoration could be calculated as an alternative form of valuation in this type of situation. Overall, cost data tend to be easier to obtain than those required to characterize production functions or supply/demand curves, but they also often have a weaker relationship with the economic values they aim to represent so need to be applied with caution (Barbier 2007).

20.4 Revealed Preference Methods

These techniques are based on observations of actual behavior and can be used to assess use values of non-market goods. They are limited in applicability to situations where individuals respond to real choices, though this also provides credibility to the valuations generated (Fisher et al. 2015).

One example is *hedonic pricing* where the ES are part of the bundle of characteristics associated with a good that can be purchased in a market. Most commonly this technique utilizes data on property sales and environmental characteristics such as proximity to greenspace or components of the view are valued by controlling for other influences such as the number of rooms, their internal condition and neighbourhood attributes (e.g. crime rates). This requires details on a sufficiently large number of properties with a variety of features, as well as the use of statistical techniques to separate out the premium people are willing to pay to live in an area with particular environmental benefits (e.g. Bockstael and McConnell 2007). The technique can also be used to estimate how much home owners need to be compensated for a potential disamenity, such as noise from a new road (Day et al. 2007).

A second example is the *travel cost method* which values the recreational benefits provided by a natural area through assessment of the costs incurred by individuals in travelling to the site. This approach requires information on expenditure (e.g. in terms of fuel, accommodation, food, entry fees, time incurred etc.) and is based on the assumption that what people pay to travel is at least how much they value the recreational benefit, otherwise they would not make the trip. Again this requires substantial amounts of information on recreational visits and their costs, as well as the use of statistical techniques to isolate the environmental benefits from other aspects of the visit (e.g. Sen et al. 2014). A particular complication is when there are multiple substitute recreational sites, both in terms of compiling information on these alternatives and correctly specifying the statistical model. Discrete choice modelling techniques (e.g. *random utility models*) are commonly used to assess the influences on consumer choices between different bundles of site attributes (e.g. quality of recreational experience, distance involved, number of intervening opportunities etc.). An example is provided by Englin et al. (2006) who used information from over 1600 route permits in Jasper National Park (Canada) to estimate the value of accessing old-growth forest to hikers. These techniques can also be employed to test the potential effects of altering environmental characteristics (e.g. introducing new amenities) on the visit rate at a particular site.

20.5 Stated Preference Methods

These techniques involve asking people hypothetical questions about how they would respond to certain choices, either in terms of willingness to pay for a particular benefit or to accept compensation in exchange for a loss. An advantage of the method is flexibility, including the range of situations where it can be applied (such as those which have not occurred previously) and the ability to assess both use and non-use values. The main drawback is the hypothetical nature of the preferences and consequent reservations regarding the robustness of the values generated (Badura et al. 2016).

Two main types of survey are commonly employed, the *contingent valuation method* and *choice experiments*. Contingent valuation involves the use of a questionnaire to construct a hypothetical market for a particular environmental change (e.g. preservation of a rare species) and directly asks participants for their willingness to pay for (or accept) this outcome. Design of the survey is critical to obtaining reliable results and needs to encompass sufficient background information to allow an informed decision as well as realistic payment options (e.g. taking account of household income). Implementation of the survey also requires attention to matters of sampling strategy and presentation in order to achieve meaningful results (Fisher et al. 2015).

Contingent valuation was the first stated preference method to be widely applied and as more experience has been reported so there has been increasing appreciation of the procedures that need to be followed (Arrow et al. 1994; Bateman et al. 2002a). Choice experiment techniques originally developed in marketing research have also been adapted for use in non-market valuation (e.g. Naidoo and Adamowicz 2005). This method is less direct in eliciting willingness to pay values for individual attributes. Instead survey participants are presented with goods that consist of multiple attributes (including environmental properties) and asked to select their preferred combination or rank several alternative combinations. Typically each participant is asked to make a number of such choices or rankings so that across all respondents every different combination of attributes is compared on multiple occasions. Each set of attributes has a price associated with it and therefore across the whole set of survey data it is possible to identify the average willingness to pay or willingness to accept for each individual attribute. This method is therefore less vulnerable to some of the 'warm glow' or 'anchoring' biases that can occur in contingent valuation studies (Bateman et al. 1999) and can also examine multiple changes in provision of goods while contingent valuation is typically restricted to a single alteration. Nevertheless, the need for large samples of respondents and careful design and implementation of surveys applies to both techniques.

20.6 Benefit Transfer

It is frequently not practical (on grounds of time or cost) to undertake specific economic valuation studies for all the ES-related benefits that need to be considered in a landscape plan. In addition, the growing extent of valuation literature and reviews (e.g. see Turner et al. 2007b in the context of wetlands) has reduced the need to undertake primary research. Instead a common solution is to apply published non-market value estimates (e.g. per hectare of woodland) to a location of interest where no studies have taken place. Examples of this approach at a global scale include the studies by Costanza et al. (1997, 2014) to estimate economic values for all types of goods and services from all biomes types across the entire earth.

Implementation of such *benefit transfer* typically occurs in two main ways. The simplest approach is to adjust an existing mean value to take account of differences in socio-economic and landscape characteristics between the location where the study was conducted and where it needs to be applied. However, this is easier said than done and there is a particular danger of uncontrolled extrapolation in a manner that generates unrealistically large benefit estimates (Bateman et al. 2006; Fisher and Naidoo 2011).

A potentially more robust approach is the use of *value functions*. Initially, this involves using multiple regression techniques to estimate a relationship between the characteristics of a set of study sites and their known non-market monetary values (e.g. for the recreational benefits provided). Subsequently the attributes of new sites are input to the derived function in order to generate corresponding value estimates. Brainard et al. (1999) present an example of such an approach in the context of woodland recreation. The advantage of this method is that it explicitly controls for certain differences between sites, although it is still dependent on the appropriate specification of the underlying function. Bateman et al. (2011b) assess the two approaches in the context of non-market benefits of water quality improvements and conclude that the more heterogeneous the set of sites the better the performance of the value function method compared to the use of mean values.

The developments in value transfer have gained appreciably from the use of GIS techniques (see Chap. 6) to take account of spatial variations in ES provision (Bateman et al. 2002b; Brander et al. 2012). This is particularly important when the proximity of substitutes has a major influence on the value of a proposed environmental change (e.g. the recreational benefit of a new woodland). Sen et al. (2014) illustrate the application of a methodology for spatially sensitive prediction of outdoor recreation visits and values in the UK which combines a GIS-based analysis of a recreation activity database with a meta-analysis of recreation values. The ability to present valuation results in map form is also an advantage in terms of the communication of ecosystem service benefits (Troy and Wilson 2006; Crossman et al. 2013).

20.7 Databases and Tools to Support Valuation

Interest in applying economic valuation techniques and the growing volume of literature on non-market valuation has stimulated a range of initiatives to support such studies. These include databases summarizing valuation study results, guidance on applying the methods and software tools for modelling changes in ES provision and undertaking monetary valuations of the outcomes. Table 20.1 provides details for a number of these resources.

The Environmental Valuation Reference Inventory (EVRI) is an example of a searchable database designed specifically to support benefit transfer applications. It currently contains over 4000 study records and, although there is some bias in content towards water resource valuations and North American research, can be easily used to screen literature for a wide range of benefit transfer purposes.

Guidance materials include both reports outlining how to undertake certain types of analyses and online resources that provide a portal to other websites and documents. The Ecosystems Knowledge Network is an example of a UK initiative whose website provides profiles of analytical tools and links to a series of guidance resources. Although these are UK orientated many of the concepts and approaches discussed have international applicability.

Table 20.1 Resources to support economic valuation studies

Description	Internet Address
Databases	
TEEB valuation database	http://naturalcapitalcoalition.org/the-teeb-valuation-database/
Environmental valuation reference inventory	https://www.evri.ca/
Guidance	
Ecosystems knowledge network	http://ecosystemsknowledge.net/resources/tools
Natural England ecosystem services transfer toolkit	http://publications.naturalengland.org.uk/publication/5890643062685696
Software tools	
ARIES	http://aries.integratedmodelling.org/
InVEST	http://www.naturalcapitalproject.org/invest/
Ecosystem valuation toolkit – SERVES	http://esvaluation.org/

Note: Compiled from information available in December 2016. The TEEB database dates from 2010

Software tools to support systematic assessments of ES provision now exist in a variety of formats ranging from relatively simple spreadsheet applications to integrated sets of spatially-explicit models. Most have the capacity to assess the impacts of climate or land-use changes on ES provision (i.e. the first two stages outlined in Sect. 20.2) and some also incorporate monetary valuation of the alternation in benefits (i.e. stage three). Bagstad et al. (2013) review 17 different tools and assess their performance against eight evaluation criteria. Seven of the tools are also applied in an assessment of the San Pedro River watershed spanning the USA/Mexican border, the results of which highlight both the complementarity of different tools and that their suitability may vary between geographical and decision contexts. This also emphasizes the need for more comparative assessments of such tools.

A further research priority will need to include investigation of how to most effectively use such tools with stakeholders (de Groot et al. 2010; Harwood et al. 2015), not least because of the amount and complexity of information that can be produced, as well as the arguments that the use of monetary valuation can generate. Ozdemiroglu and Hails (2016) provide some helpful guidance regarding the communication of economic valuation evidence, emphasizing the need to be:

- Specific about the types of decisions economic values can be used for
- Relevant to the needs of stakeholders and presenting monetary values as part of a wider assessment of possible changes
- Clear about what's included in the economic estimates and what isn't
- Open about uncertainties and assumptions, not presenting a single value to answer all questions.

20.8 Conclusions

This chapter has provided an overview of the different techniques that can be used to derive monetary values for the benefits provided by ecosystem services. It should be apparent that the range of possibilities is now considerable, so in any practical application it is worthwhile considering the following sequence of questions:

1. Is the ability to compare options in monetary terms likely to be useful given the proposed type of change being assessed?
2. Is the magnitude or controversy associated with the potential change such that original economic valuation research is required?
3. If specific research is indicated, what is the likely relative importance of use and non-use values? This will be a good guide to the suitability of different techniques.
4. If an original study is not merited, or practical, then do the types of resources listed in Table 20.1 provide guidance on the most appropriate means of conducting a benefit transfer exercise?

Whatever the approach adopted it is important to recognize that deriving monetary values is a means to an end, not an end in itself. Careful thought therefore needs to be given to how this evidence is utilized in wider decision-making processes.

References

Arrow, K., Solow, R., Schuman, H., et al. (1994). Report to the NOAA panel on contingent valuation. *US Federal Register, 58*(10), 4602–4614.

Badola, R., & Hussain, S. A. (2005). Valuing ecosystem functions: An empirical study on the storm protection function of Bhitarkanika mangrove ecosystem, India. *Environmental Conservation, 32*, 85–92.

Badura, T., Bateman, I. J., Agarwala, M., et al. (2016). Valuing preferences for ecosystem-related goods and services. In M. Potschin, R. Haines-Young, R. Fish, & R. K. Turner (Eds.), *Routledge handbook of ecosystem services* (pp. 228–242). Oxford: Routledge.

Bagstad, K. J., Semmens, D. J., Waage, S., et al. (2013). A comparative assessment of decision-support tools for ecosystem services quantification and valuation. *Ecosystem Services, 5*, e27–e39.

Barbier, E. B. (2007). Valuing ecosystem services as productive inputs. *Economic Policy, 22*, 177–229.

Bateman, I. J., & Willis, K. G. (Eds.). (1999). *Valuing environmental preferences*. Oxford: Oxford University Press.

Bateman, I. J., Langford, I. D., & Rasbash, J. (1999). Elicitation effects in contingent valuation studies. In I. J. Bateman & K. G. Willis (Eds.), *Valuing environmental preferences* (pp. 511–539). Oxford: Oxford University Press.

Bateman, I. J., Carson, R. T., Day, B., et al. (2002a). *Economic valuation with stated preference techniques: A manual*. Cheltenham: Edward Elgar.

Bateman, I. J., Jones, A. P., Lovett, A. A., et al. (2002b). Applying geographical information systems (GIS) to environmental and resource economics. *Environmental and Resource Economics, 22*, 212–269.

Bateman, I. J., Day, B. H., Georgiou, S., et al. (2006). The aggregation of environmental benefit values: Welfare measures, distance decay and total WTP. *Ecological Economics, 60*, 450–460.

Bateman, I. J., Mace, G. M., Fezzi, C., et al. (2011a). Economic analysis for ecosystem service assessments. *Environmental and Resource Economics, 48*, 177–218.

Bateman, I. J., Brouwer, R., Ferrini, S., et al. (2011b). Making benefit transfers work: Deriving and testing principles for value transfers for similar and dissimilar sites using a case study of the non-market benefits of water quality improvements across Europe. *Environmental and Resource Economics, 50*, 365–387.

Bennett, J. (Ed.). (2013). *The international handbook on non-market environmental valuation*. London: Edward Elgar.

Bockstael, N. E., & McConnell, K. E. (2007). *Environmental and resource evaluation with revealed preferences: A theoretical guide to empirical models*. Dordrecht: Springer.

Brainard, J. S., Lovett, A. A., & Bateman, I. J. (1999). Integrating geographical information systems into travel cost analysis and benefit transfer. *International Journal of Geographical Information Science, 13*, 227–246.

Brander, L. M., Brauer, I., Gerdes, H., et al. (2012). Using meta-analysis and GIS for value transfer and scaling up: Valuing climate change induced losses of European wetlands. *Environmental and Resource Economics, 52*, 395–413.

Costanza, R., d'Arge, R., de Groot, R., et al. (1997). The value of the world's ecosystem services and natural capital. *Nature, 387*, 253–260.

Costanza, R., de Groot, R., Sutton, P., et al. (2014). Changes in the global value of ecosystem services. *Global Environmental Change, 26*, 152–158.

Crossman, N. D., Burkhard, B., Nedkovc, S., et al. (2013). A blueprint for mapping and modelling ecosystem services. *Ecosystem Services, 4*, 4–14.

Day, B., Bateman, I. J., & Lake, I. (2007). Beyond implicit process: Recovering theoretically consistent and transferable values for noise avoidance from a hedonic property price model. *Environmental and Resource Economics, 37*, 211–232.

de Groot, R. S., Alkemade, R., Braat, L., et al. (2010). Challenges in integrating the concept of ecosystem services and values in landscape planning, management and decision making. *Ecological Complexity, 7*, 260–272.

Diaz, S., Demissew, S., Carabias, et al. (2015). The IPBES conceptual framework — Connecting nature and people. *Current Opinion in Environmental Sustainability, 14*, 1–16.

Englin, J., McDonald, J. M., & Moeltner, K. (2006). Valuing ancient forest ecosystems: An analysis of backcountry hiking in Jasper National Park. *Ecological Economics, 57*, 665–678.

Fezzi, C., Bateman, I. J., Askew, T., et al. (2014). Valuing provisioning ecosystem services in agriculture: The impact of climate change on food production in the United Kingdom. *Environmental and Resource Economics, 57*, 197–214.

Fisher, B., & Naidoo, R. (2011). Concerns about extrapolating right off the bat. *Science, 333*, 287.

Fisher, B., Turner, R. K., & Morling, P. (2009). Defining and classifying ecosystem services for decision making. *Ecological Economics, 68*, 643–653.

Fisher, B., Naidoo, R., & Ricketts, J. (2015). A field guide to economics for conservationists. In *Roberts and company*. Colorado: Greenwood.

Freeman, M. A., III, Herriges, J. A., & Kling, C. L. (2014). *The measurement of environmental and resource values: Theory and methods* (3rd ed.). Washington DC: RFF Press.

Garrod, G., & Willis, K. G. (1999). *Economic valuation of the environment*. Cheltenham: Edward Elgar.

Gómez-Baggethun, E., & Ruiz-Perez, M. (2011). Economics valuation and the commodification of ecosystem services. *Progress in Physical Geography, 35*, 613–628.

Goulder, L. H., & Kennedy, D. (2011). Interpreting and estimating the value of ecosystem services. In P. Kareiva, H. Tallis, T. H. Ricketts, G. C. Daily, & S. Polasky (Eds.), *Natural capital: Theory & practice of mapping ecosystem services* (pp. 15–33). Oxford: Oxford University Press.

Harwood, A., Lovett, A. A., De-Gol, A., et al. (2015). Simulating and conveying changes in ecosystem services for geodesign. In E. Buhmann, S. M. Ervin, & M. Pietsch (Eds.), *Peer reviewed proceedings of digital landscape architecture 2015 at Anhalt University of Applied Sciences* (pp. 205–212). Berlin: Herbert Wichmann Verlag. http://gispoint.de/gisopen-paper/1823-simulating-and-conveying-changes-in-ecosystem-services-for-geodesign.html. Accessed 20 Dec 2016.

Naidoo, R., & Adamowicz, W. L. (2005). Economic benefits of biodiversity conservation exceed costs of conservation at an African rainforest reserve. *Proceedings of the National Academy of Sciences of the United States of America, 102*, 16712–16716.

Norgaard, R. B., Bode, C., & the Values Reading Group. (1998). Next, the value of god, and other reactions. *Ecological Economics, 25*, 37–39.

Ozdemiroglu, E., & Hails, R. (eds) (2016). Demystifying economic valuation. *Valuing nature paper VNP04*. http://valuing-nature.net/news/. Accessed 23 Aug 2016.

Pascual, U., Muradian, R., Brander, L., et al. (2010). Chapter 5: The economics of valuing ecosystem services and biodiversity. In P. Kumar (Ed.), *The economics of ecosystems and biodiversity: Ecological and economic foundations*. London: Earthscan. www.teebweb.org.

Ricketts, T. H., Daily, G. C., Ehrlich, P. R., et al. (2004). Economic value of tropical forest to coffee production. *Proceedings of the National Academy of Sciences of the United States of America, 101*, 12579–12582.

Rolls, S., & Sunderland, T. (2014) Microeconomic evidence for the benefits of investment in the environment 2 (MEBIE2). *Natural England research reports, number 057*. Natural England: Bristol.

Sen, A., Harwood, A., Bateman, I. J., et al. (2014). Economic assessment of the recreational value of ecosystems: Methodological development and national and local application. *Environmental and Resource Economics, 57*, 233–249.

Troy, A., & Wilson, M. A. (2006). Mapping ecosystem services: Practical challenges and opportunities in linking GIS and value transfer. *Ecological Economics, 60*, 435–449.

Turner, R. K., Burgess, D., Hadley, D., et al. (2007a). A cost-benefit appraisal of coastal managed realignment policy. *Global Environmental Change, 17*, 397–407.

Turner, R. K., Georgiou, S., & Fisher, B. (2007b). *Valuing ecosystem services: The case of multifunctional wetlands*. London: Earthscan.

Part IV
Methods for Deriving Response Measures

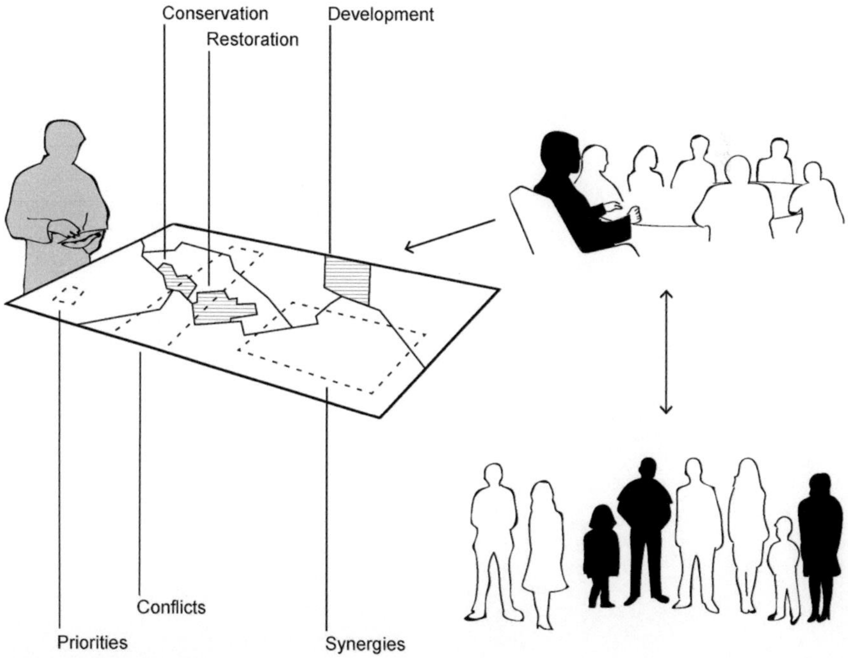

The review of responses describes landscape planning objectives and measures for mitigating existing pressures, as well as for safeguarding, restoring, and enhancing the state of biodiversity and ecosystem services. We suggest an approach to decision-support that helps identify priorities for action, resolves conflicts and utilizes opportunities regarding synergies between different ecosystem services. The latter includes using multifunctional measures as a means to enhance the investment or spatial efficiency of landscape plans. The public and other stakeholders should also participate in selecting measures for implementation within the framing conditions set by higher levels of governance.

Developing Landscape Planning Objectives and Measures

21

Christian Albert and Christina von Haaren

Abstract

The development of targets and measures is one of the core tasks of landscape planning. This chapter introduces requirements, distinguishes between different types, and provides an overview of the basic modules for developing targets and measures in landscape planning. The chapter begins with an explanation of the context dependence of objective development processes in landscape planning. In terms of requirements, the chapter proposes that landscape planning targets and measures will need to be perceived simultaneously as scientifically credible, politically salient and procedurally legitimate by users and stakeholders in order to yield influence on decisions. The chapter proposes ten modules which can be used to develop targets and measures in landscape planning. In addition, the process character of objectives in landscape planning is pointed out. The chapter concludes with remarks concerning how the principles and approaches described in this chapter can be applied in practice.

Keywords

Planning objectives · Targets · Methods for developing objectives

21.1 Landscape Planning Targets in Different Contexts

While we generally refer to landscape planning as a strong forward-looking action to enhance, restore or create landscapes (as stipulated by the European Landscape Convention), the actual understanding of landscape planning and its purpose, obligations, responsibilities and tasks differs substantially across EU member states (cf.

C. Albert (✉) · C. von Haaren
Institute of Environmental Planning, Leibniz University Hannover, Hannover, Germany
e-mail: albert@umwelt.uni-hannover.de; haaren@umwelt.uni-hannover.de

C. von Haaren et al. (eds.), *Landscape Planning with Ecosystem Services*, Landscape Series 24, https://doi.org/10.1007/978-94-024-1681-7_21

Chaps. 1, 2 and 3). Consequently, the legislative power that is associated with the targets proposed by landscape planning diverges. In countries where landscape planning is associated with spatial or territorial planning, its targets are already negotiated between different sectorial interests. In other countries, such as Germany where landscape planning is understood as a sectorial planning system oriented only towards an interest for nature and environmental conservation, the targets proposed by landscape planning are much more focused on the environmental perspective (while also taking into account other sectors' interests in order to enhance implementation). The integration with spatial planning follows as a separate step, thus allowing for a transparent process of political weighing between economic, environmental and other spatial objectives. The following description focuses primarily on the latter example of objective development emphasising nature-conservation considerations. However, the description of requirements, the types of landscape planning targets, and the procedures for developing them, can also be insightful for landscape planners in other governance contexts.

21.2 Requirements for Landscape Planning Targets

In general, landscape planning targets and measures, as with any kind of scientific or planning advice, are more likely to be considered in decision-making when they are perceived as simultaneously scientifically credible, politically salient and procedurally legitimate by users and stakeholders (Cash et al. 2003). The scientific credibility of landscape planning stems from a clear and internally logical system of targets and measures that are derived from the best available scientific knowledge as well as from overarching legal and professional prerequisites (such as the Federal Nature Conservation Act) (Termorshuizen et al. 2007). The scientific information basis must be relevant for the issues at stake and the description of the objectives needs to be transparent and understandable. In addition, the comprehension of the objective system can be enhanced if it includes a consideration of possible alternative objectives or measures and options to illustrate the decision space.

In order to achieve political salience, landscape planning targets should adequately address each of the issues considered as relevant and important by decision-makers and stakeholders. In particular, they need to relate to existing legislative objectives, standards and requirements as stipulated by laws and directives (for legitimacy of objectives see Chap. 4). Attaining a high level of salience is particularly important for landscape planning as its key mode of influencing decision-making is through providing relevant information. Landscape planning can draw, to a limited degree, from government planning instruments, economic incentives or communication to stakeholders to achieve its objective. Consequently, gaining acceptance of the proposed actions and integrating suggestions into the activities of other stakeholders and sectors needs to be taken into account. The effectiveness and implementation ability of landscape planning targets can be enhanced if the targets are clearly defined, spatially and temporally explicit and if they are measureable. Most importantly, the priorities of the targets need to be clarified (discriminating between obligatory – usually legally prescribed – and optional targets, see the following section). The targets and measures have to be described in ways that respond to the needs of different groups of addressees and their implementation abilities. For example, spatial planners will accept propositions more easily if the ES objectives have been translated into the planning categories of a regional plan. Citizens may respond positively to 3D visualisations showing the consequences of a neighbourhood development or renewable energies in the landscape. For nature conservation authorities, information about habitats and species rareness is needed in combination with a proposal on protection priorities and synergies with other environmental interests.

The legitimacy of landscape planning targets and measures can be enhanced if they are derived not only from legal documents and democratic decisions, but also if they are scrutinised in an open participatory process that systematically involves all interested and concerned parties (Albert et al. 2012; Bohnet and Smith 2007; Luz 2000). The participation of these parties should ideally begin in the early stages of planning, where the overarching targets and the scope for planning are defined. It should continue throughout the planning process until the final product is published. As the formal participation process, as seen in existing legislation, only introduces citizen participation in the later stages of planning, it can be helpful to provide additional voluntary and early participation opportunities for interested parties.

21.3 Types of Landscape Planning Targets and Measures

When developing targets and measures in landscape planning, it is useful to discriminate between different types of targets and measures for structuring landscape plans and accompanying texts. In light of this, von Haaren (2004) proposes three dimensions of objective types: (i) the level of legal relevance, (ii) the level of substantiation, (iii) and the type of content.[1]

21.3.1 Discriminating Targets by Their Degree of Legal Relevance

From a legal perspective, landscape planning targets can be discerned as obligatory or optional targets. Obligatory targets (also termed 'minimum targets') characterise the minimum level of societal responsibility for protection of nature and landscapes. They should be based on laws and legislation. Optional targets, in contrast, are more ambitious from the perspective of nature conservation, but their attainment is not formally required by law.

The fact that obligatory targets are derived from democratically legitimised laws and legislation provides landscape planners with convincing arguments for the need of implementation, even when faced with resistance from affected land users. As

[1] This section is an adapted and expanded translation from von Haaren (2004).

obligatory targets are set by definition, they essentially 'frame' the decision-space within which other alternatives can be debated. Distinguishing between obligatory and optional targets is recommended in landscape plans. This allows easy identification of targets and measures that should be implemented with high priority, and targets that are of less priority and open for debate in decision-making processes. Obligatory targets should determine the minimum threshold at which the sustainability of natural capital is threatened. Available laws and legislation from the EU level (e.g. the Water Framework Directive, Natura 2000) and from national and sub-national levels can be used. In addition, expert standards such as Red Lists can be employed as additional sources. An important challenge, however, is that this minimum level often is not clearly predefined for the regional and local scale and needs some interpretation. More detailed information on local and regional scarcities, endangerment and quality, and potential responses of ecosystems can help to downscale the general legal principles.

Optional nature conservation objectives are those that are more ambitious than obligatory targets. Their attainment is not formally required by any laws or legislation. Optional targets are thus of less official priority and do not need to be implemented in all areas in which they are potentially suitable. They are more oriented towards improving the state of the landscape and will be a major field local activities and public participation.

21.3.2 Discriminating Targets by Their Level of Substantiation

The level of substantiation provides another dimension for discriminating landscape planning targets and measures. More specifically, we can hierarchically distinguish between overarching targets (which are quite general), operational targets (which are more specific) and measures (which describe a particular action). Discriminating between these levels of targets enhances the transparency of the development of measures. It also enables a procedural and stepwise development of the planning and implementation process.

Overarching targets refer to the motivation and the purpose of landscape planning. They can be derived from guiding principles and general environmental development targets at upper levels of policy- and decision-making, i.e. from the national or federal state level. They form the highest level of targets and are addressed in the respective planning process. Overarching targets in landscape planning should not be as global or general as objectives from legislation, e.g. from the German Federal Nature Conservation Act. Instead of being generally applicable to Germany, overarching targets in landscape planning should help specify the legal targets for the study area. Overarching targets can be seen as guidelines for implementation.

Operational targets and measures represent the sub-objectives for these overarching targets. Operational targets provide greater detail and describe possible ways of reaching the overarching targets. The definition of operational targets and measures is only possible in relation to a specific overarching objective. Targets answer such questions as: "What shall be attained?" (for example, a high level of species diversity), while measures answer the question of "How can this objective be achieved" (for example, through the development of a habitat network). In the hierarchical concept of targets and measures, measures can turn into targets when the scope of the analysis becomes more detailed. For example, developing a habitat network can then become an objective, with the measure being the widening of buffer strips.

Another useful example for the hierarchical system of objectives and measures concerns the issue of soil erosion. According to the German Federal Nature Conservation Act (§1), one of the purposes of nature conservation should be to "permanently safeguard the performance and functioning of the natural balance, including the ability of natural resources to regenerate and lend themselves to sustainable use" (BMU 2009: 7). In landscape planning, the objective is to cement an overarching objective such as 'reducing soil erosion to the minimum level attainable under agricultural land use in the study area'. An operational objective could then be 'to reduce soil erosion on sites A to C to a maximum of 2 tonnes per hectare per year'. As this example shows, operational objectives should at best be described with measurable quantities to ease both the selection of appropriate measures and the monitoring of implementation effectiveness (see e.g. Jones et al. 2013).

A hierarchical system of overarching objectives, operational objectives and measures allows for flexibility in the implementation process. Periodically, in the implementation phase of a plan, it transpires that some measures cannot be executed as the context conditions have changed. In such cases, planners can go back to the overarching objectives and try to identify alternative operational objectives and measures. So that despite the changed conditions the overarching objective can still be attained.

21.3.3 Discriminating Objectives by Content

Landscape planning objectives and measures can be distinguished from one another by their content regarding conservation, protection, restoration and development objectives. The identification and description of objectives and measures can often be derived directly from the results of the analysis, assessment and evaluation of the landscape. As such, the objectives and measures represent responses to identified qualities, problems and potentials. For instance, in situations where a site is in danger of losing its capacity to provide ecosystem services, the need to develop and implement objectives and measures to protect or restore the site is obvious. In such cases, the ongoing impairment of a protected habitat needs to be halted by protecting the site from further pressure and eventually restoring it so that it recovers its full capacities.

In addition, many objectives and measures can be identified and recommended in a much more creative way. These objectives are usually development objectives that go beyond the mere attainment of obligatory objectives and are thus much more flexible. Creativity and design considerations can therefore play a particularly great role in the development of these objectives. Furthermore, a differentiated

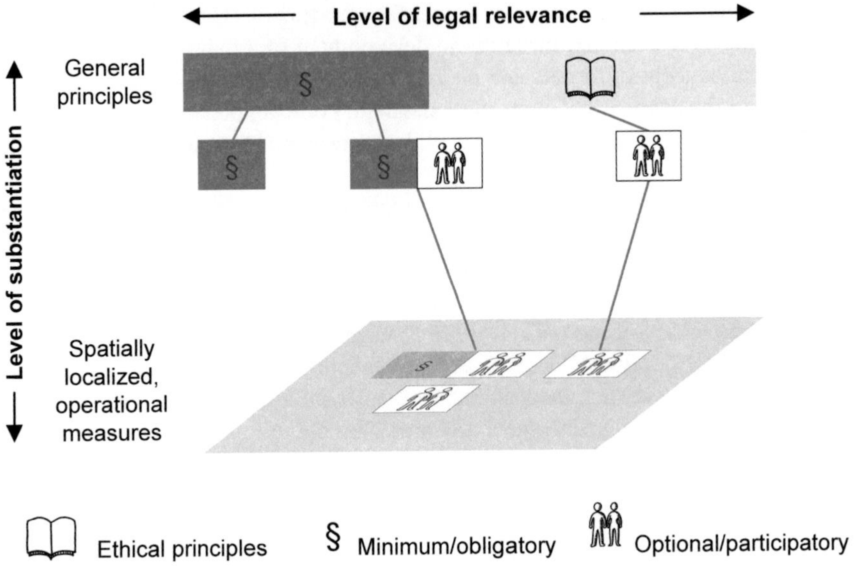

Fig. 21.1 The different types of objectives that help characterise the goals and measures in a landscape plan and thus support the prioritising of measures and defining the decision and participation space

perspective on the delivered ecosystem services compared to the actually used ecosystem services, as well as on monetised benefits of activating the potential, can trigger development measures. For instance, the aesthetic capacities of a landscape can be compared to the recreation activities, which may be impeded by the absence of paths for exploitation. Consequently, infrastructural development can be proposed.

Landscape planning practice usually does not clearly distinguish between the different types of objectives and their combination (Fig. 21.1) so as not to overload planning documents with too many detailed classification systems. Planners will usually choose one of the presented classifications to structure the response part in a landscape plan. However, the foundational categories and the hierarchical system of objectives and measures should be considered by planners as they are important for identifying priorities and syntheses between objectives, and for deciding between alternatives.

21.4 Modules and Procedures for Developing Objectives

This section describes the structure, procedures and interrelations for developing objectives and measures in landscape planning. Within this process, the different types of objectives and measures described in the preceding section are considered in addition to the requirements for attaining objective development. Our description is comprehensive; it introduces the full spectrum of options that can be included in

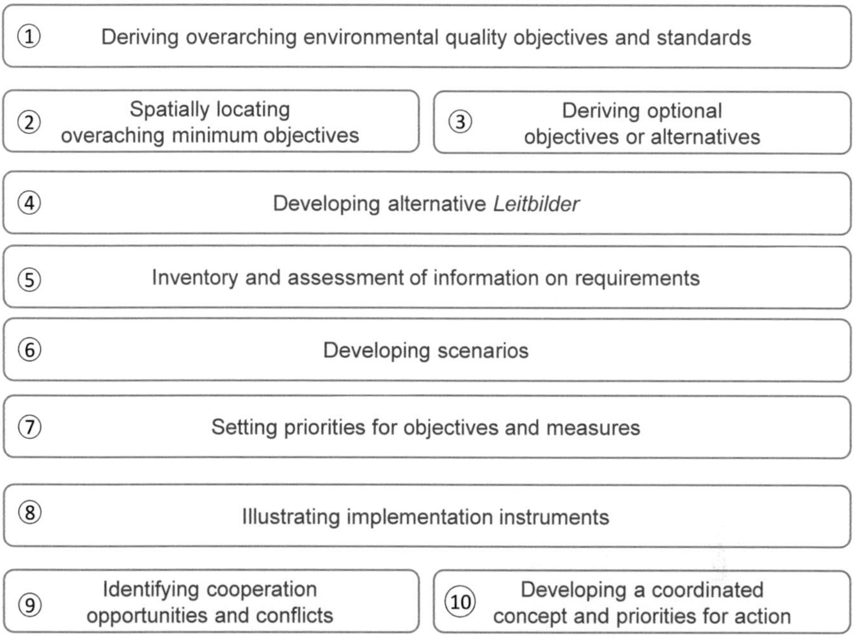

Fig. 21.2 Modules for the development of objectives in landscape planning

the objective development process (see Fig. 21.2). In practice, however, resource limitations for each case study will require adaptation and selection of a subset of the different steps in the objective development. Furthermore, the legal context of each study will further shape and influence the choice of approaches available to landscape planners.

The context conditions of the case study area influence the type of stakeholders that can participate in the process. Local studies will more strongly emphasise the participation of individual citizens, whereas larger planning areas will usually invite representatives of different actor groups to become involved as representatives. It also needs to be emphasised that the whole process is not linear but involves feedback loops as well as parallel activities. For instance, implementation options should be introduced early in the process in order to make clear that the planner is not being fanciful, but that there is a good chance for easy realisation of certain response measures.

Across all modules, the systematic involvement of decision-makers, stakeholders and citizens is crucial for the legitimacy and salience of the planning proposals. The participation opportunities are diverse and include various formats and audiences. They include, for example, scoping meetings with key decision-makers such as administration representatives and stakeholders who can help focus the aims on the most important issues at stake. Exchanges with representatives from different spatial planning sectors should begin early in the process. This enables them to be involved in the discussion and to have access to up-to-date data and plans, as well as the scope to assess any potential areas of conflict and synergies. Citizens can be involved in workshops for developing *Leitbilder* (similar to mission statements, see Chap. 27 for details). Workshops with relevant stakeholders should be organised to allow for joint discussions of objectives. Finally, landscape planners can support the development of political decisions regarding which objectives should become binding through, for example, the integration of specific measures into spatial and territorial plan designations.

21.4.1 Module 1 – Deriving Overarching Environmental Quality Objectives and Standards

Overarching guiding principles (Chap. 4) should be kept in mind in the course of the whole process in order to refer back to them when the operative objectives and measures may be disputed and alternatives have to be found. Goals and objectives provide insights into minimum objectives that are pre-set at higher governance levels and therefore not available for local discretion. Examples for such overarching objectives include the preservation of habitats of international and national importance (e.g. according to the FFH Directive). Often checklists are available, for example at the state level, summarising the overarching objectives which need to be fulfilled.

21.4.2 Module 2 – Spatially Locating Overarching Minimum Objectives

This stage of the objective development process revisits the results of the inventory and evaluation of the ecosystem services that was conducted in the assessment phase of the landscape planning process. It identifies the areas where overarching minimum objectives are present. These minimum objectives are not determined at local discretion. Minimum objectives usually refer to the preservation of the current state of important habitats. In Germany, about ten percent of the country's area can be considered as in need of such high conservation standards and thus minimum objectives.

21.4.3 Module 3 – Deriving Optional Objectives or Objective Alternatives

Based on the results from the assessment and the recognition of the current state and its development potential, the minimum objectives for a particular landscape case study should be accompanied by optional objectives. Optional objectives can illustrate the different development alternatives at a site. Therefore, they remain open to different options as long as they are acceptable from a nature conservation perspective. Keeping this in mind, firstly individual optional objectives for the sustainment, enhancement or restoration of specific ecosystem services should be identified.

Between the identified objectives, both synergies and conflicts may exist. The step-by-step process of developing an integrated objective concept, by firstly identifying objectives for each ecosystem service and then exploring potential synergies and conflicts, can enhance transparency and comprehension. In this way, planners can demonstrate arguments for their choice of particular objectives and provide proposals which can be openly discussed. In general, it is advisable to connect objectives with concrete measures, even at this stage of development, because conflicts between objectives often only become apparent when talking about the specific land use management measures to be implemented.

21.4.4 Module 4 – Developing Alternative Leitbilder

Together, the overarching objectives and the optional objectives can be jointly described in alternative *Leitbilder* (see Chap. 27) to convey a guiding vision for the future development of a landscape. Such *Leitbilder* should describe the vision in an appropriate and understandable way. At the same time, however, they should remain slightly vague in order to minimise the risk of conflicts and to expose opportunities for joint discussions and solution development. Such *Leitbild* processes may also be relevant to initiatives seeking to foster or create particular landscape identities.

21.4.5 Module 5 – Inventory and Assessment of Information on Legal, Economic, Social and Cultural Requirements of Sectors and Stakeholders

This module can help in the development of landscape planning objectives and spatial concepts that are adapted to the implementation conditions. It can help to develop potential synergies and deal with conflicts between nature conservation and landscape planning objectives and other sectorial interests. It can guide the adaptation of landscape objectives (as long as they are feasible) to suit the context conditions and can help the crafting of strategies for communication and implementation. The module can include document analyses as well as different forms of stakeholder consultation and involvement.

21.4.6 Module 6 – Developing Scenarios

In the context of developing landscape planning objectives, scenarios can help to explore the impacts of proposed or foreseeable alternative pathways. In this way, scenarios can help answer 'what-if' questions about the future (see Chap. 27 for a more detailed discussion). Scenarios allow for the development of objectives and measures in different contexts, but also the use of so-called backcasting approaches to identify the measures needed to attain a particular vision. A common use of scenarios in landscape planning is for the investigation of quantitative impacts (e.g. Bryan et al. 2011) and the costs of alternative implementation options.

21.4.7 Module 7 – Setting Priorities for Objectives and Measures from the Perspective of Nature Conservation

This module involves the definition of priorities for objectives and measures as proposed from the perspective of nature conservation and landscape planning. It thereby provides the basis for subsequent decision-making processes regarding the uptake of nature conservation objectives in spatial planning and land use decisions. It also provides important basic information for relevant government agencies and civil society stakeholders.

Priorities for nature conservation objectives and measures are usually set on the basis of the inventory and evaluation of the status quo. In cases where alternative objectives exist that are of equal importance, aspects of implementation feasibility and potential for synergies with other land use interests may also influence the decision-making process.

An important issue when setting priorities is to make the process transparent and comprehensible for different audiences. Usually objectives with the highest level of importance, for instance as derived from overarching objectives, are of greatest priority. In cases where conflicts between priorities exist, the ecosystem service of highest value may determine which objectives should be prioritised. If conflicts among alternative objectives cannot be decided upon based on nature-conservation considerations alone, further aspects such as the temporal dimension (how long does it take to achieve the objective), funding aspects and considerations of implementation practicality may also need to be considered. The spatial illustration of the different priority levels of objectives can help in the decision-making process.

21.4.8 Module 8 – Illustrating Implementation Instruments

The proposal of objectives and measures by landscape planning should be accompanied by suggestions concerning suitable instruments and who should implement them. Several types of implementation instruments can be distinguished. First, objectives and measures can be formally stipulated in spatial and territorial plans and are therefore of binding character. Second, mitigation banking can help in implementing actions. Third, protected areas can be designated. Fourth, additional funding instruments and economic incentives can be provided for the implementation of particular measures by the land users. Fifth, information can be used to initiate and support voluntary implementation of measures by land users. A very important step for implementation is communicating the objectives and the rationale for measures (e.g. economic costs and benefits) to those parties (e.g. farmers) who will decide whether to adopt them.

21.4.9 Module 9 – Identifying Cooperation Opportunities and Conflicts

Landscape planners should systematically identify opportunities for harnessing synergies from cooperation with land users, but also those areas where conflicts between different land users could persist. To do so, landscape planners should first identify those areas where no problems exist, and those where conflicts are apparent. In the next step, planners should investigate how the remaining conflicts can be minimised. A precondition for solving conflicts is the implementation of an open and transparent process of stakeholder consultation and public participation procedures. In addition, the concerned stakeholders need to be open and flexible when negotiating solutions. While the process of negotiating options will result in some solutions, there may well remain a set of conflicts that cannot be solved. These may be addressable at a later time when context conditions and actor networks have changed. Where urgent solutions are necessary, political decisions may need to be made regardless of conflicts.

21.4.10 Module 10 – Developing a Coordinated Objective Concept and Priorities for Action

The final result of the objective development process is a flexible (i.e. adaptable) procedural plan in the form of words and maps that fulfils a series of characteristics. In particular, the outcome should:

- Distinguish between minimum objectives and optional objectives
- Provide a plausible and logical derivation of objectives and measures from politically legitimised norms and additional assumptions
- Clarify uncertainties in the assessment and prognosis
- Highlight and explain temporal and spatial priorities for the proposed objectives and measures
- Emphasize areas for cooperation and remaining issues of conflict,
- Identify objectives and measures where continuous monitoring, evaluation, and possible modification are of particular importance (for example, by including what-if options)
- Elaborate on implementation options and eventually adapt objectives to those options (from a nature conservation perspective)
- Clearly address the what, who, how and when questions relevant for objectives and measures

The whole objective system should allow for continuous adaptation and further development, reacting to monitoring results, changing pressures and new legislation. This can be facilitated by storing the spatial data in accessible digital formats and processing them in a Geographical Information System (see Chaps. 5, 6). Documented changes in pressure, state or evaluation standards can automatically lead to different assessment results and highlight the need for adaptation of the plan. Since the digital revolution, landscape plans are no longer set in stone but can be designed as half-automated learning entities.

21.5 Conclusions

This chapter has provided insights into the requirements, types, and basic modules for developing objectives in landscape planning. Our concluding remarks now concern the question of how the principles and approaches described in this chapter can be applied in practice.

The substantial resource and time constraints characterising many actual cases of landscape planning limit the possibility of implementing the full spectrum of modules for objective development. Consequently, only a selected set of modules can be used to address the issues at stake in the best possible way. The choice of modules to implement, and the level of detail with which they are applied depends on several aspects of the individual case study. These factors include the quality and extent of available data, the interests and prior knowledge of the stakeholders involved, the specific problems at stake and the available resources. For example, the development of comprehensive landscape *Leitbilder* is not a necessary exercise in landscape planning. Whether or not it is performed can be left to local preferences. In addition, some cases of landscape planning do not necessarily require the creation of alternative scenarios, even though the exercise might yield interesting results. Furthermore, the identification of implementation interests is often not covered by the standard budgets or procedures for landscape planning. However, communities and counties can decide to spend additional funds in order to acquire such additional information or other strategic insights.

References

Albert, C., Zimmermann, T., Knieling, J., et al. (2012). Social learning can benefit decision-making in landscape planning: Gartow case study on climate change adaptation, Elbe valley biosphere reserve. *Landscape and Urban Planning, 105*, 347–360. https://doi.org/10.1016/j.landurbplan.2011.12.024.

BMU – German Federal Ministry for the Environment, Nature Conservation and Nuclear Safety. (2009). *Act on nature conservation and landscape management* (Federal Nature Conservation Act – BnatSchG) of 29 July 2009 – unofficial translation – https://www.bmu.de/fileadmin/Daten_BMU/Download_PDF/Naturschutz/bnatschg_en_bf.pdf. Accessed 5 July 2018.

Bohnet, I., & Smith, D. M. (2007). Planning future landscapes in the Wet Tropics of Australia: A social-ecological framework. *Landscape and Urban Planning, 80*, 137–152.

Bryan, B. A., Crossman, N. D., King, D., et al. (2011). Landscape futures analysis: Assessing the impacts of environmental targets under alternative spatial policy options and future scenarios. *Environmental Modelling and Software, 26*, 83–91. https://doi.org/10.1016/j.envsoft.2010.03.034.

Cash, D. W., Clark, W. C., Alcock, F., et al. (2003). Knowledge systems for sustainable development. *Proceedings of the National Academy of Sciences of the United States of America, 100*, 8086–8091.

Jones, K. B., Zurlini, G., Kienast, F., et al. (2013). Informing landscape planning and design for sustaining ecosystem services from existing spatial patterns and knowledge. *Landscape Ecology, 28*, 1175–1192. https://doi.org/10.1007/s10980-012-9794-4.

Luz, F. (2000). Participatory landscape ecology – A basis for acceptance and implementation. *Landscape and Urban Planning, 50*, 157–166.

Termorshuizen, J. W., Opdam, P., & van den Brink, A. (2007). Incorporating ecological sustainability into landscape planning. *Landscape and Urban Planning, 79*, 374–384.

von Haaren, C. (Ed.). (2004). *Landschaftsplanung*. Stuttgart: Eugen Ulmer.

22 Measures to Safeguard and Enhance Soil-Related Ecosystem Services

Miguel A. Cebrián-Piqueras

Abstract

This chapter introduces measures for the protection of soil-related ecosystem functions and services. It outlines available mandatory and voluntary measures regarding conservation of soil production capacity in agricultural land at a European level. It continues with an overview of specific management measures for soil erosion control and for enhancement of the greenhouse gases storage and carbon sequestration service. Finally, measures to protect geodiversity, including geosites and soils with historical and cultural value are presented.

Keywords

Soil-related ecosystem services · Measures · Soil conservation · Agriculture · CAP · Soil erosion · Geodiversity · Carbon sequestration · Production capacity

22.1 Introduction

22.1.1 The Specific Role of Soil and Geodiversity Protection

Safeguarding and sustainably managing soils and associated geodiversity is a key objective of landscape planning due to the many positive effects that such resources have on biodiversity and ecosystem services. The provision of food, materials and bioenergy, erosion prevention, soil-related greenhouse gas storage (Stolte et al. 2016), as well as geodiversity (LABO 2011), flood mitigation, filtering of nutrients or recycling of waste all depend on site-adapted land use and soil management (Dominati et al. 2010).

M. A. Cebrián-Piqueras (✉)
Institute of Environmental Planning, Leibniz University Hannover, Hannover, Germany
e-mail: cebrian@umwelt.uni-hannover.de

C. von Haaren et al. (eds.), *Landscape Planning with Ecosystem Services*, Landscape Series 24, https://doi.org/10.1007/978-94-024-1681-7_22

Given the contributions that soils provide to the delivery of a considerable diversity of ecosystem services, it comes to no surprise that soil management is often multifunctional in enhancing the provision of not just one but several ecosystem services (Hauck et al. 2014). For example, the maintenance and enhancement of appropriate soil humus content (i.e. soil organic carbon) has not only a positive impact on soil fertility and thus food production, but also on other properties and services such as water retention or greenhouse gases storage (Tilman et al. 2002; Magdoff and Weil 2004).

Against this background, the objective of this chapter is to synthesize existing knowledge on measures to safeguard and enhance soils for three key ecosystem services, namely soil production capacity, soil-related greenhouse gas storage, and geodiversity. Prevention of soil erosion is addressed in this chapter as part of measures to safeguard soil production capacities (maintenance of soil fertility). However, erosion also affects other ecosystem services (Panagos et al. 2015). Additional soil-related measures can also be found in Chap. 23 regarding water quality enhancement and Chap. 25 in terms of biodiversity.

In general, the identification of appropriate response objectives and siting of measures for soil protection and sustainable use should be based on an in-depth analysis of relevant pressures and the current state of soils in the study area. Response options can aim to minimize existing pressures, safeguard the state, or enhance the current situation (see Chap. 21 for more details). Key sources of information for the development of soil-related response options are therefore the assessment of existing pressures (Chap. 9), as well as details regarding the current state of a landscape regarding the provision of habitats (Chaps. 17, 18), food, materials, and bioenergy (Chaps. 10, 12), the regulation of water (Chap. 11) and local climate (Chap. 13), and the mitigation of greenhouse gas emissions (Chap. 14). Last, but not least, assessments of landscape capacities to safeguard geodiversity and other cultural services (Chaps. 15, 16) need to be taken into account as well.

Implementation of objectives and measures for soil-related ecosystem services can be achieved through enforcing existing legal standards, providing economic incentives to land managers and by informing relevant stakeholders. Furthermore, spatial planning has a long tradition of soil conservation through land use designations that often draw upon the landscape planning information base. In addition, Environmental and Strategic Impact Assessments also need to take impacts on soil functions into account.

The following sections provide an overview of objectives and measures for safeguarding soil production capacities, minimizing soil erosion, reducing greenhouse gas emissions and protecting geodiversity. Where appropriate, the description of measures is differentiated between mandatory measures required by laws or other regulations, and those where implementation is voluntary.

22.1.2 Good Farming Practice (GFP) and Additional Voluntary Measures

Measures for soil protection which are undertaken as part of 'good farming practice' are those the farmer has to implement without compensation because they are obligations that need to be fulfilled according to the 'polluter pays principle'. Usually such objectives and measures are defined in legal or quasi-legal norms. Within the European Union, legal obligations are defined in many national laws as well as 'quasi mandatory' requirements of the European Common Agricultural Policy (CAP). Farmers receiving direct payments from CAP are required to comply with the so-called 'cross-compliance' and 'greening' standards. Member states have some flexibility in the implementation of these standards and specific measures are adapted to different regional and climatic conditions. In principle, farmers can opt not to receive direct payments so the standards linked to them are termed 'quasi mandatory' here. They are not legally binding in a strict sense, but given the importance of direct payments to many farm incomes have, in practice, a similar or perhaps an even stronger impact on land management practices.

Cross-compliance is a mechanism that ties direct payments to farmers with accordance to basic standards termed *Statutory Management Requirements* (SMRs). These concern the environment, food safety, animal and plant health, and animal welfare, as well as the requirement of maintaining land in *Good Agricultural and Environmental Condition (GAEC)*. Cross-Compliance was introduced in 2003 and subsequently updated by Regulation (EC) No 73/2009 and Regulations (EU) No 1306/2013 and (EU) No 1307/2013. Even in European countries which are not EU member states these regulations influence farming practice because they need to be observed if agricultural products are imported to the EU.

Apart from the obligatory measures included in the first pillar of the CAP (i.e. cross-compliance and greening), the second pillar includes more targeted voluntary agri-environmental measures (AEMs) (Regulation No 1698/2005) which go beyond the basic obligations of good farming practice. AEMs are included in the rural development programs (RDPs) of member states and farmers can voluntarily apply to join such schemes for periods of five or more years and receive additional payments for implementing measures (Matthews 2013). AEMs should play a prominent role in supporting the sustainable development of rural areas and in responding to society's increasing demand for environmental services, though there is much debate as to how effective they have been (Batáry et al. 2015). The recent Regulation (EU) No 1305/2013 also introduced the idea of agri-environment-climate measures (AECMs) which aim to promote changes to agricultural practices that make a positive contribution to the environment and climate (see Sect. 25.4).

For landscape planning the distinction between GFP and additional voluntary actions is very important, because it directly influences implementation strategies, particularly engagement with landowners. Technical advice and enforcement will be needed to support the implementation of GFP, whereas voluntary measures are dependent on participation and can be combined with financial incentives. Thus the differentiation between GFP and voluntary measures relates to the separation of minimum targets from voluntary, desirable objectives proposed in Chap. 21. Due to the direct relevance for the participation process and consequences for both officials and landowners, the landscape plan should show the distinction between GFP and voluntary measures in relevant maps.

22.2 Multifunctional Objectives and Measures Regarding Soil Production Capacity and Other Services

22.2.1 Multifunctional Objectives and Measures for Good Soil Conditions According to Good Farming Practice

The most important cross-compliance measures to directly preserve soils in good conditions are the GAECs (see Table 22.1). These include measures against soil erosion (GAECs 4 and 5) and for improvement of soil organic matter and structure (GAEC 6). In addition, some of the SMRs, in particular those related to Area 1 (Environment, Climate Change and Good Agricultural Condition of Land) are expected to have a direct or indirect positive effect on soil conservation. Examples of EU legislation where implementation is supposed to be supported by Area 1 – SMRs include the Nitrates Directive, Groundwater Directive, and Sewage Sludge Directive (SMR 1 Protection of Water against Pollution caused by Nitrates), the Birds Directive (SMR 2 Conservation of Wild Birds) and Habitats Directive (SMR

Table 22.1 Good agricultural and environmental conditions directly related to soil conservation, soil production capacity and other soil-related ecosystem services

GAEC 4: Minimum soil cover. The aim of this requirement is to protect soil against erosion after harvest until the end of winter ('winter' is up to and including the last day of February).	Positive effects on soil erosion, enhancement of humus content, biodiversity, water retention, nutrients filtering.
GAEC 5: Minimum land management reflecting site specific conditions to limit erosion. The aim of these rules is to protect soil against erosion in certain situations.	Positive effects on soil erosion, production capacity, soil structure, humus content and biodiversity.
GAEC 6: Soil organic matter. The aim of this requirement is to maintain soil organic matter levels through appropriate practices.	Positive effects on soil erosion, soil structure, humus content, production capacity, nutrients filtering, water retention, and biodiversity.

References: DEFRA (2015), Tilman et al. (2002), Magdoff and Weil (2004), Dale and Polansky (2007), Zhang et al. (2007) and Power (2010)

3 Conservation of Natural Habitats and of Wild Flora and Fauna). The specific implementation of GAECs is shaped by national and regional regulations.

Greening measures were introduced in the CAP reform of 2013 (Regulation (EU) No 1306/2013 and COM (2011) 625 Final). They account for up to 30% of national direct payment budgets and seek to achieve a more environment-friendly agriculture. All the three groups of greening measures i.e. diversifying crops, maintaining permanent grasslands and preserving ecologically beneficial elements are expected to have positive effects on soil conditions (Table 22.2). However the extent of these effects depends on particular management practices (Hauck et al. 2014).

Even if the catalogues of CAP measures change in the future, the measures listed above can be expected to stay a benchmark for good farming practice – not least

Table 22.2 Greening measures and expected impacts on soil conservation, production capacity and other ecosystem services

Greening measures	Expected impacts on soil conservation, production capacity and other ecosystem services.
Dedicating 5% of arable land to **'ecologically beneficial elements': Ecological Focus Areas:** directly, such as fallow land, field margins, hedges & trees, buffer strips. Indirectly, by cutting use of inputs or better soil protection (e.g. in areas covered by catch crops (fast-growing crops grown between plantings of main crops) or nitrogen-fixing crops).	**Set aside:** Short-term negative impacts on food production. Positive impacts on soil structure, soil fertility, humus content and carbon sequestration, water regulation and pollination.
	Buffer strips: Positive impacts on soil erosion control, water regulation, soil carbon stocks and soil structure.
Maintaining permanent grassland: National governments are obliged to designate **environmentally sensitive permanent grasslands** within Natura 2000 areas. Environmentally sensitive permanent grasslands may also be designated outside Natura 2000 areas. These valuable permanent grasslands cannot be ploughed or converted to arable land. The ratio of permanent grassland to the total agricultural area cannot fall more than 5% compared to the reference year.	**Intensive grasslands:** Positive impacts on productivity, however negative impacts are found on several regulating and cultural services (i.e. biodiversity, soil organic carbon, soil structure and fertility and aesthetic value).
	Extensive grasslands: Positive impacts on production capacity and several regulating and cultural services (e.g. soil fertility, water regulation and purification, soil carbon stocks, recreation and aesthetic value).
Diversifying crops: This requirement applies to farmers with over 10 ha of arable land. Up to 30 ha: farmers have to grow at least 2 crops and the main crop cannot cover more than 75% of the land. Over 30 ha: farmers have to grow at least 3 crops, the main crop covering at most 75% of the land and the 2 main crops at most 95%.	Positive impacts on yields, soil fertility, soil carbon stocks, soil erosion control and aesthetic value.

References: Hauck et al. (2014), Hajjar et al. (2008), IEEP (2008) and Uthes et al. (2010)

Table 22.3 Agri-environmental measures expected to have a positive impact on soil conservation

Organic farming
Integrated production
Other extensification of farming systems: fertilizer and pesticide reduction, lower livestock densities
Crop rotation, maintenance of set-aside areas
Specific actions to prevent or reduce soil erosion
Upkeep of the landscape including the conservation of historical features on agricultural land
Water-related actions (apart from nutrient management) such as buffer strips, field margins, wetland management

References: Stolte (2016), von Haaren (2004) and WWF (2012)

because they support the EU directives mentioned above which have a binding character for EU member states. The measures strive to prevent soil degradation and thus contribute to safeguarding ES, in particular the production capacity of the soil but also water retention and climate protection functions.

Besides the mandatory and quasi-mandatory measures for soil protection at the EU level, national legislation often includes mandatory measures to support soil functions. Examples are the protection of grassland, standards regarding organic and inorganic contaminants in soil which limit the application of slurry and sludge, and requirements for diversity in crop rotations.

22.2.2 Voluntary Objectives and Measures

The main types of farming practices that are eligible for agri-environment payments in the EU and are expected to have a positive effect on soil conservation are listed in Table 22.3. Particular AEMs having positive effects on soil conservation and preventing impacts such as soil erosion are discussed in Sect. 22.3.

Overall, these measures are generally positive for soil conservation. However, if applied indiscriminately and in an untargeted manner, the capacity of the landscape plan to prioritize measures and financial resources will remains unused. If the measures are directed to areas where they are (most) needed or have the greatest effects, the information basis of the landscape plan could be used to increase the efficiency of implementation (see Chap. 26). To this end, the implementation of responses needs to directly refer to the assessment of state conditions.

22.3 Objectives and Measures for Maintaining Soil Production Capacity by Minimizing Soil Erosion

Among all the threats to soil, erosion has one of the major impacts on productive capacity. For example, 12.7% of European arable land has soil loss >5 t/ha annually and requires conservation measures. Among all land uses, arable and sparse vegetation have the highest soil loss rates (Panagos et al. 2015). As part of the CAP, GAEC requirements include mandatory soil-protection measures such as reduced tillage and contour farming. According to Panagos et al. (2016), implementation of GAECs on agricultural land has helped to significantly reduce soil loss rates. From 2003–2010, it reduced soil loss across the EU by 9.5% and by over 20% on arable land. Management options with the greatest impact on soil loss rates were reduced and no-tillage practices, which are currently applied to over 25% of agricultural land in the EU. In addition to the GAECs, other voluntary instruments and agri-environment measures are helping to minimize impacts on soils (Stolte et al. 2016; Panagos et al. 2015, 2016; Louwagie et al. 2009). Some national regulations also include management requirements to prevent soil erosion. For instance the German Federal Soil Conservation Act (§ 17 (2) 4 (BBodSchG) requires site-appropriate use to reduce erosion as much as possible, in particular by taking into account the slope, ground cover, water and wind conditions. This environmental quality target applies to all usable for agriculture and forestry. The remainder of this section discusses the most relevant mandatory and voluntary measures directly related to soil erosion control.

22.3.1 Good Farming Practice for Minimizing Soil Erosion

Within the framework of cross-compliance obligations, EU members have classified agricultural areas according to their risk of wind and water soil erosion. Farmers managing 'at risk' areas and receiving direct payments are obliged to implement a minimum set of measures. The regulations defining the areas in different risk categories and the associated measures are specified in national laws.

For instance, the German legislation (federal direct payments-obligation regulation – 'DirektZahlVerpflV') adopted the DIN19706 and DIN19708 methodologies to classify areas according their wind and water erosion risk. Areas classed as CCwasser0 and CCwind0 had no risk of erosion, while those categorised as CCwind1, CCwasser1 and CCwasser2 had varying degrees of risk. The federal states subsequently adapted and implemented this requirement in their state laws. An example of the requirements in the state of North Rhine-Westphalia is shown in Table 22.4.

22.3.2 Voluntary Objectives and Measures

In addition to the legally required measures, a broad range of voluntary objectives and measures exist in Europe to avoid soil erosion. Ideally, erosion should be

Table 22.4 Examples from North Rhine-Westphalia of measures to be conducted in agricultural areas classed at different levels of risk for wind and water erosion

Measures for areas catalogued as **"CCWind1" (Wind erosion risk)**
May only be ploughed when sowing before 1 March.
Ploughing from 1 March is only permitted with immediately following sowing (except for row crops). 'Immediate' includes, as far as necessary, the period for settling the soil (about 3–6 weeks) and weather-related delays.
The plough ban on row crops does not apply under specific conditions (see the federal act specifications).
The regulation specifies particular requirements and ploughing ban exceptions for maize, sugar beet and potato crops.
Measures for areas catalogued as **"CCWasser1" (Water erosion risk)**
Ploughing ban in the period between December 1 and the end of February 15.
Ploughing after harvest of the pre-crop is only allowed for sowing before 1 December.
The requirements do not apply to management across the slope. 'Management' covers all operations of soil tillage, seedbed preparation and sowing / planting. The headlands do not have to be worked across the slope. Management across the slope can only be claimed if the slope is clearly inclined in one direction.
Ploughing may be carried out before 15 February if further processing of the furrow occurs after 15 February, followed immediately by summer crops, grain legumes, summer rape, forage or maize with a row spacing of less than 45 cm (LESchV).
Measures for areas catalogued as **"CCWasser2" (High water erosion risk)**
Ploughing ban in the period from December 1 to the end of February 15.
Ploughing between February 16 and the end of 30 November 30 is only permissible in cases of immediately following sowing. Latest time of sowing is November 30.
As described under CCWasser1, plowing may be carried out before February 15, if further cultivation continues after February 15 immediately followed by summer crops, grain legumes, summer rape, forage or maize with a row spacing of less than 45 cm (LESchV).
Ploughing ban before sowing of row crops.
The regulation specifies ploughing ban exceptions for maize, sugar beet, potatoes and horticultural crops.

Source: examples taken from the Federal State Erosion Protection Act of North Rhine-Westphalia ('LESchV')

avoided, but since it cannot be totally prevented under arable use efforts must be made to minimize soil losses (Panagos et al. 2015; Louwagie et al. 2009; Stolte et al. 2016). Examples of the most common measures to limit water and wind erosion are given in Table 22.5. Some of these have been adopted by European states regions in the form of agri-environment measures or agri-environment-climate measures (UBA 2017). According to Posthumus et al. (2015), tramline management, mulching, buffer strips, high-density planting and sediment traps have been identified as the most cost-effective erosion control measures in the UK.

Table 22.5 Common measures to prevent soil erosion in agricultural areas

(a) General measures against soil erosion
On-site inspection of impacts in areas with high risk of erosion and, where appropriate, in areas where 'offsite' damage can occur (water edge strips, other biotopes), to check whether erosion actually occurs.
Erosion protection planning at the farm level to meet the tolerance limits of soil erosion by consultants and farmers.
Use of optimization models.
(b) Measures to limit soil erosion by water
Crop rotation design: Potato and maize cultivation only up to max. 14% inclination (replacement of maize with in-house produced ryegrass silage and cereal crop silage from winter barley); replacement of summer with winter cereals; sugar beet cultivation only up to max. 9% slope; catch crops, under-seeding of maize and sugar beet; direct sowing. The goal is year-round plant cover; weed control only in the wake;
Conservation tillage with **mulch sowing** as far as possible throughout the crop rotation process and without seedbed preparation; tillage only in spring instead of autumn; elimination of soil compaction to improve infiltration; timely use of machinery; promoting soil moisture on organic soils; alternation of ridge direction in potato cultivation
Other measures to **improve structural stability,** e.g. **increase of humus content of soils, liming;**
Strip use: (e.g. cultivation of cereal crops and root crops alternately); at 1 to 2% slope 40 m maximum strip width; at 17 to 20% 20 m strip width);
Erosion control strips e.g. of grass or cereals such as winter barley; length of the strips: depending on the slope 5 to 30 m, width of the strips 1 to 3 m. The main application is in maize cultivation;
Shortening of the erosive slope length, depending on erosion risk shortening of the parcel length to <300 m, especially in maize cultivation; with slope > 10%: parcel length < 200 m, with a slope of 21 to 25%: 17 m; Avoidance of downhill tramlines etc.;
Installation of drainage barriers (e.g. deep terminal furrows, shallow dams); grassland use or forest on very high risk of erosion;
Reforestation of erosion-prone areas, exclusion areas (for reasons of biotope protection, recreation and climate protection) need to be monitored (e.g. bogs, marshes, wet meadows, inland dunes, heaths, cultural landscapes, land for special recreation activities);
High-density planting to increase the number of plant/tree individuals per area (e.g. in orchards) on specific parcels or fragments with high erosion risk;
Special measures for highly endangered arable lands: great effects can be achieved by the redistribution of individual arable land parcels in the landscape (parcel); changing the land use and permanent abandonment (greening of high-risk areas);
Tramline management: Tramlines should be established only after the winter, or if possible, they should not be used until the spring. In addition, the tramlines and wheelings should be cultivated to avoid risk of erosion derived from compaction (Posthumus et al. 2015).
(c) Measures to limit soil erosion by wind
Crop production (short-time effects): change cultivation direction; adaptation of crop rotation to give the longest possible ground cover, reduce share of maize, potatoes, sugar beets in the crop rotation <25%; cultivation of catch crops; under sowing; improvement of humus supply; plant erosion protection strips.
Landscape and land improvements: (medium-term effects): wind protection planting; plant agroforestry systems; liming;

(continued)

Table 22.5 (continued)

Agricultural structure (long-term effects): land consolidation; creation of biotopes;
Other erosion control measures: straw mulch, mulch-sowing process to maize and sugar beet, no grassland change, not even as a nursing change, especially on organic soils (over-/slit-seeding method for reseeding).

References: Posthumus et al. (2015), AID (2015), Duttmann et al. (2012), Frielinghaus et al. (2002a, b), von Haaren (2004), UBA (2017) and Deasy et al. (2009)

22.4 Objectives and Measures for Greenhouse Gas Storage and Carbon Sequestration

Despite its unequivocal contribution to global climate regulation and being one of the most widely recognized ecosystem services (IPCC 2013; Lal 2004; MEA 2005), the capacity of soils and vegetation to store carbon has rarely been addressed in landscape planning (Saathoff and von Haaren 2011) and, to date, no overarching legally-binding European legislation exists in this regard.

Though more robust data are needed to assess carbon stocks at plot and landscape scales, the precautionary principle must guide actions in landscape planning towards avoiding land use transformations that can put at risk existing organic carbon stocks and the potential for future sequestration (Füssel and Jol 2012). Chapter 14 discusses spatially-explicit methods to assess the soil carbon sequestration function and highlights the importance of identifying hotspots of carbon stocks for optimization in landscape planning.

In terms of policy, important advances have been made through the EU CAP and Rural Development Programs (RDPs). Wetlands, peatlands, grasslands and agricultural areas in general have been targets of sets of measures that aim to enhance the capacity of soils to store and sequestrate carbon. Both mandatory and voluntary measures are discussed below.

22.4.1 Good Farming Practice

Several GAEC requirements are expected to positively affect the maintenance and enhancement of organic carbon stocks in soils and therefore contribute to climate change mitigation (Borrrelli et al. 2016). Particularly relevant is the GAEC 6 standard on 'Maintenance of soil organic matter level through appropriate practices'. Specific requirements of this standard vary between countries and regions, but commonly include not burning stubble or crop residues such as straw and complying with a prescribed burning code of practice where burning is permitted.

Both GAEC 4 'Minimum soil cover' and GAEC 5 'Minimum land management reflecting site specific conditions to limit erosion' are specifically designed to prevent soil erosion; however both may also have, depending on the particular conditions, positive effects on soil organic matter.

Indirectly, measures such as GAEC 1 'Establishment of buffer strips along watercourses' and GAEC 7 'Retention of landscape features' can also contribute positively to soil carbon storage, since the associated natural or semi-natural vegetation have been found to retain and enhance below-ground biomass and soil organic carbon stocks on agricultural land (Walter et al. 2003; Follain et al. 2007; Lal et al. 1999; Blanco-Canqui and Lal 2008). Further discussion of buffer strips is included in Chap. 23.

National implementations of cross-compliance include requirements to preserve and enhance the humus content in agricultural soils. For example, the German regulation (DirektZahlDurchfV) requires that if the minimum level of crop rotation at farm level is not respected (and only then) either (i) a farm humus balance for the entire arable area (including set aside) is conducted annually by 31 March of the following year, or (ii) uniform farmed parcels with similar soil properties should be tested for soil humus contents at least every six years. In any case, the humus balance must not fall below an average value of minus 75 kg humus carbon (Humus-C) per hectare per year.

Finally, the greening measures introduced in 2013 (ecologically beneficial elements, maintenance of permanent pastures and crop diversification) should positively affect soil organic carbon stocks and carbon sequestration. However, the efficacy of these requirements to support carbon stocks is likely to vary according to site conditions and the particular type of measure. Hauck et al. (2014) review existing knowledge about the expected effects of greening measures on several ecosystem services including climate regulation and carbon sequestration.

22.4.2 Voluntary Objectives and Measures

As mentioned earlier, Regulation (EU) No 1305/2013 introduced a set of voluntary agri-environment-climate measures (AECMs) to support conservation of soil organic carbon stocks and permanent pastures. The aim of measure M.10.1 'agri-environment-climate commitments' was to preserve and promote the changes to agricultural practices that make a positive contribution to the environment and climate. It is the only mandatory Rural Development Programme measure for EU member states, but it remains voluntary for farmers. Less directly, organic farming can enhance carbon stocks and measure M.11 provides support to farmers who convert to or maintain organic farming practices and methods as defined in Regulation (EC) No 834/2007. Overall, successful application of measures can benefit from targeting based on knowledge of spatial variations in carbon sequestration, as discussed in Chap. 14.

22.5 Objectives and Measures for the Protection of Geodiversity

Valuable sites for geodiversity (as identified by methods explained in Chap. 16) should be protected and sustainably managed. Policies, strategies, guidelines, general and specific measures are all part of what has been defined as geoconservation during last decades (Nieto 2001; Gray 2004). According to Gordon et al. (2018)

> Geoheritage conservation (or geoconservation) is the practice of conserving, enhancing and promoting awareness of those features and underlying processes of geodiversity that have significant scientific, educational, cultural, aesthetic or ecological value.

This section discusses conservation guidelines and examples of measures that can be applied to preserve and enhance geosites. As already noted in Chap. 16, there is no Europe-wide legislation offering assessment standards or requirements to protect geosites. However, international organisations such as IUCN, UNESCO and ProGeo have set guidelines and recommendations which are being implemented at national level.

Some national nature conservation legislation already includes legal frameworks to assure the conservation of outstanding geosites. For instance, the German Federal Soil Conservation Act specifies that soils with relevant natural and cultural history should be protected and so-called 'geotopes' (analogue term for geosite in the central European tradition) are included in the nature conservation regulations of many states. Additionally, the Federal Nature Conservation Act can be used to take account of soil properties in designating areas for their outstanding value for natural or cultural history.

In the reform of the Spanish Nature Conservation Act (Law 47/2007), the value of geodiversity is legally comparable to biodiversity and the concept of geodiversity is defined as

> the variety of geological elements, including rocks, minerals, fossils, soils, relief forms, formations and geological units and landscapes that are the product and record of the evolution of the Earth.

This legislation also includes the identification of Geoparks as

> delimited territories that present unique geological forms, of special scientific importance, singularity or beauty and that are representative of the geological evolutionary history and of the events and processes that have formed them.

Following this law and associated guidelines (National Strategic Plan for the Conservation of Nature and Biodiversity), most Spanish regions have incorporated geodiversity into their specific nature conservation regulations and subsequently identified worth protecting geosites. Information about Places of Geological Interest

(LIGs) (i.e. geosites), is easily accessible through an online GIS (Spanish Inventory of Places of Geological Interest (IELIG), http://info.igme.es/ielig/).

Suggested steps in the development of a management plan are shown in Fig. 16.2 (Chap. 16). As a general guideline, the Committee of Ministers of the Council of Europe (2004) recommended the steps shown in Table 22.6 for the management of places with geological value.

Practical experience suggests that determination of the *degradation risk* is a crucial task for the effective protection of geosites (Brilha 2016; Fuertes-Gutiérrez and Fernández-Martínez 2012). A set of *general measures* can be included in landscape plans or used by stakeholders to avoid the deterioration of geodiversity and geosites of high value (Table 22.7).

In addition, *specific measures* should be implemented according to the type and properties of the geosite. The German federal/state soil protection working group (LABO 2011) provides a good example of measures to protect soils for their natural and cultural historical value depending on these characteristics (Table 22.8). Implementation of these measures can occur in practice through nature conservation or conservation of heritage regulations, as well as the usual instruments of spatial planning and building regulation.

Table 22.6 Recommendations of the Committee of Ministers of the Council of Europe on the conservation of geological heritage and areas of special geological interest (Rec (2004) 3)

1. Recognition of the **distribution and nature of this resource** through development of national area (site) inventories;
2. **Classification of area (site)** types according to:
(a). **scientific value** (geological or geomorphological features and their scientific importance)
(b). **physical characteristics** (coastal, river valley, mountain, quarry, roadside exposure, etc.)
(c). **specific management requirements** of individual areas (sites)
3. Development of indicators to **identify threats and monitor degradation** of geological heritage
4. Implementation of **site-condition monitoring programmes** based upon management requirements of specific area (site) types; these programmes should be linked to existing biodiversity monitoring programmes where possible;
5. **Creation of national/regional databases**, to include inventory and monitoring information. Such databases are essential for management of areas (sites) and the dissemination of information relating to their scientific and educational value. Internet-based databases should be the standard, to ensure the maximum dissemination of information
6. **Linking national 'areas of special geological interest'** databases to:
(a). **regional and local planning** to ensure that planning authorities are aware of, and take into account, these special areas in creating/implementing plans
(b). **biodiversity databases** to ensure consistency of approach when managing natural heritage.

Table 22.7 General measures to protect geodiversity and valued geosites

No dismantling of rocks and earth on protected geotopes/geosites.
Avoidance of sealing, building, excavation and redistribution (risk of complete disappearance of the function).
No change in the surface relief (e.g. in the case of volcanic bogs), forest use, if necessary, natural forest parcels.
Abandonment of melioration measures as well as deep pits.
Public information (geological trails, information boards, information on hiking and cycling maps).
Inclusion in design concepts, artistic portrayals.
Preservation of information and accessibility; e.g. if necessary, preventing the filling of depleting sites.
If necessary, deforestation.
No sealing, building, excavation and rearrangement.
Avoiding deposits or the application of soil material.
Agricultural use restrictions.
Forestry restrictions.

References: LABO (2011) and von Haaren (2004)

Table 22.8 Recommendations by LABO (2011) of measures to protect soils and geotopes for their natural and cultural history value

Distinctive features	Examples	Recommendations and Remarks
Examples of soils and geotopes to be protected for their **natural history**		
Recent soil formation processes	Initial soil development, e.g. of Rendzina soils	Recommendations:
	Pronounced material transfer, degradation or enrichment processes, e.g. B. Sesquioxid shift in Podzols	Developing vegetation or biotope networking if biotope development potential exists (intervention regulation)
	Pronounced redox processes, e.g. of Gleysols	
	Sedimentation in floodplain areas, z. B. of Vega soils (Fluvisols)	
Pedogenic processes from past geological and climatic periods (paleo-soils)	Terra rossa	For covered palaeosols a justification for protection is often difficult
	Schwarzerden (i.e. Chernozems)	

(continued)

Table 22.8 (continued)

Distinctive features	Examples	Recommendations and Remarks
Information in moors of historical climate and vegetation communities	Fens	High synergy effects between nature protection and soil protection exists. Therefore, in practice most sites are simply protected as nature conservation areas.
	Raised bogs	
	Upper soil	
Periglacial processes in soils and morphological elements or landforms	Frost sample soils and cryogenic bacteria occurrence, e.g. ice wedges;	In the case of morphological landscape elements, there are high synergy effects between soil protection and nature conservation and usually the protecting status of landscape conservation area is used.
	Morphological landscape elements such as kettles, drumlins, sinkholes, dunes and "Kare" (Cirque) or end moraines	
Starting/raw materials of soil development	Soils of volcanic rocks, reef-time loose rocks or limestone sockets	For small-scale structures, such. as lime sinter
	Soils of special substrates such as silicification, calcareous sinter, fracture fillings, sea clays or shale coal.	
Formations and structures depicting earth history (Geotope)	Outcrops with striking layer sequences	Easy implementation, as geotope cadastre is managed nationwide and made available to the municipalities in most countries.
	Quarries, clay, sand and gravel pits	
	Fossil deposits	
	Source and sinter terraces	
Examples of soils and geotopes relevant for their **cultural history**		
Pedogenic records of settlement and land use history ('Kultosole') and historical usages	'Plaggenesch' (Plaggept)	Forwarding of information to conservation authority by soil protection authority necessary
	Historic uses such as historic vineyards or 'Wölbäcker'	
Relics of settlement and land use history (monuments/ archaeological sites)	Finding sites, burial grounds	Strengthening of arguments for monument protection on grounds of soil protection
	Settlement remains	
	Historical landmarks, burial mounds etc.	

22.6 Conclusions

This chapter has provided an overview of commonly applied mandatory and voluntary measures to safeguard and enhance soil-related ecosystem services such as soil production capacity, soil erosion control, greenhouse gas storage, carbon sequestration and geodiversity. It is evident that diverse measures are available. In choosing between measures, planners, farmers and relevant stakeholders should consider the opportunities for multifunctional options that help attain several objectives with just one measure. This could enhance investment effectiveness and minimize the land area required to achieve multiple objectives (see Chap. 26, Galler et al. 2015; Schindler et al. 2014).

References

AID. (2015). *Infodienst Ernährung, Landwirtschaft, Verbraucherschutz e.V.: Gute fachliche Praxis – Bodenbewirtschaftung und Bodenschutz*, 2nd edn. Bonn.

Batáry, P., Dicks, L. V., Kleijn, D., et al. (2015). The role of agri-environment schemes in conservation and environmental management. *Conservation Biology, 29*, 1006–1016.

Blanco-Canqui, H., & Lal, R. (2008). *Principles of soil conservation and management*. Dordrecht: Springer Netherlands.

Borrelli, P., Paustian, K., Panagos, P., et al. (2016). Effect of good agricultural and environmental conditions on erosion and soil organic carbon balance: A national case study. *Land Use Policy, 50*, 408–421.

Brilha, J. (2016). Inventory and quantitative assessment of geosites and geodiversity sites: A review. *Geoheritage, 8*(2), 119–134.

COE – Council of Europe. (2004). *Recommendation Rec(2004)3 on conservation of the geological heritage and areas of special geological interest*. Council of Europe, Committee of Ministers. https://wcd.coe.int/ViewDoc.jsp?id=740629&Lang=en. Accessed 24 Aug 2018.

Dale, V. H., & Polasky, S. (2007). Measures of the effects of agricultural practices on ecosystem services. *Ecological Economics, 64*(2), 286–296.

Deasy, C., Quinton, J. N., & Silgram, M. (2009). Mitigation options for sediment and phosphorus losses from winter-sown arable crops. *Journal of Environmental Quality, 38*, 2121–2130.

DEFRA (ed). (2015). *Cross compliance in England: Soil protection standards*. https://www.gov.uk/government/publications/cross-compliance-guidance-for-2015. Accessed 24 Aug 2018.

Dominati, E., Patterson, M., & Mackay, A. (2010). A framework for classifying and quantifying the natural capital and ecosystem services of soils. *Ecological Economics, 69*(9), 1858–1868.

Duttmann, R., Hassenpflug, W., Bach, M., et al. (Eds.). (2012). *Winderosion in Schleswig-Holstein – Kenntnisse und Erfahrungen über Bodenverwehungen und Windschutz*. Flintbek: Landesamt für Landwirtschaft, Umwelt und ländliche Räume Schleswig-Holstein (LLUR).

Follain, S., Walter, C., Legout, A., et al. (2007). Induced effects of hedgerow networks on soil organic carbon storage within an agricultural landscape. *Geoderma, 142*(1–2), 80–95.

Frielinghaus, M., Winnige, B., Deumlich, D., et al. (Eds.). (2002a). *Informationsheft zum landwirtschaftlichen Bodenschutz im Land Brandenburg – Teil Bodenerosion*. Potsdam: Ministerium für Landwirtschaft, Umwelt und Raumordnung Brandenburg (MLUR) und Leibniz-Zentrum für Agrarlandschaftsforschung (ZALF).

Frielinghaus, M., Deumlich, D., Funk, R., et al. (Eds.). (2002b). *Bodenerosion – Beiträge zum Bodenschutz in Mecklenburg-Vorpommern* (2nd ed.). Schwerin: Landesamt für Umwelt, Naturschutz und Geologie Mecklenburg-Vorpommern (LUNG).

Fuertes-Gutiérrez, I., & Fernández-Martínez, E. (2012). Mapping geosites for geoheritage management: a methodological proposal for the Regional Park of Picos de Europa (León, Spain). *Environmental Management, 50*, 789–806.

Füssel, H.-M., & Jol, A. (2012). *Climate change, impacts and vulnerability in Europe 2012 an indicator-based report*. Luxembourg: Publications Office of the European Union.

Galler, C., von Haaren, C., & Albert, C. (2015). Optimizing environmental measures for landscape multifunctionality: Effectiveness, efficiency and recommendations for agri-environmental programs. *Journal of Environmental Management, 151*, 243–257.

Gordon, J. E., Crofts, R., & Díaz-Martínez, E. (2018). Geoheritage conservation and environmental policies: Retrospect and prospect. In E. Reynard & J. Brilha (Eds.), *Geoheritage. Assessment, protection, and management* (pp. 213–235). Chennai: Elsevier.

Gray, M. (2004). *Geodiversity. Valuing and conserving abiotic nature*. Sussex: John Wiley & Sons.
Hajjar, R., Jarvis, D. I., & Gemmill-Herren, B. (2008). The utility of crop genetic diversity in maintaining ecosystem services. *Agriculture, Ecosystems and Environment, 123*, 261–270.
Hauck, J., Schleyer, C., Winkler, K. J., et al. (2014). Shades of greening: Reviewing the impact of the new EU agricultural policy on ecosystem services. *Change and Adaptation in Socio-Ecological Systems, 1*(1), 51–62.
IEEP – Institute for European Environmental Policy (ed). (2008). *The environmental benefits of set-aside in the EU – A summary of evidence*. Institute for European Environmental Policy, Report to DEFRA. http://archive.defra.gov.uk/evidence/statistics/foo-farm/enviro/observatory/setaside/documents/ieepfeb08.pdf. Accessed 18 June 2018.
IPCC. (2013). *Climate change 2013: The physical science basis* (Contribution of working group I to the fifth assessment report of the intergovernmental panel on climate change). Cambridge/New York: Cambridge University Press.
LABO – Bund/Länder-Arbeitsgemeinschaft Bodenschutz (ed). (2011). *Archivböden – Empfehlungen zur Bewertung und zum Schutz von Böden mit besonderer Funktion als Archiv der Natur- und Kulturgeschichte*. www.labo-deutschland.de Accessed 4 Sept 2018.
Lal, R. (2004). Soil carbon sequestration impacts on global climate change and food security. *Science, 304*, 1623–1627.
Lal, R., Follett, R. F., Kimble, J., et al. (1999). Managing US cropland to sequester carbon in soil. *Journal of Soil and Water Conservation, 54*(1), 374–381.
Louwagie, G., Gay, S. H., & Burrell, A. (2009). *Addressing soil degradation in EU agriculture, relevant processes, practices and policies: Report on the project "Sustainable Agriculture and Soil Conservation" (SoCo)*. EUR-OP.
Magdoff, F., & Weil, R. R. (2004). *Soil organic matter in sustainable agriculture*. Boca Raton: CRC Press.
Matthews, A. (2013). Greening agricultural payments in the EU's common agricultural policy. *Bio-based and Applied Economics, 2*(1), 1–27.
MEA – Millennium Ecosystem Assessment. (2005). *Ecosystems and human well-being*. Washington, DC: Island Press.
Nieto, L. M. (2001). Geodiversidad: propuesta de una definición integradora. *Boletín Geológico y Minero, 112*(2), 3–11.
Panagos, P., Borrelli, P., Poesen, J., et al. (2015). The new assessment of soil loss by water erosion in Europe. *Environmental Science & Policy, 54*, 438–447.
Panagos, P., Imeson, A., Meusburger, K., et al. (2016). Soil conservation in Europe: Wish or reality? *Land Degradation and Development, 27*(6), 1547–1551.
Posthumus, H., Deeks, L. K., Rickson, R. J., et al. (2015). Costs and benefits of erosion control measures in the UK. *Soil Use and Management, 31*, 16–33.
Power, A. G. (2010). Ecosystem services and agriculture: Tradeoffs and synergies. *Philosophical Transactions of the Royal Society of London. Series B, Biological Sciences, 365*(1554), 2959–2971.
Saathoff, W., & von Haaren, C. (2011). Klimarelevanz der Landnutzungen und Konsequenzen für den Naturschutz. Biosphärenreservat Niedersächsische Elbtalaue als regionaler Testfall. *Naturschutz und Landschaftsplanung, 43*(5), 138–146.
Schindler, S., Sebesvari, Z., Damm, C., et al. (2014). Multifunctionality of floodplain landscapes: Relating management options to ecosystem services. *Landscape Ecology, 29*(2), 229–244.
Stolte, J., Tesfai, M., & Øygarden, L. et al. (eds). (2016). *Soil threats in Europe*. EUR 27607 EN. Publications Office doi:https://doi.org/10.2788/488054 (print); doi:https://doi.org/10.2788/828742 (online).
Tilman, D., Cassman, K. G., Matson, P. A., et al. (2002). Agricultural sustainability and intensive production practices. *Nature, 418*(6898), 671.
UBA – Umwelbundesamt (Ed.). (2017). *Bodenerosion durch Wind. Sachstand und Handlungsempfehlungen zur Gefahrenabwehr*. Dessau-Roßlau: UBA.

Uthes, S., Matzdorf, B., Müller, K., et al. (2010). Spatial targeting of agri-environmental measures: Cost-effectiveness and distributional consequences. *Environmental Management, 46*(3), 494–509.

von Haaren, C. (Ed.). (2004). *Landschaftsplanung*. Stuttgart: Eugen Ulmer.

Walter, C., Mérot, P., Layer, B., et al. (2003). The effect of hedgerows on soil organic carbon storage in hillslopes. *Soil Use and Management, 19*, 201–207.

WWF – World Wildlife Fund. (2012). *Baltic sea eco region programme – Sorting out the goods – Agri-environment measures in the Baltic sea member states*. https://wwf.fi/mediabank/2003.pdf. Accessed 18 June 2018.

Zhang, W., Ricketts, T. H., Kremen, C., et al. (2007). Ecosystem services and dis-services to agriculture. *Ecological Economics, 64*(2), 253–260.

Mitigation Measures for Water Pollution and Flooding

23

Richard J. Cooper, Kevin M. Hiscock, and Andrew A. Lovett

Abstract

This chapter discusses the range of measures that can be used to mitigate the impacts of water pollution and flooding. It makes a distinction between source measures which aim to reduce the amount of water or pollutant initially mobilised, pathway interventions which seek to slow the flow of pollutant enriched water once it has become mobilised and methods to protect receptor water bodies which are intended to reduce peak flows or prevent pollutants moving further through a catchment. In many European countries the policies and programmes used to increase the adoption of such measures are heavily influenced by EU obligations stemming from the Floods, Nitrates and Water Framework Directives. Typical approaches used involve a combination of regulation, financial incentives and advice provision. There are also a range of tools that can be used to model the potential effects of mitigation measures and a number of research programmes generating findings that may be of value to the landscape planner.

Keywords

Mitigation measures · Source-pathway-receptor paradigm · Water framework directive · Regulation · Financial incentives · Advice

23.1 Introduction

Intensification of agriculture and extensive urbanisation have resulted in environmentally-sensitive freshwater systems across Europe becoming degraded by nutrient and sediment enrichment, pesticide contamination, overexploitation,

R. J. Cooper (✉) · K. M. Hiscock · A. A. Lovett
School of Environmental Sciences, University of East Anglia, Norwich, UK
e-mail: richard.j.cooper@uea.ac.uk; k.hiscock@uea.ac.uk; a.lovett@uea.ac.uk

C. von Haaren et al. (eds.), *Landscape Planning with Ecosystem Services*, Landscape Series 24, https://doi.org/10.1007/978-94-024-1681-7_23

the introduction of invasive species and through a simplification of hydromorphology. With human resource demands exerting pressure on both water quality and water quantity, catchment water resources experience an array of detrimental ecological and economic impacts which threaten the sustainable ecosystem functioning of this essential natural resource. Under national and international legislation, such as the EU Water Framework Directive (2000/60/EC), governments have legal obligations to ensure that water bodies achieve good ecological and chemical status. Nevertheless, many freshwater systems across Europe are still failing to achieve recommended water quality standards due to continuing poor land management practices contributing to the delivery of contaminants from the terrestrial environment. Mitigation measures are therefore required to help reduce land-to-water pollutant transfers, however for these to be targeted effectively, it is essential to understand catchment functioning and the provenance of pollutants. This chapter builds upon the catchment water resource concepts presented in Chap. 11 by exploring a range of commonly applied mitigation methods for tackling water pollution and flooding, considering both the physical performance of these options as well as the policy and economic drivers to incentivise uptake. It focuses heavily on mitigation measures employed in agricultural settings due to the dominant role of agriculture in contributing to the degradation of European freshwater environments (Box 23.1).

Box 23.1: Definitions and Concepts

Mitigation measure: Term used to describe any process or feature designed to prevent, reduce and/or remediate the impact of pollution upon a water body. Measures are classified via the source-pathway-receptor paradigm (see Chap. 11) and largely seek to minimise the terrestrial-to-freshwater transfer of nutrients, sediments, pesticides, heavy metals and organic contaminants. In other chapters of this book the overarching term **response measure** is used in a similar sense.

Critical source area (CSA): An area within a catchment where elevated *pollutant availability* and good *hydrological connectivity* coincide to facilitate the rapid and efficient land-to-water transfer of pollutants. This term can refer to transfer into surface water bodies or leaching of pollutants into groundwater. CSAs are most commonly discussed in the context of soil erosion, where there exists high antecedent soil moisture conditions and an abundance of readily mobilised nutrient-rich soil. These CSAs include silage storage areas, field gateways, infield tramlines, compacted headlands, intensive pig and poultry units, road and river crossings, livestock paths, farmyard hardstanding and animal feeding stations. It is typically more cost-effective to target mitigation efforts on CSAs that cover a small part of the catchment yet are responsible for a majority of the pollution than to distribute mitigation efforts across the entire catchment (Thompson et al. 2012).

(continued)

Box 23.1 (continued)

Pollution swapping: A term used to describe the paradox when a land management measure introduced to mitigate one type of pollution inadvertently results in an increase in another type of pollution, thus *swapping* one pollutant for another. This necessitates the adoption of a holistic approach to the implementation of mitigation measures to ensure the most effective site-specific options are chosen from both an economic and environmental perspective (Stevens and Quinton 2009).

Hydromorphology: a WFD legislative term that encompasses fluvial geomorphology and hydrology and which describes the physical factors that govern river ecosystems.

Green infrastructure: A network of new or existing green space (i.e. vegetation) in rural or urban areas that supports the natural functioning of ecosystem processes and is integral to the health and wellbeing of communities. An example would be the use of *sustainable urban drainage systems (SUDs)* to reduce surface water flood risk by increasing the infiltration rate of rainwater into the soil in towns and cities (Ellis et al. 2002), as well as possibly contributing to urban biodiversity and recreation. This contrasts with *grey infrastructure* which entails artificial ecosystem modifications to control natural processes for human needs (e.g. the building of concrete dams to reduce downstream flood risk and provide hydroelectric power).

23.2 Types of Mitigation Measure

A wide range of mitigation options are available to address the threats of flooding and water pollution to ecosystem services and these can be classified according to their primary function with respect to the source-pathway-receptor paradigm. Source measures are options which aim to reduce the amount of water or a pollutant initially mobilised (e.g. by reducing soil erosion). Pathway measures are options which seek to slow the flow of pollutant enriched water once it has become mobilised (e.g. through intercepting surface runoff). Lastly, receptor measures are options deployed in or around water bodies which aim to reduce peak flows or prevent pollutants entering and moving further through the catchment. Examples of commonly used mitigation measures are presented in Fig. 23.1 and Table 23.1. A number of studies have sought to compile inventories of measures, including details of their applicability, cost and effectiveness, with examples including Kania et al. (2014), GWP/INBO (2015) and NWRM (2017). Selected measures are discussed in more detail in the following paragraphs.

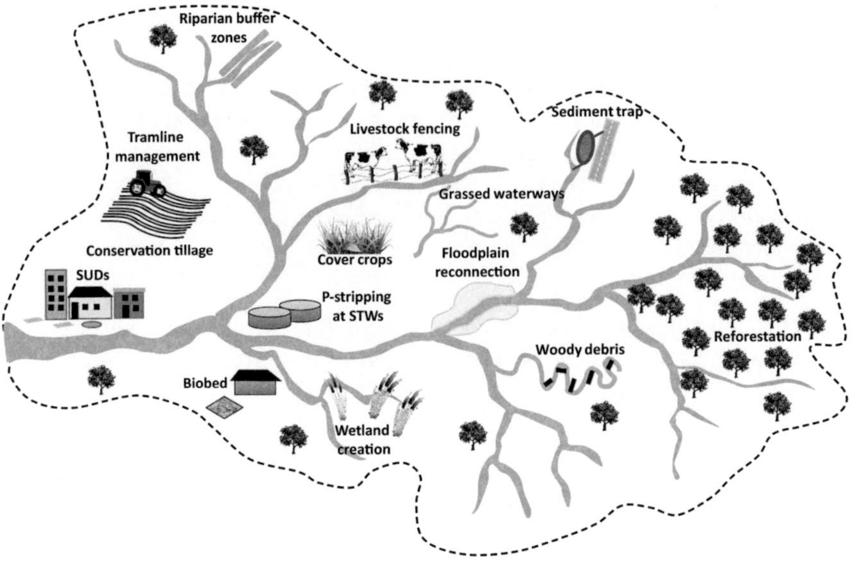

Fig. 23.1 Conceptual model of example land management interventions available to mitigate the impacts of water pollution and flooding to ecosystem services in river catchments

23.2.1 Source Measures

23.2.1.1 Cover Crops

In many conventional farming systems, arable fields are typically cultivated in the early autumn to destroy crop residues and weeds and to prepare the land for sowing of the subsequent crop by loosening compacted soil, incorporating oxygen and bringing nutrients to the surface. Where spring cropping is practiced, this can result in fields being left fallow and devoid of vegetation for 4–5 months during the winter. Under these circumstances, the absence of roots to bind the soil together or leaves to intercept rainfall mean the risk of soil erosion is significantly elevated, resulting in the enhanced transport of sediments and nutrients from the land into surface water courses threatening ecosystem services (see Section 11.4). To mitigate against this issue, a *cover crop* (or *catch crop*) can be sown in the autumn to provide winter ground cover and soil protection (Fig. 23.2). A range of species can be grown, including nitrogen fixing leguminous (e.g. clover, vetch and pea) and non-leguminous (e.g. rye, sorghum and brassicas) varieties. Cover crops have primarily been used to minimise nitrate (NO_3^-) fertiliser leaching into groundwater by scavenging highly soluble residual soil NO_3^- and converting it into relatively immobile organic nitrogen (Snapp et al. 2005). Reported reductions in nitrate leaching under cover crops range from 38–70% (Hooker et al. 2008), 25–60% (Valkama et al. 2015) and 75–97% (Cooper et al. 2017). Cover crops have also been shown to provide a range of other ecosystem service benefits including protecting soils from erosive surface flows, increasing soil organic matter content, improving soil

Table 23.1 Example mitigation measures employed to reduce the impacts of pollution and flooding on the water resources of river catchments

Type	Example	Primary objective	Main impacts on water resources
Source	Cover crops	Soil protection	Reduce nutrient leaching
	Conservation tillage	Soil stabilisation	Reduce soil erosion, lower turbidity
	Biobed	Pesticide degradation	Reduce pesticide concentrations
	Phosphorus stripping	Improving STW effluent	Lower P concentrations
	Reforestation	Water retention	Reduce downstream flood risk
	Rain gardens/soakaways (SuDS)	Increase infiltration	Reduce peak river flows, recharge groundwater
	Green roofs (SuDS)	Increase evapotranspiration	Reduce peak river flows
Pathway	Grassed waterways	Intercept surface runoff	Reduce soil erosion; lower turbidity
	Tramline management/ contour ploughing	Disrupt surface flow path	Reduce soil erosion, lower turbidity
	Controlled traffic farming	Reduce number of flow paths	Reduce soil erosion, lower turbidity
	Sediment traps (swales)	Capture mobilised soil	Lower turbidity; lower P concentrations
	Road crossing redesign	Disrupt surface flow path	Lower turbidity; reduce organic contaminants
Receptor	Buffer strips	Intercept surface runoff	Reduce turbidity; reduce N + P concentrations
	Livestock fencing	River bank protection	Reduce turbidity; reduce FIOs; improve morphology
	Floodplain reconnection	Improve water retention	Reduce downstream flood risk
	Woody debris	Meander creation	Improve morphology; slow water flows
	Riverbank stabilisation	Reduce bank erosion rates	Reduce turbidity; improve morphology
	Wetland creation	Water purification	Reduce turbidity; reduce N + P concentrations

STW sewage treatment works, *SuDS* sustainable drainage systems, *P* phosphorus, *N* nitrogen, *FIO* faecal indicator organism

structure, suppressing weeds and enhancing the soil moisture balance (Dabney et al. 2007; Stevens and Quinton 2009). However, some negative aspects of cover crops have also been reported and include the cost of establishment, difficulty in destroying the cover crop prior to sowing the subsequent cash crop, the harbouring of insect and mollusc pests and the complexity of predicting the release of mineralised nitrogen as the cover crop residues degrade (Deasy et al. 2010).

Fig. 23.2 Example mitigation options to reduce water pollution and flood risk. From top: a winter oilseed radish cover crop (*source reduction*); an on-farm biobed (*source reduction*); a U-shaped sediment trap to intercept road runoff (*pathway interruption*); and grassed riparian buffer strips (*receptor protection*)

23.2.1.2 Conservation Tillage

In conventional tillage systems, autumn cultivations typically see the soil inverted to a depth of 10–30 cm using a mouldboard plough prior to secondary cultivation with harrows and rollers to create a seedbed into which the subsequent cash crop is sown. However, such practice damages the soil structure, breaking up soil aggregates and disturbing the natural soil horizons which increases the likelihood of erosion and the transport of soil and associated nutrients into water bodies. The main objective of conservation tillage systems is to improve soil structure and stability by either disturbing the soil to a lesser degree (e.g. shallow non-inversion tillage to a depth of <10 cm using discs or tines) or not disturbing the soil at all, with sowing occurring directly into the residue of the previous crop (e.g. direct drilling) (Morris et al. 2010). By improving soil structure, conservation tillage methods have been shown to reduce soil erosion, improve drainage and water holding capacity, reduce incidences of soil crusting and compaction (thus increasing infiltration and reducing surface runoff), and increase microbial and earthworm activity by preserving the habitat of soil organisms (Holland 2004; Soane et al. 2012). Conservation tillage can also increase soil organic carbon content, an important determinant of both soil fertility and structural stability, by retaining crop residues on the soil surface and reducing the exposure of organic matter to oxygen deeper in the soil profile and thereby limiting aerobic decomposition and its conversion to carbon dioxide. Nevertheless, the lack of soil inversion can increase pest populations in conservation tillage systems as weed seedlings are not mechanically destroyed and surface organic residues provide food to support larger populations of molluscs. These issues can lead to higher pesticide inputs (pollution swapping) or reduced crop yields, both of which have financial implications for the farmer. Under favourable conditions, however, there is increasing evidence that conservation tillage can be financially competitive with conventional farm practice (Kertész and Madarász 2014).

23.2.1.3 Biobeds

Pesticide pollution threatens the sustainable ecosystem functioning of rivers draining agricultural catchments and therefore mitigation measures are required to reduce pesticide transfer into freshwater environments. Whilst diffuse pesticide pollution sources can in part be reduced by behavioural changes, such as timing spraying operations to avoid periods of inclement weather to limit pesticide mobility, biobeds have emerged as an important mitigation strategy for dealing with point source pollution arising from contaminated machinery washings and accidental spillages during sprayer filling (Castillo et al. 2008; Torstensson 2000). The biobed concept originated in Sweden in the 1990s as a way of using microbial activity to degrade waste pesticide residues. A biobed is essentially a moderately sized pit (typically tens of cubic metres in volume) which can be lined or unlined and is filled with a 1:2:1 matrix of compost, straw and topsoil. The surface is covered with grass and onto this the waste pesticide residues are deposited. In principle, microorganisms (e.g. bacteria and fungi) within the biobed matrix chemically and physically interact with the pesticides leading to structural changes and/or complete degradation. To work effectively, the biobed mixture needs to have a high pesticide absorption capacity and be able to facilitate high rates of microbial activity. Therefore, the content of straw, soil and compost is carefully controlled to maximise biobed performance. In lined biobed systems, the leachate is typically collected from the bottom of the biobed and re-used for either irrigation, sprayer washing or as a carrier for further pesticide applications. Biobed pesticide removal efficiencies of 52–100% have been recorded for a wide range of herbicides, fungicides and insecticides in studies conducted across Europe (Cooper et al. 2016; De Wilde et al. 2007), thus demonstrating the success of biobeds as a management tool for protecting the ecosystem services of water resources.

23.2.1.4 Phosphorus Stripping

The effluent discharged into rivers at sewage treatment works (STWs) is rich in biologically available soluble reactive phosphorus (SRP) and is a major cause of downstream freshwater eutrophication. Discharged sewage effluent typically has SRP concentrations of 1–20 mg L^{-1}, values well in excess of the 0.02–0.07 mg L^{-1} river water quality standard considered *'Good'* under the EU WFD (Withers and Jarvie 2008). Due to the continuous nature of sewage effluent discharges, SRP concentrations tend to display a highly seasonal pattern with higher concentrations during summer low flows and lower concentrations during winter high flows due to dilution. Consequently, phosphorus concentrations peak during the ecologically sensitive summer season when the rate of primary production and eutrophication risk are greatest. In order to reduce the toxicity of the effluent, wastewater undergoes numerous stages of processing at STWs, including screening through filters to remove coarse material (*pretreatment*), holding in settling tanks to encourage sedimentation of suspended fines (*primary treatment*) and promoting the degradation of organics through biological oxidation (*secondary treatment*). However, even after these treatment stages the effluent remains rich in phosphorus and requires further treatment to mitigate the pollution risk. Phosphorus stripping is a form of *tertiary treatment* increasingly being installed at STWs in which the effluent is dosed with a precipitant (e.g. iron ammonium sulphate) which causes the phosphorus to precipitate out and accumulate at the bottom of settling tanks where the sludge can be recovered and used as a P-rich fertiliser for agriculture. Such tertiary P-stripping is capable of removing up to 95% of the phosphorus within STW effluent, but the technology is expensive and its application has largely been limited to larger STWs where the benefit-cost ratios are higher.

23.2.1.5 Reforestation

Forests currently cover 32% (211 million ha) of Europe's land surface, with coverage varying from >50% in Scandinavia to <15% in Ireland and the United Kingdom which have historically high *deforestation* rates (EEA 2015). The clearance of permanent forest to make space for seasonal cultivated crops and intensively stocked livestock pasture has greatly accelerated the degradation of freshwater environments across Europe. Without the protection of above ground vegetation or stabilising subsurface root networks, soil erosion rates increase significantly, enhancing the transport of nutrient rich sediment into surface water bodies and thus promoting the development of eutrophic conditions. The loss of native forest cover also removes the valuable ecosystem services of flood prevention and drought resilience. Although dependent upon the expanse of forest cover, the tree composition, tree density, length of the growing season and complexity of the vegetation structure, forests have the potential to retain excess rainwater, prevent extreme surface runoff during storm events and to reduce peak river flows, thereby mitigating flooding. Research has shown that water retention potential in catchments with 30% and 70% forest cover is 25% and 50% higher, respectively, than in catchments with just 10% forest cover (EEA 2015). Forests also play a key role in buffering catchments against the effects of drought by enhancing soil infiltration, reducing evaporation, restricting soil desiccation and increasing water storage capacity. Overall, *reforestation* can serve as an effective means of enhancing regulatory ecosystem services, but in the context of flood prevention it is important to locate new tree planting quite carefully so as to differentially slow flows in tributaries in a way that reduces downstream peaks rather than just delaying them (Dixon et al. 2016).

23.2.1.6 Sustainable Drainage Systems (SuDS)

In urban areas, the majority of the land is covered with artificial impervious surfaces such as concrete and asphalt as houses, factories, car parks and roads have replaced the natural permeable vegetation cover. These impermeable areas reduce rainwater infiltration into the soil and increase the amount of surface runoff generated, significantly increasing the risk of *flash flooding* during storm events. *Sustainable drainage systems* mitigate this by attempting to replicate, as closely as possible, the natural drainage from a site before it was developed. SuDS are typically designed such that they are able to capture rainfall and/or surface runoff, retain it for a period of time, and increase both water infiltration into the soil and evapotranspiration into the atmosphere (Ellis et al. 2002). The net result of the regulatory services provided by SuDS is a reduction in surface water flood risk. Examples of SuDS include small, landscaped, vegetated areas used to increase infiltration (*rain gardens*); plants grown on the roofs of building to increase evapotranspiration (*green roofs*); detention basins to capture and store surface water (*swales, retention ponds*); and the substitution of impervious materials for permeable surfaces (*porous pavements, gravel car parks*). A welcome side effect of such water-related mitigation measures in urban areas is the additional support for biodiversity and urban recreation.

23.2.2 Pathway Measures

23.2.2.1 Tramline Management

Tramlines (or 'wheelings') are unvegetated tracks made within arable crops for farm machinery to travel along during fertiliser and pesticide spraying operations without damaging the surrounding crop. Typically around 30–40 cm wide and spaced 18–24 m apart depending on the width of the farmers' pesticide sprayer boom, tramlines become heavily compacted under the weight of farm machinery, significantly reducing infiltration rates and depressing the soil relative to the surrounding land. With no vegetation cover to intercept rainfall, compacted tramlines can channel erosive surface runoff during precipitation events and act as preferential pathways for the rapid land-to-river transport of nutrient-rich and pesticide contaminated soils (Silgram et al. 2010; Withers et al. 2006). Mitigating this issue is typically focused on disrupting the flow pathway by using tines to loosen tramline soil structure behind machinery wheels and thereby enhance infiltration and reduce incidences of surface runoff. This approach has been shown to reduce sediment and phosphorus concentrations in surface runoff by 72–99% in plot trials (Deasy et al. 2009). Farmers can also fit low pressure tyres to farm vehicles to dissipate the weight and thereby reduce the severity of soil compaction. Furthermore, in areas with steeper slopes, crop management operations can be adjusted to the contour lines, following them instead of ploughing downhill. This measure effectively disrupts flow pathways under conditions of moderately inclined hills and non-extreme rainfall (see Chap. 22).

23.2.2.2 Sediment Traps

Sediment traps, also known as *settling ponds*, *swales* or *constructed wetlands*, are artificial ponds dug to intercept and capture erosive surface runoff before it enters into a water body. Located along a dominant flow pathway, such as the end of field tramlines or next to an impermeable metalled road, fast moving surface runoff is directed into the ponds where it encounters a stationary body of water. The reduction in kinetic energy encourages entrained sediments to settle out of suspension and accumulate on the bottom of the trap. In an *open system*, an outflow then syphons the cleaner water from the top of the pond and discharges it to a neighbouring water course. Conversely, *closed system* traps have no outflow and the captured water is retained and allowed to slowly evaporate and infiltrate down into the soil. The decision on whether to construct an open or closed system, and on the size of the trap required, is dependent upon the volume of surface runoff generated, with larger open systems required to efficiently process high runoff volumes. How effective an open system trap is at capturing and retaining sediments will in large part be determined by the speed at which water passes through the pond, which in turn will partly depend upon the type and amount of vegetation growing within the pond. In general, the higher the plant density, the higher the flow resistance and thus the greater the settling rate. More plants also promotes higher biotic assimilation of nutrients thus reducing eutrophication risk, however too many plants will reduce trap capacity. Retaining 43–88% (69% on average) of sediment inflows, sediment traps and other type of constructed wetland have been shown to be highly effective at removing suspended sediments (Stevens and Quinton 2009), although they can be expensive to construct and maintain (e.g. removing material to prevent over siltation). Where possible, the nutrient-rich sediment should be dug out to maintain trap capacity and used as a source of fertiliser on arable fields, thus supporting crop productivity.

23.2.3 Receptor Measures

23.2.3.1 Riparian Buffer Zones

One of the biggest threats to surface water resources is erosive runoff during heavy rainfall events transporting nutrient-enriched sediment via overland flow paths off agricultural land and directly into streams, rivers and lakes. Riparian buffer zones (RBZs) are strips of permanent vegetation grown alongside river channels to protect the water course from the impacts of agricultural activities on the adjacent land. Vegetated with grasses, scrubby bushes or trees, RBZs provide a rough, high-friction surface which intercepts surface runoff and slows down the flow of the water. As the flows decrease, entrained sediments are encouraged to settle out and are deposited on the RBZ, whilst the water infiltrates down into the soil. RBZs have been shown to be highly effective at mitigating surface runoff pollution, on average reducing sediment loads into water courses by ~75% and with it ~60% of phosphorus and ~78% of pesticides (Stevens and Quinton 2009). As well as supplying the provisioning service of clean water, RBZs also increase biodiversity by providing ribbons of riparian habitat for species that have been forced out of the surrounding agricultural land. Ultimately, however, the success of RBZs at mitigating water pollution is dependent upon the buffer design, with wider and longer buffers covered in denser vegetation having the greatest potential to inhibit overland flow before it reaches the river. The siting of the RBZ is crucial to ensure it intercepts the dominant flow paths, whilst management may be required to prevent sediment build-up within the strip from reducing longer-term retention ability (Dorioz et al. 2006).

23.2.3.2 Livestock Fencing

The outdoor rearing of livestock, particularly at high stocking densities, can have significant implications for water quality when the animals have free access to a water course. As the animals come down to a river to drink their hooves damage the channel banks in a process termed *poaching*, causing the banks to collapse and rapidly erode, releasing sediment into the river and increasing water turbidity. The problem is particularly acute on dairy and beef farms due to the heavy weight of cattle (500–1000 kg) contributing to a high ground pressure that is capable of causing serious structural damage to riparian soils. Furthermore, livestock defecation within the river can contribute to faecal contamination of the water body and the growth of microorganisms toxic to human health, thus threatening drinking water provisioning services. To protect the riparian zone and mitigate against soil erosion, pastured livestock can be relatively inexpensively fenced off (e.g. using barbed wire) from water courses to prevent unrestricted access and instead be provided with an alternative drinking water source within the field.

23.2.3.3 Floodplain Reconnection

A floodplain is a low lying area of land bordering a river channel that is formed by the lateral erosion of a meandering river within the confines of a river valley (Fig. 23.3). During high-flow conditions, a river may overtop its banks and flood out onto the surrounding floodplain, depositing mounds of coarse sands and gravels close to the river channel (*levees*) and fine silt and clay at a greater distance. This periodic breaching of the river channel is part of a natural process which allows the fluvial system to absorb excess water, dissipating the energy of high flows and helping to transport fertile sediments out of the channel and onto the surround land. Inundation of the floodplain helps to reduce downstream flood risk, increase the fertility of the valley floor, provides a diverse habitat for wetland species, cleans the river of excess sediment and nutrients, decreases riparian erosion and contributes cultural, aesthetic and recreational benefits (e.g. wildlife tourism, wildfowling). However, historically, rivers have been extensively deepened and straightened (i.e. *channelization*) through dredging to speed up the flow of water and enable the floodplain to be more efficiently drained for agricultural use. A direct consequence is that the rivers become disconnected from their floodplains with the river water surface several metres below the height of the surround land and thus preventing overbank flows from occurring. A similar situation arises in towns and cities where, to protect buildings built on the floodplain, authorities install unnaturally high artificial levees (typically made of concrete) to reduce the incidences of overbank flow and thereby mitigate local flood risk. Without this floodplain connection, valuable

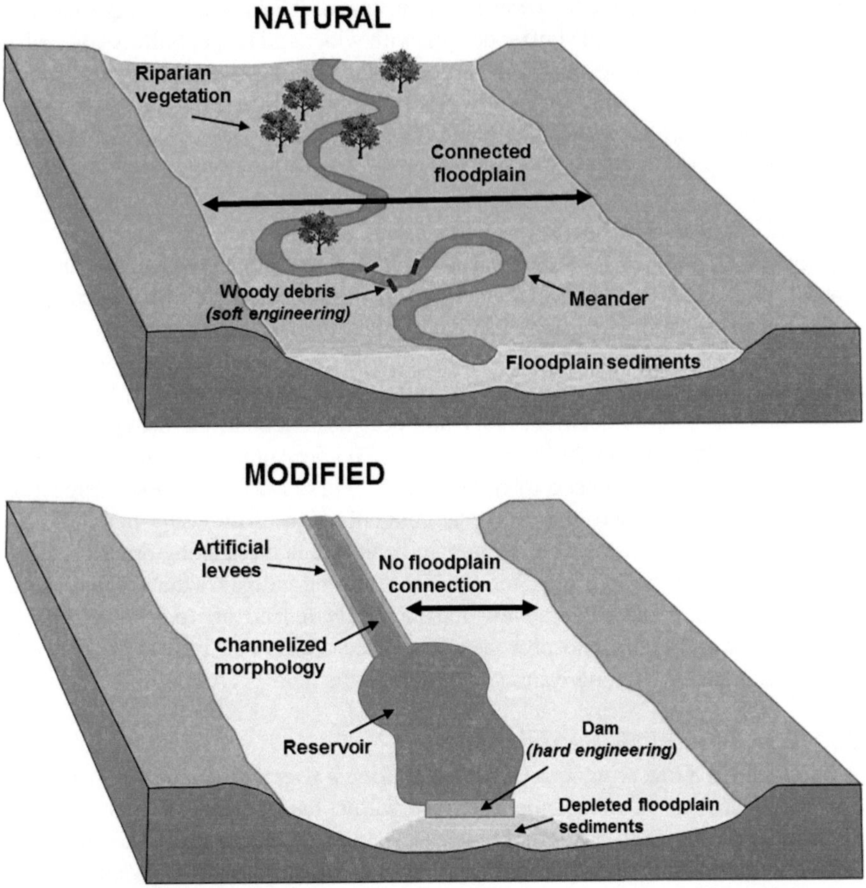

Fig. 23.3 Conceptual diagram of a natural and human-modified river valley. Anthropogenic modifications to a river channel to drain the surrounding land for agriculture or house building can result in the disconnection of the river from its floodplain, even under high-flow conditions

wetland habitat is lost and the main river channel is forced to transport more water during high flows, exacerbating flooding downstream in areas not protected by artificial levees.

Floodplain reconnection aims to restore natural processes by removing artificial flood defences, raising the height of the riverbed and breaching gaps in the river banks to facilitate overbank flow and floodplain retention. Floodplain reconnection is just one example of numerous *soft engineering* mitigation options available termed *natural flood management*, which sees natural processes favoured over *hard engineering* solutions to mitigating flood risk (Fig. 23.3). Another example includes the use of *woody debris* (e.g. felled trees, branches, log piles) strategically placed perpendicular to the direction of flow within straight, homogeneous river sections which acts as a baffle, deflecting the river sideways, increasing flow diversity and encouraging the river to *meander*. The increased *sinuosity* of a meandering river increases its length and reduces its gradient, which in turn slows down the flow of the river and delays peak flows during flood events, thus helping to alleviate flood risk (ECRR 2017). Berms can be used to create narrower sections with faster flowing water and gravel glides can be installed to create a pool-and-riffle type channel morphology. Examples of river restoration features on the River Wensum in eastern England are shown in Fig. 23.4. The left-hand and right-hand columns of photographs are, respectively, prior to (June 2012) and after (October 2012) the implementation of the scheme. From top to bottom in the right-hand column, the river restoration features include: a filled berm to narrow the river width in order to increase flow velocity and the cleaning of river bed sediment; the positioning of woody debris and a gravel glide to decrease the water depth and deflect the river flow in order to increase flow velocity and create a pool-and-riffle type channel morphology; a channel plug to remove a previously straightened section; and a reinstated meander loop following diversion of the river due to the channel plug. Further design information is contained in Natural England (2009, 2012).

23.3 Methods to Incentivise the Adoption of Mitigation Measures

In many European countries the policies and programmes used to increase the adoption of mitigation measures are heavily influenced by EU obligations stemming from a number of EU Directives. These include the Floods (2007/60/EC), Nitrates (91/676/EEC) and Water Framework Directives (2000/60/EC, see Section 11.6). Given the important relationship between agriculture and water resources another key factor is the implementation of the EU Common Agricultural Policy (CAP). All of this means that there is greater commonality across countries in the management of water resources than exists for some other types of natural capital.

Although there are similarities arising from EU-wide policies, there are also differences between countries in the manner that EU Directives and CAP requirements are implemented. In most cases a mixture of approaches has been adopted, commonly with a pyramid of mechanisms (see Fig. 23.5), starting with nationally-applied baseline regulations and codes of good practice, then more regional or local variation in the use of advice schemes or financial incentives. Further legally-enforced restrictions may exist in local water resource protection areas (e.g. around public water supply abstraction points or boreholes).

23.3.1 Examples of Baseline Regulations

CAP Cross-Compliance Financial support to farmers under Pillar 1 (direct payments based on area farmed) of the CAP is linked to *cross-compliance* obligations regarding environmental, animal welfare and food safety standards. If farmers are found not to be meeting these standards during inspections then they can be penal-

Fig. 23.4 Examples of river restoration features at Swanton Morley on the River Wensum, eastern England

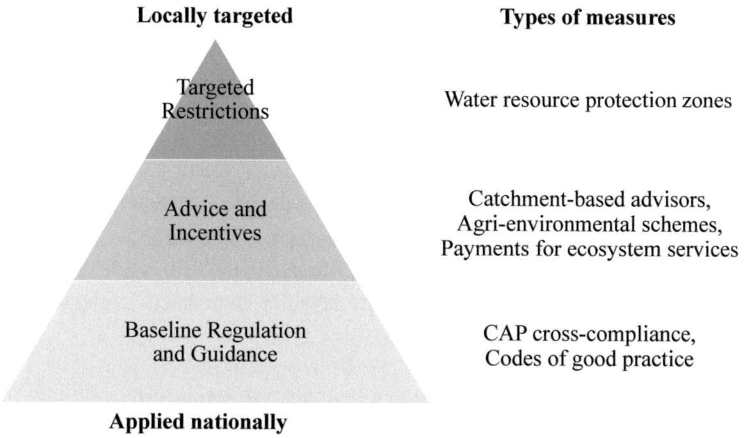

Fig. 23.5 Policy delivery mechanisms for measures to mitigate impacts on catchment water resources. (Source: modified from McGonigle et al. (2012, Fig. 3)

ised a proportion of their Pillar 1 payment. Under the 2013 CAP reforms further *greening* requirements were introduced and linked to 30% of the direct payments (European Commission 2017). The motive for this change was to strengthen the environmental sustainability of agriculture through requirements for:

- diversifying crops (to make soil and ecosystems more resilient)
- maintaining permanent grassland (to conserve soil carbon and grassland habitats)
- dedicating 5% of arable land to 'ecologically beneficial elements' ('ecological focus areas') in order to protect water and habitats.

Nitrate Vulnerable Zones (NVZs) The Nitrates Directive (91/676/EEC) aims to protect water quality across Europe by preventing nitrates from agricultural sources polluting ground and surface waters and by promoting the use of good farming practices. Implementation includes the designation as Nitrate Vulnerable Zones (NVZs) of areas of land which contribute to nitrate pollution and establishment of action programmes of measures which must be implemented by farmers within such zones. In some countries (e.g. Denmark) the entire territory has been designated as an NVZ, whilst in other cases specific zones have been defined (e.g. 58% of England as of September 2017). In England farmers in NVZs are required to meet several obligations including limiting the amount of farmyard manure and inorganic fertiliser applied to fields; keeping records of all nitrate applications within the past 5 years; having closed periods (3–5 months) when fertiliser application is prohibited; not applying organic manure within 10 m of a surface water body or 50 m of a groundwater source (i.e. spring, well or borehole); and providing at least 6 months storage capacity for poultry manures and pig slurry.

Floods Directive (2007/60/EC) The 2007 EU Floods Directive requires member states to assess if all water courses and coastlines are at risk from flooding, to map the flood extent, assets and humans at risk in these areas, and to take adequate and coordinated measures to reduce the flood risk. The directive encourages a coordinated and integrated approach to implementing flood risk measures throughout the entire river catchment to increase their effectiveness, meaning that suites of measures addressing flood risk in upland (e.g. reforestation) through to lowland (e.g. floodplain reconnection) environments are preferred. The directive is implemented in coordination with the Water Framework Directive (WFD), with flood risk management plans being incorporated into the broader river basin management plans (see Section 11.6).

In some European countries, such as Germany, the minimum standards for agricultural practice are defined in environmental laws. Often these make the European Directives more specific at the national level. According to the *polluter pays principle* (PPP) farmers cannot be paid for observing these standards, which may include maximum rates of fertilizer input or limits to pesticide use. Remuneration for water services on farmland will therefore be restricted to – often voluntary – activities beyond the legally-prescribed good practice.

23.3.2 Advice and Voluntary Measures

In addition to complying with the legal standards, an important tool for mitigating threats to water resources, is the establishment of professional *Codes of Good Agricultural Practice* which can be implemented by landowners on a voluntary basis. Such codes aim to provide practical guidance to help farmers and growers to minimise the risk of causing pollution whilst still allowing economic growth within the agricultural sector. Codes typically include advice such as the optimum application rates for fertilisers and pesticides to minimise the risk of unnecessarily applied excess chemicals entering into water courses; guidance on when agrochemicals should and should not be applied in relation to weather conditions to restrict mobility in the environment; and advice on the timings of in-field cultivations to minimise damage to soil structure and reduce the risk of soil erosion. Support given to farmers can also be delivered through government-funded training events (e.g. workshops, demonstrations, farm visits) and access to farm advisers.

Advice schemes exist in many countries and can be funded by central government, local government or industry (e.g. water supply or agri-chemical businesses). In England the *Catchment Sensitive Farming (CSF)* initiative was established by central government in 2006 to raise awareness of diffuse water pollution from agriculture and improve the environmental performance of farms by providing free training and advice to farmers in high priority areas for water quality where WFD targets are not being achieved. The *Voluntary Initiative* (http://www.voluntaryinitiative.org.uk/) is a UK industry-led programme to promote the responsible use of pesticides in order to protect water and the wider environment. There is also an

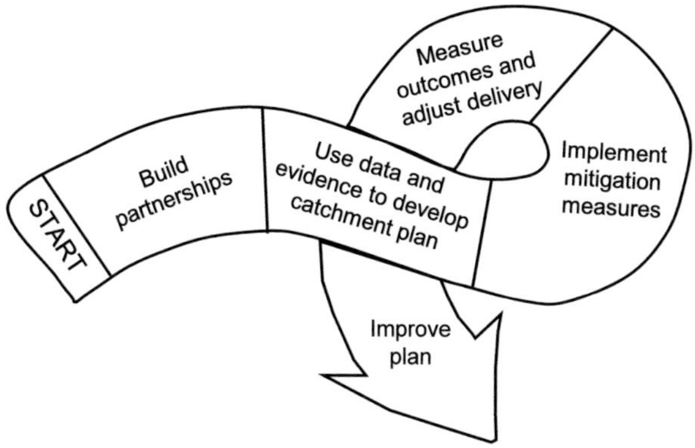

Fig. 23.6 The adaptive management cycle provides a framework for sustainable catchment management and the implementation of mitigation measures to protect water resources at the catchment-scale (US EPA 2008)

extensive literature on factors influencing farmer uptake of advice, see Inman et al. (2017) for a recent review.

Another important change in recent years has been for national governments seeking to implement WFD river basin management plans to shift focus away from national-scale thinking onto a more targeted local-scale '*Catchment Based Approach*' (CaBA 2017) in order to improve the effectiveness of delivery. This catchment-based approach is community led, engaging public, private and charitable organisations from across society to improve water resources, both quality and quantity, through the development of a holistic catchment-specific management strategy. Seeking to integrate economic, environmental and social issues into water resource planning, CaBA adopts the collaborative principles of the adaptive management cycle as a means of incorporating an appropriate combination of regulation, advice, land use measures, incentives and voluntary action to protect water resources (Fig. 23.6).

23.3.3 Financial Incentives

Agri-environment schemes funded under Pillar 2 (rural development) of the CAP provide financial incentives for land managers to look after the environment through activities such as conserving and restoring wildlife habitats, implementing flood risk management, reducing widespread water pollution from agriculture, maintaining the character of the countryside, preserving features important to the history of the rural landscape and encouraging educational access. An example is the *Countryside Stewardship Scheme*, the main CAP-funded agri-environment programme in England. Administered by central government (*Natural England*), Countryside Stewardship is a targeted, competitive scheme with a particular emphasis on biodiversity, water quality and flood management for which land managers must submit funding applications. With a budget of £925 million (€1 billion) for 2015–2020, the scheme is split into main three elements:

- **Higher Tier (£380 million):** covers management of the most environmentally significant sites such as ancient woodland, wetlands, wildflower meadows and Sites of Special Scientific Interest;
- **Mid Tier (£412 million):** simple but effective environmental measures carried out on ordinary agricultural land;
- **Capital Grants (£85 million):** larger sums of money available for capital projects such as the installation of biobeds, building settling ponds, improving manure storage facilities or creating new woodlands.

In total, there are 238 agri-environmental options eligible for funding under Countryside Stewardship, with the amount of money available to land managers dependent upon the extent, nature and effectiveness of the scheme. In 2014, 62% of UK agricultural land (10.6 million ha) was registered under some form of agri-environmental scheme. However, it is not just the EU or national governments that fund measures to protect water resources. *Water companies* are increasingly becoming involved in financially supporting pollution and flood risk mitigation measures as a way of protecting water supplies for consumers as part of their asset management programmes and *paying for ecosystem services (PES)*. One example is the Upstream Thinking initiative (http://www.upstreamthinking.org/) run by South West Water in the UK.

23.4 Modelling the Effects of Mitigation Measures

For land-use planners developing on-farm mitigation strategies to reduce water pollution and flood risk, it is useful to consider eight important factors which will ultimately determine the degree of success of measures deployment (Newell Price et al. 2011). These are:

(i) the nature of the problem being targeted (e.g. nutrient enrichment, pesticide contamination);
(ii) the land-use typologies to which the measures are applicable (e.g. intensive arable, lowland dairy);
(iii) the mechanism of mitigation action (i.e. how does the measure reduce pollution/flood risk);
(iv) the potential for applying the measure (i.e. spatial assessment of the area to which the measure could be applied);
(v) the practicality of deployment (e.g. ease of adoption, impact on farm business, resistance from landowners);
(vi) the likely uptake rate (e.g. percentage of farms on which the measure could be adopted given existing economic and legislative drivers);
(vii) the costs of measure deployment (e.g. € per km^2 or € per unit);
(viii) the likely effectiveness of the measure (e.g. percentage reduction in nitrate concentrations based on published research or expert knowledge).

The economic evaluation of mitigation options (stage vii), is a key determinant of whether measures to protect water resources will be pursued. Such evaluation either takes the form of a *cost-effectiveness analysis (CEA)*, where a specified water quality objective is given and the aim is to identify the cheapest set of measures for achieving it; or via *cost–benefit analysis (CBA)*, where the overall costs and benefits of a set of measures are assessed to determine if it should be carried out. In the context of the practical implementation of the WFD, applications of CEA are much more common than CBA. To assist in the assessment process, land-use planners can take advantage of decision-support tools, such as FARMSCOPER (FARM Scale Optimisation of Pollutant Emissions Reduction) or SWAT (Soil & Water Assessment Tool; see Box 11.3), which can estimate baseline pollutant losses and then quantify the effectiveness of combinations of mitigation measures at reducing pollutant losses at the farm- or catchment-scale (Gooday et al. 2014) (Box 23.2).

Box 23.2: Research Programmes for Mitigation Schemes

DTCs (**D**emonstration **T**est **C**atchments): UK government funded initiative to assess the extent to which on-farm mitigation measures can cost-effectively reduce the impact of agricultural pollution on river ecology whilst maintaining food production capacity (http://www.demonstratingcatchmentmanagement.net).

ECRR (**E**uropean **C**entre for **R**iver **R**estoration): pan-European network of national centres, organisations, institutions and individuals linked together to support the development of best management practices for restoring Europe's rivers (http://www.ecrr.org).

NWRM (**N**atural **W**ater **R**etention **M**easures): expert network established to develop a structured knowledge base on the application of natural water retention measures which can be disseminated through the development of web-based practical manuals for supporting the design and implementation of new NWRM schemes (http://www.nwrm.eu/).

REFORM (**RE**storing rivers **FOR** effective catchment **M**anagement): EU-wide project aimed at providing a framework for improving the success of hydromorphological restoration measures to achieve improved ecological status of rivers in a cost-effective manner (http://www.reformrivers.eu).

RESTORE (**R**ivers **E**ngaging, **S**upporting and **T**ransferring kn**O**wledge for **R**estoration in **E**urope): EU-funded project led by the Environment Agency (England) to encourage the restoration of European rivers towards a more natural state that delivers increased ecological quality, flood risk reduction and social and economic benefits (https://www.restorerivers.eu).

RRC (**R**iver **R**estoration **C**entre): a UK-based organisation promoting best-practice river restoration, habitat enhancement and catchment management through knowledge exchange, technical advice and assessment, and training and guidance (http://www.therrc.co.uk/rrc).

References

CaBA. (2017). *Catchment based approach*. https://www.catchmentbasedapproach.org/. Accessed 14 June 2018.

Castillo, M. D. P., Torstensson, L., & Stenstrom, J. (2008). Biobeds for environmental protection from pesticide use – A review. *Journal of Agricultural and Food Chemistry, 56*, 6206–6219.

Cooper, R. J., Fitt, P., Hiscock, K. M., et al. (2016). Assessing the effectiveness of a three-stage on-farm biobed in treating pesticide contaminated wastewater. *Journal of Environmental Management, 181*, 874–882.

Cooper, R. J., Hama-Aziz, Z., Hiscock, K. M., et al. (2017). Assessing the farm-scale impacts of cover crops and non-inversion tillage regimes on nutrient losses from an arable catchment. *Agriculture, Ecosystems and Environment, 237*, 181–193.

Dabney, S. M., Delgado, J. A., & Reeves, D. W. (2007). Using winter cover crops to improve soil and water quality. *Communications in Soil Science and Plant Analysis, 32*, 1221–1250.

de Wilde, T., Spanoghe, P., Debaer, C., et al. (2007). Overview of on-farm bioremediation systems to reduce the occurrence of point source contamination. *Pest Management Science, 63*, 111–128.

Deasy, C., Quinton, J. N., Silgram, M., et al. (2009). Mitigation options for sediment and phosphorus loss from winter-sown arable crops. *Journal of Environmental Quality, 38*, 2121–2130.

Deasy, C., Quinton, J. N., Silgram, M., et al. (2010). Contributing understanding of mitigation options for phosphorus and sediment to a review of the efficacy of contemporary agricultural stewardship measures. *Agricultural Systems, 103*, 105–109.

Dixon, S. J., Sear, D. A., Odoni, N. A., et al. (2016). The effects of river restoration on catchment scale flood risk and flood hydrology. *Earth Surface Processes and Landforms, 41*, 997–1008.

Dorioz, J. M., Wang, D., Poulenard, J., et al. (2006). The effect of grass buffer strips on phosphorus dynamics – A critical review and synthesis as a basis for application in agricultural landscapes in France. *Agriculture, Ecosystems and Environment, 117*, 4–21.

ECRR. (2017). *European centre for river restoration*. http://www.ecrr.org/. Accessed 14 June 2018.

EEA. (2015). *Water-retention potential of Europe's forests: A European overview to support natural water-retention measures* (Technical report No. 13/2015). Copenhagen: European Environment Agency. https://doi.org/10.2800/790618.

Ellis, J. B., D'Arcy, B. J., & Chatfield, P. R. (2002). Sustainable urban-drainage systems and catchment planning. *Water Environment Journal, 16*, 286–291.

European Commission. (2017). *Greening*. https://ec.europa.eu/agriculture/direct-support/greening_en. Accessed 14 June 2018.

Gooday, R. D., Anthony, S. G., Chadwick, D. R., et al. (2014). Modelling the cost-effectiveness of mitigation methods for multiple pollutants at farm scale. *Science of the Total Environment, 468*, 1198–1209.

GWP/INBO. (2015). *The handbook for management and restoration of aquatic ecosystems in river and lake basins*. International Network of Basin Organisations, Paris. http://www.inbo-news.org/. Accessed 14 June 2018.

Holland, J. M. (2004). The environmental consequences of adopting conservation tillage in Europe: Reviewing the evidence. *Agriculture, Ecosystems and Environment, 103*, 1–25.

Hooker, K. V., Coxon, C. E., Hackett, R., et al. (2008). Evaluation of cover crop and reduced cultivation for reducing nitrate leaching in Ireland. *Journal of Environmental Quality, 37*, 138–145.

Inman, I., Vrain, E., Jones, I., et al. (2017). An exploration of individual, social and material factors influencing water pollution mitigation behaviours within the farming community. *Land Use Policy*. in press.

Kania, J., Vinohradnik, K., & Knierim, A. (2014). *Prospects for farmers' support: Advisory services in European AKIS (PRO AKIS) – Synthesis report*. http://www.proakis.eu/synthesis-report. Accessed 14 June 2018.

Kertész, A., & Madarász, B. (2014). Conservation agriculture in Europe. *International Soil and Water Conservation Research, 2*, 91–96.

McGonigle, D. F., Harris, R. C., McCamphill, C., et al. (2012). Towards a more strategic approach to research to support catchment-based policy approaches to mitigate agricultural water pollution: A UK case-study. *Environmental Science & Policy, 24*, 4–14.

Morris, N. L., Miller, P. C. H., Orson, J. H., et al. (2010). The adoption of non-inversion tillage systems in the United Kingdom and the agronomic impact on soil, crops and the environment – A review. *Soil and Tillage Research, 108*, 1–15.

Natural England. (2009). *River wensum restoration strategy*. Natural England Commissioned Report NECR010. http://www.naturalengland.org.uk. Accessed 14 June 2018.

Natural England. (2012). River Wensum restoration strategy: Swanton Morley restoration scheme – Reach 14a.

Newell Price, J. P., Harris, D., Taylor, M., et al. (2011). *An inventory of mitigation methods and guide to their effects on diffuse water pollution, greenhouse gas emissions and ammonia emissions from agriculture* (Final report for Project WQ0106). London: Department for Environment/Food and Rural Affairs.

NWRM. (2017). *Natural water retention measures*. http://nwrm.eu/. Accessed 14 June 2018.

Silgram, M., Jackson, D. R., Bailey, A., et al. (2010). Hillslope scale surface runoff, sediment and nutrient losses associated with tramline wheelings. *Earth Surface Processes and Landforms, 35*, 699–706.

Snapp, S. S., Swinton, S. M., Labarta, R., et al. (2005). Evaluating cover crops for benefits, costs and performance within cropping system niches. *Agronomy Journal, 97*, 322–332.

Soane, B. D., Ball, B. C., Arvidsson, J., et al. (2012). No-till in northern, western and south-western Europe: A review of problems and opportunities for crop production and the environment. *Soil and Tillage Research, 118*, 66–87.

Stevens, C. J., & Quinton, J. N. (2009). Diffuse pollution swapping in arable agricultural systems. *Critical Reviews in Environmental Science and Technology, 39*, 478–520.

Thompson, J. J. D., Doody, D. G., Flynn, R., et al. (2012). Dynamics of critical source areas: Does connectivity explain chemistry? *Science of the Total Environment, 43*, 499–508.

Torstensson, L. (2000). Experiences of biobeds in practical use in Sweden. *Pesticide Outlook, 11*, 206–211.

US EPA. (2008). *Handbook for developing watershed plans to restore and protect our waters*. Washington, DC: United States Environmental Protection Agency. EPA 841-B-08-002.

Valkama, E., Lemola, R., Känkänen, H., et al. (2015). Meta-analysis of the effects of undersown catch crops on nitrogen leaching loss and grain yields in the Nordic countries. *Agriculture, Ecosystems and Environment, 203*, 93–101.

Withers, P. J., & Jarvie, H. P. (2008). Delivery and cycling of phosphorus in rivers: A review. *Science of the Total Environment, 400*, 379–395.

Withers, P. J. A., Hodgkinson, R. A., Bates, A., et al. (2006). Some effects of tramlines on surface runoff, sediment and phosphorus mobilization on an erosion-prone soil. *Soil Use and Management, 22*, 245–255.

Measures for Landscape Aesthetics and Recreational Quality

24

Christian Albert, Thiemen Boll, Peter Haus, Johannes Hermes, and Christina von Haaren

Abstract

This chapter introduces approaches for developing targets and measures to enhance the visual quality and the recreation potential of landscapes. It begins by introducing recommendations for the enhancement of visual quality as the context within which all recreational activities take place. Subsequently, options for enhancing recreation potential are explored.

Keywords

Cultural ecosystem services · Aesthetics · Visual quality · Recreation potential · Conflict minimisation

24.1 Introduction

Relevant stakeholders to be considered in the process of developing targets and measures for cultural ecosystem services (CES) include decision-makers and stakeholders which influence, or are affected by, changes in the provision of such

C. Albert (✉) · J. Hermes · C. von Haaren
Institute of Environmental Planning, Leibniz University Hannover, Hannover, Germany
e-mail: albert@umwelt.uni-hannover.de; hermes@umwelt.uni-hannover.de; haaren@umwelt.uni-hannover.de

T. Boll
Dezernat für Natur- und Landschaftsschutz, Fischerei, Bezirksregierung Münster, Münster, Germany
e-mail: thiemen.boll@bezreg-muenster.nrw.de

P. Haus
Grün Stadt Zürich, Zürich, Switzerland
e-mail: peter.haus@zuerich.ch

C. von Haaren et al. (eds.), *Landscape Planning with Ecosystem Services*, Landscape Series 24, https://doi.org/10.1007/978-94-024-1681-7_24

services. People influencing change in the delivery of CES include policy makers, planners and land managers. They cause changes by influencing land cover and land uses in various ways, from measures at the international scale (e.g. trade agreements) to local alterations in farming practices. Local residents will be one group particularly affected by changes in CES, as well as anyone who appreciates aesthetically appealing landscapes and enjoys them in recreational activities. In addition, the visual landscape provides the background to a diverse range of nature-based recreation and tourism activities. As such, the beneficiaries of beautiful landscapes are not only the citizens who enjoy them, but also tourism and recreation companies along with the hospitality sector. Optimising landscapes for better provision of CES may occur at the expense of, or indeed enhance, other ecosystem services (see Chap. 26 on multifunctionality). In implementing the general objective of safeguarding and developing aesthetic landscape qualities for recreation purposes, landscape planning must therefore also take into account potential conflicts, synergies and trade-offs and address them in a site-specific and locally adapted manner.

The appropriate spatial level for the development of targets and measures for CES is usually the regional and local level. This is due to their design being highly dependent on the specific characteristics of the landscapes, the regional or local demand for recreation and other cultural ecosystem services, and existing recreational infrastructure.

In general, the authority of landscape planning is limited to implementing measures addressing the preservation or the improvement of landscape aesthetics and recreational qualities. Sometimes proposed targets and measures can be implemented by including them in the plans of spatial planning authorities or other agencies. Private sector stakeholders such as local businesses, agricultural interests, private associations or landscape users themselves could similarly include measures into their management plans. A list of stakeholders and agencies is given in Table 24.1.

Table 24.1 Stakeholders and agencies with an interest in the implementation of measures for landscape aesthetics and recreational quality

Regional level	Local level
Regional sectoral organisations (e.g. for water, energy or agricultural administration)	Local authorities
Regional planning agencies	Local spatial planning authorities
Agricultural federations	Farmers
Tourist associations	Catering and hospitality establishments
Regional societies (e.g. hiking, biking, fishing, culture, history)	Local clubs (e.g. biking, fishing, boating, historical)
Regional business associations	Local businesses and cooperatives
Regional infrastructure businesses (e.g. railways, boat services)	Landscape users

The implementation of measures to safeguard and enhance landscape aesthetics and recreational quality can take various forms. The following list contains commonly employed approaches, either as single measures or in combination:

- Integration into spatial or sectoral planning designs
- Integration into protected area designations
- Integration into other forms of restrictions at the local or regional level approved by competent local or regional authorities
- Implementation through habitat banking or biodiversity offset measures
- Implementation through projects using subsidies for regional businesses, tourism, rural development or environmental protection
- Integration into agricultural land management practices, i.e. through agri-environmental measures as part of Common Agricultural Policy (CAP) implementation or in cooperation with local or regional agricultural associations,
- Implementation through action agreements on a voluntary basis
- Integration through cooperative activities with businesses, i.e. marketing of nature conservation products (Albert et al. 2009), regional development cooperation or sponsorship (e.g. financing of outdoor barbeques or benches).

24.2 Developing Targets

Targets endeavor to either safeguard the landscape's aesthetic quality or its usability for recreational purposes. As is the case for other types of measures, the aesthetic and recreational qualities of landscapes should be developed in accordance with overarching objectives from planning documents at higher levels of decision making. However, these overarching targets have to be adapted to take into account local circumstances. In regard to the current condition and demand for recreation, different kinds of targets should be addressed as described in Table 24.2.

Table 24.2 Priority actions with different levels of recreation opportunities and demand

	Low demand for recreation	High demand for recreation
Low level of recreation opportunities	Focusing on other environmental goods	Protecting areas of particularly high visual quality
	Priority to other land-uses	Safeguarding or increasing visual quality
		Enhancing recreation opportunities
		Minimising potential conflicts
High level of recreation opportunities	Focusing on other environmental goods	Protecting areas of particularly high visual quality
	If necessary, decommissioning of unused infrastructure	Safeguarding or increasing visual quality
		Minimising potential conflicts
		Diverting intense recreation demand from sensitive areas

The derivation of objectives in this table draws upon the distinction in landscape assessment between the delivered aesthetic quality and the demand for or actual use of this service. The following overview of measures is based upon and expanded from Wöbse and Ott (2004).

24.3 Measures for Safeguarding and Enhancing Visual Quality

Proposing generally applicable measures for enhancing the visual quality of landscapes is difficult. The visual character of landscapes is very place-specific and dependent upon the quality, location and combination of various landscape components. Given this situation, the chapter makes some general suggestions as inspiration. The measures presented in Table 24.3 are mainly derived from selected landscape plans (Stadt Königslutter 2005, Stadt Freiburg 2006, Kreis Lippe 2005, Region Hannover 2013).

Table 24.3 Possible measures to enhance landscape aesthetics

Examples of measures	Effects of measures
Designation of protected areas or other orders and prohibitions	Safeguarding aesthetic landscape quality
Development of fruit trees, woodland, tree lines, shrubs, groves and hedges for recreation. Indigenous planting material should be used	Increase of structural diversity
	Contribution to spatial orientation. Habitats created to enhance visual and acoustic diversity. Concealment of technical structures such as roads to increase perceived naturalness
Restoration of rivers, brooks and banks of waters, wetlands, moorlands, dry grasslands and heathlands	Create habitats to enhance visual and acoustic diversity as well as landscape naturalness
Support of extensive agriculture, increase in crop diversity	Habitat creation: enhanced visual and acoustic diversity
	Enhancement of landscape diversity
Support of diversified forestry management	Increasing diversity as well as natural appearance of forests
Creation of a natural forest edges	Preserving forestry elements contributing to uniqueness of landscape
Preservation of special (old) types of forest management (e.g. coppicing)	
Increasing share of grassland or of coppice and forest	Enhancement of structural landscape diversity
Preservation and development of typical and structured settlement boundary as transition zone between developed area and open landscape	Hiding technical structures
	Uniqueness
Using characteristic building materials	Conservation of typical landscape character
Limiting deterioration of existing scenery	Increasing natural appearance of landscape
Conservation of historical sites, objects and constructions	Preserving landscape uniqueness and historical identity
Improvement of visual relations	Contribution to spatial orientation

The selection of appropriate measures should be based on a spatially explicit assessment and evaluation of the landscape character, for example through one of the methods presented in Chap. 15. Furthermore, the selection of specific measures should take into account the history of the landscape. Where possible, potential synergies with the provision of other ecosystem services should be exploited (Chap. 26) and trade-offs should be avoided.

Sources: based upon examples from Stadt Königslutter (2005), Stadt Freiburg. (2006), Kreis Lippe (2005), Region Hannover (2013)

Areas identified as having a relatively high visual landscape quality should be preserved by avoiding changes that could potentially impair their current value. Such potentially damaging changes include, amongst others, the development of infrastructure such as pylons for electricity transmission. In areas of high aesthetic quality, additional measures could relate to the careful continuation or extensification of farming, the enhancement of recreational access, and planting of additional trees and hedges.

Landscape areas evaluated as having medium visual quality should exploit existing opportunities for enhancement. These could include, for example, the restoration of rivers and the development of new trees or woodland patches. Further impairments of visual quality, for example through the development of new roads, should be avoided as much as possible. Landscape areas with low or very low visual quality urgently require measures for improvement – especially if they are located near settlements or used for recreational purposes. Historical remains such as ancient roads, watercourses and field borders should be considered in the design of improvement measures. Existing visual connections or perspectives of particular importance should also be preserved.

24.4 Measures for Enhancing Recreational Quality

Landscapes may serve as a platform for a large number of different nature-based recreational activities. In many cases, an aesthetically attractive landscape enhances the value of such recreation potential. In light of this, most of the measures contributing to the improvement of aesthetic landscape quality (as described in the previous section), also enhance the recreational qualities of landscapes.

Landscape planning can further enhance recreation potential by introducing or designating specific sites and routes where activities can be undertaken. For example, routes for activities such as walking, cycling, inline skating or horse riding. While the specific requirements concerning the route surface, elevation and length differ, the route system as a whole will benefit from the implementation of some general measures:

- Establishing and maintaining a continuous and eventually circular network, possibly with different route alternatives that offer variations in length and difficulty.
- Careful maintenance and furnishing of the routes with benches, waste containers and other elements to raise the quality of the visit, especially in areas near to settlements (Stadt Königslutter 2005).

- Measures to improve accessibility for people with disabilities.
- Installing appropriate signposts and information boards (Stadt Freiburg 2006) to improve orientation and visitor knowledge about the local area.
- Incorporating highlights such as historical buildings, viewpoints, lakes or gastronomic facilities within the network or initiating their establishment.
- Designing routes in such a way that they pass through a large variety of aesthetic experiences such as open landscapes, complex land use patterns, and forests. Sections of different curvature further enhance the quality of the experience. Additionally, edge strips should be established with diverse vegetation types.
- Facilitating accessibility with private and public transport e.g. by locating parking places and bus stops at access points to the network and equipping them with information boards.

While some activities (e.g. walking) can be performed in any landscape, others such as fishing and swimming are geographically bound to the occurrence of open waters. Landscape planning should allow for bathing in parts of suitable water bodies (cf. LP Stadt Freiburg 2006: 217, 246). Therefore, landscape planning in cooperation with other authorities (e.g. spatial planning authority) should allow for access by providing paths to the water body edges as well as limiting the establishment of private property next to water bodies. In addition, the installation of changing rooms and sunbathing lawns can improve bathing experiences.

Some activities like alpine skiing, sailing or golf depend on more elaborate infrastructure such as prepared ski slopes, cable cars, marinas or golf courses as well as gastronomic facilities. Landscape planning can suggest whether such infrastructure should be encouraged at a particular locality or designate areas where they should be banned due to their high conflict potential.

As described in Chap. 15, landscapes are not only of importance for recreation but also as a source for education and spiritual activities. Appropriate signage of such places can help convey their particular importance (Stadt Freiburg 2006: 219). Educational trails and open air museums (cf. Stadt Königslutter 2005: 105 f.) can increase knowledge and awareness of ecology or landscape history.

Sufficient open green spaces should be provided in relation to settlement areas (see Region Hannover 2013: 677). Within settlement areas, the planting of trees and shrubs on open space as well as green roofs or walls can extend nature experience. Expanding existing urban greens and the restauration of water bodies may have additional positive effects (cf. Stadt Königslutter 2005: 103, Bowen and Lynch 2017).

24.5 Approaches for Minimising Conflicts

The use of landscapes for recreational purpose can lead to various conflicts. These can be caused by the execution of the activity, by the necessary infrastructure, or by the travel involved to get to the place of the activity. Sometimes a simple ban on certain activities is inevitable but often other solutions can allow both the protection of environmental goods and the enabling of leisure activities.

Visitor guidelines have the potential to enable landscapes in sensitive areas to be experienced while keeping visitors away from sites of extreme sensitivity. Such guidance can include elements such as the creation of attractive paths, observatories, or the provision of information about the location, its characteristics and its environmental value.

Particular sites, routes or areas can be specified for recreational activities such as camp fires, bathing, climbing or downhill biking. Necessary infrastructure can be restricted to defined sites so that these activities can occur while protecting sensitive sites elsewhere. The designation of areas and the installation of infrastructure should be carried out in cooperation with the prospective users (e.g. mountain biking clubs). This will increase the acceptance of implemented measures while also raising awareness of potential conflicts. Some sites are only sensitive to recreational activities during particular periods of the year so that temporary prohibition of their use may be sufficient.

Furthermore, voluntary agreements can be established between sports clubs and nature conservation authorities to avoid conflicts between recreational activities and nature conservation. On the more regional and state levels, policies for integrating outdoor recreation and nature conservation objectives can be developed and guidelines for identifying and mitigating local conflicts can be agreed upon (Wolf and Appel-Kummer 2009: 258 f.).

Conflicts may also occur between different recreational activities, e.g. between hiking and mountain biking, or between fishing, bathing or boating. Such conflicts can be avoided by anticipatory planning in cooperation with the clubs and associations that may be involved.

References

Albert, C., Aurbacher, J., von Haaren, C., et al. (2009). Ökonomische Auswirkungen zukünftiger Agrarentwicklungen auf die Landschaftspflege und mögliche Beiträge der Aufpreisvermarktung von Naturschutzprodukten im Landkreis Diepholz. *Berichte über Landwirtschaft, 87*(3), 357–379.

Bowen, K. J., & Lynch, Y. (2017). The public health benefits of green infrastructure: the potential of economic framing for enhanced decision-making. *Current Opinion in Environmental Sustainability, 25*, 90–95.

Kreis Lippe. (2005). *Landschaftsplan Nr.* 7 "Lemgo". Kreis Lippe, Lippe

Region Hannover. (2013). *Landschaftsrahmenplan der Region Hannover*. Region Hannover

Stadt Freiburg. (2006). *Landschaftsplan 2020*. Stadt Freiburg i.Br., Freiburg

Stadt Königslutter. (2005). *Landschaftsplan der Stadt Königslutter am Elm*. http://koenigslutter.entera-online.com/IL.php?DOC_ID=190. Accessed 9 Feb 2018

Wöbse, H. H., Ott, S. (2004). Ziele und Maßnahmen zu Landschaftserleben und Erholungsfunktion. In: von Haaren C (ed) *Lanschaftsplanung* (pp. 358–365). Ulmer.

Wolf, A., & Appel-Kummer, E. (Eds.). (2009). *Naherholung in Stadt und Land*. Norderstedt: Books on Demand GmbH.

Measures for Biodiversity

25

Christoffer Lange-Kabitz, Birte Bredemeier, Christina von Haaren, and Jan Barkmann

Abstract

Response measures for safeguarding and increasing biodiversity can be derived from biodiversity assessments. They may aim to maintain and protect the valuable well-functioning areas, to mitigate or eliminate damaging influences or develop or restore the sites with the best potential for success. The choice of these measures depends on the habitat development potential of the site, the value and endangerment of the existing biodiversity present and the habitat connectivity. This chapter introduces the basic principles of these different approaches and presents examples of measures to maintain or develop the biodiversity state of field and grassland habitats in agricultural landscapes. In addition, we outline the case for habitat banking as a specific example of an economic means of implementation. Overall, the chapter gives an initial insight into the variety of response measures and starting points for their derivation.

Keywords

Biodiversity assessment · Habitat development potential · Agricultural landscapes · Habitat banking

C. Lange-Kabitz (✉) · B. Bredemeier · C. von Haaren
Institute of Environmental Planning, Leibniz University Hannover, Hannover, Germany
e-mail: lange-kabitz@umwelt.uni-hannover.de; bredemeier@umwelt.uni-hannover.de; haaren@umwelt.uni-hannover.de

J. Barkmann
Fachbereich Gesellschaftswissenschaften, University of Applied Sciences Darmstadt, Darmstadt, Germany
e-mail: jan.barkmann@h-da.de

C. von Haaren et al. (eds.), *Landscape Planning with Ecosystem Services*, Landscape Series 24, https://doi.org/10.1007/978-94-024-1681-7_25

25.1 Introduction: Definitions and Concept

Biodiversity refers to diversity at multiple scales of biological organisation (genes, populations, species, ecosystems and ecological processes (Noss 1990)) and can be considered at any geographic scale (local, regional or global) (Millennium Ecosystem Assessment 2005). In the present context, we will refer to biodiversity as the diversity of species and habitats including habitat networks (see Chap. 18).

Safeguarding biodiversity requires conservation activities. These activities may include developing new habitats, restoring impaired biodiversity or maintaining the present state. The development of appropriate and effective measures and resultant management plans requires an initial assessment of (i) the state of biodiversity, (ii) the impacts of human activities on biodiversity and (iii) the habitat development potential (see Chaps. 17 and 18) (Fig. 25.1).

The assessment needs to identify the key features of conservation importance (targets) and attributes that define the conditions detected (cf. Tucker et al. 2005). For management and then ongoing monitoring, objectives have to be defined. Priorities should be set in terms of the timeline and importance of implementation, the targets of conservation and the scale considered.

Targets of conservation actions may be species, species groups, habitats or their connectivity. For example, prominent species may be used as 'target species'. They can increase the acceptability of measures and can be used for communicating benefits of the measures (cf. Bakker et al. 2000; Walpole and Leader-Williams 2002). Well-known examples of this strategy are bees which are often used for demonstrating the economic impact of biodiversity degradation. First and foremost, however, endangered species (e.g. those listed in the annexes of the FFH-Directive on a European scale or in national legislation) are used as target species, as their needs should be the most important objectives for protection and development. Therefore, the conservation of these legally protected species and their habitats is a primary

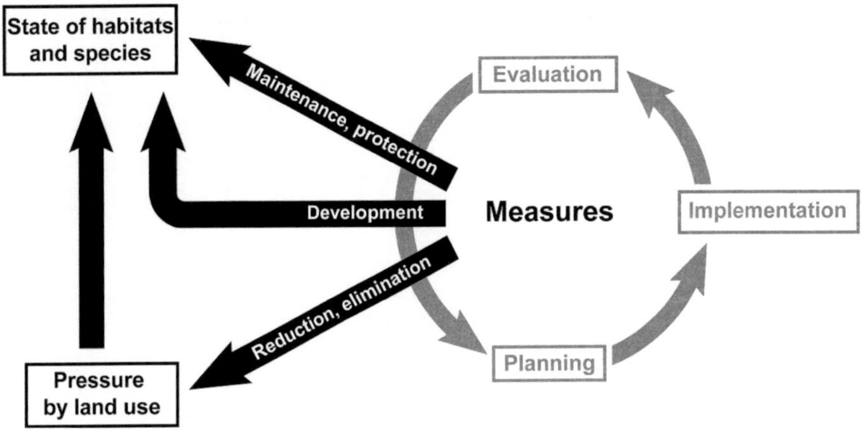

Fig. 25.1 Pressure-state-response framework complemented by a simplified management and monitoring cycle. (cf. Tucker (2005))

and minimum goal for deriving measures. In addition, certain other species are suitable as targets if they are representatives of their habitat (−complex) and the species composition occurring within it. Therefore, species with high ecological demands, with high space requirements or a close link to key structures within the ecosystem are well suited as target species. Some measures will create multifunctional benefits for both target species and other species living in the habitat. In these cases target species simultaneously function as 'umbrella species' (cf. Roberge and Angelstam 2004). For instance, the dusky large blue (*Maculinea nausithous*) is a highly endangered butterfly species and thus protected by European (FFH-Directive Annex II and IV) and national law (e.g. German species protection law). *M. nausithous* lives in wet meadows where the metamorphosis of caterpillars occurs exclusively on flowers of the great burnet (*Sanguisorba officinalis*), and in symbiosis with certain host-ants (*Myrmica*-species). These highly specific habitats require customised measurements, such as one-time mowing before or after the flight period of *M. nausithous* (Johst et al. 2006). At the same time, there are several other species that benefit from this adapted mowing regime (e.g. meadow birds such as the lapwing (*Vanellus vanellus*), see Sect. 25.2).

The most important fauna species groups which could function as target species for measures in their typical habitats are listed in Table 25.1. By way of example, some amphibian and dragonfly species may be relevant target species for wetlands, whilst reptile or hymenoptera species could be used as target species for dry open habitats such as heathlands.

The scale of conservation actions refers to different dimensions of landscape: the vertical dimension (habitat-based assessment, see Chap. 18), the horizontal dimension (assessment of habitat networks, see Chap. 18) and the temporal dimension. Development and restoration measures are usually limited in time until the desired state is reached. Maintenance measures may be needed permanently, especially if the habitat represents an earlier state of succession that is conserved by human activities for the sake of habitat and species protection.

In general, measures should be planned across different spatial and temporal scales to assure that there will be habitats in different stages of the successional cycle (leading to high patchiness). This creates a mosaic landscape structure, which may provide benefits for several species (e.g. amphibians cf. Pope et al. 2000, butterflies cf. Dennis 2001) and may even promote higher species diversity (e.g. birds, amphibians, reptiles and butterflies (Atauri and De Lucio 2001)). This general approach, which is less targeted to particular species, is complementary to species-oriented measures.

Sites for the implementation of measures (both restoration and maintenance) to improve the connectivity of the landscape should be prioritised according to their importance for habitat connectivity on the local or regional scale. An important criterion for improving habitat connectivity is not only the ability of species to overcome barriers such as streets, villages or fields (dispersal mobility, see Chap. 18), but also the specific habitat requirements of species in different seasons (e.g. seasonal habitats of bats, cf. Russ et al. 2003). Measures for the improvement of habitatconnectivity should focus on the development and maintenance of stepping-stone

Table 25.1 Possible target species groups for different habitat types

Species groups / Habitats	Medium and large mammals	Bats	Small mammals	Birds	Reptiles	Amphi bians	Fish and cyclos tomes	Macro zoobenthos	Dragonflies	Grass hoppers	Butterflies and burnet moths	Moths	Spiders	Ground beetles	Xylobiont beetles	Mussels	Snails	*Aculeate hymenoptera*
Forests and -edges	●	●	●	●								●	●	●	●		●	
Shrubs, single trees			●												●			
Springs								●										
Streams	●		●	●			●	●	●							●		
Standing waters	●		●	●		●		●	●								●	
Bogs and fens, shore			●	●		●			●	●	●	●	●	●			●	
Peatland				●					●		●	●						
Rocks and bare soils					●					●				●				●
Heathland and dry grassland					●					●	●	●	●				●	●
Grassland and -edges				●						●	●		●	●			●	●
Fields and -edges														●				
Ruderal sites																		●
Buildings		●																
Caves		●																

According to Bernotat et al. (2002), and Brinkmann (1998)

habitats or corridors. They should also include the removal or reduction of the negative effects of barriers, e.g. with wildlife crossing or fish ladders. In agricultural landscapes set-aside land can serve as a refuge for species (van Buskirk and Willi 2004). Additionally, establishing a heterogeneous landscape at the farm scale, e.g. through the cultivation of diverse field crops (Hawes et al. 2010) or the preservation of near natural habitats (Rusch et al. 2012), will lead to multifunctional benefits for habitats, their connectivity, species and even for cultivation itself. This is evident, for example, in terms of higher crop yields due to higher abundance of pollinators (Klein et al. 2007) or better pest control (Bianchi et al. 2006).

25.2 Objectives and Measures for Habitats and Species

25.2.1 Restoration and Development Measures for Habitats

25.2.1.1 General Approach

Measures for restoring, enhancing and developing biodiversity primarily entail changing or abandoning intensive land use. The specific measures, as well as their prospects of success, can be derived from the species and habitat assessment, in parallel with the assessment of land use impacts and the habitat development potential (HDP) (see Chaps. 17, 18).

The actual success of measures will strongly depend on the HDP, associated with the site conditions (relief, soil type, nutrients, humidity, micro climate). If these conditions have been changed fundamentally beforehand, restoration will not be possible. The potential area for development measures will usually consist of sites with no particularly valuable state and, simultaneously, have scope for improvement. Sites with the highest development potential can be identified using the HDP. Here, investments in payments for conservation are likely to produce the highest outcome in terms of the habitat value (von Haaren and Bathke 2008).

For each site with its specific conditions of soil, climate, morphology, and seed bank, there are different development alternatives. Besides considering the best added value for biodiversity, the decision about these alternatives will also have to take into account implementation options such as costs and willingness of land users to comply with obligations. As so many factors have to be considered and as key information is often incomplete (e.g. data about seedbanks are usually missing), projections of potentially emerging habitats cannot be precise. Habitat projections will usually be given in the form of a roughly classified habitat type. Usually, this information will suffice for providing a capacity to act. Under this premise, also an uncertain projection is useful (cf. Neuendorf et al. 2018). If the consequences of wrong or imprecise projections may be very harmful or costly, preliminary measures which will test the projection (e.g. targeted examinations of seedbanks, injection of regional seeds) should be performed.

In spite of uncertainty concerns, sufficient detail of the objectives and projections is needed in communication with stakeholders and the public (e.g. in order to illustrate possible results for decision makers). If the objectives of habitat development

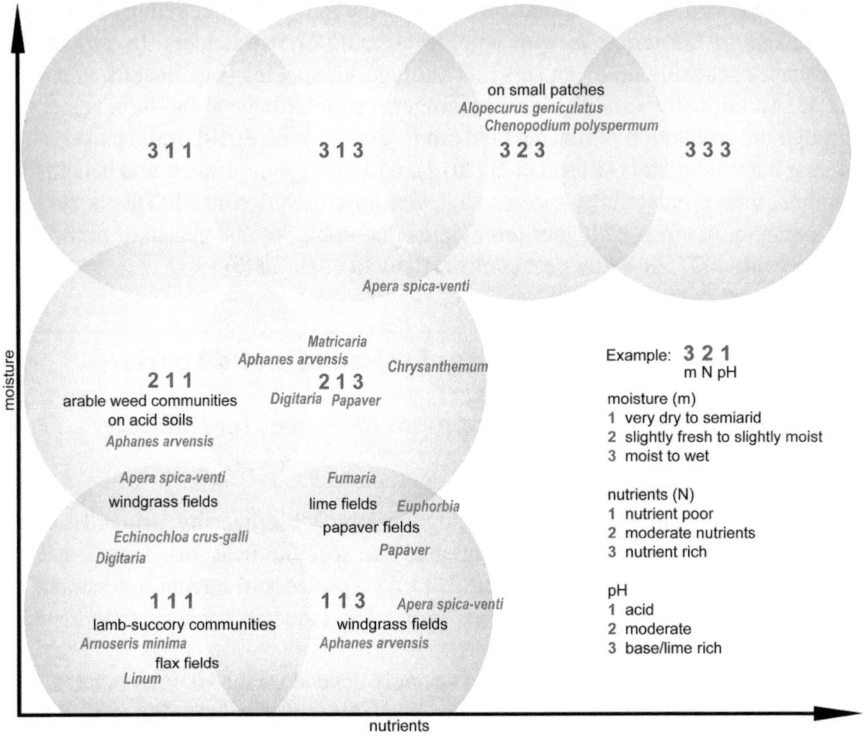

Fig. 25.2 Field habitats – Development options for target habitats on cropland under different site conditions assuming adapted nutrient and pesticide management (with examples of characteristic plant species); based on simplified ecogram (Chap. 17) focused on target habitats

remain abstract or without spatial specification, they will be unlikely to garner public and political support. Therefore, the projections should be concrete enough to give a picture of what is to be expected but general enough, and in the form of good examples, to include the uncertainties.

The following paragraphs provide examples of habitat development projections under different site conditions and management options (see Figs. 25.2, 25.3 and 25.4).

25.2.1.2 General Habitat Development Potential Approach

Due to the uncertainty of the prognosis for specific plant communities, development options are rough projections of the potentially emerging habitats. While the habitat potential ecograms (see Chap. 17) help to select the sites with the highest potential value, the response ecograms support the selection of measures and illustrate the vegetation that may be restored. More precise models of the biodiversity state of certain habitats (e.g. Bredemeier et al. 2015; Sybertz et al. 2017) and, in particular, site-specific mapping of the actual soil conditions, seed bank, landscape context,

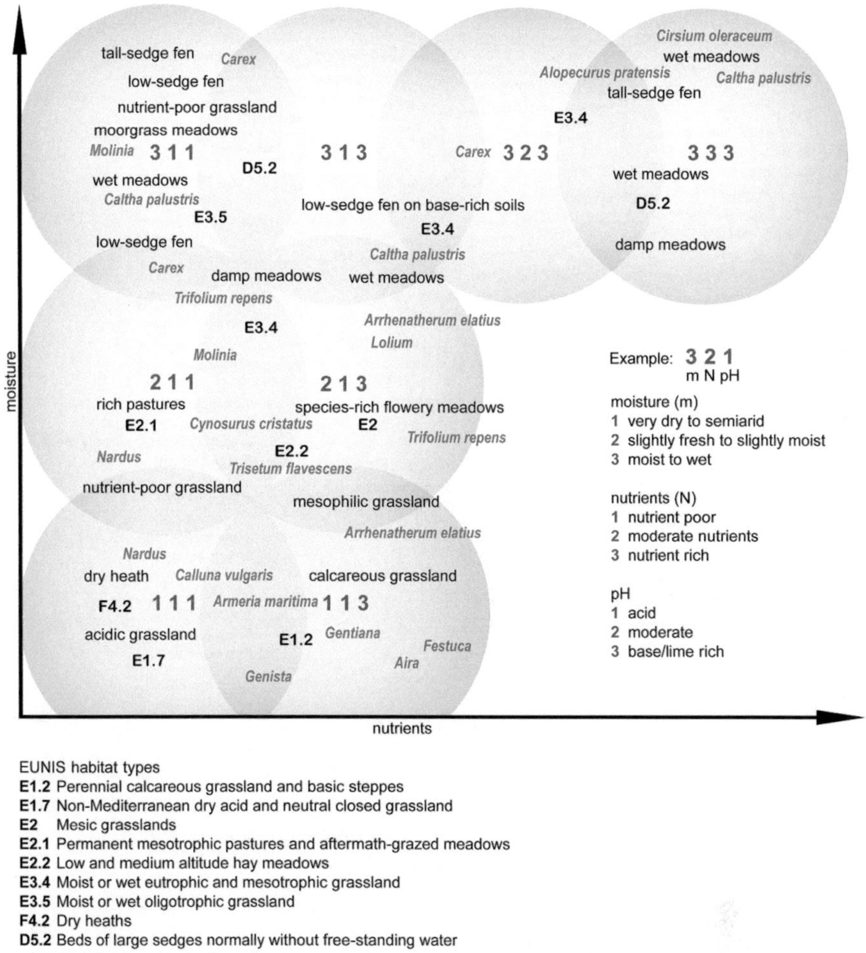

Fig. 25.3 Grassland habitats – Development options for target habitats under different site conditions and grassland management with nutrient and mowing management (with EUNIS habitat types and examples of characteristic plant species); based on a simplified ecogram (Chap. 17) focused on target habitats

fauna, and relicts of the natural vegetation, can considerably improve the projection of the options for habitat development (cf. Bredemeier et al. 2015).

For example, even under the most favourable soil conditions, the potential for species-rich endangered field or grassland habitats can only be developed if the soil seed bank is still in good condition. This may be the case, for example, on grasslands with a short history of improved cultivation (Bekker et al. 1997), particularly if specific nature conservation management had been carried out on the plot for a long period of time prior to intensification.

Habitat development measures may be particularly applicable if land use, for example farming, is restricted by various impediments such as:

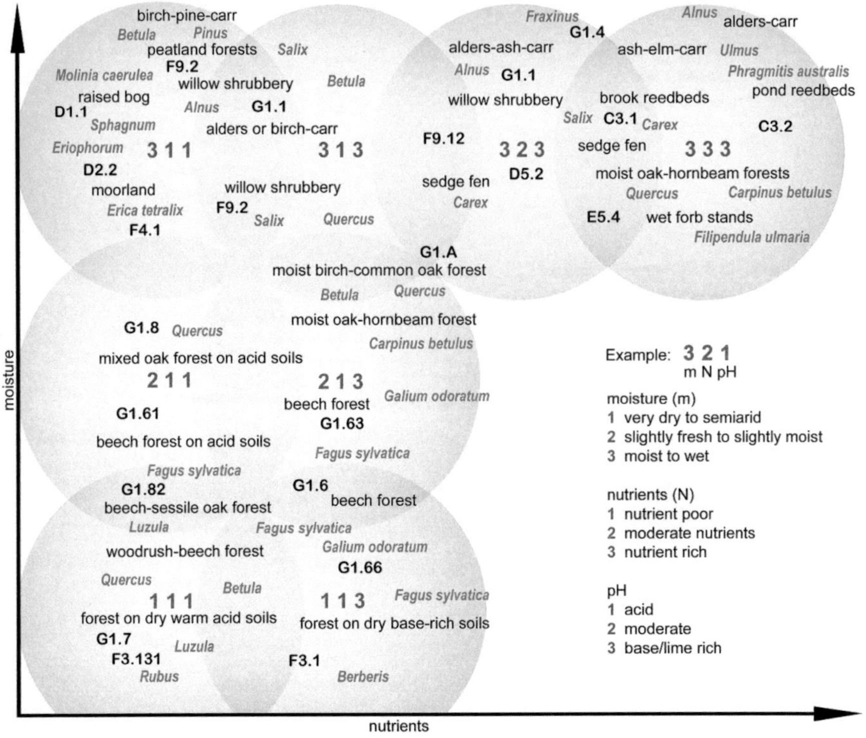

EUNIS habitat types

C3.1 Species-rich helophyte beds
C3.2 Water-fringing reedbeds and tall helophytes other than canes
D1.1 Raised bogs
D2.2 Poor fens and soft-water spring mires
D5.2 Beds of large sedges normally without free-standing water
E5.4 Moist or wet tall-herb and fern fringes and meadows
F3.1 Temperate thickets and scrub
F3.131 Bramble thickets
F9.12 Lowland and collinar riverine Salix scrub
F9.2 Salix carr and fen scrub
F4.1 Wet heaths
G1.A Meso- and eutrophic Quercus, Carpinus, Fraxinus, Acer, Tilia, Ulmus and related woodland
G1.1 Riparian and gallery woodland, with dominant Alnus, Betula, Populus or Salix
G1.4 Broadleaved swamp woodland not on acid peat
G1.6 Fagus woodland
G1.61 Medio-European acidophilous Fagus forests
G1.63 Medio-European neutrophile Fagus forests
G1.66 Medio-European limestone Fagus forests
G1.7 Thermophilous deciduous woodland
G1.8 Acidophilous Quercus-dominated woodland
G1.82 Atlantic acidophilous Fagus - Quercus forests

Fig. 25.4 Succession – Development options for target habitats on different sites without any land use or management or requiring only initial measures (with EUNIS habitat types and examples of characteristic plant species); based on a simplified ecogram (Chap. 17) focused on target habitats

- Sandy or heavy clay soil on slopes
- Sites with unfavourable shapes or remote from the farmstead
- Water edges with legal restrictions on pesticides and fertiliser use
- Areas close to shrubs and trees which compete with crops for water, nutrients and light
- Verges of tracks, roads, railway
- Areas beneath power lines

Depending on the site conditions, land use and target habitats, different development measures are required. These measures may be applied to the whole site but also to small patches within, especially if there are adverse or specific site conditions. After setting target habitats, the development measures should aim at constraining negative impacts on the restoration sites (e.g. reduce nutrient loads, reverse drainage). In some cases, the initial measures should also seek to establish favourable conditions for future habitats (e.g. initial planting or modification of topology).

25.2.1.3 Field Habitat Examples

In the case of field habitats, the highest development potential for endangered and diverse biocoenoses is on dry calcareous soils (cf. Bredemeier et al. 2015). However, on dry sandy soils a very diverse plant community can also be restored under favourable conditions (e.g. with the highly endangered lamb succory (*Arnoseris minima*) see Fig. 25.2). In contrast, very wet areas are not as relevant for field habitat plant communities as these sites are usually not suitable for arable farming (exception: small wet patches e.g. with marsh foxtail (*Alopecurus geniculatus*) within fields).

Restrictions on the use of fertiliser and plant protection products (PPP) are usually a crucial factor for restoring (and also maintaining) valuable and species-rich field habitats (Hyvönen and Salonen 2002; Kleijn et al. 2006). Even a total relinquishment may be essential (Kaule 1986; Wetterich and Köpke 2003) to conserve remnants of the prior field flora. Sparse crop stands (e.g. through wider row spacing or reduced seed rates) are recommended to support the weakly competitive species of the field flora against the more competitive field crops (Stein-Bachinger et al. 2010); this is even possible without great loss of crop yield.

Additionally, on sites with marginal yields, a high species inventory, and in the direct neighbourhood of other extensively managed areas, the creation of field margins is a widely accepted measure to develop species-rich field habitats (Oesau and Henke 2002; Schacherer 2007). Despite farmers' fears of weed invasion, positive effects of field margins or fallows on pest control have been reported, e.g. reducing problematic weeds such as cleavers (*Galium aparine*) (Moonen and Marshall 2001). Field margins also assist in enhancing natural enemies of pest species (Thies and Tscharntke 1999) leading to higher predation or parasitism rates (Bianchi et al. 2006).

25.2.1.4 Grassland Habitat Examples

In the case of grassland management, wet sites offer an important development potential for valuable habitats (e.g. for wet meadows with marsh marigold (*Caltha palustris*) or tall-sedge fens, see Fig. 25.3). In a similar manner to field habitats, endangered, diverse and species-rich grassland habitats can also be restored on dry, calcareous or acid soils (e.g. butterfly species diversity is usually the highest in habitats on calcareous soils (cf. van Swaay 2002)). Calcareous grassland habitats (e.g. with gentian species (*Gentiana* sp.)) or acidic sand grasslands (e.g. with mat grass (*Nardus stricta*)) will have a chance to occur on these sites if the soil seed bank is still in good condition or the seeds can be imported from nearby habitats. Most of the valuable grassland habitats grow on more or less nutrient poor sites (e.g. peatlands, heathlands). In contrast, on sites with excessive nutrient inputs (present or past), success can only be expected if nutrient loads are reduced. This can be achieved by top soil removal, which is an expensive but very fast effective measure (Patzelt et al. 2001; Hölzel and Otte 2003; Klimkowska et al. 2007; Klimkowska et al. 2010). Extensive mowing or grazing (also in combination) is a less expensive way to reduce nutrients on grassland sites by removing vegetation (Rosén and van der Maarel 2000; Walker et al. 2004). However, effects are only visible after many years.

On peat soil sites, it is necessary to regulate the water regime in combination with top soil removal in order to restore the ancient water and nutrient regime (van Dijk et al. 2007). This may enhance biodiversity but also generate benefits for other ES too (e.g. soil or water properties, see Chaps. 22 and 23).

If the previous seed bank potential is no longer available, the re-establishment of the seed bank and specific vegetation can be important on grassland sites. This may be achieved by spreading seed-containing autochthonous hay (Hölzel and Otte 2003; Kiehl et al. 2006) or regional seeds (Jongepierová et al. 2007; Conrad and Tischew 2011). For a more comprehensive and detailed outline of grassland measures see, for example, Török et al. (2011).

25.2.1.5 Succession Examples

Refraining from intensive management or land use is often a preferable option for habitat development if management is difficult to organise or prohibitively expensive (cf. Prach and Pyšek 2001). Nevertheless, in some cases (especially open grasslands) abandonment of land use may lead to loss of typical and, in particular, endangered species (Diemer et al. 2001). Therefore, such 'hands-off' management should be favoured for less valuable sites where succession or 'natural' habitats are end targets. One example is riparian forests (alder carr) for which Schrautzer et al. (2007) describe a model of successional restoration. Therefore, depending on different grades of human impact (e.g. drainage or fertilisation) and the consequence of different initial stages, the abandonment of land use may lead to different successional stages (seres) and output habitats (see Fig. 25.5).

In general, initial modification of the terrain may be helpful when restoring habitats by successional sere (e.g. recreating floodplains or otherwise altering the hydrological regime, cf. Orczewska 2009). It will increase micro habitat diversity and

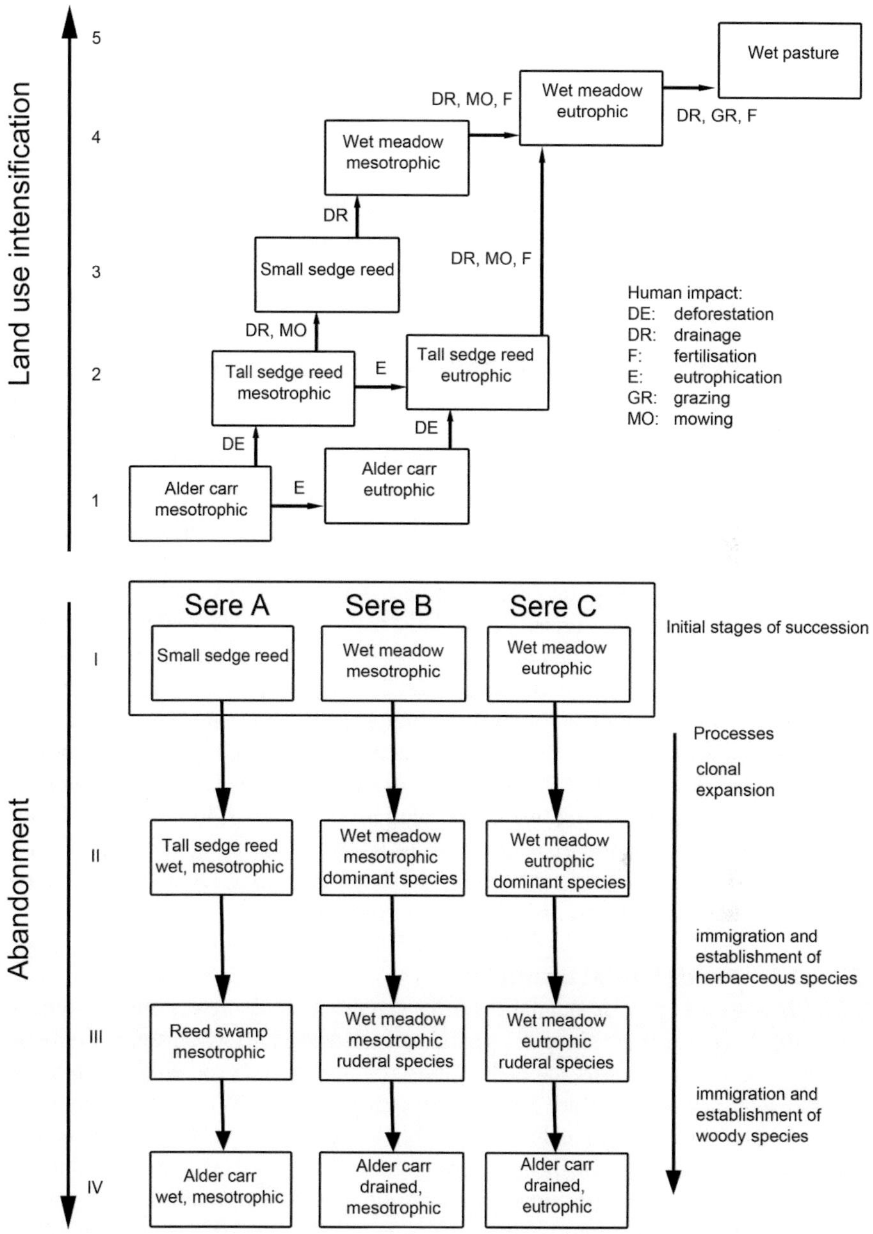

Fig. 25.5 Models describing successional stages: stages 1 to 5 in retrogressive succession by land use intensification and stages I to IV in secondary progressive succession by abandonment in seres A, B and C on fen sites. The initial stages of the abandonment seres differ in drainage and land use intensity. (Adapted from Schrautzer et al. 2007)

slow down or prevent succession in pioneer habitats, which constitute a favourable contribution to the overall habitat diversity of the restoration site.

On sites where extreme drought or extreme moisture prevail and hamper the succession to dominant forest, particularly valuable habitats can emerge when nature is left to itself (e.g. raised bogs see Fig. 25.4). However, endangered and rare habitats can also be developed through succession on sites with very fertile top soil (e.g. successional development of alder carr on sites with formerly intensive agricultural use (cf. Douda et al. 2009)). These sites, e.g. on loess soils, are historically and nearly always used for agriculture, which has made nearly natural habitats very rare.

25.2.2 Conservation and Maintenance Measures for Habitats

25.2.2.1 General

Usually the highest priority in nature conservation is to preserve the existing valuable habitats and endangered species. This can be achieved through maintenance measures, sometimes required in combination with one-time development measures.

Maintenance measures may include the protection of very valuable habitats, for example nature protection areas – especially if the present state is endangered by recently changed land uses or impending risks. Orientation is needed to choose the right measures or land use restrictions in order to develop or maintain the present state by active management or by reducing harmful pressures. Such measures mostly aim at interrupting or reversing the successional cycle to maintain a specific valuable stage (Prach et al. 2007). In many cases there is a need to remove the established vegetation (e.g. cutting trees or burning broom heather) to restart succession (Luken 1990). If flora and/or fauna seem to be in a continuing suboptimal state, there are several development measures which can be used as recurring procedures to maintain habitats (e.g. extensive grazing or mowing).

25.2.2.2 Field Habitat Examples

Field flora species rely on regular disturbances due to their weak competitiveness. Thus, the soil needs to be ploughed periodically. However, to support field flora species that bear fruit towards the end of the growing season (e.g. *Kickxia* sp.), late stubble working is necessary (in central Europe after the 15th of September). Additionally, sparse crop stands are advisable to support the weak competitive species of the field flora against the more competitive field crops. This can be achieved, for example, by a wider row spacing of two or three sowing widths or by sowing gaps at two or three locations per hectare at a length of up to 10 m (Fuchs and Stein-Bachinger 2008). Due to the increasing risk of weed infestation, such measures should not be implemented at sites with problematic weeds such as couch grass (*Elymus repens*). On specific field flora conservation sites, cultivators should not be used at depths of more than 20 cm to enable bulbous plants like the Orange lily (*Lilium bulbiferum*) to reproduce (Wicke 1998). For habitat development the use of PPP and fertiliser also needs to be reduced (see Sect. 25.2.1). The input of fertiliser should range between 60 to 80 kg of nitrogen per hectare per year; the initial fertilisation during spring should be discontinued (Oesau and Henke 2002; Jedicke 1994). If quantities of 100 kg of nitrogen or more per hectare per year are applied, species-rich inventories of field flora are not possible (Wetterich and Köpke 2003). Correspondingly, organic farming requires adapted measures like minimising tillage on the entire field or at least on subplots.

Overall, for the maintenance of species-rich field habitats, a small-scale mosaic of crops with different sowing and harvesting times, as well as different growth patterns, are necessary to providing favourable food and habitat for several species (Glemnitz et al. 2008). For a detailed outline of field habitat measures see e.g. Andrews and Rebane (1994), Hill et al. (1995), Fuchs and Stein-Bachinger (2008) or Berger and Pfeffer (2011).

25.2.2.3 Grassland Habitat Examples

Maintenance measures mostly include grazing or/and mowing with intensity being the crucial factor for biodiversity. In the case of grazing, intensity can be described both by the stocking rate and the duration of grazing. A reduced stocking rate may lead, for example, to a higher diversity of invertebrates (Eschen et al. 2012). A higher stocking rate leads to greater nutrient loads via excrement and effects of trampling. In Germany about 0.3–0.5 GLU/ha (grazing livestock unit) on less productive land and about 0.8–1.5 GLU/ha on high productive land is recommended as extensive grazing management (Steidl 2002). Additionally, the choice of livestock species plays an important role. Different breeds have different grazing behaviours, e.g. selection of plants for feeding (Ausden and Treweek 1995). For example, goats are able to feed on woody and even thorny plant species, while other livestock species prefer herbaceous plants (Celaya et al. 2007). Therefore, different species promote diverse plant structures. Moreover, mowing after the grazing period is often necessary. Otherwise specific plants, such as thistles or tussock grass (*Deschampsia cespitosa*), will proliferate because they are rejected by the grazing animals.

Key factors for species conservation on predominantly mown grasslands are the cutting dates and frequency, as well as the type of mowing machine (Andrews and Rebane 1994). Especially on nutrient-poor soils, it is important for plants to accumulate nutrients during the vegetation period. A late cut (after the first of October) provides this opportunity to plant species (e.g. Stammel et al. 2003).

Specific measures often have to be applied to improve conditions for the typical fauna of a habitat. For example, typical grassland fauna such as ground-nesting birds, grasshoppers and butterflies need an adapted mowing regime. In general, a cutter bar mower and a cutting height of at least 10 cm should be preferred because of its reduced impact on grassland fauna (Humbert et al. 2009). Machines such as rotary mowers cause injury or death to fauna species because of the rotating blades (cf. Liczner 1999). The timing of cutting also has to be adjusted to the hatching dates of ground-nesting meadow birds like the lapwing (*Vanellus vanellus*) or the common redshank (*Tringa tetanus*), as described by Kruk et al. (1996), to avoid damaging eggs or nestlings.

25.2.3 Species-Supporting Measures

In some cases, additional species-supporting measures are required, such as (nest) boxes for bats or birds and maybe even ex situ reproduction of nearly extinct species. Nest boxes for the cavity-nesting european roller (*Coracias garrulous*) (Rodríguez et al. 2011) or the hoopoe (*Upupa epops*) (Weber 2011) serve as examples. For the species under protection of European law, there are specific guidelines for planning. These are recommended by the nature conservation authorities of the Member States, e.g. for the dusky large blue (*Maculinea nausithous*) (see Sect. 25.1) guidelines have been produced by the nature conservation authority of Lower-Saxony, Germany (cf. NLWKN 2011).

The first step in planning species-supporting measures should be the clarification of the viability of populations, since this needs to be the objective of the measures. Sites with remnant populations should gain priority for species measures or should at least be preferred in decision making.

It is also important to realise that planning for one species may have negative effects on another. Therefore, the objectives should be clear and well considered. This is also crucial for monitoring the success of measures and for judging whether any adjustments need to be made (von Haaren et al. 1997). The management option chosen as the best for the restoration or maintenance objective is defined, in part, by previous land use or the site conditions. It will also be influenced by its acceptability to land users (and NGOs), costs, public preferences, and implementation options.

25.3 Implementation of Measures

In general, defining measures is a cross-sectoral task. It needs to take the principles of multifunctionality (see Chaps. 19 and 26), legal obligations, cost-efficiency, sustainability and societal acceptability into account (cf. Louette et al. 2015). The development and application of specific response measures is a necessity to reach the desired status for a given habitat or species. Important issues in planning include, e.g. the protection of endangered habitats and species, the development of the Natura 2000 network or Environmental Impact Assessments and the compensation of impacts.

Commonly, the implementation of measures for conserving and supporting biodiversity is part of legal restrictions such as regulations in protected areas or the compensation for impacts. According to the 'polluter pays' principle, the source of impacts should be obliged to compensate or substitute the impairment of function which occurred as result of an intervention in an ecosystem. In some European states, this is regulated by mandatory offset mechanisms such as the German intervention regulation. For valuable parts of the landscape, outside of protected areas, the authorities have the choice of 'buy or borrow' to safeguard these sites (cf. Schöttker et al. 2016).

The purchase of land is the most obvious, sustainable but expensive way to safeguard valuable parts of the landscape. On these sites, nature conservation organisations are important stakeholders for implementing measures. They can often manage valuable areas as representatives of the responsible authorities. Besides these organisations, other NGOs (e.g. hiking or water sport clubs) can commit themselves to implement measures for habitats and species on a voluntary basis.

Time-limited contracts are a useful instrument for implementation, especially if the area of interest is not in municipal possession and the purchase of land is out of the question due to high prices or the landowner's unwillingness to sell. On the basis of such contracts, landowners such as farmers, hunting or fishing cooperatives are obliged to implement measures for biodiversity (mostly restrictions on land use) for a defined period of time. Depending on the objectives, the use of shorter or longer lasting contracts is recommended (Lennox and Armsworth 2011).

Within agri-environmental schemes (e.g. EU-supported schemes), farmers can receive financial support as compensation for restricting cultivation on grassland or arable land, e.g. reduced usage of fertiliser or PPP. These measures can be taken for the period of a CAP programme and can be result-oriented. Agri-environmental payments (AEP) to farmers can be set on the basis of the present or the potential conservation value of a site (von Haaren and Bathke 2008). Information obtained from the HDP and exemplary development options can support landowners in assessing how biodiversity-friendly the created habitat may be. The results of the agri-environmental measures do not only depend on management activities but also on the site conditions (cf. Bredemeier et al. 2015). For instance, it will be much easier and faster to successfully establish a high-nature value farmland habitat, such as a moist grassland habitat, on a site with a correspondingly high HDP than on a well-drained site with a good soil nutrient supply.

Furthermore, the HDP can support spatial targeting of agri-environmental programs. AEPs could be made conditional upon achieving a certain minimum HDP or the payment could be scaled according to the HDP of the contracted sites. In these cases, the HDP could represent an intermediate step between result-oriented AEP and traditional agri-environmental programs.

However, each implementation of measures for biodiversity should be checked for multifunctional benefits regarding other ecosystem services. The integration of measures in higher-level planning such as management plans for the water framework directive or urban land use plans could be a way to achieve synergies.

25.3.1 Habitat Banking Example

Habitat banking is a form of offset mechanisms and aids in implementing the 'no net loss' principle. If a habitat of a certain size and value is destroyed or impaired by infrastructure development (e.g. settlement expansion, road or industrial development), conservation or planning law in some states mandates *in kind* compensation. For example, a habitat of comparable size and value needs to be restored in order to safeguard public goods such as biodiversity. Habitat banking is an institution that simplifies the task for a developer to prove that respective restoration is actually provided. It can also reduce the direct costs of providing compensatory habitats. To be systematically acceptable, from a conservation perspective, the habitat value provided must be at least equal to the habitat value lost. This task can be improved by using the HDP.

Landowners can use the HDP concept to make a first assessment of their site. For example, an intensively-used grassland on a drained fen or a heavily-manured maize stand on a very sandy soil both have a very low current conservation value. Given a suitable landscape matrix, their HDP may be considerable as fen and inland dune habitats could be developed.

With their knowledge of actual agricultural profits from these sites, the landowners may choose to register the land with a habitat banking organisation. This will require regularly updated evidence regarding the conservation value of the current situation and, potentially, certification that the land was not degraded just before registration. For example, let us assume that the landowners carry out the habitat development themselves. This involves terminating the use of fertilisers and re-establishing a near-natural hydrologic regime at a fen site. Depending on the success of habitat development and the details of the habitat banking scheme, the land is assigned habitat banking credits based on calculations of the difference between the old and the projected (after restoration) conservation value of the site. The habitat bank can either (i) purchase the land and pay a price according to the market value of the habitat credits to the landowner, or (ii) enter into a contractual agreement with the landowner to ensure the long-term stability of the developed habitat, and market the credits as a broker on behalf of the land owner. In both cases, the habitat credits can be sold to investors interested in a speedy approval of their environmentally destructive projects. Properly-endowed nature conservation interests could also purchase the credits.

For private landowners, the difference in farming profits with and without habitat banking will often be a decisive factor in deciding whether to adopt such a practice. This profit difference can be substantially influenced by farming-related transfer payments. Ignoring such incentives, the sites of most interest for habitat banking tend to be those with a comparably low agronomic or forestry value. This is due to the fact that the profit lost by taking such sites out of production is low. At the same time, the potential to improve agriculturally highly-profitable sites is often limited. Even with conservation measures in place, relatively common plant communities and species are likely to develop. In contrast, on low productivity sites endogenous habitat development is more promising and/or externally initiated development less costly. Additionally, these sites are often embedded in less degraded landscapes where the chance of recolonization by a diversity of species is higher than in highly productive areas.

Additionally, the price of the habitat credits can be subject to market forces or the habitat credits may be sold by a monopolistic habitat bank at politically fixed prices. Fundamental economic considerations suggest that too high a fixed price will lead to a lower utilisation of this tool by developers than economically optimal. Too low a fixed price will result in less supply of habitat credits than is useful. The stricter the spatial or special in-kind restrictions, the more imbalances are likely to occur as the market for habitat credits is fragmented into a number of submarkets. It is generally more useful, from an economic point of view, to implement a common market for habitat credits, but to allow for the generation of additional credits in areas with especially high conservation demand.

25.4 Conclusion

This chapter has discussed how response measures for safeguarding and increasing biodiversity can be derived from biodiversity assessments. Selecting appropriate measures typically depends on the habitat development potential of the site, the existing value and endangerment of biodiversity present and the habitat connectivity. Several examples have been discussed, focusing on field and grassland habitats. In addition, the potential role of habitat banking as a means of implementation has been outlined.

References

Andrews, J., & Rebane, M. (1994). *Farming & wildlife: A practical management handbook.* Bedfordshire: The Royal Society for the Protection of Birds.

Atauri, J. A., & De Lucio, J. V. (2001). The role of landscape structure in species richness distribution of birds, amphibians, reptiles and lepidopterans in Mediterranean landscapes. *Landscape Ecology, 16*(2), 147–159.

Ausden, M., & Treweek, J. (1995). Grasslands. In W. J. Sutherland & D. Hill (Eds.), *Managing habitats for conservation* (pp. 197–229). Cambridge: Cambridge University Press.

Bakker, J. P., Grootjans, A. P., Hermy, M., et al. (2000). How to define targets for ecological restoration? – Introduction. *Applied Vegetation Science, 3*(1), 3–6.

Bekker, R. M., Verweij, G. L., Smith, R. E. N., Reine, R., Bakker, J. P., & Schneider, S. (1997). Soil seed banks in European grasslands: Does land use affect regeneration perspectives? *Journal of Applied Ecology, 34*(5), 1293–1310.

Berger, G., & Pfeffer, H. (2011). *Naturschutzbrachen im Ackerbau: Praxishandbuch für die Anlage und optimierte Bewirtschaftung kleinflächiger Lebensräume für die biologische Vielfalt,* 1 Aufl. Natur & Text, Rangsdorf.

Bernotat, D., Schlumprecht, H., Brauns, C., et al. (2002). Gelbdruck "Verwendung tierökologischer Daten". In H. Plachter, D. Bernotat, R. Müssner, et al. (Eds.), *Entwicklung und Festlegung von Methodenstandards im Naturschutz: Ergebnisse einer Pilotstudie. 2 Aufl* (pp. 109–217). Bonn-Bad Godesberg: Bundesamt für Naturschutz.

Bianchi, F., Booij, C., & Tscharntke, T. (2006). Sustainable pest regulation in agricultural landscapes: A review on landscape composition, biodiversity and natural pest control. *Proceedings of the Royal Society B: Biological Sciences, 273*(1595), 1715–1727.

Bredemeier, B., von Haaren, C., Rüter, S., et al. (2015). Evaluating the nature conservation value of field habitats: A model approach for targeting agri-environmental measures and projecting their effects. *Ecological Modelling, 295*, 113–122.

Brinkmann, R. (1998). Berücksichtigung faunistisch-tierökologischer Belange in der Landschaftsplanung. *Informationsdienst Naturschutz Niedersachsen, 18*(4), 58–128.

Celaya, R., Oliván, M., Ferreira, L., et al. (2007). Comparison of grazing behaviour, dietary overlap and performance in non-lactating domestic ruminants grazing on marginal heathland areas. *Livestock Science, 106*(2–3), 271–281.

Conrad, M. K., & Tischew, S. (2011). Grassland restoration in practice: Do we achieve the targets? A case study from Saxony-Anhalt/Germany. *Ecological Engineering, 37*(8), 1149–1157.

Dennis, R. L. (2001). Just how important are structural elements as habitat components? Indications from a declining lycaenid butterfly with priority conservation status. *Journal of Insect Conservation, 8*(1), 37–45.

Diemer, M., Oetiker, K., & Billeter, R. (2001). Abandonment alters community composition and canopy structure of Swiss calcareous fens. *Applied Vegetation Science, 4*(2), 237–246.

Douda, J., Cejková, A., Douda, K., & Kochánková, J. (2009). Development of alder carr after the abandonment of wet grasslands during the last 70 years. *Annals of Forest Science, 66*(7), 712.

Eschen, R., Brook, A. J., Maczey, N., et al. (2012). Effects of reduced grazing intensity on pasture vegetation and invertebrates. *Agriculture, Ecosystems and Environment, 151*, 53–60.

Fuchs, S., & Stein-Bachinger, K. (2008). *Naturschutz im Ökolandbau: Praxishandbuch für den ökologischen Ackerbau im nordostdeutschen Raum*, 1 Aufl. Mainz: Bioland-Verl.

Glemnitz, M., Platen, R., & Saure, C. (2008). Auswirkungen des Anbaus von Energiepflanzen auf Biodiversität: Bewertungsmethodik und Einfluss des Anbauverfahrens. In Kuratorium für Technik und Bauwesen in der Landwirtschaft e.V (Ed.), *Ökologische und ökonomische Bewertung nachwachsender Energieträger* (pp. 136–150). Münster-Hiltrup: Landwirtschaftsverlag.

Hawes, C., Squire, G. R., Hallett, P. D., et al. (2010). Arable plant communities as indicators of farming practice. *Agriculture, Ecosystems and Environment, 138*(1–2), 17–26.

Hill, D., Andrews, J., Sotherton, N. W., et al. (1995). Farmland. In W. J. Sutherland & D. Hill (Eds.), *Managing habitats for conservation* (pp. 230–266). Cambridge: Cambridge University Press.

Hölzel, N., & Otte, A. (2003). Restoration of a species-rich flood meadow by topsoil removal and diaspore transfer with plant material. *Applied Vegetation Science, 6*(2), 131–140.

Humbert, J.-Y., Ghazoul, J., & Walter, T. (2009). Meadow harvesting techniques and their impacts on field fauna. *Agriculture, Ecosystems and Environment, 130*(1–2), 1–8.

Hyvönen, T., & Salonen, J. (2002). Weed species diversity and community composition in cropping practices at two intensity levels – A six-year experiment. *Plant Ecology, 159*(1), 73–81.

Jedicke, E. (1994). *Biotopverbund: Grundlagen und Maßnahmen einer neuen Naturschutzstrategie*. Stuttgart: Ulmer.

Johst, K., Drechsler, M., Thomas, J., et al. (2006). Influence of mowing on the persistence of two endangered large blue butterfly species. *Journal of Applied Ecology, 43*(2), 333–342.

Jongepierová, I., Mitchley, J., & Tzanopoulos, J. (2007). A field experiment to recreate species rich hay meadows using regional seed mixtures. *Biological Conservation, 139*(3–4), 297–305.

Kaule, G. (1986). *Arten- und Biotopschutz*. Stuttgart: Eugen Ulmer.

Kiehl, K., Thormann, A., & Pfadenhauer, J. (2006). Evaluation of initial restoration measures during the restoration of calcareous grasslands on former arable fields. *Restoration Ecology, 14*(1), 148–156.

Kleijn, D., Baquero, R. A., Clough, Y., et al. (2006). Mixed biodiversity benefits of agri-environment schemes in five European countries. *Ecology Letters, 9*(3), 243–254.

Klein, A.-M., Vaissiere, B. E., Cane, J. H., et al. (2007). Importance of pollinators in changing landscapes for world crops. *Proceedings of the Royal Society B: Biological Sciences, 274*(1608), 303–313.

Klimkowska, A., van Diggelen, R., Bakker, J. P., et al. (2007). Wet meadow restoration in Western Europe: A quantitative assessment of the effectiveness of several techniques. *Biological Conservation, 140*(3–4), 318–328.

Klimkowska, A., Kotowski, W., van Diggelen, R., et al. (2010). Vegetation re-development after fen meadow restoration by topsoil removal and hay transfer. *Restoration Ecology, 18*(6), 924–933.

Kruk, M., Noordervliet, M., & ter Keurs, W. J. (1996). Hatching dates of waders and mowing dates in intensively exploited grassland areas in different years. *Biological Conservation, 77*(2–3), 213–218.

Lennox, G. D., & Armsworth, P. R. (2011). Suitability of short or long conservation contracts under ecological and socio-economic uncertainty. *Ecological Modelling, 222*(15), 2856–2866.

Liczner, Y. (1999). Auswirkungen unterschiedlicher Mäh- und Heubearbeitungsmethoden auf die Amphibienfauna in der Narewniederung (Nordostpolen). *RANA Sonderheft, 3*, 67–79.

Louette, G., Adriaens, D., Paelinckx, D., et al. (2015). Implementing the habitats directive: How science can support decision making. *Journal for Nature Conservation, 23*, 27–34.

Luken, J. O. (1990). *Directing ecological succession* (1st ed.). London: Chapman and Hall.

Millennium Ecosystem Assessment. (2005). *Ecosystems and human well-being: Biodiversity synthesis*. Washington, DC: World Resources Institute.

Moonen, A. C., & Marshall, E. J. P. (2001). The influence of sown margin strips, management and boundary structure on herbaceous field margin vegetation in two neighbouring farms in southern England. *Agriculture, Ecosystems and Environment, 86*(2), 187–202.

Neuendorf, F., von Haaren, C., & Albert, C. (2018). Assessing and coping with uncertainties in landscape planning: An overview. *Landscape Ecology, 33*, 861.

NLWKN (Niedersächsischer Landesbetrieb für Wasserwirtschaft, Küsten- und Naturschutz) (2011). *Vollzugshinweise zum Schutz von Wirbellosenarten in Niedersachsen.* – Wirbellosenarten des Anhangs II der FFH-Richtlinie mit höchster Priorität für Erhaltungs- und Entwicklungsmaßnahmen: Schwarzer Moorbläuling (Dunkler Wiesenknopf-Ameisenbläuling) (*Maculinea nausithous*).

Noss, R. (1990). Indicators for monitoring biodiversity: A hierarchical approach. *Conservation Biology, 4*(4), 355–364.

Oesau, A., & Henke, W. (2002). Ackerwildkräuter schützen. *Ökologische Bausteine der Kulturlandschaft – Leben braucht Vielfalt, 1445.*

Orczewska, A. (2009). The impact of former agriculture on habitat conditions and distribution patterns of ancient woodland plant species in recent black alder (Alnus glutinosa (L.) Gaertn.) woods in south-western Poland. *Forest Ecology and Management, 258*(5), 794–803.

Patzelt, A., Wild, U., & Pfadenhauer, J. (2001). Restoration of wet fen meadows by topsoil removal: Vegetation development and germination biology of fen species. *Restoration Ecology, 9*(2), 127–136.

Pope, S. E., Fahrig, L., & Merriam, H. G. (2000). Landscape complementation and metapopulation effects on leopard frog populations. *Ecology, 81*(9), 2498.

Prach, K., & Pyšek, P. (2001). Using spontaneous succession for restoration of human-disturbed habitats: Experience from Central Europe. *Ecological Engineering, 17*(1), 55–62.

Prach, K., Marrs, R., Pyšek, P., et al. (2007). Manipulation of succession. In L. R. Walker, J. Walker, & R. J. Hobbs (Eds.), *Linking restoration and ecological succession* (pp. 121–149). New York: Springer.

Roberge, J., & Angelstam, P. (2004). Usefulness of the umbrella species concept as a conservation tool. *Conservation Biology, 18*(1), 76–85.

Rodríguez, J., Avilés, J. M., & Parejo, D. (2011). The value of nestboxes in the conservation of Eurasian Rollers Coracias garrulus in southern Spain. *Ibis, 153*(4), 735–745.

Rosén, E., & van der Maarel, E. (2000). Restoration of alvar vegetation on Öland, Sweden. *Applied Vegetation Science, 3*(1), 65–72.

Rusch, A., Valantin-Morison, M., Roger-Estrade, J., et al. (2012). Using landscape indicators to predict high pest infestations and successful natural pest control at the regional scale. *Landscape and Urban Planning, 105*(1–2), 62–73.

Russ, J. M., Briffa, M., & Montgomery, W. I. (2003). Seasonal patterns in activity and habitat use by bats (Pipistrellus spp. and Nyctalus leisleri) in Northern Ireland, determined using a driven transect. *Journal of Zoology, 259*(3), 289–299.

Schacherer, A. (2007). 20 Jahre Ackerwildkrautschutz in Niedersachsen: Entstehung eines Förderprogramms. *Informationsdienst Naturschutz Niedersachsen, 27*(2), 79–85.

Schöttker, O., Johst, K., Drechsler, M., et al. (2016). Land for biodiversity conservation – To buy or borrow? *Ecological Economics, 129*, 94–103.

Schrautzer, J., Rinker, A., Jensen, K., Müller, F., Schwartze, P., & Dierßen, K. (2007). Succession and Restoration of Drained Fens: Perspectives from Northwestern Europe. In L. R. Walker, J. Walker, & R. J. Hobbs (Eds.), *Linking restoration and ecological succession* (pp. 90–120). New York: Springer.

Stammel, B., Kiehl, K., & Pfadenhauer, J. (2003). Alternative management on fens: Response of vegetation to grazing and mowing. *Applied Vegetation Science, 6*(2), 245–254.

Steidl, I. (2002). Beweidung von Feuchtgrünland – Ökologische, naturschutzfachliche und betriebsökonomische Aspekte im Landschaftspflegekonzept Bayern (LPK). *Laufener Seminarbeiträge, 1*, 67–83.

Stein-Bachinger, K., Fuchs, S., & Gottwald, F. (2010). Naturschutzfachliche Optimierung des Ökologischen Landbaus: „Naturschutzhof Brodowin". Ergebnisse des F+E-Projektes „Naturschutzhof Brodowin". *Naturschutz und biologische Vielfalt, 90.*

Sybertz, J., Matthies, S., Schaarschmidt, F., et al. (2017). Assessing the value of field margins for butterflies and plants: how to document and enhance biodiversity at the farm scale. *Agriculture, Ecosystems and Environment, 249*, 165–176.

Thies, C., & Tscharntke, T. (1999). Landscape structure and biological control in agroecosystems. *Science, 285*(5429), 893–895.

Török, P., Vida, E., Deák, B., et al. (2011). Grassland restoration on former croplands in Europe: An assessment of applicability of techniques and costs. *Biodiversity and Conservation, 20*(11), 2311–2332.

Tucker, G. (2005). *A review of biodiversity conservation performance measures*. https://www.cbd.int/doc/case-studies/suse/cs-suse-024-earthwatch-en.pdf. Accessed 21 Mar 2019.

Tucker, G., Hill, D., & Fasham, M. (2005). Introduction to planning. In D. Hill, M. Fasham, G. Tucker, et al. (Eds.), *Handbook of biodiversity methods: Survey, evaluation and monitoring* (pp. 3–5). Cambridge: Cambridge University Press.

van Buskirk, J., & Willi, Y. (2004). Enhancement of farmland biodiversity within set-aside land. *Conservation Biology, 18*(4), 987–994.

van Dijk, J., Stroetenga, M., van Bodegom, P. M., et al. (2007). The contribution of rewetting to vegetation restoration of degraded peat meadows. *Applied Vegetation Science, 10*(3), 315–324.

van Swaay, C. A. M. (2002). The importance of calcareous grasslands for butterflies in Europe. *Biological Conservation, 104*(3), 315–318.

von Haaren, C., & Bathke, M. (2008). Integrated landscape planning and remuneration of agri-environmental services: Results of a case study in the Fuhrberg region of Germany. *Journal of Environmental Management, 89*(3), 209–221.

von Haaren, C., Janßen, U., Haubfleisch, E., et al. (1997). Naturschutzfachliche Erfolgskontrollen von Pflege- und Entwicklungsplänen: Erfahrungen im Rahmen einer beispielhaften Durchführung an den Eifelmaaren. *Natur und Landschaft, 72*(7/8), 319–327.

Walker, K. J., Stevens, P. A., Stevens, D. P., et al. (2004). The restoration and re-creation of species-rich lowland grassland on land formerly managed for intensive agriculture in the UK. *Biological Conservation, 119*(1), 1–18.

Walpole, M. J., & Leader-Williams, N. (2002). Tourism and flagship species in conservation. *Biodiversity and Conservation, 11*(3), 543–547.

Weber, M. (2011). Starke Bestandszunahme und hohe Siedlungsdichte des Wiedehopfes (Upupa epops) in der Vorbergzone des nördlichen Ortenaukreises. *Naturschutz südl. Oberrhein, 6*, 43–49.

Wetterich, F., & Köpke, U. (2003). *Indikatoren für ein nationales Monitoring der Umwelteffekte landwirtschaftlicher Produktion. Band 2: Biologische Vielfalt und Landschaftsästhetik*. Projekt im Auftrag des Bundesministeriums für Umwelt, Naturschutz und Reaktorsicherheit (FKZ 200 12 118).

Wicke, G. (1998). Neue Entwicklung im Ackerwildkrautschutz. *Natur und Landschaft, 73*(3), 91.

Methods for Increasing Spatial and Cost Effectiveness of Measures Through Multifunctionality

26

Carolin Galler and Peter Stubkjær Andersen

Abstract

This chapter presents a quantitative approach for valuing the multifunctional effects of environmental measures in relation to the delivery of diverse ecosystem services. The approach consists of five methodological steps: (i) Evaluating the delivery of ecosystem services (ii) Estimating likely effects of environmental measures or land use changes on ecosystem services delivery (iii) Defining regional environmental quality objectives (valuation benchmarks) for each ecosystem service (iv) Quantifying multifunctional effects and (v) Assessing the added value with regard to the fulfillment of regional objectives through multifunctional compared to monofunctional measures. The approach also sheds light on the quantity *and* quality of multifunctional effects of land use options on ecosystem services, including trade-offs between individual services.

Keywords

Landscape multifunctionality · Ecosystem services · Trade-offs

C. Galler (✉)
Niedersächsisches Ministerium für Umwelt, Energie und Klimaschutz, Hannover, Germany
e-mail: carolin.galler@mu.niedersachsen.de

P. S. Andersen
Department of Geosciences and Natural Resource Management, Landscape Architecture and Planning, University of Copenhagen, Frederiksberg, Denmark
e-mail: stub@ign.ku.dk

C. von Haaren et al. (eds.), *Landscape Planning with Ecosystem Services*,
Landscape Series 24, https://doi.org/10.1007/978-94-024-1681-7_26

26.1 Quantifying and Optimising Multifunctionality

The development of integrated landscape development strategies necessitates the consideration of multiple ecosystem services and their trade-offs (Raudsepp-Hearne et al. 2010). Assessing such multifunctional environmental effects is challenging and is not yet systematically accounted for in planning and decision making processes (Selman 2009, 2012).

This chapter presents a method for assessing the likely effects of landscape management measures on diverse ecosystem services. It builds on insights from Chap. 19 of this book which introduced methods for evaluating multifunctionality and focused on identifying and assessing multifunctional areas.

In general, landscape management measures can be differentiated according to their degree of multifunctionality (Bennett et al. 2009). Monofunctional measures, i.e. measures to optimize a single ES, may either be neutral with respect to other ES or possibly in conflict with them. Whilst implementing several spatially overlapping monofunctional measures, multiple ES benefits may be achieved on the same piece of land. The other group of management measures are those which create synergies for multiple ES. That is, two or more ES benefit from one measure. However, these multifunctional measures may vary in terms of their effectiveness for each individual ES.

When responses affect multiple ecosystem services this may – in general – enhance their effectiveness and the efficient use of land resources. However, multifunctional measures are not necessarily more efficient than monofunctional measures. Hence it is an important question for landscape planning to find the best option to maximize multifunctional effects while investing the least money or utilizing the smallest land area (Torralba et al. 2018).

The assessment method presented here is indicator-based and considers the area-specific conditions for delivering ES. As the available area or the cost of management are often decisive factors when deciding whether to implement environmental measures, mono- and multifunctional management options should be assessed in terms of their cost- and area efficiency. This requires quantification of the effects on the target ES as well as of the costs and land area needed to reach the desired objectives.

The method is illustrated with a case study which is described in more detail in Galler et al. (2015) and Galler (2016). It can be applied for planning at either regional or local levels. Both multifunctional effects of land use or management concepts (e.g. within landscape planning) or of single measures (e.g. agri-environmental measures) can be quantified. Planning tasks that can be handled by applying the method include:

- where to apply a certain type of measure in order to maximize multifunctional effects for selected ES,
- how to avoid or minimize trade-offs between different ES,
- evaluating the level of multifunctionality that can be attained and rating the multifunctional effects that can be achieved by a certain measure.

26.2 General Conditions and Prerequisites

Unlike approaches which classify the multifunctionality of measures or land uses (e.g. Burkhard et al. 2012), the method presented here is based on spatially-explicit assessments and evaluations of particular ecosystem services. Such information is rarely generated by one discipline alone, but dispersed across the various policy sectors and disciplines of environmental management and planning including, for example, soil management, water management, and nature conservation. In this respect, multifunctionality analysis is an interdisciplinary and cross-sectoral exercise. To value multifunctionality, planners should integrate indicators and valuation methods that are established in the respective disciplines in order to make use of existing data. Given limited capacities within the environmental planning sectors, multifunctionality analysis should not require additional data collection.

In Germany and other European countries, a synopsis of information about different landscape functions can usually be derived from landscape plans. However, these data tend to describe the supply of ES (delivered ES) rather than demand (utilised ES). The approach presented here therefore focuses on delivered ES and measures for their management.

The complexity of the assessment will increase with the number of ES included. The set of ES which should be considered in the multifunctionality analysis needs to be identified with respect to the specific planning task (see Chap. 19). Furthermore, both the selected ES and those excluded from analysis need to be transparently stated.

26.3 Methodology

The case study focuses on four landscape functions: natural capacity for crop yields (including soil erosion prevention), water resources (water quality), climate change mitigation and biodiversity. Multifunctionality is quantified by considering synergies and trade-offs of individual ES management measures to generate two kinds of information: i) The number of (delivered) ES that are affected by the management measures and ii) the sum of (physical) effects of the management measures on multiple ES.

The actual effects of measures vary according to the conditions of the sites where they are implemented. Therefore, the method follows the principle of an environmental impact analysis, following the Driving force-Pressure-State-Impact-Response (DPSIR) model (Smeets and Weterings 1999; Tscherning et al. 2012). Similar to the approach of an ecological risk analysis, the assessment takes into account the general effects on delivered ES and the results of the spatial-functional process analysis.

A key indicator of the effectiveness of measures is the increase in delivered ES. In areas where the natural conditions only allow for limited functionality, the demonstrated effects of measures may be limited. Where potential high functionality is reduced by existing impairments in conditions then the (multifunctional) effects of measures may be realizable to a higher extent.

The assessment consists of five steps which are summarized in Fig. 26.1. The site-specific effects of applying environmental measures for each ES are derived from an evaluation of present landscape conditions (landscape capacities, specific sensitivities, actual impacts and underlying pressures) on the one hand (see Step 1) and from an estimation of the maximum possible effects of environmental measures on the other hand (see Step 2). The quantified physical effects for each ES can be related to regional quality objectives derived from legislation (e.g. legally binding thresholds), political decisions (e.g. long term objectives) or scientifically (e.g. critical load) (see Step 3). Multifunctional effects can then be quantified in terms of the proportional fulfillment of these objectives (see Step 4). Hence, the environmental quality objectives (EQO) for the study region serve as benchmarks for quantifying multifunctional effects. In Step 5 the cost- and area-efficiency of the measures is calculated. This allows an appraisal of their contribution to efficient landscape management.

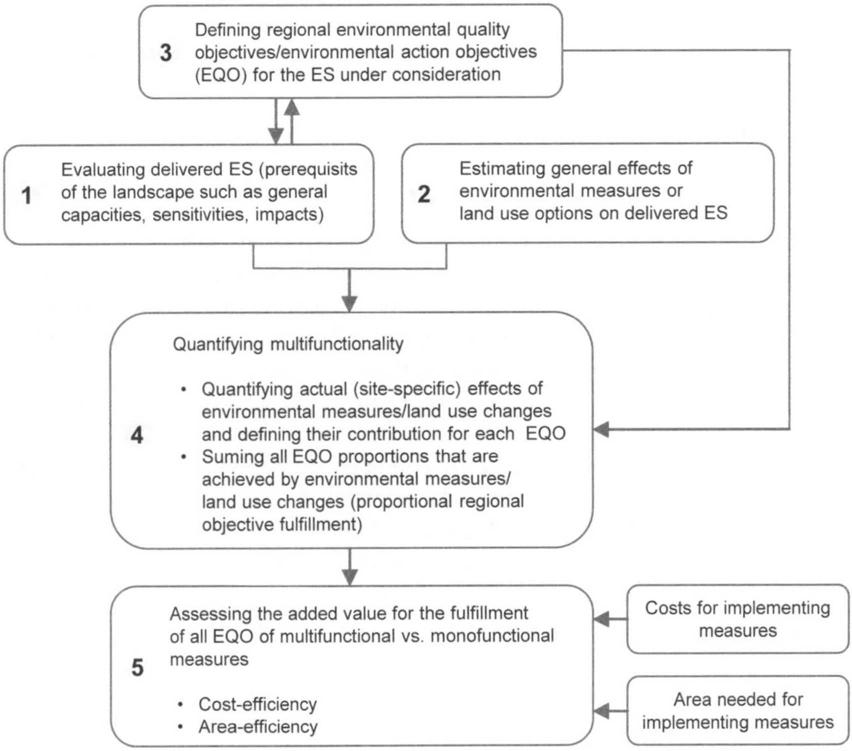

Fig. 26.1 Flowchart of the working steps

26.3.1 Step 1: Evaluating Delivered ES

A geographical information system (GIS) was used to assess the scope for locations within a region to provide the relevant ES, both in terms of existing delivery and the potential for increases. An overview of possible state and impact indicators is given in Albert et al. (2016). In this case study, which focused on the county of Verden in Lower Saxony, four main landscape functions were examined. The indicators used to assess these are listed in Table 26.1 and information extracted from regional sectoral and landscape plans was used to quantify them.

These assessment results were then overlaid in the GIS to provide an overview of the number of landscape functions that could be improved in each area (for more details, see Galler et al. (2015) and Galler (2016)). The map in Fig. 26.2 shows the result, with the darker shadings representing greater scope for multifunctional delivery.

26.3.2 Step 2: Estimating Effects of Measures on Delivered ES

Environmental measures vary in the number of ES they influence. However, the extent to which the effects actually occur can also depend on the natural conditions of the sites where the measures are implemented. Information from literature (Osterburg and Runge 2007; Saathoff et al. 2012; von Drachenfels 2012) and regional plans was used to construct look-up tables summarizing the effects of different measures on the indicators listed in Table 26.1. These look-up tables also took

Table 26.1 ES analysis indicators used in the case study

Purpose of indicator	Water quality	Crop yield	Climate protection	Biodiversity
Indicators for spatial differentiation of general capacity to deliver ES	Chemical status (nitrate) of ground water bodies	Risk of wind or water erosion and flooding	CO_2-sequestration of soils (based on soil and land use)	Value of biotope according to von Drachenfels 2012
Indicators for locating and grading areas with specific sensitivities	No spatial differentiation	Erosion prone sites (water erosion: Universal Soil Loss Equation, wind erosion: soil type), floodplains	Soil type with CO_2 retention potential	Potential for upgrading biotope value
Indicators for quantifying the impacts of measures	N-input (kg N/a)	Soil erosion (erosion protection)	CO_2-sequestration (kg CO_2-emission from soil)	Upgrade in biotope value points (biotope value x area size)

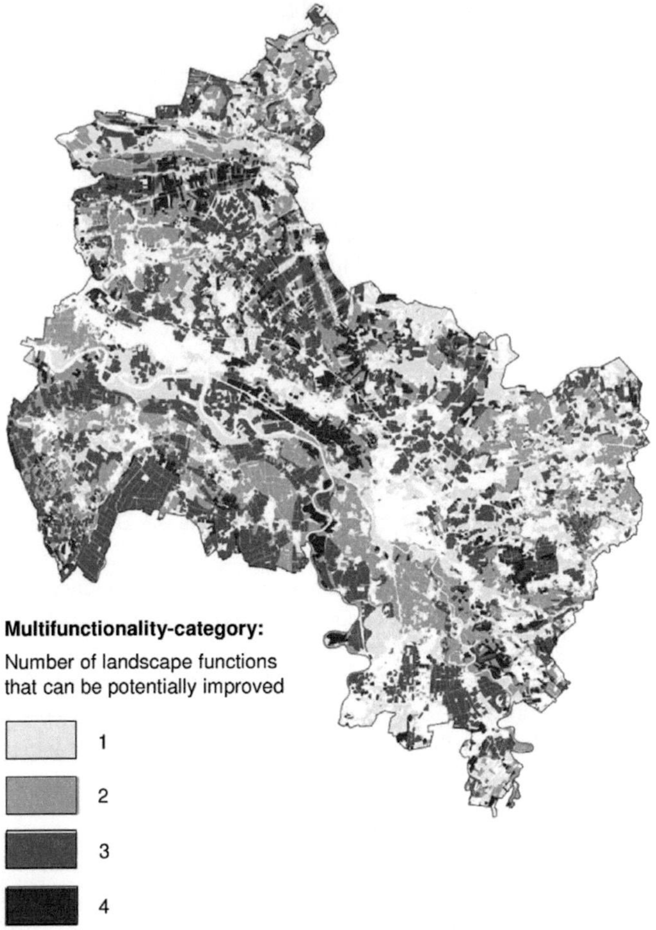

Fig. 26.2 Scope for improved multifunctional delivery in the agricultural areas of Verden county. (Source: Galler et al. 2015)

account of any interactions between measures and site-specific conditions and their effects are classified for respective site-conditions. For example, the effects of converting cropland to extensive grassland in terms of climate protection vary according to the soil type of the field involved.

Estimating the general effects of measures also provides information about their potential multifunctionality. For example, reducing nitrogen fertilizer applications is a monofunctional measure for water protection with no effects on erosion protection, decrease of soil-based CO_2-emissions and, hence, to this respect no effects on climate protection and – at least on fields with very low habitat potential – for biodiversity. In contrast, conversion of land to extensive grassland is a measure with potentially high multifunctionality. However, the degree to which it effects soil protection, climate protection and biodiversity depends on other characteristics of the site involved. For example, large increases in CO_2-sequestration can only be achieved on hydromorphic soils.

26.3.3 Step 3: Defining a Regional Environmental Quality Objective (EQO) for Each ES

A regional environmental quality objective (EQO) was determined for each ES. This drew upon knowledge of relevant legislation, targets in the landscape plan, and the landscape capacities in the relevant region. Table 26.2 lists and explains the EQO defined for the Verden study area.

The EQO serves as a benchmark for each ES and as reference levels for quantifying multifunctional effects of environmental measures. Within each area, the average proportional fulfillment of objectives with respect to different ES can be used as a measure of the multifunctional effects of different measures (see Fig. 26.3). Using proportional fulfillment of EQO in this way makes the different physical effects on various ES comparable.

26.3.4 Step 4: Quantifying Multifunctionality

The information on landscape conditions and effects of measures compiled in Steps 1 and 2 was applied to specific sites within the study region by using a GIS to combine and intersect attributes for each polygon. As illustrated in Fig. 26.3 the effects on delivered ES (originally measured in specific physical units) were then re-scaled as proportional fulfillment of regional objectives. This made the assessment results for different ES comparable.

Multifunctional effects were quantified as total environmental objective fulfillment in percentages, divided by the number of objectives included in the assessment. Hence, if all the specified environmental objectives were attained, the total objective fulfillment would be 100%. This approach makes it possible to quantify multifunctional effects either for particular measures or management concepts that are intended for specific sites or planning regions.

26.3.5 Step 5: Comparing the Impacts of Multifunctional vs. Monofunctional Measures

In the final stage, individual measures are compared in relation to their contribution to regional quality objectives. It is also possible to compare monofunctional against multifunctional measures in terms of their overall environmental effectiveness. Cost- and area-efficiency can be calculated by relating measure costs or the area required to implement them to the proportional objective fulfillment achieved in order to derive cost-benefit-ratios.

Table 26.2 Environmental quality objectives in the study area

Landscape function	Quality objective	Explanation
Soil erosion prevention	Preventive action on all prone sites (18,983 ha)	Erosion of soil by wind, water and floods is considered a major factor that impacts yields. The regional landscape plan indicates erosion prone sites, where preventative action is necessary. These account for 18,983 hectares of cropland.
Water quality conservation	Reduction of 1500 tonnes annual N-input	The environmental quality objective refers to the EU Water Framework Directive and Groundwater Directive that require a maximum 50 mg NO_3/l concentration in percolate water below all agricultural land for groundwater and maximum 3 mg/l N-concentration for inputs to surface water. For the river basin in which the case study region is located, the total amount of N-reduction (in tons per year) on agricultural land required to achieve good status of water bodies was calculated by Kreins et al. (2009). The environmental quality objective for the case study region follows this limit and amounts to a proportional N-reduction of about 1500 tonnes a year.
Climate change mitigation	Reduction of 10,604,017 tonnes land use related CO_2-emissions	The objective for this landscape function considers the theoretical potential for carbon sequestration of soils in the case study region and the possible reduction of soil based CO_2-emissions caused by agricultural use. To estimate the potential soil based CO_2-emission from the utilized agricultural area in the region we used a method from Saathoff et al. (2012). The calculation is based on the soil organic carbon content and considers the total amount of CO_2 emitted over an indefinite period of time. The estimated reduction of land use related CO_2-emissions amounts to 10,604,017 tonnes in the study area.
Safeguarding biodiversity	Increase of 47,909 habitat value points	The objective of safeguarding biodiversity is represented in the regional landscape plan through the spatially explicit habitat concept. The targeted habitat types on agricultural land constitute the environmental quality objective. For all habitat types their value has been estimated in habitat value points (VP) by the Lower Saxony State Office (von Drachenfels 2012). These VP were used to quantify the environmental quality objective. The differences between the VP of the status quo and targeted status were multiplied by the size of each area (this is similar to the procedure used to assess the need for compensation measures within German impact mitigation regulation). Meeting the EQO for the utilized agricultural area (UAA) in the case study region required upgrading the habitat value points total to 47,909 VP. The biodiversity effects of environmental measures within the scenarios were quantified by the increase in VP.

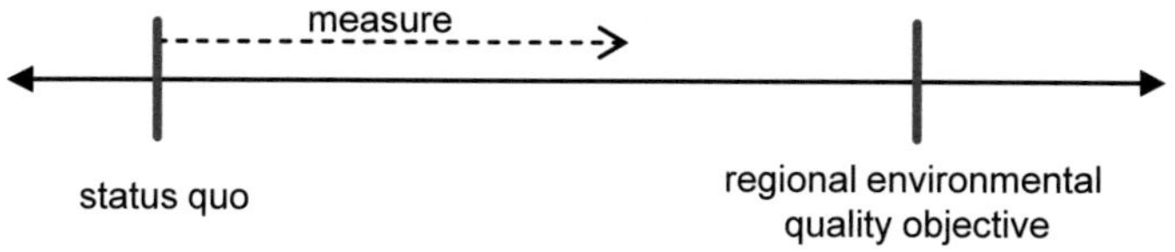

Fig. 26.3 Using the regional EQO as a benchmark to value multifunctional effects of measures

Table 26.3 Outcome of four scenarios involving different combinations of measures

	Scenarios			
	S1	S2	I1	I2
Soil erosion prevention (%)	22.4	17.2	9.6	43.9
Water quality conservation (%)	45.9	19.9	6.5	47.3
Climate change mitigation (%)	5.7	9.2	8.5	5.7
Safeguarding biodiversity (%)	12.4	13.8	12.8	47.5
Overall objective fulfillment (%)	21.6	15.0	9.4	36.1
Spatial efficiency (proportion of total objective fulfillment per 1000 hectares)	0.9	1.8	4.9	1.8
Cost efficiency (implementation costs for 10 years in € per 1% total objective fulfillment)	1,850,742	2,665,398	4,268,903	1,108,517
Area of measures on UAA (ha)	23,064	8496	1912	19,539
% of total UAA	43.8	16.1	3.6	37.1

The results for individual measures can be compared to generate different management scenarios for regions. Table 26.3 shows the outcomes and results of the following four scenarios in the study region.

- S1: uncoordinated sectoral scenario, equal distribution of financial resources to the sectoral implementation programs
- S2: uncoordinated sectoral scenario, unequal distribution of financial resources for a balanced fulfillment of environmental objectives
- I1: integrative scenario 'optimizing spatial efficiency'
- I2: integrative scenario 'optimizing cost efficiency'

Fig 26.4 illustrates the results of the scenarios for optimizing spatial and cost efficiency.

Overall, the results suggest that the integrative scenarios perform better in a number of important respects that the purely sectoral ones. In particular, the integrative scenario that is spatially optimized (I1) achieves the highest environmental benefit per hectare. None of the other three scenarios achieve even half the level of benefit (0.9% or 1.8% compared to 4.9% in Table 26.3). However, although this scenario involves measures on the smallest proportion of UAA it is also by far the most expensive (i.e. poorest in terms of cost efficiency). This implies that when the budget is limited, optimizing spatial efficiency can lead to a low total objective fulfillment (i.e. the bars in Fig. 26.4 are generally lower for scenario I1 than I2).

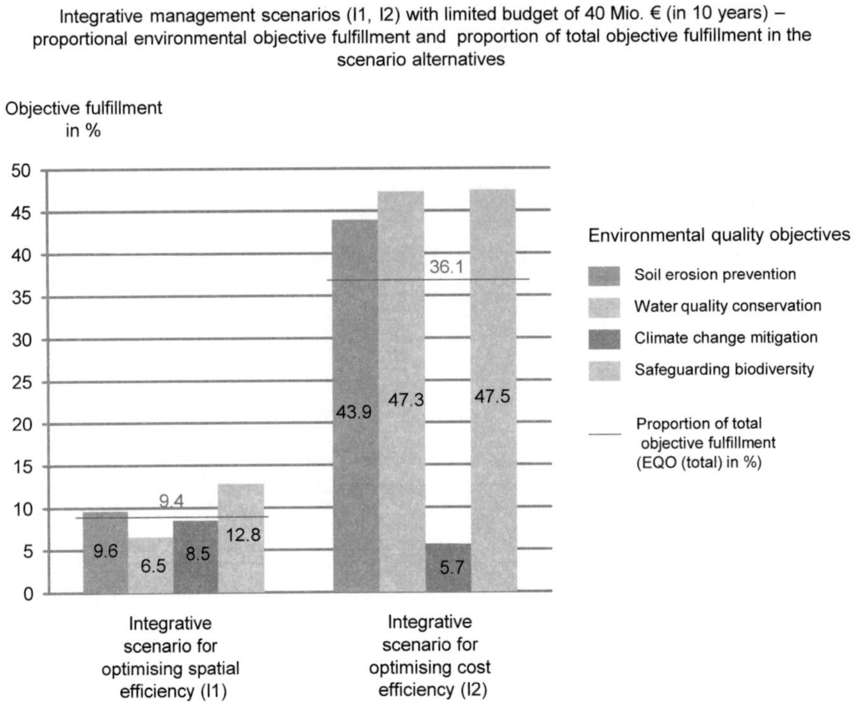

Fig. 26.4 Comparison of multifunctional effects of two management scenarios, oriented towards optimizing spatial efficiency (I1) and towards cost efficiency (I2), in the utilized agricultural area of the county of Verden, Germany

A second important point is that there were some significant trade-offs for environmental objectives. Measures with greater multifunctional effects usually involve more restrictions on land use and higher implementation costs. For example, measures for climate change mitigation and safeguarding biodiversity are generally multifunctional. They often simultaneously contribute to both water quality conservation and erosion prevention. However, these measures are the most costly. In contrast, many typical measures for water quality conservation generally only provide a few additional benefits to other landscape functions. They are comparably inexpensive, but have a lower spatial efficiency. These effects can be seen in some of the results where, for example the spatially optimized integrative scenario (I1) had a much lower objective fulfillment for water quality conservation. Another feature was that none of the scenarios were especially effective in terms of climate change mitigation. See Galler et al. (2015) for further discussion of the results.

26.3.6 Methodological Challenges

The approach discussed above allows for a quantitative valuation of multifunctional effects of environmental measures. Hence, for landscape planning purposes it helps to indicate where multiple effects can be achieved. However, it differs fundamentally from ordinally-scaled classification of multifunctionality (cf. Galler et al. 2015). Other than in ordinally-scaled methods, qualitative parameters are generally inadequate to use and can be included only in addition to quantitative indicators.

Defining regional EQOs plays a key role within the approach. This highlights the normative background of ES valuation, but it should be noted that the degree to which objectives are obtained depends on the targets that are set. In other words, specifying more ambitious environmental objectives leads to a lower proportional fulfillment while the effects of measures are the same. Having benchmarks to highlight regional differences is nevertheless desirable, because areas will vary in their potentials for delivering ES. If required, quality objectives may be differentially weighted using multiplication factors in order to respond to contrasts in landscape capacities.

26.4 Added Value for Planning and Governance

Landscape planning is particularly important for coordinating responses to environmental issues. It is the most inclusive form of environmental planning because it integrates different environmental objectives. Fundamentally, a main task of landscape planning is the overall assessment of landscape functions, including specifying and prioritizing of environmental objectives.

The method described in this chapter can be used within environmental planning to assess the implications of plans or projects for different landscape functions in terms of delivered ES. Alternative management actions can be compared concerning the quantity *and* quality of (presumed) multifunctional effects. Quantification of multifunctional effects allows for a standardised consideration of ES synergies and trade-offs.

The assessment results from the case study show that the effectiveness of the same set of measures can be increased when multifunctional effects are optimized (cf. Galler et al. 2015; Galler 2016). This can be achieved by allocating measures with potentially high multifunctionality on sites which have the conditions to generate these multifunctional effects. Hence, the efficiency of management can be increased when cost-intensive measures with potential effects on diverse ES are located on sites with relevance for these ES, whereas less cost-intensive (but monofunctional) measures are applied only where there is scope to improve the relevant single ES.

This also implies that multifunctional management measures are not always the most efficient option. Furthermore, there may be situations in which there is a high priority for a certain ES (e.g. species protection), and the recommended specialized management may be in conflict with other ES. In such circumstances monofunctional measures may be inevitable.

For the assessment of multifunctionality (as well as to handle ES trade-offs in general), regional EQOs for safeguarding and enhancing delivered ES need to be defined for the planning or administrative region. Landscape planning offers the possibility to develop a spatially-explicit integrated management concept, including harmonized environmental quality objectives that other sector-administrations can refer to. However, defining these regional quality objectives is challenging because it requires cross-sectoral and vertical/cross-scale coordination. An added value of relating the assessment of ES to regional environmental quality objectives is that the results demonstrate the relevance of multifunctional effects in achieving wider policy aspirations.

References

Albert, C., Galler, C., Hermes, J., et al. (2016). Applying ecosystem services indicators in landscape planning and management: The ES-in-Planning framework. *Ecological Indicators, 61*(1), 100–113.

Bennett, E. M., Peterson, G. D., & Gordon, L. J. (2009). Understanding relationships among multiple ecosystem services. *Ecology Letters, 12*(12), 1394–1404.

Burkhard, B., Kroll, F., Nedkov, S., & Müller, F. (2012). Mapping ecosystem service supply, demand and budgets. *Ecological Indicators, 21*, 17–29. https://doi.org/10.1016/j.ecolind.2011.06.019.

Galler, C. (2016). Multifunktionalität von Umweltmaßnahmen: Quantifizierung multipler Umweltwirkungen und ihre Berücksichtigung in der Umweltplanung. Dissertation, Hannover.

Galler, C., von Haaren, C., & Albert, C. (2015). Optimizing environmental measures for landscape multifunctionality: Effectiveness, efficiency and recommendations for agri-environmental programs. *Journal of Environmental Management, 151*, 243–257.

Kreins, P., Behrendt, H., Gömann, H., et al. (2009). Analyse von Agrar- und Umweltmaßnahmen im Bereich des landwirtschaftlichen Gewässerschutzes vor dem Hintergrund der EG-Wasserrahmenrichtlinie in der Flussgebietseinheit Weser. *Landbauforschung vTI Agriculture and Forestry Research, 336*, 1–308.

Osterburg, B., & Runge, T. (2007). Maßnahmen zur Reduzierung von Stickstoffeinträgen in Gewässer – eine wasserschutzorientierte Landwirtschaft zur Umsetzung der Wasserrahmenrichtlinie. *Landbauforschung Völkenrode – FAL Agricultural Research. Special Issue, 307*, 1–302.

Raudsepp-Hearne, C., Peterson, G. D., & Bennett, E. M. (2010). Ecosystem service bundles for analyzing tradeoffs in diverse landscapes. *Proceedings of the National Academy of Sciences of the United States of America, 107*(11), 5242–5247. https://doi.org/10.1073/pnas.0907284107.

Saathoff, W., von Haaren, C., Dechow, R., et al. (2012). Farm-level assessment of CO2 and N2O emissions in Lower Saxony and comparison of implementation potentials for mitigation measures in Germany and England. *Regional Environmental Change, 13*, 825–841. https://doi.org/10.1007/s10113-012-0364-8.

Selman, P. (2009). Planning for landscape multifunctionality. *Sustainability: Science, Practice, and Policy, 5*(2), 45–52.

Selman, P. (2012). *Sustainable landscape planning: The reconnection agena*. London: Routledge.

Smeets, E., & Weterings, R. (1999). Environmental indicators: Typology and overview. Technical report No 25. EEA, Copenhagen.
Torralba, M., Fagerholm, N., Hartel, T., et al. (2018). A social-ecological analysis of ecosystem services supply and trade-offs in European wood-pastures. *Science Advances, 4*(5), eaar2176. https://doi.org/10.1126/sciadv.aar2176.
Tscherning, K., Helming, K., Krippner, B., et al. (2012). Does research applying the DPSIR framework support decision making? *Land Use Policy, 29*, 102–110.
von Drachenfels, O. (2012). Liste der Biotoptypen in Niedersachsen mit Angaben zu Regenerationsfähigkeit, Wertstufen, Grundwasserabhängigkeit, Nährstoffempfindlichkeit und Gefährdung (Rote Liste). *Infdienst Natschutz Niedersachs, 32*, 1–60.

Leitbilder and Scenarios in Landscape Planning 27

Christian Albert, Christina von Haaren, Hermann Klug, and Raphael Weber

Abstract

This chapter introduces the concepts of 'Leitbilder' and scenarios in landscape planning. Leitbilder can be understood as descriptions of target states that diverse stakeholders can agree on. Scenarios represent plausible descriptions of pathways of change that can help explore resulting future land use changes (alternative futures) and their respective impacts.

Keywords

Leitbild · Scenarios · Alternative futures

27.1 'Leitbilder' in Landscape Planning

The term 'Leitbild' is used in various ways in landscape planning and nature conservation in Germany. Based on an extensive review of relevant literature, (Potschin et al. 2010: 657) define Leitbilder as follows:

> A Leitbild (pl. Leitbilder) is a summary statement describing a desired and releasable future state for a specific issue or spatial unit, which takes account of the primary objectives and drivers in a holistic and integrated way. All present knowledge is used to balance future constraints and demands from social, economic, cultural, political and environmental perspectives. Therefore, a commonly accepted Leitbild projects a specified trajectory for

C. Albert (✉) · C. von Haaren · R. Weber
Institute of Environmental Planning, Leibniz University Hannover, Hannover, Germany
e-mail: albert@umwelt.uni-hannover.de; haaren@umwelt.uni-hannover.de

H. Klug
Interfaculty Department of Geoinformatics – Z_GIS, University of Salzburg, Salzburg, Austria
e-mail: hermann.klug@sbg.ac.at

C. von Haaren et al. (eds.), *Landscape Planning with Ecosystem Services*, Landscape Series 24, https://doi.org/10.1007/978-94-024-1681-7_27

the future spatial structure, distribution, utilisation, condition and development of the socio-natural system. It provides a set of guidelines that shape actions, and a framework within which the impact of particular developments can be judged and socially negotiated.

While some authors interpret a Leitbild as a sectoral activity, others use the term to describe a cross-sectoral, integrated vision for a particular study area. In addition, a Leitbild can be of varying levels of detail – ranging from vague ideas of the general vision to spatially explicit, detailed concepts with locally specified development targets (cf. von Haaren 2004).

A Leitbild includes a future vision as well as guidelines on how an identified target condition can be reached. The objectives are identified and deduced on the basis of the fundamental norms and principles of sustainable development (Chap. 4), which also guide the assessment of ecosystem services. As described in Chap. 21, the derivation of objectives and measures often results in a complex system of mandatory and desirable objectives, which may be hard to communicate to the public and decision makers. The key role of a Leitbild is to put the complex public and possibly individual objectives in the context of other social, economic and environmental information and to transfer this into an easily comprehensible but holistic-integrated view of the future conditions (including functional relationships) of the landscape to be attained. This makes the Leitbild development process particularly important for communication between different stakeholders. For an example of the Leitbild approach see the paper by Klug (2012).

Leitbild prepare for, and complement, detailed landscape development targets and corresponding mid-term actions. Developments and resulting consequences need to be reviewed continuously and considered carefully in the planning process. Thus, a Leitbild is not a static and final concept but illustrates the spectrum between minimum and optimal developments (Klug 2006). It emphasizes different alternative solutions within planning objectives and priorities set, e.g. a Leitbild on 'sustainable use' or 'optimizing habitat and species protection'. A Leitbild can consist of text, maps, or images created by the planning team. Increasingly, a Leitbilt includes spatio-temporal analysis and visualization using computer-generated maps and 3D Models (Schroth et al. 2011; Wissen-Hayek 2009; Shaw et al. 2009).

A recurring theme in the debate about Leitbilder is the question of whether historic, cultural landscapes or a multifunctional landscape should be used as the preferred end point. While historic cultural landscapes often seem to be the aspired option for many citizens, their development is often not feasible within the current governance and implementation context (von Haaren 1988, 1991). Instead, landscapes designed for functionality, for delivering ES, including aesthetics, are usually considered much more appropriate. Such multifunctional landscapes can, however, include small-scale targets for conserving, restoring or creating historic forms of landscape development, but only within a broader fabric of multifunctional use.

As described in depth by Potschin et al. (2010), using a case study of Leitbild development from the Mondsee in Austria, the creation of scenarios of future changes can be a useful approach to explore diverging perspectives and to come to common ideas for a Leitbild. While Leitbilder focus primarily on the potential target situation, scenarios emphasize the pathways of change and potential impacts on different objectives.

27.2 Scenarios in Landscape Planning

Scenarios have been formally used at least since the end of World War II in the field of war game analysis (Shoemaker 1993; van der Heijden 1996). Civilian application of the scenario technique in planning was pioneered by Herman Kahn (1967) and others and has been further developed and applied in business planning (e.g. Wack 1985a, b; von Reibnitz 1987; Gausemeier et al. 1995; Georgantzas and Acar 1995; Schwartz 1996; van der Heijden 1996). At least since the 'Limits of Growth' study by Meadows et al. (1972), scenarios have been applied to numerous long-term environmental challenges of public concern, ranging from global to regional and local scales (Gallopin et al. 1997; Nakicenovic et al. 2000; Raskin et al. 2002).

The field of landscape and environmental planning has seen the application of scenarios for several decades. These have received increasing attention in recent years, which is reflected in a growing number of publications in relevant journals (e.g. Albert et al. 2012, 2016; Fritsch 2002; Theobald and Hobbs 2002; Steinitz et al. 2003; Tress and Tress 2003; Baker et al. 2004; Hulse et al. 2004; Nassauer and Corry 2004; Santelmann et al. 2004; Shearer 2005; Sisk et al. 2006; Bohnet and Smith 2007; Stock et al. 2007; Walz et al. 2007; Grêt-Regamey et al. 2008; Schroth et al. 2009). Excellent reviews of the history and use of scenarios in spatial planning are available in Shearer (2005) and Xiang and Clarke (2003).

Within landscape planning, scenarios can be used to develop storylines of future landscape change, e.g. modelling the potential land use and land cover changes (LUCC) resulting from them (alternative futures), and assessing their consequences (cf. Hulse et al. 2004). Given that the terms 'scenarios', 'alternative futures' and 'modelling' are interpreted differently in the literature, depending on context and discipline, some further clarifications is needed. According to the Millennium Ecosystem Assessment (MEA) scenarios are:

> "plausible and often simplified descriptions of how the future may develop based on a coherent and internally consistent set of assumptions about key driving forces and relationships" (Carpenter et al. 2005: 148).

Scenarios describe potential futures and the ways of attaining them, as well as potential impacts on diverse targets. While scenarios describe potential pathways of change, alternative futures are understood as possible end states. The latter illustrate (for example) the land use and land cover configurations of the landscape that may result from the changes within a particular scenario at a specified point of time in the future (Fig. 27.1, cf. Steinitz et al. 2003, Shearer 2005). Modeling LUCC change

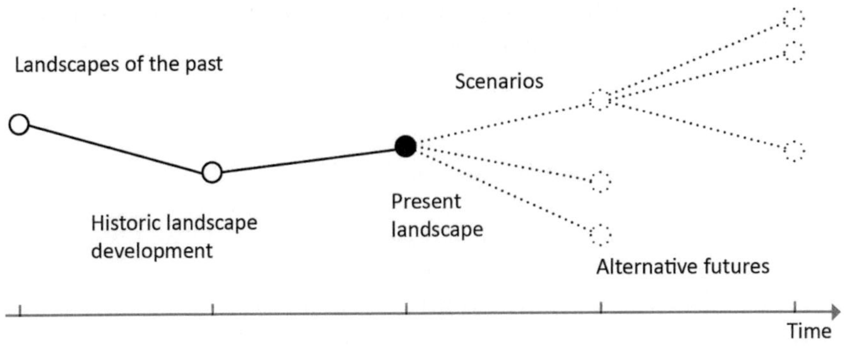

Fig. 27.1 Scenarios and alternative futures. (Source: Albert 2011, based on Steinitz et al. 2003)

is a means of producing a simulation of potential future landscape patterns, which may be based on a formal computer model or intuitive reasoning, based on simple decision rules.

27.2.1 Types of Scenarios

Scenarios in landscape planning can be of various types and embody various approaches. For example, Bradfield et al. (2005) describe the evolution of three 'schools' of scenario development, while Bishop et al. (2007) summarise the different techniques for developing scenarios. Van Notten et al. (2003, 2005) present a scheme consisting of three main themes (project goal, process design, and scenario content) with various additional parameters.

To answer the question of which kind of scenario is best to use in landscape planning, Börjeson et al. (2006) propose a particularly useful typology. The authors distinguish between predictive, explorative and normative scenarios with two sub-categories for each type (see Table 27.1).

The first group of scenario types is predictive and responds to the question *'what will happen?'* in the more or less near future. One kind of predictive scenario is forecasts, dealing with the results of a possible event occurring with high probability. The second kind of predictive scenario, what-if, predicts the impact based on a set of preconditions. For example, what-if scenarios may reflect several possible outcomes based on the different impacts of a participation initiative.

A second group is explorative scenarios which concentrate on the question *'what can happen?'*. These focus on the far future horizon and can be divided into external and strategic scenarios. External scenarios observe the outcome of external events on the local setting or policy. In contrast, strategic scenarios explore the impact of an implemented action or policy on an event.

The third group are normative scenarios, asking *'how a specific target can be reached?'* and include preserving and transforming scenarios. Preserving scenarios focus on internal decisions such as measures and policies which could be implemented and unfold their effects to reach a set target. Transforming scenarios,

Table 27.1 Scenario types based on Börjeson et al. (2006)

Scenario type	Description	Targeted time frame	Influencing factors	Landscape planning example
Predictive				
Forecast	Predicts the result after a foreseeable development (*What will happen?*)	Near future	External events	Forecasting a single natural phenomenon
What-if	Predicts the result based on certain preconditions (*What will happen?*)	Near future	External events	Effects of an external event (e.g. Brexit) on local landscape change
Explorative				
External	Explores possible effects of an external event (*What can happen?*)	Far future	External events	Impacts of climate change on habitats
Strategic	Explores the impact of an implemented action (*What can happen?*)	Far future	Internal decisions	Impacts of a policy choice
Normative				
Preserving	Focuses on the impact of policies or measures on the current situation (*How can a specific target be reached?*)	Near future	Internal decisions	Introducing new fertiliser application limits
Transforming	Examines the weaknesses of internal structure that could hinder reaching a specific target (*How can a specific target be reached?*)	Past (backcasting from far future)	Internal decisions	Setting up an agenda or vision

in contrast, set up an agenda or vision of the future and cast back from that point towards the present. This analysis aims to identify weak points of the internal structure which may be hindrances to achieving a certain goal. Another widely-cited approach to the characterization of scenarios is that proposed by Van Notten and colleagues (2003, 2005), addressing alternative objectives, processes, and outputs.

The objectives of using scenarios in landscape planning can range from exploration to decision support. Explorative scenarios aim at awareness raising, facilitating creative thinking, and studying the complex interactions of different processes over time. Decision support-oriented planning uses scenarios that are more or less desirable. The vantage point of the scenarios may be either forecasting or backcasting (Haslauer et al. 2012). Forecasting scenarios start from the present and explore how the future might evolve. Backcasting scenarios assume a specific future situation and explore the range of actions or developments necessary to attain (or not) the projected condition.

27.2.2 Implementing Scenario-Based Planning

The process of scenario-based planning varies according to the degree of quantitative and qualitative data and approaches used, the choice of methods, and the level of involvement of decision makers and stakeholders. At one end of the range is the intuitive approach that relies strongly on qualitative methods. The approach may use narrative outlines, texts, storylines, diagrams, pictures and/or collages to describe future developments with high levels of complexity and uncertainty. It may include non-quantifiable, normative aspects like values, mental maps, and expectations. At the other end of the spectrum is the formal approach, consisting of a rather rational and analytical exercise and often employing quantitative methods and formalized computer models. The latter approach offers structural consistency and scientific rigor through explicit assumptions. Both approaches have their advantages and recent efforts increasingly aim at combining them (e.g. Alcamo 2008; van Vliet et al. 2010). Other developments emphasise an 'Automated Geosynthesis' where standardised (real-time) data offerings are combined with open modelling interfaces for real-time spatio-temporal scenario building with stakeholders (Klug and Kmoch 2015).

The involvement and input of stakeholders and decision makers in scenario-based planning varies on a gradient from citizen-driven to expert-driven approaches (Hulse et al. 2004). The gradient can be further classified into five different levels of involvement (Arnstein 1969; Pahl-Wostl 2008; Volkery et al. 2008). At the lowest level, stakeholders and decision makers are only *informed* about the process and results of a scenario exercise. Transdisciplinary intensive participation occurs when non-scientific stakeholders are *consulted* during the exercise to provide input. *Co-thinking*, the third level, means that participants are actively involved in the development of the scenarios but do not make decisions. At the *co-designing* stage, participants are furthermore engaged in the structuring of the scenario process and the joint definition of 'game rules' for collaboration. Finally, participants can *co-decide* and assume responsibility for the scenario process design, the analysis, and the recommendations derived from them.

Over the last two decades, a range of approaches and frameworks for implementing scenario-based planning have been proposed (cf. Horlitz 1998; Wollenberg et al. 2000; von Haaren 2004; Ahern 2006; Scholles 2008). While the approaches differ, most of them consist of steps to define scenario assumptions. The scenario assumptions then shape pattern-process relationships to impact the modelled land use/land cover changes and resulting consequences.

One of the most prominent approaches to scenario-based landscape planning is the Alternative Futures Framework developed by Carl Steinitz (1990, 1993, 2003). It has been employed in many projects around the world and has recently been re-interpreted as a concept for Geodesign (Steinitz 2012). The framework consists of six questions that need to be addressed in any landscape planning study (Fig. 27.2).

The framework should be passed through three times. The first cycle defines the context and scope. Within the second cycle, the methods are specified. The last

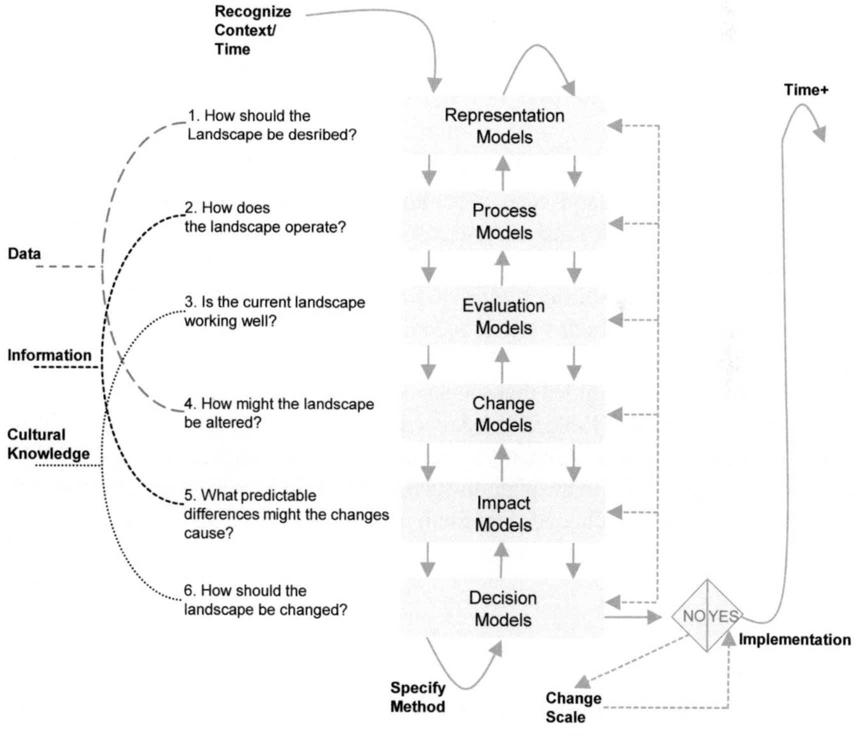

Fig. 27.2 The framework for alternative futures studies (Steinitz 1990, 1993; Steinitz et al. 2003). Scenarios are created here in the change models component, thereby answering the question of "How might the landscape be altered?". In the decision-models phase, different targets, resulting scenarios and respective alternative futures are explored in order to inform the decision-making process

cycle contains the implementation of the methods to conduct the study. The six questions that should be addressed are:

- Question 1, Representation Models: How should the state of the landscape be described? This includes considerations of the location and extent of the study area, its history and geography.
- Question 2, Process Models: How does the landscape operate? Here processes and their interactions are assessed.
- Question 3, Evaluation Models: Is the current landscape working well? This question refers to current problem issues and their location.
- Question 4, Change Models: How might the landscape be altered? This refers to which changes are foreseen for the region, which policies and actions might be developed.
- Question 5, Impact Models: What predictable differences might the changes cause? This refers to the evaluation of the foreseeable changes and an assessment of their seriousness.
- Question 6, Decision Models: How should the landscape be changed? This refers to the types and interests of major stakeholders.

27.3 Conclusions

This chapter has introduced Leitbilder and scenarios as two similar but different approaches to inform landscape planning. Leitbilder have been proposed as descriptions of target states that diverse stakeholders can agree on. Scenarios, on the other hand, can be understood as plausible descriptions of pathways of change that can help in exploring resulting future land use states (alternative futures) and their corresponding impacts. In this sense, scenarios can be regarded as a part of a larger Leitbild generation process.

The chapter has highlighted that scenarios can be used in various ways to support landscape planning (see Table 27.1). A particularly important function of scenarios can be to explore the land use changes needed to fulfill both mandatory and desirable targets. Scenarios in landscape planning can also help in exploring the consequences of different pathways of change. And finally, they can facilitate participation and aid communication about desirable goals.

References

Ahern, J. (2006). Theories, methods and strategies for sustainable landscape planning. In B. Tress, G. Tress, G. Fry, et al. (Eds.), *From landscape research to landscape planning: Aspects of integration, education and application*. Dordrecht: Springer.

Albert C (2011) *Scenario-based Landscape Planning – Influencing decision-Making through Substantive Outputs and Social Learning*. Dissertation. Leibniz Universität Hannover.

Albert, C., Zimmermann, T., Knieling, J., et al. (2012). Social learning can benefit decision-making in landscape planning: Gartow case study on climate change adaptation, Elbe valley biosphere reserve. *Landscape and Urban Planning, 105*, 347–360.

Albert, C., Galler, C., Hermes, J., et al. (2016). Applying ecosystem services indicators in landscape planning and management: the ES-in-Planning framework. *Ecological Indicators, 61*(1), 100–113.

Alcamo, J. (Ed.). (2008). *Environmental futures: The practice of environmental scenario analysis.* Amsterdam: Developments In Integrated Environmental Assessment. Elsevier.

Arnstein, S. R. (1969). A ladder of citizen participation. *Journal of the American Planning Association, 35*(4), 216–224.

Baker, J. P., Hulse, D. W., Gregory, S. V., et al. (2004). Alternative futures for the Willamette River Basin, Oregon. *Ecological Applied, 14*(2), 313–324.

Bishop, P., Hines, A., & Collins, T. (2007). The current state of scenario development: An overview of techniques. *Foresight, 9*(1), 5–25.

Bohnet, I., & Smith, D. M. (2007). Planning future landscapes in the Wet Tropics of Australia: A social-ecological framework. *Landscape and Urban Planning, 80*(1–2), 137–152.

Börjeson, L., Höjer, M., Dreborg, K. H., et al. (2006). Scenario types and techniques: Towards a user's guide. *Futures, 38*(7), 723–739.

Bradfield, R., Wright, G., Burt, G., et al. (2005). The origins and evolution of scenario techniques in long range business planning. *Futures, 37*, 795–812.

Carpenter, S., Pingali, P., Bennett, E. et al. (eds.). (2005). *Ecosystems and human well-being* (vol. 2, pp. 145–172). Scenarios. Oxford: Island Press.

Fritsch, U. (2002). *Entwicklung von Landnutzungsszenarien für landschaftsökologische Fragestellungen*. Dissertation, Universität Potsdam.

Gallopin, G. C., Hammond, A., Raskin, P., et al. (1997). *Branch points: Global scenarios and human choice – A resource papaer of the global scenarios group*. Stockholm: Stockholm Environmental Institute.

Gausemeier, J., Fink, A., & Schlake, O. (1995). *Szenario-Management: Planen und Führen mit Szenarien*. München: Carl Hanser Verlag.

Georgantzas, N. C., & Acar, W. (1995). *Scenario-driven planning: learning to manage strategic uncertainty*. Westport: Quorum Books.

Grêt-Regamey, A., Bebi, P., Bishop, I. D., et al. (2008). Linking GIS-based models to value ecosystem services in an Alpine region. *Journal of Environmental Management, 89*(3), 197–208.

Haslauer, E., Biberacher, M., & Blaschke, T. (2012). GIS-based backcasting: An innovative method for parameterisation of sustainable spatial planning and resource management. *Futures, 44*(4), 292–302.

Horlitz, T. (1998). Naturschutzszenarien und Leitbilder Eine Grundlage für die Zielbestimmung im Naturschutz. *Natschutz Landschplan, 30*, 327–329.

Hulse, D. W., Branscomb, A., & Payne, S. G. (2004). Envisioning alternatives: Using citizen guidance to map future land and water use. *Ecological Applications, 14*(2), 325–341.

Kahn, H., & Wiener, A. J. (1967). *The year 2000*. New York: Macmillan.

Klug H (2006) *The Leitbild concept: A holistic transdisciplinary approach for landscape planning*. PhD Thesis, Salzburg University.

Klug, H. (2012). An integrated holistic transdisciplinary landscape planning concept after the Leitbild approach. *Ecological Indicators, 23*, 616–626.

Klug, H., & Kmoch, A. (2015). Operationalizing environmental indicators for real time multi-purpose decision making and action support. *Ecological Modelling, 295*, 66–74.

Meadows, D. H., Meadows, D. L., Randers, J., et al. (1972). *The limits to growth*. New York: Universe Books.

Nakicenovic, N., Alcamo, J., Davis, G., et al. (2000). *Special report on emissions scenarios: A special report of Working Group III of the intergovernmental panel on climate change*. New York: Cambridge University Press.

Nassauer, J. I., & Corry, R. C. (2004). Using normative scenarios in landscape ecology. *Landscape Ecology, 19*(4), 343–356.
Pahl-Wostl, C. (2008). Participation in building environmental scenarios. In J. Alcamo (Ed.), *Environmental Futures: The practice of environmental scenario analysis* (Vol. 2, pp. 105–122). Amsterdam: Elsevier.
Potschin, M. B., Klug, H., & Haines-Young, R. H. (2010). From vision to action: Framing the Leitbild concept in the context of landscape planning. *Futures, 42*(7), 656–667.
Raskin, P., Banuri, T., Gallopin, G. C., et al. (2002). *Great transition: The promise and Lure of the times ahead*. Boston: Stockholm Environmental Institute.
Santelmann, M. V., White, D., Freemark, K., et al. (2004). Assessing alternative futures for agriculture in Iowa, USA. *Landscape Ecology, 19*(4), 357–374.
Scholles, F. (2008). Szenariotechnik. In D. Fürst & F. Scholles (Eds.), *Handbuch Theorien und Methoden der Raum- und Umweltplanung* (pp. 380–392). Dortmund: Vertrieb für Bau- und Planungsliteratur.
Schroth, O., Pond, E., Muir-Owen, S. et al. (2009). *Tools for the understanding of spatio-temporal climate scenarios in local planning: Kimberley (BC) case study*. SNSF Report PBEZP1–122976.
Schroth, O., Hayek, U. W., Lange, E., et al. (2011). Multiple-case study of landscape visualizations as a tool in transdisciplinary planning workshops. *Landscape Journal, 30*(1), 53–71.
Schwartz, P. (1996). *The art of the long view*. New York: Doubleday.
Shaw, A., Sheppard, S., Burch, S., et al. (2009). Making local futures tangible--Synthesizing, downscaling, and visualizing climate change scenarios for participatory capacity building. *Global Environmental Change, 19*(4), 447–463.
Shearer, A. W. (2005). Approaching scenario-based studies: three perceptions about the future and considerations for landscape planning. *Environment and Planning. B, Planning & Design, 32*, 67–87.
Shoemaker, P. J. H. (1993). Multiple scenario development: Its conceptual and behavioral foundation. *Strategic Management, 14*(3), 193–213.
Sisk, T. D., Prather, J. W., Hampton, H. M., et al. (2006). Participatory landscape analysis to guide restoration of ponderosa pine ecosystems in the American Southwest. *Landscape and Urban Planning, 78*(4), 300–310.
Steinitz, C. (1990). A framework for theory applicable to the education of landscape architects (and other environmental design professionals). *Landscape Journal, 9*(2), 136–143.
Steinitz, C. (1993). A framework for theory and practice in landscape planning. *GIS Europe, 2*(6), 42–45.
Steinitz, C. (2012). *A framework for Geodesign: Changing geography by design*. Redlands: ESRI Press.
Steinitz, C., Arias, H., Bassett, S., et al. (2003). *alternative futures for changing landscapes: The Upper San Pedro River Basin in Arizona and Sonora*. Washington D.C: Island Press.
Stock, C., Bishop, I. D., & Green, R. (2007). Exploring landscape changes using an envisioning system in rural community workshops. *Landscape and Urban Planning, 79*(3–4), 229–239.
Theobald, D. M., & Hobbs, N. T. (2002). A framework for evaluating land use planning alternatives: Protecting biodiversity on private land. *Conservation Ecology, 6*(1), 5.
Tress, B., & Tress, G. (2003). Scenario visualisation for participatory landscape planning—a study from Denmark. *Landscape and Urban Planning, 64*(3), 161–178.
van der Heijden, K. (1996). *Scenarios: The art of strategic conversation*. Chichester: Wiley.
van Notten, P. (2005). *Writing on the wall: Scenario development in times of discontinuity*. Maastricht: Universiteit Maastricht.
van Notten, P. W. F., Rotmans, J., van Asselt, M. B. A., & Rothman, D. S. (2003). An updated scenario typology. *Futures, 35*(5), 423–443.
van Vliet, M., Kok, K., & Veldkamp, T. (2010). Linking stakeholders and modellers in scenario studies: The use of Fuzzy Cognitive Maps as a communication and learning tool. *Futures, 42*(1), 1–14.

Volkery, A., Ribeiro, T., Henrichs, T., et al. (2008). Your Vision or My Model? Lessons from Participatory Land Use Scenario Development on a European Scale. *Systemic Practice and Action Research, 21*(6), 459–477.

von Haaren, C. (1988). Beitrag zu einer normativen Grundlage für praktische Zielentscheidungen im Arten- und Biotopschutz. *Landschaft + Stadt, 20*(3), 97–106.

von Haaren, C. (1991). Leitbilder oder Leitprinzipien? *Garten + Landschaft, 101*(2), 29–34.

von Haaren, C. (2004). Szenarienmethode. In C. von Haaren (Ed.), *Landschaftsplanung* (pp. 287–291). Stuttgart: Eugen Ulmer.

von Reibnitz, U. (1987). *Szenarien – Optionen für die Zukunft*. Hamburg: McGraw-Hill.

Wack, P. (1985a). Scenarios: Uncharted waters ahead. *Harvard Business Review, 63*(5), 73–89.

Wack, P. (1985b). Scenarios: Shooting the rapids. *Harvard Business Review, 63*(6), 139–150.

Walz, A., Lardelli, C., Behrendt, H., et al. (2007). Participatory scenario analysis for integrated regional modelling. *Landscape and Urban Planning, 81*(1–2), 114–131.

Wissen-Hayek, U. (2009). *Virtuelle Landschaften zur partizipativen Planung*. Zürich: vdf Hochschulverlag AG.

Wollenberg, E., Edmunds, D., & Buck, L. (2000). Using scenarios to make decisions about the future: anticipatory learning for the adaptive co-management of community forests. *Landscape and Urban Planning, 47*(1–2), 65–77.

Xiang, W. N., & Clarke, K. C. (2003). The use of scenarios in land-use planning. *Environment and Planning. B, Planning & Design, 30*(6), 885–909.

Part V
Communication in Landscape Planning

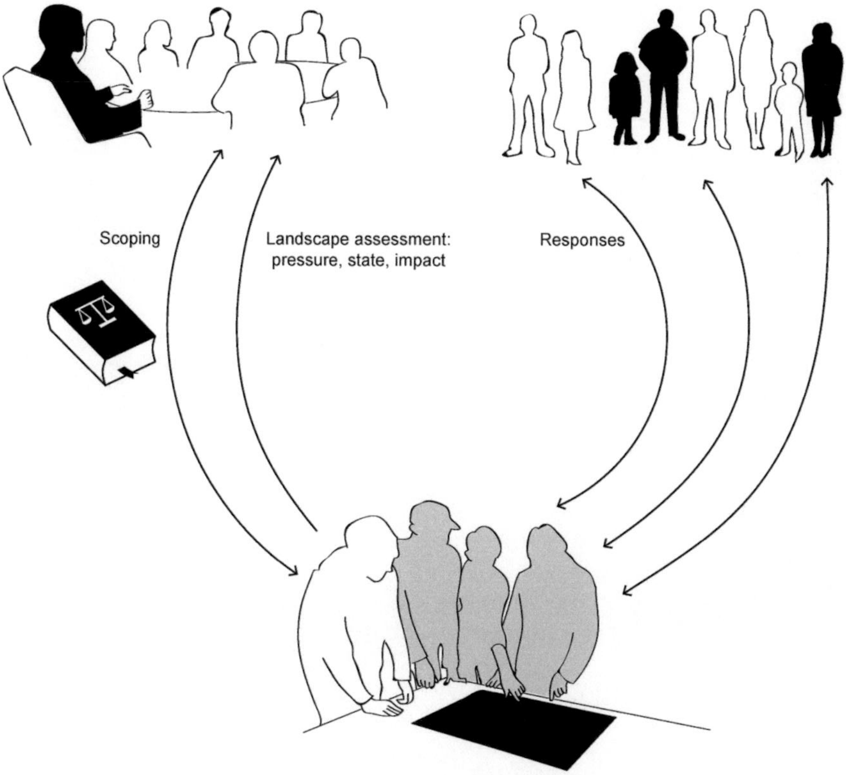

This part of the book introduces different methods for engaging the public and interest groups in debates and decisions regarding landscape planning issues in the context of a democratic and constitutional state. Furthermore, we discuss the role of design in landscape planning and the communication of options.

Techniques for Participatory Approaches 28

Sebastian Krätzig, Carolin Galler,
and Bartlett Warren-Kretzschmar

Abstract

A challenge of comprehensive landscape planning communication processes is to integrate digital information technology into planning processes by matching the present technological capabilities to the specific requirements and tasks of the different planning phases. Web 2.0 technologies hold the potential for supporting landscape planning tasks such as scoping, landscape analysis, assessment, planning concepts, assigning objectives and measures, implementation concepts, implementation support and continuous updates and monitoring. Application fields of crowd sourcing and social media/networks are discussed and recommendations are made for deciding which E-tools best fulfil the interactive functions requested in the planning process.

Keywords

Participation requirements · Digital technologies · Crowd sourcing · Social media · E-tools

S. Krätzig (✉)
Institute of Environmental Planning, Leibniz University Hannover, Hannover, Germany
e-mail: kraetzig@umwelt.uni-hannover.de

C. Galler
Niedersächsisches Ministerium für Umwelt, Energie und Klimaschutz, Hannover, Germany
e-mail: carolin.galler@mu.niedersachsen.de

B. Warren-Kretzschmar
Department of Landscape Architecture and Environmental Planning, Utah State University, Logan, UT, USA
e-mail: barty.warren@usu.edu

C. von Haaren et al. (eds.), *Landscape Planning with Ecosystem Services*,
Landscape Series 24, https://doi.org/10.1007/978-94-024-1681-7_28

28.1 Requirements and Options for Participation in Landscape Planning

28.1.1 Public Involvement

Decisions in environmental planning must be supported by public participation and the environmental information must be accessible (UN/ECE Aarhus convention). This right to environmental information and participation is implemented in European (Directive 2003/35/EC, Directive 2003/4/EC) and national laws such as the German 'Umweltinformationsgesetz' and 'Öffentlichkeitsbeteiligungsgesetz'. Additionally, it is also recognised in administrative procedures. Legal minimum standards regulate the participation of various agencies and the public, specifying (formal) participation procedures (Fürst and Scholles 2008). The European Landscape Convention also requires that landscape planning should be accompanied by public participation as a core component of the planning process. However, there are countries, such as Germany, who have yet to ratify the Convention and where public participation in landscape planning is not mandatory. Nevertheless, participation, as in Germany, is usually standard in the landscape planning practice. However, it is often restricted to a few meetings covered by the agreement between the planner and local or regional authorities.

An important precondition for successful participation is that the planner prepares information in a way which indicates the aspects that are open for local public input. It is frustrating for participants if, after extensive consultation in which they have invested participatory efforts, they find out that the subject is not open to local discretion because a European interest is at stake, for instance in preserving an endangered species. Therefore, the objectives in the landscape plan should differentiate between mandatory objectives, which are not negotiable in the participation process, and those open for local discretion and input (see Chap. 21).

Participative planning addresses a range of non-governmental organisations or individuals – e.g. stakeholders, affected persons or interested citizens, environmental groups, citizens' initiatives, bodies with a statutory consultative role. Their roles vary greatly in the planning process because they have different institutional and organisational backgrounds as well as varying understandings of environmental information and accesses to a technical infrastructure. Thus, they may require different opportunities for participation (von Haaren and Galler 2012).

28.1.2 Addressing Citizen Groups and Optimising Availability of Information

The internet and smart phone technology provide new ways to communicate and discuss planning objectives and measures with local citizens. These new channels of communication are independent of time and space and enable users to access information and respond at their convenience. Web 2.0 and social media open opportunities and networks for citizens, stakeholders to exchange opinions and knowledge about the planning process. Furthermore, the threads of the discussion are documented and provide a transparent and comprehensive overview of the salient arguments.

However, the internet and social media and their role in making planning processes more transparent are, alone, not enough to motivate citizens to engage in conversations about their landscape. Often, landscape planning addresses relatively abstract concepts and targets experts in the administration and NGOs with specific information that may not interest the public. In other words, landscape planning is concerned with abstract legal values and standards as well as species or soil and water problems, which are unknown or not understood by the public. However, specific issues and interests that do concern citizens are not addressed.

One approach to increase citizen interest in landscape planning may be to raise citizens' awareness and understanding of the benefits of ES in their local landscape (Harwood et al. 2015). When citizens appreciate the value of the ES that are provided in their landscape, then landscape planners can inform citizens about their supply and vulnerability in the local environment. Furthermore, when citizens are informed about the use and potential endangerment of ES then they can actively participate in decisions about the priority of nature conservation measures.

28.1.3 Interactive Functions Within the Planning Process

Although landscape planning is an iterative process, it has distinct phases (Galler et al. 2014). In each planning phase participation has its place and can be supported by specific interactive functions (see Fig. 28.1). These facilitate different landscape planning tasks that follow general objectives, in particular, transparency, consolidating democratic procedures, improving environmental information base and education.

- **Scoping:** Following the example of environmental impact assessments, the planning process starts with a scoping phase (see Fig. 28.1) to define the planning issues of specific concern, identify recent problems and determine the assessment framework. Public agencies, environmental NGOs and citizens are requested to contribute information and ideas for the landscape plan. In this way, planners and responsible authorities can ensure that the plan will focus on current and pressing issues, without neglecting to give an overall picture of the state of the environment in the municipality or region. Furthermore, the scoping exercise helps to provide a framework for participation within the planning process (who, when, how, the decision space and possible areas of co-decisions). This initial step should be implemented in planning practice to clearly frame the subjects in which, and the ways how, the public will be involved in the planning and decision-making process (State Ministry Baden-Württemberg 2014). E-tools with interactive functions such as a web discussion platform can complement face-to-face events, for example in town hall meetings. They should be integrated

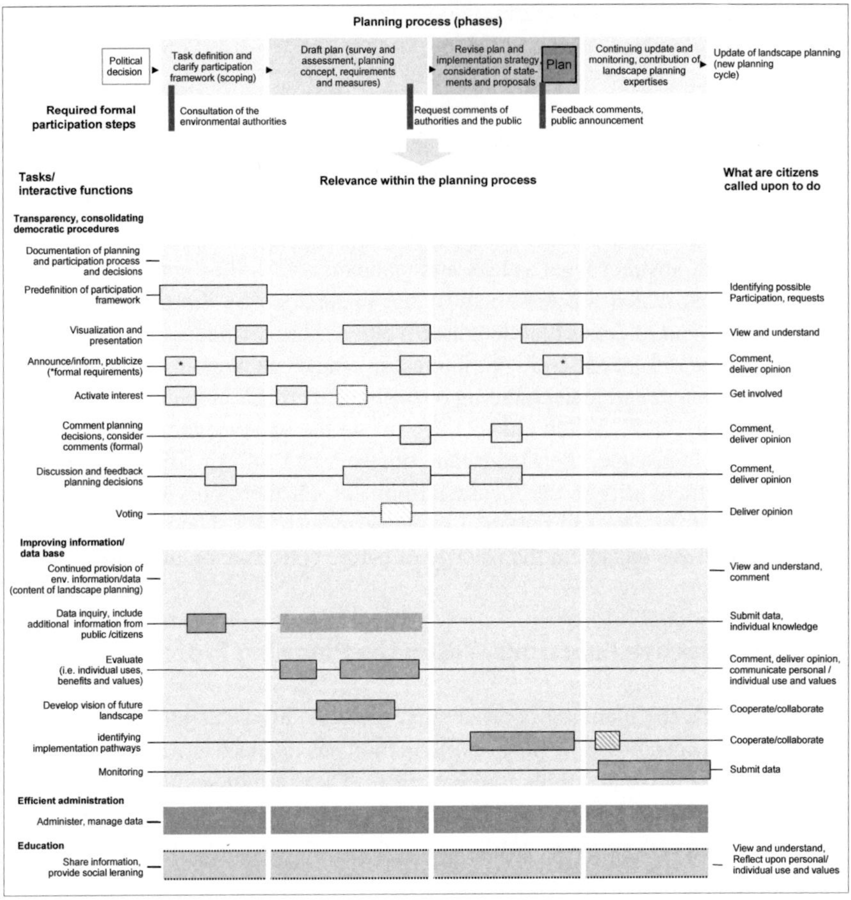

Fig. 28.1 The landscape planning process is closely linked to interactive functions that respond to the requirements of the addressees. (Galler et al. 2014)

in this initial phase of the process to encourage citizens and NGOs to contribute their views and local knowledge in the planning.

- **Landscape analysis, assessment:** Landscape planning is based on digital environmental information about the status quo, historical status and forecasts of the prospective state of the environment. Various public (and private) authorities maintain this data (Galler and Gnest 2011). To some degree, database portals (e.g. the Environmental Portal of Lower Saxony, www.umweltkarten-niedersachsen.de) integrate this data about the landscape. Furthermore, non-governmental organisations or citizens can provide additional data, for example about species (Ardini 2012) or assessments. In this way, a user-driven assessment, based on landscape preferences of the local population or actual recreational use, can complement the non-user-related landscape aesthetic assessment of the existing CES (see Chap. 15). The relevant information must be consolidated and processed for case-related analyses. Furthermore, the landscape planning process, as well as the planning content, should be documented and made accessible to the public. These requirements are achieved preferably with a web-based information system that offers interactive functions for information exchange and integration of user-based values about the landscape (georeferenced preferences/feedback).

- **Development/planning concept, assigning objectives and measures:** In many respects, landscape planning allows for alternatives in the specification of nature conservation objectives and their resulting spatial and thematic prioritisation. Citizens and local stakeholders can and should be involved in decisions about alternative objectives and measures. For this, feedback functions as well as interactive scenario development and visioning provide important tools for collaborative planning.
- **Implementation concept and implementation support:** Policy makers, together with relevant agencies and the public, should draw up an agreement on priorities in terms of objectives and timing of environmental measures (that are recommended in the landscape plan, see Chap. 21). This is the basis for an implementation strategy. In this planning phase, interactive functions need to be included that allow involvement and feedback from relevant parties.
- **Continuous update and monitoring:** Increasingly, environmental monitoring plays an important role in ensuring the targeted outcomes of the landscape plan or adapting objectives and measures if success appears uncertain (Chap. 21). Crowd-sourcing functions offer inexpensive opportunities to survey and monitor the landscape and its development.

Participants in the landscape planning process may have different requirements regarding interactive functions. For example, public agencies must respond to planning proposals with a formally documented comment, e.g. written comments, while citizens may respond more informally. In order to decide which E-tools best fulfil the interactive functions the following criteria should be considered: (i) phase of the planning, (ii) target groups and stakeholders involved, (iii) institutional background/procedural requirements (formal or informal participation), (iv) desired outcomes for participants, (v) desired feedback and input for the planning process.

An array of E-tools and applications that offer such interactive functions are available to support landscape planning tasks (Table 28.1).

Many of the tools mentioned in Table 28.1 are already employed in landscape planning whilst others, especially new approaches based on Web 2.0 technology, remain untested.

Table 28.1 E-tools that support the tasks and functions in the landscape planning (grouped along overall objectives); for references to good practice see footnotes at the end of the chapter (Galler et al. 2014)

Task of landscape planning	Applications, E-tools: Web 1.0 (Websites + Email)	webGIS	Scenario planning software	Crowd sourcing apps	Augmented reality	Visualization	Virtual globes	Discussion groups	Social networks	Votings, Polls	E-file (www)	Best practice examples
Transparency												
Predefinition of participation framework								X	X	X		
Scope of the assessment framework	X			X				X				
Continuing supply of environmental information/ data (content of LP)		X		X			X				X	Portal U – Environmental Portal Germany[a]; Environmental register Münster[b]
Documentation of planning and participation process and decisions	X					X					X	Environmental information system Bremen[c]
Announce/ inform, publicise	X	X					X	X	X			Mapserver FNP Bremen[d]; Citymap Dresden[e]; Interactive maps Hamburg[f]

Consolidating democratic procedures												
Activate interest	X			X	X	X	X	X	X	X		Participatory budget Weimar on Facebook[g]; Conversion of form. US-space in Heidelberg (YouTube)[h]
Feedback/ comment, discussion about planning decisions (formal, informal)	X	X	X	X				X	X	X		National park Rheinland-Pfalz[i]; BOB-SH[j]
Develop vision of future landscape			X		X	X				X		Future vision Region Hannover[j]; Wasatch Choice for 2040[k]
Improving information/data												
Include additional information from public/citizens	X	X		X		X	X	X	X		X	Fix my Street[l]; Ardini[m]; OpenElm[n]; Noise action plans[o]; Cultural landscapes-Wiki KLEKS[p]

(continued)

Table 28.1 (continued)

Task of landscape planning	Applications, E-tools											Best practice examples
	Web 1.0 (Websites + Email)	webGIS	Scenario planning software	Crowd sourcing apps	Augmented reality	Visualization	Virtual globes	Discussion groups	Social networks	Votings, Polls	E-file (www)	
Monitoring		X		X							X	
Evaluate (i.a. individual uses, benefits and values)	X	X		X		X	X	X	X	X		GeoVisualizer (Tracking of roots and emotions (stress))[q]; Experience natural heritage (Coesfeld, Ger.)[r]
Education												
Share information, provide social learning	X	X	X		X	X	X	X	X			Climate city Bremerhaven[s]; Experience natural heritage (Coesfeld, Ger.)[t]; Eye on Earth (Information portal of the European Environmental Agency)[u]

Efficient administration												
Administer, manage data											X	
Respond									X		X	Regional plan for the Brunswick Region[v]

[a]http://www.portalu.de/kartendienste
[b]http://www.geo.stadt-muenster.de/umweltkataster
[c]http://www.umwelt.bremen.de/
[d]http://www.fnp-bremen.de/kartenuebersicht/
[e]http://www.stadtplan2.dresden.de/%28S%28x5ktt5wrcjaipksqqyaqayss%29%29/spdd.aspx;
[f]http://www.hamburg.de/schutzgebietskarte/
[g]https://www.facebook.com/pages/B%C3%BCrgerhaushalt-Weimar/164357443579622
[h]http://www.heidelberg.de/Konversion,Lde/Startseite+Konversion/Mediathek.html
[i]http://www.nationalpark.rlp.de/
[j]http://www.bob-sh.de/app.php/
[k]http://hannover.zukunftsbild.net/sieben-mal-zukunft/bild1/
[l]http://www.envisionutah.org/about-wc2040/item/198-vision-map-brochure
[m]http://www.fixmystreet.com/
[n]http://www.ardini.de
[o]https://www2.domap.de/wps/portal/laermaktionsplan/buerger_forum; http://www.machsleiser.de/karte
[p]http://www.kleks-online.de
[q]Zeile et al. (2013)
[r]http://www.erlebnis-naturerbe.de; http://www.air-verband.de/fileadmin/template/img/Download/Veranstaltungen/6.AIR_BF/AIR_BF_6_Apps_Tourismus_EFTAS_NZ.pdf
[s]http://www.klimastadt-bremerhaven.info/category/projekte/
[t]http://www.eyeonearth.org
[u]Schulze-Wolf and Menzel (2007)
[v]http://www.openelm.org.im

28.2 Potentials of new Approaches Based on Web 2.0 Technologies Within the Landscape Planning Process

28.2.1 Crowd Sourcing

Public authorities and NGOs are increasingly interested in using crowd sourcing to incorporate citizens' knowledge and preferences into planning. The EU Directive on Environmental Noise (2002/49/EG) requires cities and regions to map traffic and industrial noise. Crowd sourcing has been used successfully in action plans to reduce noise in cities. For example, the city of Dortmund/Germany has used crowd sourcing to examine how citizens subjectively perceive noise in their city. Participants were asked to locate areas with disturbing noise or pleasant silence onto a map on the internet and to describe the situation. The administration was able to connect and compare this data with official noise mapping data, and the results were used to develop and prioritise measures to reduce noise sources, noise propagation, or noise perceptions among citizens.

Today, apps for crowd sourcing not only support information exchange in landscape planning (see Table 28.1), they allow for input into a database from mobile end-user devices (tablets, smartphones) in real time. In landscape planning, citizens and stakeholders have helped to update and expand the environmental information base using the bird mapping app 'ARDINI', the biodiversity mapping app 'Anymals+plants' (2011) or iNaturalist for different observations. These projects represent good practice for coordinating the local knowledge of volunteers/NGOs with the methodological, technical and administrative requirements of official data management. Furthermore, these applications can collect spatial information about the citizens' needs and perceptions, such as noise and olfactory perception or aesthetical experiences in the landscape. However, the collected data must be compatible with administrative data management systems and the accuracy of the data, as well as the uncertainties, must be well documented.

28.2.2 Social Media and Social Networks

Social media and social networks offer different levels of participatory involvement to support the objectives and tasks of planning (Krätzig and Warren-Kretzschmar 2014). Social media offers the following five potential levels of participation in environmental planning (administrative objectives are in brackets): 'listen' (know what is said online about environmental issues), 'promote' (raise awareness of environmental planning programs or opportunities), 'participate' (join a conversation with citizens about environmental measures), 'share content' (share information or results of environmental measures) and 'build community' (develop relationships online, nurture community, engage people, encourage them to take action) (see Table 28.2).

Social media and social networks, such as Facebook, offer the possibility to access and incorporate citizens' opinions and suggestions through examining

Table 28.2 2 Levels of participation in social media (Krätzig and Warren-Kretzschmar 2014)

	Listen	Promote	Participate	Share content	Build community
Objectives	Know what is said online about environmental issues	Raise awareness of environmental planning programs, opportunities, planning activities	Join a conversation with citizens about sustainability or environmental measures	Share information or results of environmental measures	Build relationships online, engage people to take action or adapt sustainable life styles
Technical possibilities	Google Alerts, Twitter, Socialmention, and RSS readers, etc.	Twitter, Share This, and Digg, etc.	Facebook, Twitter and LinkedIn, encourage bloggers to write about environmental issues	Blogging, podcasts, You Tube, video sharing sites, Instagram, Flickr, etc.	Ning, LinkedIn for knowledge and skill sharing, Facebook to engage new supporters
Tasks or skills needed	Pattern analysis (link listening and analysis to decisions or actions)	Build trust, credibility and a relationship with interested participants	Participatory competencies like listening, interpreting, opinion sharing skills	Create content and use buzz tools to share. Motivate others to share their information/ideas about the issues	Engage participants to reach out to their personal social networks, create their own content, or activities to share
Level of involvement	Passive engagement	Broadcast or share	Low engagement	Content intensive	High engagement

comments and group discussions. Further, they allow administrators to inform a large number of citizens about environmental issues with relatively little effort. Administrators could use, for example, Facebook groups to improve transparency about existing ES and to publicise information about new planning measures and monitoring activities (see Fig. 28.2). Finally, social media offers the opportunity to engage citizens in group discussions, possibly activating their interest in the planning process and building a community of involved citizens.

In a social network a citizen receives information as it happens by networking with people who are involved in similar or related issues, instead of explicitly searching for information from a specific person or institutional source. Observations of Facebook indicated that users often receive answers to questions or obtain additional information much faster than when waiting for answers from official institutional sources. In addition, statements from non-governmental sources often contain more details or local knowledge, special tips or alternative interpretations of particular issues. This immediate and dynamic exchange of information can also be spread very quickly when it goes 'viral' (Kanter and Fine 2010). In contrast to formal participation procedures with strict requirements, social networks offer the opportunity to engage citizens in an informal way without formal regulations.

However, social media also has limitations for use in planning practice. Its use is difficult to direct and may not always follow the intended participatory objective. For example, participants on Facebook may not enter into a group discussion, or they may use the platform to express their opinion without reading the comments of other participants (Krätzig and Warren-Kretzschmar 2014). Self-managed discussion platforms are therefore usually much more capable of supporting a qualified discussion about the planning proposals. The question remains whether social media and networks could be a permissible and representative form of communication for formal and informal participation processes. For formal participation, they must fulfil requirements such as time-limits for participation, social equity and usability as well as binding character and reliability, privacy and the right of use (Martini and Fritzsche 2013). Presently, further development of social media or proprietary software is needed in order to reach a permissible and representative (formal or informal) form of communication.

28.3 Coloured, Faster, Better? Options and Recommendations for Interactive Landscape Planning

The 'Interactive Landscape Plan Königslutter am Elm', which was developed from 2002 to 2005, used web-based information with map server technology and allowed (georeferenced) feedback functions (von Haaren et al. 2005). The project also capitalised on the potential of the internet to disseminate information. Today the internet is a primary source of environmental and planning information, and it has become an accepted method for publishing information and announcements to the public. However, E-tools have not been broadly applied and the innovations of Web 2.0

Fig. 28.2 Posted information and photos, regarding a landscape planning measure, in a citizen's discussion group about Königslutter on Facebook. The post was made under the name "research team environmental planning Hannover" (Forschungsteam Umweltplanung Hannover) and information about our relationship to the city and involvement in the particular measure were also posted. Facebook group members were invited to respond to two questions about the planning measure: Do you know about the measure? Did you use the pond after renaturalisation? Nineteen citizens commented. (Krätzig and Warren-Kretzschmar 2014)

have not yet been broadly used in landscape planning – at least in Germany. Applications within the wider context of environmental planning – such as in urban land use planning (e.g. land development plan of Bremen) or noise action plans (e.g. for the city of Dortmund) – exemplify how today's IT/web-solutions could contribute to landscape planning practice. However, the diversity of E-tools illustrates the need for standardisation in order to promote their use. Tools offered by service providers and federal or national initiatives help to standardise participation systems (e.g. 'BOB-SH', http://www.bob-sh.de). This enables a uniform processing of formal participation in urban land-use planning. Such applications could be expanded to environmental and landscape planning.

The German situation illustrates a long-standing landscape planning challenge. On the one hand, there is a wealth of data and information, though dominated by governmental and public welfare perspectives. On the other hand, the individual perspectives of citizens are neglected, illustrating the structural deficits in participation.

Web 2.0 and the use of social media and networks offer planners new forms of two-way communication with citizens that go beyond simply providing information on the internet. In addition, these tools show great potential in helping individual citizens voice their preferences in the planning process. In turn, this can support the citizen's active identification with the surrounding landscape and their involvement in the planning decisions, making the decisions more transparent and understandable.

In addition to the development of social media and networks, the development of different end user technology, such as smart phones and tablets, provides the opportunity not only to access information but also to provide information through crowd sourcing applications. They offer a new and exciting opportunity for planners to update and expand information with the help of citizens and stakeholders. Planners must therefore increasingly formulate their information needs so that applications can be developed specifically to support planning purposes.

Different visualisation techniques, from hand sketches to VRML interactive virtual worlds, were used in the Interactive Landscape Plan Königslutter and evaluated with regard to their usefulness for the citizens (von Haaren and Warren-Kretzschmar 2006). Since then, the technology has become more powerful and more intuitive (Lovett et al. 2015). Augmented reality increasingly offers participants the possibility to see simulations of change in the landscape where it is happening (Lange 2011; Gill and Lange 2015). GeoDesign provides powerful tools to evaluate the impacts of decisions (Abukhater and Walker 2010; Warren-Kretzschmar et al. 2012) and software such as CommunityViz or Envision Tomorrow Plus enables citizens to be part of the process of developing scenarios for the future development of their community and landscape (Kwartler and Longo 2008; Walker and Daniels 2011). At present, however, the use of such software in the planning process is still the exception rather than the norm.

28.4 Conclusion

Citizen participation is an indispensable part of landscape planning. It has the potential to focus planning activities on issues which are of relevance to the people, include local knowledge and preferences, and enable co-decision within the limits set by mandatory objectives. The forms of participation, as well as the support of participation by new media, needs to take into account the specifics of different audiences, the demands in different phases of the planning process, uncertainties and possible biases in citizen-generated data, and the IT capabilities of participants.

References

Abukhater, A., & Walker, D. (2010). Making smart growth smarter with GeoDesign. In: *Changing geography by design: Selected readings in GeoDesign*. Available via ESRI. http://www.esri.com/library/ebooks/GeoDesign.pdf. Accessed 4 Feb 2016.

Anymals+plants. (2011). *Artenvielfalt für ihr Smartphone*. https://www.anymals.org. Accessed 8 Feb 2016.

Ardini. (2012). *Artenerfassung digital in Niedersachsen*. http://www.ardini.de. Accessed 8 Feb 2016.

Fürst, D., & Scholles, F. (2008). Partizipative Planung. In D. Fürst & F. Scholles (Eds.), *Handbuch Theorien und Methoden* (pp. 161–178). Dortmund: Rohn.

Galler, C., & Gnest, H. (2011). Datenmanagement und Monitoringsysteme im sektoralen Verwaltungsaufbau. Zukunftsfähiger Umgang mit Wasser im Raum. *ARL Forschungs- und Sitzungsberichte, 234*, 118–122.

Galler, C., Krätzig, S., Warren-Kretzschmar, B., et al. (2014). Integrated approaches in digital/interactive landscape planning. In U. Wissen Hayek, P. Fricker, & E. Buhmann (Eds.), *Peer reviewed proceedings of digital landscape architecture 2014 at ETH Zurich* (pp. 70–83). VDE VERLAG GMBH, Berlin/Offenbach: Herbert Wichmann Verlag.

Gill, L., & Lange, E. (2015). Getting virtual 3D landscapes out of the lab. *Computers, Environment and Urban Systems, 54*, 356–362.

von Haaren, C., & Galler, C. (2012). *Landschaftsplanung. Grundlage nachhaltiger Landschaftsentwicklung*. Available via Bundesamt für Naturschutz. https://www.bfn.de/fileadmin/MDB/documents/service/Landschaftsplanung_2012.pdf. Accessed 8 Feb 2016.

Harwood, A. R., Lovett, A. A., & Turner, J. A. (2015). Customising virtual globe tours to enhance community awareness of local landscape benefits. *Landscape and Urban Planning, 142*, 106–119.

Kanter, B., & Fine, A. H. (2010). *The networked nonprofit. Connecting with social media to drive change*. San Francisco: Jossey-Bass.

Krätzig, S., & Warren-Kretzschmar, B. (2014). Using interactive web tools in environmental planning to improve communication about sustainable development. *Sustainability, 6*(1), 236–250.

Kwartler, M., & Longo, G. (2008). *Visioning and visualization: People, pixels and plans*. Cambridge, MA: Lincoln Institute of land Policy.

Lange, E. (2011). 99 volumes later: We can visualise. Now what? *Landscape and Urban Planning, 100*, 403–406.

Lovett, A., Appleton, K., Warren-Kretzschmar, B., et al. (2015). Using 3D visualization methods in landscape planning: An evaluation of options and practical issues. *Landscape and Urban Planning, 142*, 85–94.

Martini, M., & Fritzsche, S. (2013). Zwischen Öffentlichkeitsauftrag und Gesetzesbindung: zum Dilemma deutscher Behörden bei der Einbindung privater Social-Media-Werkzeuge und Geodatendienste in ihre Internetangebote. *Verwaltungsarchiv, 104*(4), 449–485.

Schulze-Wolf, T., & Menzel, A. (2007). Neue Wege der Öffentlichkeitsbeteiligung in der Raumplanung. Hintergründe, Konzepte und Erfahrungen. In Initiative eParticipation & Stiftung Mitarbeit (Ed.), *E-Partizipation. Beteiligungsprojekte im Internet* (pp. 120–143). Bonn: Stiftung Mitarbeit.

State Ministry Baden-Württemberg. (2014). *Leitfaden für eine neue Planungskultur*. https://mitwirkung.bw21.de/Downloads/140304_Planungsleitfaden.pdf. Accessed 8 Feb 2016.

von Haaren, C., & Warren-Kretzschmar, B. (2006). The interactive landscape plan – Use and benefits of new technologies in landscape planning, including initial results of the interactive Landscape plan Koenigslutter am Elm, Germany. *Landscape Research, 31*, 83–105.

von Haaren, C., Oppermann, B., Friese, K. I. et al. (2005). *Interaktiver Landschaftsplan Königslutter am Elm: Ergebnisse aus dem E+E-Vorhaben "Interaktiver Landschaftsplan Königslutter am Elm" des Bundesamtes für Naturschutz*. Bundesamt für Naturschutz, Bonn–Bad Godesberg.

Walker, D., & Daniels, T. (2011). *The planners guide to communityviz*. London: Routledge.

Warren-Kretzschmar, B., Haaren, C. V., Hachmann, R. et al. (2012, May 31–June 2). *The potential of GeoDesign for linking landscape planning and design*. Paper presented at the Digital Landscape Architecture Conference 2012, Anhalt University of Applied Siences, Bernburg/Dessau.

Zeile, P., Exner, J. P., Bergner, B. S., et al. (2013). Humansensorik und Kartierung von Emotionen in der räumlichen Planung. In E. Buhmann, S. M. Ervin, & M. Pietsch (Eds.), *Peer review proceedings of digital landscape architecture 2013 at Anhalt University of Applied Sciences. Bernburg, 6–8 June 2013* (pp. 129–141). Berlin/Offenbach: Herbert Wichmann Verlag, VDE VERLAG GMBH.

Design in Landscape Planning Solutions 29

Bartlett Warren-Kretzschmar and Christina von Haaren

Abstract

Supplementing planning measures with design approaches is a way of generating both new measures as well as communicating nature conservation to decision makers and the public. This chapter characterises the design approach with a focus on features that are different from planning. Furthermore, a framework is provided that identifies opportunities for integrating design approaches into landscape planning. Finally, examples are given, which illustrate how design can be incorporated in landscape planning.

Keywords

Landscape design processes · Design opportunities · Communication · Participation

29.1 Introduction: Specific Tasks Foster a Landscape Design Approach

Landscape design is defined here as a more intuitive and artistic approach to changing the landscape than the more deductive and scientific landscape planning approach represented in the majority of this book. Combining both approaches has great potential for communicating environmental objectives to decision makers and

B. Warren-Kretzschmar
Department of Landscape Architecture and Environmental Planning, Utah State University, Logan, UT, USA
e-mail: barty.warren@usu.edu

C. von Haaren (✉)
Institute of Environmental Planning, Leibniz University Hannover, Hannover, Germany
e-mail: haaren@umwelt.uni-hannover.de

C. von Haaren et al. (eds.), *Landscape Planning with Ecosystem Services*, Landscape Series 24, https://doi.org/10.1007/978-94-024-1681-7_29

the public. Thus, landscape design approaches hold the potential to compliment and improve the effectiveness and implementation of landscape planning objectives. If seen from a behavioural science perspective (e.g. Kahneman 2011) landscape design can be regarded as an intervention to translate complex or invisible environmental issues into a form which speaks directly to people's intuitive feelings. However, both 'planners' and 'designers' need to understand the differences in the approaches and make them productive for a complementary outcome (von Haaren et al. 2014).

For instance, in general, landscape design projects start with the consensus that change is desirable (Ogrin 1994). In contrast, for environmental planning the impulse for change usually stems from an actual, identified problem. Furthermore, support for landscape change is not a precondition of the plan but has to be generated, sometimes with great difficulties. This difference must be acknowledged and accepted as a challenge by the designer, as well as the legal constraints of some of the landscape plan objectives.

The capacity of landscape design approaches to emphasise creativity, functional requirements and aesthetic meaning offers opportunities to convey ideas and solutions to environmental issues that are not ordinarily considered in the landscape planning framework. For example, an attractive site design that incorporates landscape planning objectives is a persuasive way to communicate planning solutions to the local community or stakeholders (e.g. when the need for nature protection or habitat networks conflicts with the economic interest of local farmers), or when local policy and decision-makers require more tangible results than e.g. the assignment of a protected area.

Landscape planning offers the scope to incorporate landscape design approaches when change or development of a site is expected or required. This assumes that few conservation restrictions are associated with the site, and that land ownership and rights have been clarified and present no conflicts. Such development situations give design approaches the freedom to seek creative solutions that emphasis aesthetic and functional user-related criteria.

29.2 Landscape Planning and Landscape Design Processes and Methodologies

Landscape design and landscape planning processes and methodologies offer different approaches that can complement each other (see Fig. 29.1). Landscape planning involves an analytical process that uses scientific methodologies to evaluate landscape functions and assess land use impacts (Leitão and Ahern 2002). The design process combines knowledge and intuition in a way that translates complex information into coherent designs (Stokman and von Haaren 2011). Landscape design approaches emphasise the subjective, creative dimension (Swaffield 2002) and are reflected in a cyclical design process (Halprin 2002; Murphy 2005). While the landscape design process tends to be more flexible, there is a growing tendency in landscape planning to standardise assessments and base decisions and objectives

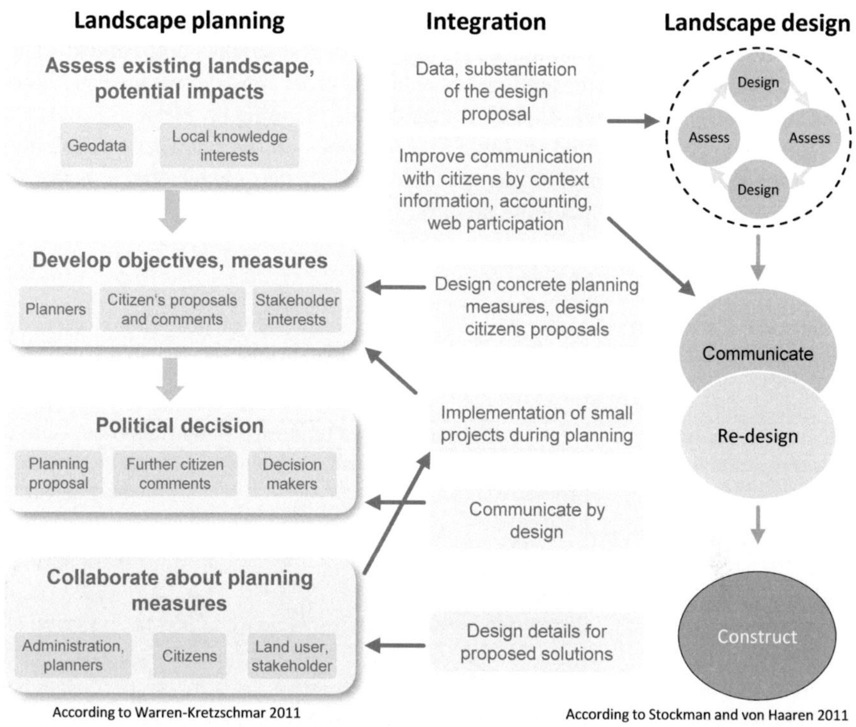

Fig. 29.1 Integration of landscape planning and landscape design processes – opportunities for the integration of the two approaches to improve outcomes of the different phases of the design/planning process. (Warren-Kretzschmar et al. 2012)

on scientific methods and information. In this way, planners must justify consequences of decisions for land use by making the process transparent, objective and comprehensible.

The approaches and underlying values of landscape design and landscape planning can be explained by the different contexts of their tasks and applications. However, the integration of landscape design and landscape planning approaches begins by identifying 'design situations' or opportunities for design to improve landscape planning results.

29.3 Framework for Design Opportunities in Landscape Planning Tasks

The characteristics of planning situations determine the potential uses of design approaches. Landscape planning projects without legally mandated objectives offer opportunities for landscape design approaches. For example, publicly owned land with no valuable natural or cultural assets are places where new landforms and designs can be developed. Furthermore, many planning tasks offer opportunities for hybrid design and planning approaches, i.e. tasks such as post-industrial sites, contaminated land, or infrastructure planning. The characteristics and context of the project should determine the approaches, which is more constructive than an 'either-or' classification of landscape design or landscape planning projects.

The framework presented in Table 29.1 provides an overview of possible task and application situations that are suitable to integrate characteristic landscape design approaches in the planning process.

Table 29.1 Context of landscape planning and landscape design and their characteristic approaches – a framework for identifying design opportunities in landscape planning (von Haaren et al. 2014)

Characteristics of the task/application situation	Characteristics of Landscape Design Approach		
	Flexibility to choose planning and design priorities	**Form**, aesthetics especially important	**Creativity**, starting with creative ideas
A Change, development expected or welcome	X	X	X
B Project site permits change (few conservation restrictions)	X	X	X
C Defined site, no property conflicts	X	X	X
D Participation		X, C	
E Planning for public interests on private land; diverse implementation options to be considered		C	
F Complex conditions in large planning area		C	
G Focus on conservation/rehabilitation of land		C	

X	Situations for suitable approach of prototypical applications
(dark shading)	"Design situations"; When they occur in landscape planning, a design approach should be considered and may take the lead.
C (light shading)	Additional landscape planning situations, in which conservation objectives can be communicated with design approaches

29.4 Examples of Design Approaches in Landscape Planning

The characteristics of the tasks represented in rows A to C of Table 29.1 are typical in design situations. However, such situations can also be found on specific sites in a landscape planning context, and, when they exist, a landscape design approach could be appropriate and beneficial. For example, design approaches can be used to brand a local community's landscape identity in a visually compelling way, or to redesign a mining site into an inviting recreational landscape. Another opportunity is the restoration of urban rivers. A plan to improve aquatic habitats and water quality can incorporate access and recreational use of the river. The environmentally informed design of trails and recreational opportunities along such rivers can help stimulate public interest in dynamic river systems and their protection.

The application situations described in rows D to G of Table 29.1 involve legal requirements that typically demand a landscape planning solution. However, the landscape design approach can complement or support the legal objectives by handling restrictions in a creative way or by visibly interpreting the underlying issues. For example, Joan Iverson Nassauer (2002) proposes to 'frame messy ecosystems' with 'cues for care', such as mown strips around wild patches of vegetation. In this way, the 'messy' habitats become more acceptable in residential or urban environment.

Design approaches offer important opportunities to include stakeholder input without the precondition of expert knowledge. Stakeholder ideas and proposals can be illustrated by design in participatory situations. In Geodesign workshops (see Fig. 29.2) a hybrid between planning and design is applied. Interactive software helps stakeholders to collaboratively create future alternatives for landscape development (Steinitz 2012). Participants can design projects and create policies that are used to develop landscape planning proposals. In an iterative process, different stakeholder groups can negotiate priorities and visualise the impacts of their proposals.

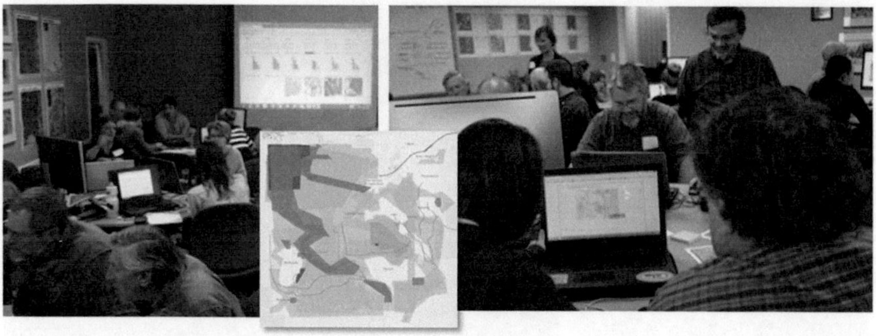

Fig. 29.2 South Cache Valley Geodesign Workshop at Utah State University. Six stakeholder groups develop proposals for future development. Insert: Farmers propose projects for the protection of agricultural land (green) as well as water quality (blue) and locations for residential (orange), commercial (red), and solar park (yellow) development. (Photos: C. McGinty)

The creative expression of design elements embedded within a landscape can also communicate a particular message or invite visitors to explore a landscape in an intentional way. The Red Ribbon by Kongjian Yu uses a bright red footpath to draw attention to and direct the public though a wetland ecosystem with high biodiversity (Fig. 29.3a). It exemplifies the potential of a design feature to communicate nature conservation with an attractive medium that is more exciting than a standard wooden trail (Fig. 29.3b).

Similarly, in the Schöneberger Südgelände in Berlin, design and nature conservation have been merged in a landscape that holds clues to the historical past of the cultural landscape, but also illustrates the potential of natural succession in the landscape (see Fig. 29.4).

29.5 Conclusion

The characteristics of landscape design and landscape planning approaches reflect the different contexts of their tasks and applications. The heterogeneous conditions in a planning area may offer situations that are typical for the design context and suitable for applying a design approach. For example, landscape planning can incorporate design approaches when:

- Land is not exclusively zoned for conservation
- The land owner wishes to improve his/her land in a visible way
- The area is owned by a municipality or a foundation with the wish to develop the land in a visually pleasing way.

Fig. 29.3 Promote acceptance of landscape protection through creative design intervention. (**a**) "The Red Ribbon" by Kongjian Yu. (Photo by Kongjiang Yu Turenscape, published by courtesy of the author). (**b**) Typical trail though wetland. (Photo C. v. Haaren)

Fig. 29.4 Marking nature and culture – communicate the hidden meaning (Schöneberger Südgelände Berlin). (Photo: C. v. Haaren)

The proposed framework, as well as some examples for integrating the two approaches, may improve landscape planning by extending the methodological spectrum and creating more convincing and acceptable results for stakeholders. A design approach may also help initiate the planning discussion about models for future landscapes or new, highly technically modified landscapes, e.g., former mining or energy landscapes. Finally, the integration of landscape design approaches can help provide solutions for contemporary issues in landscape planning by exploring approaches that make hidden ecological process visible, raise consciousness about land degradation problems, or reconcile people with new features in the landscape, e.g. wind turbines or other energy infrastructure.

References

Halprin, L. (2002). The RSVP cycles. In S. R. Swaffield (Ed.), *Theory in landscape architecture: A reader* (pp. 43–49). Philadelphia: University of Pennsylvania Press.

Kahneman, D. (2011). *Thinking, fast and slow*. New York: Farrar Straus and Giroux.

Leitão, A. B., & Ahern, J. (2002). Applying landscape ecological concepts and metrics in sustainable landscape planning. *Landscape and Urban Planning, 59*(2), 65–93.

Murphy, M. T. (2005). *Landscape architecture theory – An evolving body of thought*. Long Grove: Waveland Press.

Nassauer, J. I. (2002). Messy ecosystems, orderly frames. In S. R. Swaffield (Ed.), *Theory in landscape architecture: A reader* (pp. 196–207). Philadelphia: University of Pennsylvania Press.

Ogrin, D. (1994). Landscape architecture and its articulation into landscape planning and landscape design. *Landscape Urban Planning, 30*, 131–137.

Steinitz, C. (2012). *A framework for geodesign: Changing Geography by design*. Redlands: Esri Press.

Stokman, A., & von Haaren, C. (2011). Integrated science and creativity for landscape planning and design of urban areas. In M. Richter & U. Weiland (Eds.), *Applied urban ecology: A global framework* (pp. 170–185). Chichester: Wiley.

Swaffield, S. (2002). *Theory in landscape architecture. Reader*. Philadelphia: University of Pennsylvania Press.

von Haaren, C., Warren-Kretzschmar, B., Milos, C., et al. (2014). Opportunities for design approaches in landscape planning. *Landscape Urban Planning, 130*, 159–170.

Warren-Kretzschmar, B., von Haaren, C., Hachmann, R., et al. (2012). The potential of GeoDesign for linking landscape planning and design. In E. Buhmann, S. Ervin, & M. Pietsch (Eds.), *Digital landscape architecture 2012* (pp. 168–179). Bernburg/Berlin: Keynote/Wichmann.

Part VI
Global Context and Conclusion

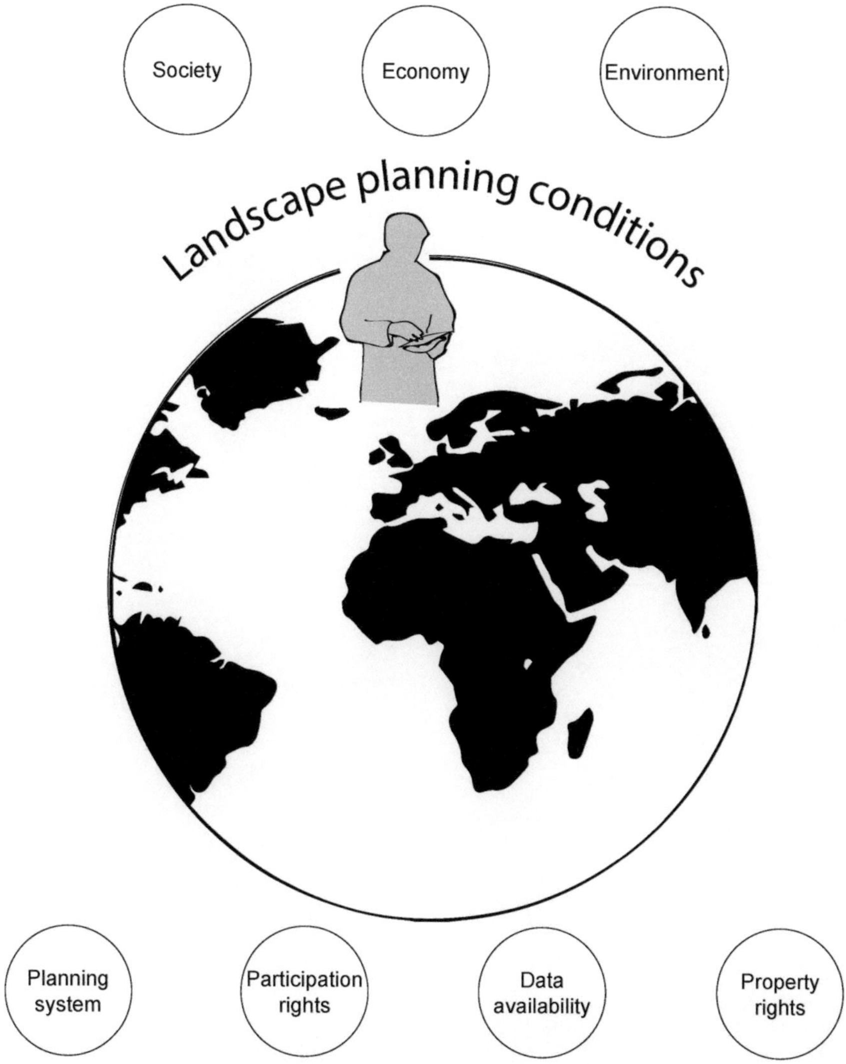

While this book focusses on landscape planning in the European setting a chapter has been included to review the context for landscape planning elsewhere in the world. Examples from the USA and Japan illustrate the implications of different framing conditions for planning, and what this may mean for the application of methods introduced in this book. We close the book with a perspective on landscape planning in the future.

Perspectives From Outside the EU: The Influence of Legal and Planning Frameworks on Landscape Planning

30

Vivek Shandas, Christina von Haaren, Hiroyuki Shimizu, Rachelle Alterman, and Andrew A. Lovett

Abstract

The legal and governance context for landscape planning in countries outside the EU can differ greatly from their EU counterparts. We propose a framework for characterizing that context in order to enable readers from non-EU countries to relate their planning systems to the European baseline for landscape planning. Methodologies for the assessment of ES in landscape planning, such as presented in this book, can be applied in principle in most countries. However, their planning context often will be very different. Legal, political, economic, demographic, cultural and physical-environmental conditions define whether comprehensive environmental planning is possible at all, or whether incremental actions are the only feasible strategy. The context also influences the role of citizen

The original version of this chapter was revised: The Fig. 30.1 has been updated now. The correction to this chapter is available at https://doi.org/10.1007/978-94-024-1681-7_32

V. Shandas (✉)
Portland State University, Portland, OR, USA
e-mail: vshandas@pdx.edu

C. von Haaren
Institute of Environmental Planning, Leibniz University Hannover, Hannover, Germany
e-mail: haaren@umwelt.uni-hannover.de

H. Shimizu
Nagoya University, Nagoya, Japan

R. Alterman
Technion – Israel Institute of Technology, Samuel Neaman Institute for National Policy Research, Haifa, Israel
e-mail: alterman@technion.ac.il

A. A. Lovett
School of Environmental Sciences, University of East Anglia, Norwich, UK
e-mail: a.lovett@uea.ac.uk

C. von Haaren et al. (eds.), *Landscape Planning with Ecosystem Services*, Landscape Series 24, https://doi.org/10.1007/978-94-024-1681-7_30

participation and different spatial or political tiers at which particular planning tasks take place. The methodologies applied in any kind of landscape planning must also be adapted to the quality and availability of data, and particularly to the evaluation standards and roles of citizen preferences in different legal and political systems. We present two particular examples from advanced economy countries – Oregon in the USA and Japan. These examples illustrate the different governance contexts for environmental planning in the selected jurisdictions and their possible consequences for managing ecosystem services.

Keywords
Landscape planning context · International · Japan · Oregon · Planning systems

30.1 Introduction: Why Look Outside Europe?

Planning systems are embedded within a broader social and legal context. By enhancing our understanding of this context for advanced economy countries outside Europe (those with broadly similar governance and planning systems) we gain three advantages:

1. a better recognition of the factors that impact planning systems as a whole and landscape planning in particular, allowing us to contextualize the European experience;
2. an ability to evaluate the role of information in landscape planning and environmental assessment as related to the legal and social contexts;
3. enhancement of the value of the chapters in this book for mutual learning and application across multiple contexts.

To that end, we first explain the relevance of the key framing conditions for planning, namely environmental legislation and property rights regimes, decision hierarchy, and public participation. In the following sections we describe the environmental aspects of the planning systems in the state of Oregon in the USA and of Japan. These examples differ significantly in framing conditions, but share with EU countries an acceptance of the need to protect the environment through a systematic planning approach. This common ground offers a valuable perspective on the role that the type of theoretical and methodological approach to landscape planning described in this book could play in countries outside the EU.

30.2 The Relevance of Framing Conditions for Landscape Planning

Legal, political and cultural conditions shape the planning system of a country. They have to be taken into account when considering whether and how the assessment of ES can be conducted and the type of planning that would be most appropriate.

30.2.1 The Political and Legal Context of Landscape Planning

The methods presented in this book are tailored for the European context. Although within the EU, the planning systems of individual countries may differ but there is also much common ground for landscape planning. For instance, spatial planning is performed in most countries, the ELC has been widely ratified, the precautionary principle is established as a common ethical standard, and environmental evaluation standards stemming from EU Directives are identical for all EU member states. Furthermore, all EU states have subscribed to the Aarhus Convention and the ensuing EU directives about public participation, disclosure of environmental information and the rights of the public to sue or initiate petitions on environmental matters.

In other parts of the world there can be very different framing conditions for environmental planning. The over-arching question that we address in this chapter is, therefore, how the methods presented in this book may be applied in such countries. Methods for mapping and assessing ES are discussed globally (Egoh et al. 2008; Nelson et al. 2009; Maes et al. 2012) for different applications such as planning, offsetting mechanisms or environmental impact studies (e.g. Waage et al. 2008). Basically, the broad approach presented in this book could be applied in other countries, or at least be used as a starting point for developing methods adapted to the needs and data availability in particular national contexts. In some countries with limited data, it may be wise to start with relatively simple methods. Furthermore, the results of such methods may often suffice to identify initial priorities for environmental planning.

However, whether in Europe or other parts of the world, the different legal, political and cultural framing conditions will be very relevant to the following types of decisions:

- a strategic choice as to whether a systematic planning approach is possible or if small incremental steps must be taken first;
- determining which spatial scales, political tiers or subject areas are most appropriate for landscape planning;
- choosing the evaluation approach (e.g. legally-based, preference-driven or a combinations of the two);
- determining the appropriate legal status of policies; e.g. should they be prescribed in detail and universally applied, or discretion allowed to permit differential implementation in response to stakeholder negotiations or public participation;
- deciding whether, and to what extent, public participation will have the ability to influence or generate the responses to environmental challenges.

In our view, the most important legal, political, or cultural factors which define how landscape planning is performed are: a) the property rights regime and the degree to which environmental standards set by legislation are comprehensive and binding; b) the distribution of power within the institutional hierarchy of the planning system and the scope for discretionary action; c) the legal opportunities for public participation and the depth of engagement with such processes. These can vary greatly both between and within countries. A fourth factor is not addressed here because it is our common denominator – economically advanced countries with a relatively solid governance base.

To date, there is not enough systematic comparative research about these framing conditions to support generalizations to other countries. Our own modest aim is to provide planners in countries not included here with an external perspective for comparing their situation with others. For this purpose, we have developed a schematic three-dimensional triangle where the hypothesized positions of individual countries may be plotted. Figure 30.1 demonstrates how different countries may be characterized by plotting them on the three dimensions linked to the frame conditions and then linking these points to form a triangular shape. For our demonstration, we have selected five countries: Germany and the UK in Europe; the US – with Oregon as a better-practice example state; Japan, as another OECD country but with

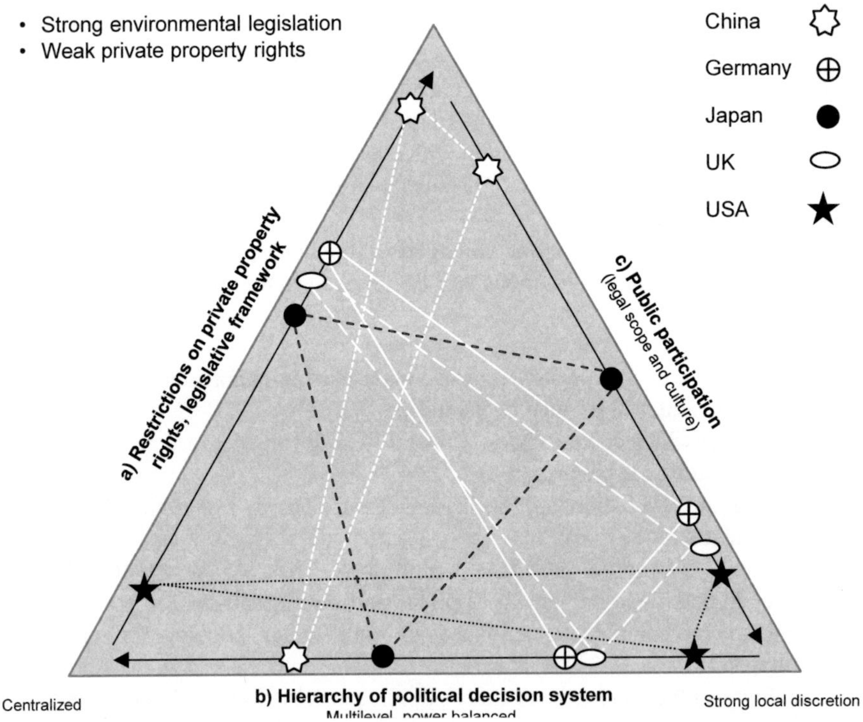

Fig. 30.1 A schematic classification of framing conditions for environmental planning in five example countries. The conditions may be decisive for the importance of economic evaluations, the binding character of objectives, the scope and role of public participation and the scale and delineation of the scope of planning. We are aware of the internal differentiation in the USA, where the preconditions may differ a lot among the states. Nevertheless there are some general frame conditions shared by all US states based on the US legal system and on federal policies

a different culture; and one non-OECD country – China. In the absence of systematic comparative research, the triangles represent the political, legal and economic dimensions within which planning systems operate and highlight, for example, the large degree of overlap between the triangles for Germany and UK. In contrast, there is very little overlap between the shapes for China and the USA, while that for Japan is more equilateral than that for the other examples. Diagrammatically, therefore, this highlights the contrasts in framing conditions for landscape planning.

An ideal system achieving the goals of landscape planning (Chap. 4) probably does not exist in reality anywhere, but its attributes would fit somewhere in the centre of the triangle. The influence of the different characteristics can be described as follows.

Legally anchored standards for landscape preservation often entail restrictions on landowners' rights to use their land or build on it. A high degree of private property rights and the comprehensiveness of binding environmental standards are thus negatively associated. In some legal systems, restrictions on land use might be regarded as tantamount to indirect expropriation or 'regulatory takings' (Alterman 2010). Whether such expropriation is possible, practiced or impeded (e.g. by high compensation rights/obligations) depends on the specific property rights regime prevalent in each country. Even where planning bodies do not intend to impose the heaviest types of intervention permitted by law the basic property rights regime is nevertheless important in (quite frequent) situations where governments prefer to achieve public goals through negotiations with private landowners.

The comprehensiveness of the environmental legislation and standards is also very relevant for deciding whether a socio-economic (e.g. preference) or a legally based evaluation method is more suitable for ES evaluations in landscape planning. One may conjecture that in countries with strong environmental legislation and limited individual property rights, it would be possible to conserve the most important ES by means of legally binding regulations. In such countries the evaluation of ES can be based primarily on legal standards and objectives accompanied by expert standards. This can lead to the full range of scales of assessment results – nominal, ordinal or cardinal. Planning targets and responses can be deduced quite easily from the outcomes of such legally-based evaluations. Resulting responses do not have to be mandatory, but may include different alternatives or priorities for action. The mandatory limits of development depend on the specifics of the law and may be defined very differently between countries. Planning has to take them into account for defining non-negotiable objectives and setting the context for decision making and participation.

The biggest advantage of strong environmental legislation – one that concentrates particularly on market failures – is its high level of legitimacy to represent public interests. Furthermore, legally predefined solutions may be much more efficient; for example, fewer economic resources are needed to purchase the 'right to pollute' from the polluter. This could even mean lower transaction costs, such as where economic instruments (e.g. emission trading) are used, thus saving on institutional and staff budgets. However, a precondition for the implementation of legal standards is a well-functioning administrative and enforcement system, and sufficient legal, economic and communication tools.

In such legal frameworks, social and economic valuations will serve as add-on in order to identify where the ES delivery in a landscape does not satisfy demand, to clarify implementation conditions (costs, social trade-offs) or to support conservation priorities through supplementary economic arguments (e.g. monetary value of the ES). Public participation in such debates may be prescribed by law (such as in the EU) and often encouraged. However, the influence of participation is also limited to a certain degree because its outcomes cannot overrule strict legal objectives and thresholds. Participation processes will often be more citizen driven on the local-level because the subjects are more specific.

In cases with very rudimentary environmental legislation or strong private property rights, at least some internationally agreed goals such as climate change targets and the Sustainable Development Goals (SDGs) (see Chap. 4) can be taken as a normative basis for identifying priorities for the most important ES. Beyond that, economic valuation may be much more important for assessment and determining objectives than reliance on legally-prescribed objectives. This occurs because economic valuation is directly linked to the interests of the decision makers and thus has greater persuasive power.

Similarly, in systems with only rudimentary evaluation standards, citizen preferences expressed during participatory processes may play a greater role in defining planning objectives than the legal framework. However, a strong role for public participation is only possible where it is mandated, or at least permitted by law, where it is politically encouraged or there is a culture of participation. Otherwise, participation may be dominated by individual stakeholders that do not represent the broad interests of all sectors of the local community. The role of public participation is also important as a checks-and-balances mechanism to ensure that administrative decision-making is ethical and conforms to the law. When public participation is not vibrant enough, other control mechanisms may be needed in order to check that the public administration is properly performing its role and not being corrupted.

Finally, political and planning systems differ concerning the number of decision levels and the distribution of powers between them. It makes a big difference for landscape planning whether the legal powers are allocated primarily to the local level or – in more differentiated systems – to several authorities in a hierarchical arrangement and with particular areas of responsibility. Landscape planning has to adapt its objectives and focus to what can realistically be implemented at the specific decision level. Information for other decision levels should be presented separately. If there is no higher political or administrative decision level which can take responsibility for transboundary environmental problems, concerns such as river pollution or migrating species have to be dealt with at the local levels of landscape planning. As a consequence, special and possibly complicated efforts will have to be made to coordinate decisions and resolve conflicts with neighbouring jurisdictions instead of their being resolved by the higher planning levels.

30.2.2 Factors Explaining the Landscape Planning Context

Very basic factors may influence the planning context and system that has evolved in different countries. Preconditions may be a result of history, culture, geographical features and can be best illustrated by comparing data across example countries (Table 30.1). Relevant variables include the population density, economic productivity (expressed by the GDP), and share of urban and rural population (Fig. 30.1). These characteristics can help to generate hypotheses about some of the contrasts between the planning systems in different countries. For instance, Germany, Japan and the UK with limited spatial resources and high population densities have developed mandatory systems of spatial planning whereas the United States with – on average – much less pressure on land has no federal planning system at all. Nevertheless, the local or regional densities in some parts of the US are as high as in some densely populated countries. As an example, Oregon took early planning action to define legally-based growth boundaries, even though its average level of population density was not especially high. Furthermore, countries with ample spatial resources such as China are in the process of developing landscape or environmental planning because a lot of land-based resources such as fertile agricultural land, drinking water, recreation landscapes or biodiversity are concentrated in specific, higher density regions. Thus, the land uses compete with each other in terms of the use of the natural resources.

The above discussion illustrates how political and socio-economic characteristics can influence the natures of the planning system in a country. However, the situation in individual countries is often much more complex than can be illustrated by a simple classification or statistical data. Planning cultures are affected by a myriad of historical and cultural forces. In order to understand more about these influences, and for a deeper understanding of the 'perspective from outside Europe', we continue by taking a closer look to at the situations in Japan and the United States. These are both industrialised, developed countries, yet have very different spatial resources, planning cultures and legal systems which belong to different 'legal families' (Siems 2016).

30.3 USA with Particular Focus on Oregon

Vivek Shandas, Christina von Haaren

Before proceeding with our analysis of the USA context, we should point out some linguistic differences. The term 'landscape planning' as used in the USA often refers to the appearance or view of an area, whereas the closest equivalent to landscape planning as used here may be 'environmental planning' or 'open-space regulation'. The US equivalent to 'spatial planning' is land use planning (urban or rural).

The USA is the birth place of many environmental innovations which have served as models for many other countries, including the EU. These include national parks, environmental impact assessments, species protection, economic approaches

Table 30.1 Geographical, demographic and economic characteristics in 2016 relevant to landscape planning for selected jurisdictions

	China	Germany	Japan	UK	USA as a whole	Oregon	Relevance for planning
Land area (sq. km)	9,388,211	348,900	364,560	241,930	9,147,420	248,607	The size of the country, in relation to the amount of inhabitants, influences the degree of pressure from land use change
Total population (millions)	1378.67	82.67	126.99	65.64	323,13	4.03	
Population density per sq. km	147	237	348	271	35	15	Higher density increases land use conflicts and a greater need to 'plan'
Annual % population growth	0.5	1.2	−0.1	0.8	0.7	1.5	Higher population growth increases development pressures and the need for restrictive planning. Declining populations provide options for developing green space
% Urban population	57	76	94	83	82	82	Fast growing urban populations makes urban landscape planning particularly important
% Rural population	43	24	6	17	18	18	Problems such as rural depopulation or intensification of agriculture need special emphasis in rural landscape planning
GDP in million current US$	11,199,145	3,466,756	4,939,383	2,618,885	18,569,100	228,900	GDP indicates the economic power of a country and thus characterizes its potential capacity to care for the environment

Sources: All data taken from the World Bank (http://data.worldbank.org, accessed 1 Aug 2017), except for data on Oregon (Department of Land Conservation and Development, US Bureau of Economic Analysis, US Census, 2015, www.bea.gov, accessed 13 Dec 2017)

to compensatory mitigation, agri-environmental measures with bidding procedures and result-oriented remuneration – to name but a few. Yet today, some countries have overtaken the US in environmental protection policies.

The innovations listed above were achieved within a context laden with legal and institutional constraints faced by comprehensive planning. At the US federal level there is a lack of overall comprehensive spatial or environmental planning legislation. By tradition, most of the legal powers to regulate private land rest with the states, not the federal government. Most of the 50 states try to minimize their degrees of intervention in the planning and development policies of local government (Alterman 2005; Cullingworth and Caves 2009: 17 f.). While many historical and cultural reasons may help to explain the limited role of federal intervention in local land use planning decisions, one plausible explanation may go back to US history. Given the country's vast land area (9 million km^2) and the relatively low population density of 35 inhabitants/km^2 there was a sense that there is ample space for further development (Kayden 2001). Indeed, a possible partial explanation for the ideological and legal status of private property – which is enshrined in the US Constitution – is the rapid settlement of Europeans in the western USA during the nineteenth century on land formerly held by indigenous tribes (for a more detailed understanding of the legal history see Levy 2001). The US government encouraged and legally enforced claims made by the settlers. The homesteading ethos elevated property rights to a high level and thus it is difficult to introduce regulation over property. Political trends in the US federal government during the Trump administration further promote this ethos by promising to roll back some of the achievements of environmental regulations and by reviving the legal debates concerning environmental protection versus private property. As a result, the ethic of landscape protection in the US is still largely governed by these early precedents and generally assigns higher reverence for private property than many EU countries. These legal-ideological factors provide the background to the current challenges facing the incorporation of ecosystem services into the market system.

Despite the emphasis on local landscape planning and the importance of private property, federal land-use controls have been the dominant means of achieving many environmental objectives, including emission and waste treatment standards, the establishment of the national parks system, procedural innovations such as environmental impact assessments, and mitigation regulations. Two federal laws, in particular, have been the focus of the debate over environmental conservation and private property: first the Endangered Species Act (ESA[1]) and Section 404 of the Clean Water Act (CWA[2]) – the two main regulations limiting the development of wetlands.

Alongside the increasing breadth and influence of environmental regulations, an ideological-political counter-reaction has emerged in recent years. Viewing some of the environmental controls as encroachment on property rights and, in some cases, as 'regulatory takings' (Bosselman et al. 1973; Roberts 2010), twenty-three state

[1] Endangered Species Act of 1973; 16 U.S.C. 1531 et seq.

[2] Clean Water Act of 1977. Pub.L. 95-217, December 27, 1977.

legislatures have enacted special laws intended to enhance property rights protection (Brady 2017). Although these legislative initiatives vary in scope and effectiveness, they share the view that the general constitutional protection of property rights is too ambiguous, and seek to provide greater certainty for landowners seeking compensation for regulatory takings. These proposals have had varying impacts on the capacity of local municipalities to undertake landscape planning.

On the municipal level, zoning is the most widespread instrument for influencing development and the role of private property in society. Zoning did not evolve from the motive to restrict land consumption or protect the environment. Rather, zoning evolved 'bottom up', incrementally, surviving many legal challenges. There were two main types of motivation. One was grounded in the desire of the affluent to protect their property values by reducing nuisances from conflicting land uses such as industry and commerce (and at the same time, distance some lower-income families). The second motivation reflected early recognition of the problems created by urban sprawl and poor coordination among land uses, resulting in high infrastructure costs and health risks (Hirt 2015; Alterman 2005; Cullingworth and Caves 2009:65 ff.). Zoning finally gained legal clearance when one of the many court cases reached the U.S. Supreme Court in 1926, and zoning was ruled as not unconstitutional (Cullingworth and Caves 2009: 72). Nevertheless, the Supreme Court also ruled, in several subsequent decisions, that when a regulation goes 'too far' it will be regarded as a 'taking' (also known as 'regulatory taking'), even though the land is not expropriated but remains in the owner's hands. In such cases, the regulation – zoning, environmental regulations or otherwise – is unconstitutional and void or, in some situations, the owners may have a right to receive compensation (Roberts 2010; Alterman 2011).

Despite the ambiguity and uncertainty caused by American law regarding regulatory takings, in reality the fear that zoning or environmental regulation would indeed be ruled unconstitutional is not as significant as the discussion of property rights makes it seem. This is demonstrated in Alterman's comparative international research (Alterman 2011). Today, zoning is deeply anchored in planning and land management in the USA, and all states allow cities to zone (except Texas, where counties are not allowed to zone, but even there most cities do choose to adopt zoning). Nevertheless, there continue to be some significant differences in planning and zoning laws and practices among the 50 states. It thus makes sense to examine one specific case in depth, so as to gain insights about the interaction between local, state and federal laws. We have chosen to focus on Oregon – a state regarded by many as a USA 'best practice' example in several aspects of planning and environmental regulation (Sullivan 2011). Within the US legal-political context, how feasible is it to enact laws, adopt programs, policies, or plans that emphasize ecosystem services?

30.3.1 Oregon Case Study

According to Oregon state law, land use planning and zoning are mandatory for local government. They are carried out by cities, counties and the Portland metropolitan region, which regulate housing and other urban uses of land. These decisions are governed by a binding set of state-wide planning goals (http://www.oregon.gov/LCD/Pages/goals.aspx#Statewide_Planning_Goals).

Oregon mandates that local governments set binding urban growth boundaries. Together with additional measures for urban densification, the growth boundaries are a powerful tool to curtail urban sprawl of residential subdivisions, shopping malls or office parks (Liberty 2009). Thus, we conclude that Oregon's planning laws and policies are protective of ES.

Among the land use tools used in Oregon is 'purchase of development rights', something practiced especially outside urban planning zones in order to contain development. In addition, designated Exclusive Farm/or Forest Use zones (EFU) and Rangeland Zones are used to protect farming and forestry areas. These tools may also (indirectly) contribute to the conservation of other ES. For this purpose, local governments in Oregon have the regulatory authority to severely limit development rights in case of proposed unbridled expansion of urban development.

The state-level governing body is the LCDC (Oregon Department of Land Conservation and Development). It is charged with initiating and monitoring the zoning and planning processes, Evaluation of the planning outcomes is mandatory and carried out as a post-plan approval process or a periodic review (Geißler 2008).

Furthermore, on federal public lands, place-based management objectives are specified through Resource Management Plans (RMP). Characteristic of Oregon, and the US in general, is the relatively large share (compared with Europe) of public lands that are designated for intensive area protection and conservation. RMPs are prepared by the Bureau of Land Management (BLM) and used for a variety of purposes such as wilderness areas, off-highway vehicle traffic, grazing management, designation and management of significant caves, and recreation area management (see for example: Lakeview District, Bureau of Land Management (BLM) 2003). The plans are prepared under the authority of the Federal Land Policy Land Management Act (FLPMA) and the National Environmental Policy Act (NEPA). They can cover large areas as – in the example of the Lakeview district – approximately 3.2 million acres of BLM-administered public lands.

Additional sectorial plans cover specific topics such as the Forest Resource Plan, a state transportation program (which manages transportation and land use planning grants for local government), or water catchment area plans such as the Oregon Plan for Salmon and Watersheds. Notably, Oregon also has a federally approved state Coastal Management Plan[3] tailored to fulfil the state planning goals.[4] Environmental planning objectives are implemented by national, state or local bodies. There are offset obligations in case of loss of wetlands (based on national regulations see Environmental Protection Agency 2002: Regulations affecting wetlands[5]) or impacts on water bodies (e.g. for rise in water temperature). Any landscape or land use planning action that is eligible for federal support requires an environmental impact assessment, and the outcomes of the relevant plan will be evaluated in the context of

[3] https://www.oregon.gov/LCD/OCMP/pages/index.aspx

[4] http://www.oregon.gov/LCD/Pages/goals.aspx#Statewide_Planning_Goals

[5] https://www.epa.gov/cwa-404 https://www.epa.gov/cwa-404

the ESEE (Economic, Social, Environmental and Energy) assessment framework (Oregon Department of Land Conservation and Development 2018).

30.3.1.1 Public Participation and Data Needs

One of the major contributions of landscape planning in the USA is the inclusion of public involvement in the planning process. Oregon's admired land use planning system has as Goal 1 'Citizen Involvement', which requires any land use planning action to enable a process for citizen inputs. This enhanced position for public participation, combined with Oregon's additional participation rights, indicates the political importance of the participation goal. Citizen involvement in Oregon is important in several ways. First, it is a specific way of implementing environmental objectives by citizens who act as agents of environmental improvement. Oregon invests in education or motivation for public action (see for example, NW Habitat Institute[6]). Second, citizen participation is a mandatory part of the planning processes. Citizens may participate either by expressing their own interests or as spokespersons for the broader public interest. Citizens thus also function as informants and control agents over government actions. Relative to Europe, the engagement and influence of NGOs and individuals seems to be stronger in the US in general and in Oregon in particular (Geißler 2008; example: http://cedarmill.org/news/UrbanNeeds/). Thirdly, citizens also have an important role in monitoring plan implementation and the achievement of the stated objectives. Indeed, participation is not only an official requirement; projects that do not involve citizens may encounter political opposition. Furthermore, in the USA legal system, the rules of legal standing are relatively liberal, and land use decisions are often challenged in the courts.

The attention given to citizen involvement in Oregon is probably partly cultural, partly due to official encouragement by the authorities, and partly reflects the constitutional rights for direct democracy available in Oregon (but not in all US states). Under that right, called the 'citizen ballot initiative', if supported by enough people, citizens can initiate legislation, vote on legislation approved by the legislature and even dismiss officials. Also, Oregon residents have broad rights to challenge government land use decisions as well as some environmental decisions in the courts.

But Oregon's story also harbours an unexpected drama. The very same state that has pioneered important environmental planning achievements, and where citizen action has been such an important stimulus, is also the state where (other) citizens undertook the most extreme counter-action to protect property rights against environmental regulations. In 2004, through a citizen-ballot legislative initiative, Oregon adopted Measure 37 (Sullivan 2007; Putter 2010) This legislation granted Oregon landowners – especially farmers – extensive protection of property by obliging the government to compensate fully for diminution of values associated with multi-generation held property due to any land use or environmental regulations. Such compensation rights are the most extreme reported anywhere in the world (see

[6] Managing Transportation and Land Use Planning Grants for Local Government wetland and forest wildlife guides. http://www.nwhi.org/index/publications, www.Oregon.gov.

Alterman 2010, 2011). Claims amounting to huge amounts of money were submitted by landowners. This was an obvious counter-revolution to Oregon's achievements in environmental protection which, inevitably, also restricted development rights. The drama had a relatively happy ending. Three years later, a counter citizen ballot initiative – Measure 49 – reversed most (but not all) of the excessive compensation rights (Putter 2010). Oregon's citizen majority demonstrated that it does want the state to continue with its good environmental planning policies.

Oregon's achievements in planning as well as participation are supported by public availability of environmental information. Although the available data are inherently fragmented due to the localized nature of landscape planning and the need to rely on inputs from many municipalities, in comparative terms the amount of available data is impressive. Furthermore, there are now initiatives towards creating a centralized data repository on the federal level. One example is EnviroAtlas (https://www.epa.gov/enviroatlas), which is a compilation of thousands of local and other datasets. It provides interactive tools and resources for exploring the benefits that people gain from nature at the municipal scale. However, at present the coverage of some data sets is partial and details are not all standardised.

The Oregon case offers several insights about how the integration of public participation with public access to information can advance the assessment of ecosystem services. Especially instructive is the example of the Portland Metropolitan Region (PMR) with its 24 cities and the urban portions of three counties. Although individually administered, these cities and counties are also coordinated through a metropolitan government. This institutional format is an innovation in the USA context. Metro standardizes all the data for the Portland Metropolitan Region, and access to their land use data is provided through a publicly available online portal named Regional Land Inventory System (RLIS, http://rlisdiscovery.oregonmetro.gov/). Through the Regional Land Inventory System, community groups, researchers, and others can access land use, land cover, transportation, wetland, river, soil, and other datasets. With Metro's curating of information on land cover, land uses, as well as administrative and political boundaries, individuals and organizations can access the full range of data about provisioning and regulating ecosystem services. Beyond the data captured at the Metro level, through the public participation process, communities can provide insights about their cultural ecosystem services. Together, these two approaches help provide an increasingly comprehensive view of the ecosystem services and biodiversity in the Portland Metropolitan Region. Once federal standards for national ecosystem services data become available, as is already the case with specific agencies (e.g. US Geological Survey, Forest Service, Land Management), Oregon could serve as a model for the integration of diverse data sets for assessing ecosystem services.

At the level of achieving spatial environmental objectives, we therefore see that the basic tools and options in Oregon are surprisingly similar to those available in many European countries, particularly with regard to the importance of spatial (land use) planning. This refers to the extensive regulation of urban sprawl by municipal planning/zoning, which even surpasses many European regulations (Liberty 2009). Methodological approaches such as provided in this book are either already applied in Oregon or could easily be integrated into planning and policymaking. The excellent data standards and easy access available to experts and citizens alike, coupled with the institutionalized modes of citizen inputs into data generation, could serve as a model not only to other US states but also for the European context.

30.4 Japan

Hiroyuki Shimizu

30.4.1 Introduction

Like the other countries discussed in this chapter, Japan's evolution of ecosystem-service governance reflects aspects of the country's history. However, Japan's story is substantively different. Japan was the first non-western (Asian) country to evolve at unprecedented speed from a developing country into the epitome of an advanced economy-plus-democracy. Indeed, the term 'developmental state' was first coined to depict Japan's unique trajectory (Johnson 1995; Sasada 2008). The term applies to the society and economy at large, but land-planning issues are an important part of the story. In a book devoted to comparative planning cultures, Sorensen (2005) entitles his chapter about Japan – *"The developmental state and the extreme narrowness of the public realm"*.

The task of nation-building placed the development of urban areas at center stage (Sorensen 2002). In such a policy context, environmental issues, along with promotion of public participation, received low priority (Shibata 2007, 2008a). Recognition of the value of promoting and preserving ES has therefore been slow, and has only become more prominent recently. This section discusses the evolution of ES governance in Japan – laws, institutions and policies. Some of the difficulties encountered are endemic to land-use planning in general and ES governance anywhere. But in many ways, as described below, the path taken in Japan on the way to institutionalizing ES seems to have been especially bumpy.

30.4.2 Framework of Legislation on Land Use, Planning, and Management, Planning Instruments and Standards for Evaluation of ES in Japan

Currently there is no legislation in Japan that requires nationwide landscape planning to ensure the preservation or promotion of ES. However, several legislative acts provide opportunities for fostering the management of ES.

The Japanese government system consists of three levels. The nation is divided into 47 prefectures and 1741 cities, towns, and villages. In the field of land use planning and management, the national level makes the policy, the prefectures plan and implement it over a wide area, and the cities, towns, and villages control implementation of the legislation in specific locations.

Japanese land use governance is rather fragmented (as occurs in other countries as well; Alexander 1993; OECD 2017: 29–30). Land use is managed by the following five separate acts:

- the Forest Act under the responsibility of the Forestry Agencies
- the Act on Establishment of Agricultural Promotion Regions (AEAPR) by the Ministry of Agriculture, Forestry, and Fisheries
- the City Planning Act (CPA) by the Ministry of Land, Infrastructure, Transport and Tourism (MLIT)
- the National Parks Act (NPA) by the Ministry of the Environment (ME)
- the Nature Conservation Act (NCA) by the Ministry of the Environment

In 1974 the National Land Use Planning Act was enacted as an additional umbrella measure to enable comprehensive land use management (see Fig. 30.2).

In the 1974 Act five major zoning types were defined – which approximately parallel to the five specialized Acts: urban areas, agricultural areas, forest areas,

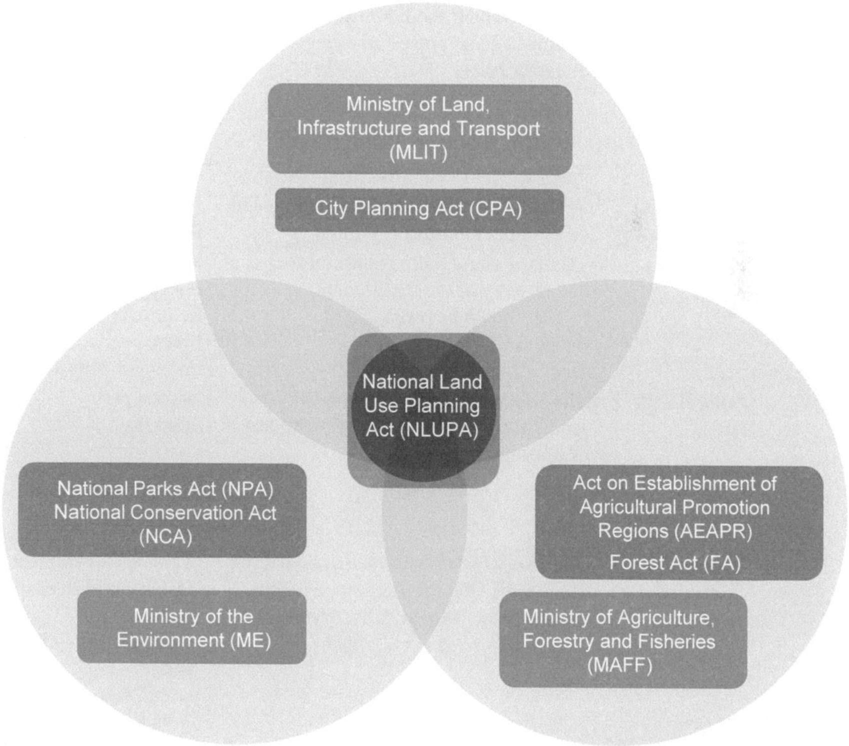

Fig. 30.2 Agencies and acts influencing land use management in Japan

natural park areas, and nature conservation areas. However, only the boundaries and sizes of these zones are designated, there are no mandatory landscape planning or spatial planning rules linked to the zones. In practice, each Ministry formulates its own spatial land use almost independently, with poor coordination among the agencies. The lack of coordination is further exacerbated by the fact that the five zones partially overlap. This issue of poor coordination may be more problematic, in comparative terms, than in many other OECD countries (Alterman 2001; OECD 2017). The important zoning issues in the five Acts are outlined in Table 30.2.

Below we discuss each of the acts in somewhat greater detail. The Forest Act is among the most significant due to its large land coverage. Designation of forest areas is seen as important for the protection of the public interest. Designated forest covers 66% of Japanese land. Within this mass, land designated as Forest Protection is 47% of the total forest area (Forest Agency 2014). The protection of forests is strongly regulated and means that the areas cannot be freely developed. This policy was originally developed for the purpose of land preservation prior to the introduction of the ES concept. Today, it still acts as a powerful tool for preservation of ES

Table 30.2 Zoning issues in five important Acts

Acts		Definition	Subdivided zonings
CPA	Urban Area	The area which should be developed, established and preserved as an integrated city	Urbanization Promotion Area Urbanization Control Area Urban Park Scenic District
AEAPR	Agricultural Area	The area where there are useful agricultural lands and agriculture should be developed integrally	Agricultural Land Area Agricultural Promotion Area
FA	Forest Area	The area where there are useful forests and forestry or the multifunction of forests should be promoted or maintained	National Forest Forest Plan Required Private Forest Protection Forest
NPA	Natural Park Area	The area with excellent natural landscapes, which should be protected or usefully promoted	Natural park Semi-Natural Park Special area Ordinary Area Special Protection District
NCA	Natural Protection Area	The area which has formed good natural landscapes, and the area where natural environments should be conserved	National Government Designated Wilderness Area Nature Conservation Area Special Area Ordinary Area
	White Area	The area which does not belongs to any zonings	

in forest areas. ES related to forest protection encompass water recharge, sediment discharge, landslide prevention, erosion prevention, protection against wind, flood protection, salt damage prevention, drought prevention, snow prevention, fog reduction, avalanche prevention, rock fall prevention, fire protection, fish reserve, promotion of health, and scenic amenities.

The Agriculture Promotion Act has a similar role regarding agricultural land. It defines Agricultural Promotion Areas and Agricultural Land Areas. The former are zones necessary for the comprehensive development of agriculture. The latter are zones with stricter protection, where only agricultural land use is allowed. Agricultural Land Areas account for 13% of the land in Japan (National Land Numerical Information Download Service 2015). In Japan zones for hunting or recreation are excluded from agricultural lands. Creating areas where hunting is permitted requires approval of the Minister of the Environment based on the Wildlife Protection Act. The most important issue regarding farmland in Japan is soil remediation and improvement of nutrients, because many parts of agricultural land consist of volcanic ash that is poor in nutrients. On the other hand, soil erosion is not a serious problem, as long as the appropriate management of paddy fields is maintained (Japan Country Section 2008). Conservation of land for drinking water is regulated mainly as part of the protection of forests in the Forest Acts.

Although urban areas in Japan, as elsewhere, do not account for a large proportion of total land, regulation of urban development is very important in such a densely inhabited country. The City Planning Act (CPA) defines a number of different zones: Urban Areas, Urbanization Control Areas, Urbanization Promotion Areas, Urban Parks, and Scenic Districts. According to the National Land Numerical Information Download Service (2015), Urban Areas, Urbanization Control Areas, and Urbanization Promotion Areas account for 20.2%, 6.9% and 3.0% of Japanese land respectively. Urban land use (UrLU) increased at a high rate between 1975 and 2010 in the overlapping agricultural and urban zone as well as in the Urbanization Promotion Areas (derived from data provided by the National Land Numerical Information Download Service 2015).

10.3% of the total Urban Area has no designated land use and is labelled as a 'White Area'. This land exists mainly in the countryside (suburban areas) and has very weak control against sprawl. Another problem is that since the declaration of the Agricultural Promotion and the Agricultural Land Areas about 50 years ago, the Urbanization Control Areas, the White Areas and the Agricultural Promotion Areas overlap. This overlap is supervised by two different ministries and causes confusion in land management of suburban areas.

For landscape planning, the designation of Scenic Districts is one of the important tools offered by the City Planning Act to preserve and maintain scenic beauty in cities. In a Scenic District the planning authorities must designate between 10% and 60% of the area as green space for residents. There are 105,744 urban parks in Japan with a total area of 122,839 ha, 0.3% of Japan's whole country (Ministry of Land, Infrastructure, Transport and Tourism 2015). This yields an average of 10.2 m^2 of urban parks per Japanese person. The Natural Park Act (NPA) controls the National Parks, Semi-National Parks, and any Prefectural Natural Parks. There are 32 National Parks, which occupy 5.6% of Japanese territory. Semi-National Parks account for 3.8% and Prefectural Natural Parks are 5.2% of the territory (Ministry of the Environment 2016). One special characteristic related to the parks is that land ownership of National Parks and Semi-National Parks is not restricted to public bodies but can include private individual and corporate owners as well. Park conservation is therefore strongly dependent on the consciousness and cooperation of private landowners.

The Nature Conservation Act (NCA) designates National Wilderness Areas and Nature Conservation Areas. The former are areas without human interference. The latter are areas with special natural assets – such as alpine and sub-alpine vegetation, excellent natural forest, unique topography, geology, natural phenomena, lakes and coasts, marshes, rivers, seas, etc. that require special care in management. Five National Government Designated Wilderness Areas and 10 Nature Conservation Areas were designated by 2013 and the total combined area is 27,224 ha. This accounts for only 0.07% of the national territory. Additionally, Prefectural Natural Conservation Areas encompass 76,403 ha, 0.2% of national land (Ministry of the Environment 2012).

In addition to the five major acts concerning land use, there are several specific pieces of legislation that enable local governments to control open space in urban areas. These are the Urban Green Space Act (UGSA), Urban Park Act (UPA), and the Act for the Preservation of Trees to Maintain the Scenic Beauty of Cities (APTMSBC). The most important of these is the UGSA which allows local governments to designate green spaces in urban areas. It also enables the authorities to apply different degrees of preservation to designated areas: Green Space Conservation Districts, Special Green Space Conservation Districts, Greening Districts, and Citizen Green Sites.

The Special Green Space Conservation District is a highly restrictive zone where no development at all is allowed. Such zones can be declared as growth boundaries to prevent urban sprawl or for disaster prevention purposes. A Special Conservation District can also be designated to protect areas of historical or cultural importance, scenic beauty, or important habitats of fauna and flora. In 2014 there were 528 such areas, amounting to 2571 ha in 79 cities (Ministry of Land, Infrastructure, Transport and Tourism 2014). The Greening District is another type of regulation. It enables local governments to impose a minimum green-area ratio on private development. So far, Greening Districts have been designated in only four cities and amount to 60,625 ha. For instance, in Nagoya City, any new building projects with plots of more than 300 m^2 must be vegetated on at least 10% of the site. The Citizen Green Site is an interesting private-public land designation where private land owners, local government, and community-based organization enter into an agreement to make private green space available to the general public.

Recent awareness of the importance of urban agriculture has led to the Productive Green Land Law (PGLL) intended to encourage agricultural production within or near urban areas. This Act incentivizes agricultural production in the Urbanization Promotion Areas by reducing property taxes. Conservation of Productive Green Lands has gained a new and important meaning in the shrinking and depopulating

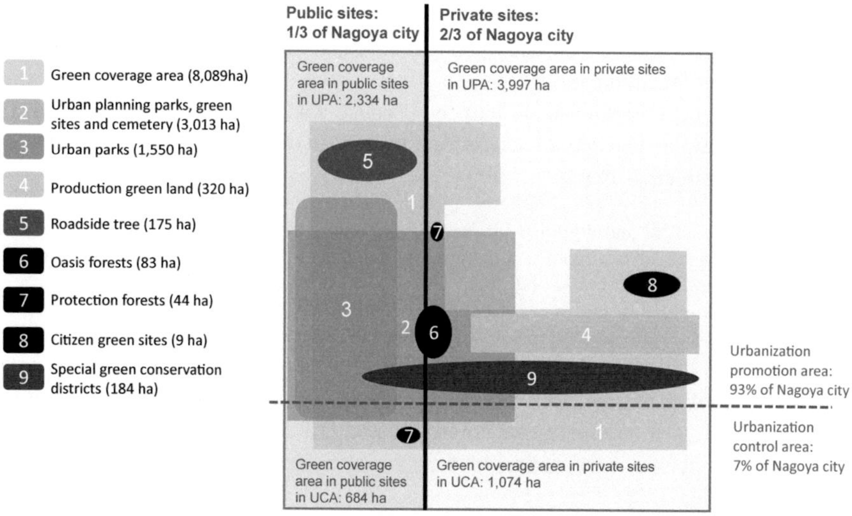

Fig. 30.3 Types and sizes of green sites in Nagoya city

urban areas. It is these areas that can provide multiple ES functions in high density urban areas. In Fig. 30.3, types and sizes of green sites in Nagoya city are shown. Importantly, green coverage areas on private property account for 63% of the total green sites (Nagoya City 2014). Furthermore, the designations of Special Green Space Conservation Districts, Greening Districts, Citizens Green Sites, and Production Green Land are mainly used to preserve green areas on private land. However, while all these regulatory tools empower governments to control the extent of green spaces, they do not influence the character and quality of the green sites. Such legal tools, necessary for good ES management, are yet to be developed in Japan.

It is also important to note that the UGSA empowers both local and prefecture governments to adopt a Basic Plan for Conservation and Promotion of Greening (Green Basic Plan). In recent years, the adoption of the Green Basic Plan at the prefectural level has been promoted by the national government. By 2016 23 of the 47 prefectures had adopted such plans and eight were in the process of doing so. These prefecture Green Basic Plans could be one means of promoting landscape planning in Japan. However, this tool is limited to urban areas because it is under the jurisdiction of the Ministry of Land, Infrastructure and Transport.

Landscape planning should not only aim at protection, conservation, development, and management of green spaces but also take into consideration wild animals, water, soil, and climate. These elements are partially addressed in Japan by the Wildlife Protection and Hunting Act, the Water Pollution Protection Act, Soil Contamination Countermeasures Act, and Air Pollution Control Act. However, landscape management issues are still not adequately addressed in these acts. The Japanese Landscape Act (see Ministry of Land, Infrastructure, Transport and Tourism 2006) was enacted in 2004, relatively late compared to many other countries.

The contents of this Act are far from a comprehensive approach to landscape planning (e.g. as envisaged by the European Landscape Convention). The idea of integrating policies for urban and rural areas did not exist in Japan when the Landscape Act was enacted. Nevertheless, in part the purpose was to promote integrative management of complex landscapes across the rural-urban divide. In article 1 the purpose of the Act is defined as being

> to create a beautiful and dignified landscape, create an attractive and comfortable living environment and realize vibrant communities with distinct personalities by taking comprehensive measures to develop good urban and rural landscapes such as formulating landscape plans, in order to improve the quality of life of the people of Japan and contribute to the growth of the national economy and sound development of society.

Unlike some countries, Japanese landscape planning is not based on a Nature Conservation Act or an integrative environmental law. Although the Landscape Act does not prevent the conservation of the natural environment, this is not its main purpose.

In another respect, the Japanese Landscape Act does have an important advantage. The Act is jointly supervised by three national ministries – IT, ME and the MAFF. This means that the act has the potential to integrate existing tools relevant to landscape planning, which are currently under the fragmented management of each authority. The Landscape Act empowers local governments and municipalities to control landscape planning and to designate Landscape Planning Areas. Furthermore, the Landscape Act has strong affinity to the conservation of cultural landscapes, which is mainly under the authority of the Ministry of Education, Culture, Sports, Science and Technology (MECST). In many municipalities, there are connections between the Landscape Planning Area (LPA) and the Preservation District for Groups of Historic Buildings of the MECST. While the Landscape Act is presently not associated with the Urban Green Space Conservation Act, collaboration between the agencies in charge of these two Acts has a high potential for landscape conservation. In order to activate this potential it is very important to enhance the environmental aspects in the Landscape Act. Even though landscape planning is optional, 713 local government and municipal authorities (out of 1741) were designated as landscape planning bodies in 2018 and 558 had formulated landscape plans (Ministry of Land, Infrastructure, Transport and Tourism 2018).

Another avenue for landscape planning is the Environmental Impact Assessment Act of 1997. An Environmental Impact Assessment (EIA) is required in certain types and sizes of developments by municipalities and individual enterprises. In such projects, there are procedures for evaluation, avoidance or mitigation of negative environmental impacts. The first attempt to carry out EIA was in the planning of the Aichi Expo in 2005. The most important issue concerning Japanese EIAs are that they are focused mainly on procedures and the targeted environmental status to be realized remains unclear. For instance, ‘no net loss’ is a core principle of the EU Biodiversity Strategy (European Commission 2016). However, in the Japanese Environmental Impact Assessment Law, no such concept is clearly articulated.

Consequently, there is currently no good example in either local or prefecture spatial plans where an EIA has enabled the 'no net loss' principle to be applied.

Furthermore, the introduction of Strategic Environmental Assessment (SEA) has yet to take place. Early introduction of the SEA into spatial planning is necessary for the achievement of holistic land management based on the precautionary principle, which is well accepted in environmental policies in Europe. This should be an urgent issue in Japan, however, the legal system has so far been hesitant about the introduction of the precautionary principle into Japanese law.

As illustrated above, the various Acts and regulations related to land use planning and management in Japan are highly fragmented under a plethora of ministries and government agencies. Complaints about institutional fragmentation and insufficient coordination are not unique to Japan, and indeed, are characteristics of the complexities inherent in land-use planning (Alexander 1993; de Roo and Silva 2010). However, the Japanese degree of fragmentation in land use planning may well be more extreme than in many other advanced economies (in the absence of systematic comparative research, we can only conjecture). Consequently, the overall vision necessary to evaluate, conserve and foster ES in Japan is absent.

Nevertheless, in recent years there are more optimistic signs. Some independent initiatives related to ES have been undertaken in a variety of projects. The common theme is to create better integration of land use planning and ES. The most pragmatic approach under the current regulatory regime is to utilize the national Landscape Act as expansively as possible by enhancing the ES aspects embedded in it. Thus, the landscape plans could be promoted at the municipal level as a means to integrate the regular spatial plans into a regional-wide assessment under the authority of the prefectures.

30.4.3 Tools for Implementing Plans for Ecosystem-Service Management: Land Readjustment, Agricultural Land Improvement, and the New Forestation Movement

Land ownership in Japan has been largely in private hands since 1873. Thus, implementation of land use policy in Japan must rely on legal tools relevant to private land. In this section we discuss two major legal instruments and one movement used to implement ES management.

Japan has a well-established system of land reallocation which was central to post-World War II reconstruction and the subsequent development boom, especially in housing. This system is still used sometimes for urban and rural development purposes. The Land Readjustment Act enables reparcelling of private land and at the same time allocates land and financing for public facilities. These can include parks and protected natural areas (see Fig. 30.4) (Urban Development and Improvement Division 2015). Under this system, the designation of land for public purposes does not incur any loss to private owners because the value of the remaining parcels is increased. However, as Japanese cities shrink and the pulse of development has weakened, the benefits to the land owner have been greatly reduced.

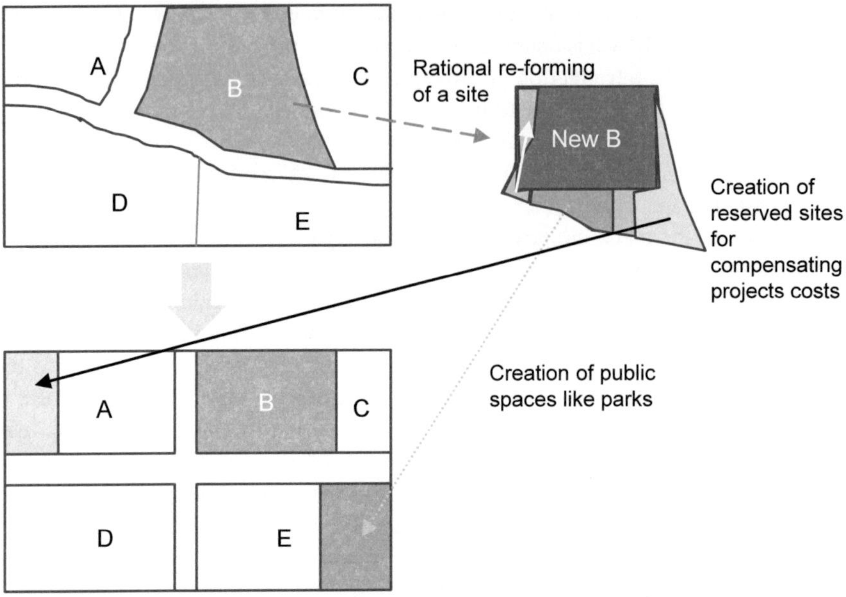

Fig. 30.4 The concept of reallocation under the Land Readjustment Act

The Land Readjustment Act mainly regulates the reallocation process, but there is no control with respect to the environmental quality of the land use created. Thus, the quality of any project greatly depends on the awareness of the municipality, the land owners, and the project management company. For example, during the high development period of 1970–80, some projects were led by companies with noble visions. They succeeded in creating high-quality residential neighbourhoods with rich green environments. However, when it comes to low-growth periods with no expected increases in land prices, it is more difficult to expect generous allocation of green spaces in private developments. In the current economic environment, a new regulatory system is needed – one that could harness some of the advantages of land adjustment projects. For example, ecosystem services could be reconceptualised as economic value and could be integrated into the calculations for land readjustment. As yet, however, there is no detailed discussion of new legislation.

A second tool for ES enhancement is through ‘agricultural land improvement projects’. This instrument, similar to land consolidation in other countries, aims to address inefficient field shapes, improvement of soil quality and drainage systems, and construction of farm roads. Such projects have been carried out all over Japan, and have succeeded in expanding yield per land unit and reducing management costs. However, unfortunately, over the years this approach has had negative impacts on farmland biodiversity, especially in paddy fields.

More positively, a change in policy is on the horizon regarding both land readjustment and agricultural promotion, shifting from a narrow production perspective towards increasing recognition of multifunctional benefits. In 2001, the Land

Readjustment Act was revised, with environmental considerations being included in Paragraphs 1 and 2 where the purpose of the Act is set out. A prefecture governor is now authorized to deny permission for a project that is not environmentally friendly. In 2014, the Act of Promotion of Fulfilment of Multifunctional Roles of Agriculture was enacted. The concept of ES has been redefined to include the conservation of land, water source recharge, conservation of the natural environment, good landscape formation, and cultural lore. For example, in some advanced projects, environmental considerations have led to the creation of fish ladders in irrigation channels or escape puddles for fish in winter. These, however, are unusual examples, since economic factors still dominate both land readjustment and agricultural enhancement projects. Thus, it is difficult to require that ecological considerations be integrated into such initiatives. Better understanding and cooperation of local residents could be a driving force for change.

The third, and most recent, way of integrating ES management in planning policy is a new Forestation Movement. This promotes that forests should serve multiple ecological functions. In 2001, the Forestry Basic Act (an important piece of legislation for forest management), together with the Forest Act, were amalgamated into the Forest and Forestry Basic Act. This modified Act introduced the concept of ecosystem services together with the concept of multiple-functions of forests. In 2009, a national Forest and Forestry Regeneration Plan was adopted. There, the notion of multifunctional forests was incentivized by means of grants to forestry associations and private enterprise bodies for promoting multiple ES. Multiple functions of forests imply, for example, the prevention of global warming by storing greenhouse gases, erosion prevention, conservation of water resources, maintenance of habitats, and connected habitat complexes, formation of landscapes with cultural ecosystem services, and wood production. In practice, a master plan for a multi-function forest would entail conversion from a conifer-dominated to mixed plantation to support the above mentioned functions. Forests are classified into two categories: productive forests and environmental forests. The Forest Agency provides subsidies to forestry associations and private enterprise bodies who wish to transform existing plantation forests that are no longer profitable (such as in mountainous areas) into environmental forests. Financing helps to cover the costs of cutting, thinning, weeding etc.

To sum up, whereas the legal recognition of the value of ES in agricultural land has risen in recent years, institutional fragmentation still slows down progress. The measures available in individual Acts remain dissociated from each other. Collaboration across government agencies is not yet sufficient for adequate management and maintenance of ES in units of complex landscape or biotopes. An example is the traditional multifunctional Satoyama landscape, which is endangered by lack of management and by urbanization. A comprehensive ES planning system is needed but it is not yet on the horizon.

30.4.4 Economic and Fiscal Implementation Options

Several of the more advanced prefectures have introduced forest environmental taxes, targeted to finance sustainable forest conservation. The tax rates – which differ among prefectures – are typically 300–1000 Yen per person per annum and 5–10% of the prefectural income tax. The tax is used for the preservation, maintenance and revitalization of forests, conservation of biodiversity and wetlands, promotion of marketing of timber from the region, and the development of environmental preservation or conservation activities such as tree management, and public education activities. In addition, there are taxes that fund water resource compliant forest management, which also differ between prefectures.

The use of these tax revenues varies across municipalities and the overall policy is not well articulated. For instance, 75% of the forest tax in the Mie Prefecture is delivered to municipalities according to their population size and their forest area. About 25% of the revenue is used for projects such as wood construction of large-scale public facilities and reforestation for water source protection, allocated at the discretion of the prefecture (Mie Prefectural Government 2015). Recently, a committee composed of academic experts has been appointed by the prefecture to evaluate the use of the tax.

At the end of 2017, the Japanese cabinet decided to establish a national environmental forest tax which will come into effect in 2024. A rate of 1000 yen per taxpayer will be collected person for environmental improvements of forests.

Regarding agricultural land, there is a newly established direct payment system based on the Act for Promotion and Fulfilment of Multifunctional Roles of Agriculture. Under this system, there are three categories of direct payments for agriculture. One is a multifunctionality payment, the second applies in hilly and mountainous areas, and the third is for environmental conservation. The distribution of these payments is made on a project by project basis directly to farmers from farmer organizations. They apply for projects to municipalities, and if reviewed positively receive grants.

Japanese agricultural policy urgently needs a more systematic method for direct allocation of financial support for preferred types of ES. Such a system is expected to promote the conservation and preservation of forest and agricultural land, particularly through the internalization of economic externalities and preventing 'free rider' actions.

30.4.5 Citizen Involvement and Participation

Citizen participation in Japan presents a paradoxical picture. On the one hand, compared to some other advanced economies, planning and environmental laws do not grant strong rights for citizens to challenge government actions (Alterman 2001). The courts too, do not readily respond to citizen appeals (Shibata 2007, 2008b) On the other hand, some forms of citizen activism are widely developed, especially in 'Machizukuri' – town management – through volunteer activities (Sorensen and

Funck 2007). Such activities take place spontaneously by groups of citizens rather than through government programmes. One particular example is that many citizens volunteer to manage the traditional Satoyama landscapes. The Ministry of the Environment supports these activities. However, there is no overall programme to efficiently integrate such citizen action with the legal system. Thus, in the future too, environmental conservation activities in Japan may have no option but to continue to rely on volunteers.

Citizen protests have been influential in blocking several public-private large scale land development projects that would have had negative environmental impacts. One such famous project – which has not yet gone ahead – is the landfill of coral reefs on the Henoko coast, Okinawa for the construction of a US military base.[7] In Japan it is quite difficult to stop a national project based on environmental arguments. It can be even more difficult to stop approved projects on private land due to strong individual property rights.

Overall, therefore, on the formal, legally mandated level, citizen participation is not well developed where environmental issues are concerned. For instance, in EIA procedures, the legal requirement to prepare a scoping document and make it available for public inspection (Article 7) is not well enforced. A similar deficiency applies to the requirement to make a draft EIS available for public inspection (Article 16).

30.4.6 Outlook: Research on Landscape Planning

In 1997, Japan chaired the COP3 meeting where the important Kyoto Protocol for the United Nations Framework Convention on Climate Change was adopted. Japan has been proactive in global policy to control CO_2 emissions. In addition, a Japanese supercomputer has contributed to the analysis of climate change, especially regarding the technique of downsizing from global scale to urban scales. Japan is thus providing data of importance for monitoring urban climates. Despite the advanced nature of Japan's research capabilities, in relation to providing the appropriate scientific data for policy decisions regarding the allocation of land uses and distribution of green sites, research contributions fail to be recognised because there is not enough political will. In addition, the right to land ownership is very strong in Japan, and there is a strong opposition to regulating land use from the standpoint of environmental conservation.

In 2010, the 10th Conference of the Parties to the Convention on Biological Diversity (COP10) was held in Nagoya. That conference triggered more research on biodiversity in Japan. For example, in Aichi Prefecture, where Nagoya is located, the prefecture carried out a comprehensive survey of endangered species and created a habitat potential map of these species. However, this specific survey did not integrate other ES.

[7] https://www.japantimes.co.jp/news/2017/12/27/national/okinawa-sit-protest-futenma-relocation-hits-5000-days/#.WsOxUExuLa0

In recent years, research evaluating ES has been on the increase. Other research on the habitat requirements of various species and habitat potential assessments has begun, mainly at a local scale. Unfortunately, this research has not yet been integrated into the legal tools for land use planning and control. Many of these studies are based on voluntary initiatives by researchers, and although academic outcomes have been obtained they have not been integrated into a comprehensive framework that integrates with national policies and the legal system.

As yet, compared with other fields of natural science in Japan, research on ES is not adequately developed. This academic gap must be addressed urgently. Furthermore, the general public is not yet acquainted with the concept of ES, and thus citizen participation processes and actions are not yet informed by it. Development of such awareness will take time.

In contrast, landscape planning methods based on ecological land evaluation have a long history in Japan. They were introduced by Takeuchi (1983) and Ide and Takeuchi (1985) during the late 1970s and developed during the 1980s. This occurred in parallel with the development of landscape planning methodology in Germany. Ide and Takeuchi first introduced a landscape planning methodology based on the evaluation of potential natural vegetation and terrain. This method has been incorporated in the land use plans of several progressive municipalities. However, this approach has not evolved into a broader and more generally accepted integrated land use planning. There are two reasons for this. One lies in the fact that the research supporting this method focused mainly on the agricultural sector, and was thus not easily extendable into an integrated land management system. Such extension would have required overcoming the fragmentation of responsibilities between many different ministries, as noted above. The second reason is that the 1970s and 1980s were an era of rapid growth in the Japanese economy, and at that time the economic forces and policies supporting land development took precedence over nature preservation.

Between the later 1990s and the early 2000s, a second stream of research focused on environmental assessment. In 1997 the Environmental Impact Assessment Law (EIAL) was prepared and was enacted in 1999. This development was an important milestone because it increased the understanding of nature conservation among citizens and entrepreneurs alike. Since then, various ecological surveys have been used as diagnostic tools to assess proposals for land developments. A representative example is the application for the development of the Toyota test course and research centre in Aichi Prefecture. The environmental assessment took place from 2007 to 2012, resulting in measures to support the regeneration of native vegetation.

30.4.7 A New Land Management and Evaluation System Is Needed in the Era of Shrinking Population

The days of the 'developmental state' – the driving force of Japan's governance for the past several decades – are over. Today, Japan's population is shrinking. While a low birth rate is not unique to Japan, the country's reluctance to receive immigrants in significant numbers makes population shrinkage a more dramatic reality than in many other OECD countries. One can hope – or even assume – that this new context will gradually elevate environmental values and considerations so as to override developmental policies that unduly harm the environment.

Nevertheless, there is a long distance to travel. Japan still lacks an adequate legal system that enables the integrated spatial planning necessary to ensure integrated management of biodiversity and ES. Even today, environmental measures are applied in an ad hoc manner to individual projects, rather than as comprehensive policy. Such measures are supported by independent grants, and carried out under the initiatives of different and fragmented governmental agencies. The lack of an integrated planning system is a major problem that should be addressed by decision makers. Research findings about ES should be better recognized and utilized by political decision makers and citizen groups.

In addition, it is urgent to reconsider ES in the context of shrinking populations. Since 2008, the Japanese population has been on the decrease. For thousands of years Japan's beautiful landscapes, such as the Satoyama and paddy fields, have been maintained by humans who protected their ES. A shrinking population threatens this harmonious relationship. Population depletion is especially severe in the countryside. The number of abandoned fields and forests is rising. Satoyama paddy fields spread out into a multi-layered terrace and form a unique landscape which is important as a spawning place for frogs or fishes and as a feeding place for birds. The maintenance of paddy fields requires a great deal of labour every year, and as soon as this stops they will disappear. Many Japanese Satoyama forests are conifer plantations. When people do undertake thinning the timber quality decreases, the ecosystem becomes weak, and it can also cause landslides. There are initiatives to change from conifer to broadleaf forest, but this also requires manual work.

In recent years increasing numbers of urban residents have visited Satoyama for the purpose of recreation or recuperation, but the inhabitants of Satoyama landscapes have not received sufficient benefit from the ES their activities support. The issue of providing compensation is not high enough on the policy agenda. Given the pace of population shrinkage and its environmental impacts, Japanese decision makers should urgently consider how to integrate offsetting or compensation measures either in the form of direct payments or by other means. The most important thing is that sufficient numbers of people continue to live in (and manage) Satoyama, and for that purpose it is necessary to develop policies that can attract people to Satoyama and enable them to continue to reside there. If people remain, the ES in Satoyama will be preserved.

The path of Japan's ES policies is thus at a junction. The reduction and aging of the population is 'bad news' in some ways, including the possible rise of social-welfare issues and a decline in the number of citizen activists. At the same time, the subsiding of Japan's developmental era allows more room for environmental issues to occupy a higher place in public policy. Greater appreciation of the value of ES will be a prime indicator of such a transformation.

30.5 Conclusions

Christina von Haaren, Rachelle Alterman

In the previous sections of this chapter we explored the implementation options for landscape planning within different governance contexts. A framework for characterizing the context for planning was proposed in order to help readers from other countries compare their situation with those discussed here.

The two examples that were examined more closely offer many insights about how the practice of landscape planning and the [potential] integration of ecosystem services assessments may differ across national contexts. They also demonstrate the diversity of national characteristics which can enhance or impair the context for landscape planning. Furthermore, the two cases demonstrate aspects of good practice, such as Oregon's exemplary public participation and data availability.

The contrasts in data availability and costs between countries also have implications for landscape planning. In the USA there is a long tradition of data collected by public agencies being made freely available to other prospective users whereas in Europe there have generally been more restrictions and sometimes high charges. However, this contrast has diminished appreciably in the past decade due to the international trend towards open data (see Chap. 5). It is also apparent from the Oregon case study that free data and extensive participation do not necessarily lead to pro environmental decisions. This situation emphasises that in landscape planning we must include the principle of participation as a goal in its own right (see Chap. 28) and not only as an instrument for achieving better environmental planning.

In the USA there has also been a trend for several federal agencies to start using an economic approach to the ES concept. The US Geological Survey (USGS) and US Forest Service, two large federal agencies, have developed ecosystem frameworks and administrative structures which aim to integrate the ES concept into broader practices of planning (USGS 2018a, US Forrest Service 2018). The social aspects of ecosystem services are also embodied in readily available tools offered by federal agencies. The SolVES project, for example, which was initiated by the USGS, allows users to characterize social values such as aesthetic and recreation as part of landscape planning projects (USGS 2018b). While similar programs exist to some degree in Europe (e.g. European Environment Agency 2018), the broad public availability of such participation initiatives, enhanced by online systems, is ahead of many countries, and can help promote new forms of public engagement. Furthermore, in the USA we see greater application of cost-benefit analysis for monetizing ecosystem services. These include examples related to outdoor recreation, ecological restoration and wildlife management (USGS 2018a). Such techniques, while also applied in Europe and Japan, are probably less commonplace than in the USA.

On the other hand, the types of regional planning widely used in Germany and other parts of Europe are not common practice in the USA, notwithstanding exceptions such as Portland (Seltzer et al. 2010). Ebenezer Howard's moto "survey before you plan" is alive and well in parts of Europe at the regional level, but in the USA, it is (probably) practiced mostly at the local, municipal level.

The Japanese example illustrates the effects of a combination of demographic and economic driving forces, a strong culture of individual property rights and an extreme fragmentation of governmental responsibilities for environmental issues. Japan experienced major environmental catastrophes earlier than many other OECD countries and therefore, public concern for the environment goes back to the 1960s. However, this long tradition has focused predominantly on technical environmental protection (e.g. emission control) and this may have diverted government and public attention from the need to adopt an integrative comprehensive perspective on ecosystems and their functions. The lack of an integrated planning system is a major problem for the preservation of features such as the traditional multifunctional Satoyama landscape. The Japanese perspective suggests that a compartmentalization of the legal framework for planning and land management in general, may adversely impact on the capacity to undertake good landscape planning. Concepts such as conservation, aesthetics, greening of the landscape and other environmental concerns are codified separately in the law and lack an adequate coordinated governance. Although legal and governance fragmentation is also common in European countries, it seems that in Japan it may be more extreme. Despite these shortcomings, the Japanese example does illustrate that even in a fragmented system, the concept of landscape planning can serve as a means of drawing together the various government sectors relevant for ES. Adoption of landscape plans at the municipal level can help integrate the regular local spatial plans with regional-wide perspectives. This approach could contribute to bridging the gap in the hierarchical structure of planning responsibilities in Japan. Additionally, the Japanese example shows, how in the absence of integrative landscape planning, a sectoral planning instrument like forest planning can be transformed into a means of supporting multiple ecosystem functions and services.

To sum up, our 'outside the EU' perspective suggests that the benefits of using a differentiated ES approach in landscape planning – as advocated in this book – can have added value for environmental planning. Decision-makers, scholars, and citizens may still be grappling with the meaning of this topic, its usefulness, and what methods can make it more effective. Current global and country-specific environmental challenges are altering land use conditions and are precipitating the need to consider how to design better governance frameworks for conserving ecosystem services. Emphasis on the value of nature for people and the economy, without neglecting the sound analysis and appraisal of the current state and capacities of ecosystems, can speak to decisions makers as well as to the affected citizens. Methodologies for the assessment of ES in landscape planning can be applied in principle in most countries and help foster public participation as part of the planning system. Adoption of legal standards as a basis for assessing ES is preferable because it enables comparisons across place and time and enhances the legitimacy of the evaluation results. A well-developed citizen participation framework is also an important pre-condition for better-informed decision making as well as for public monitoring of any deviations from the law. Finally, a tiered and well-integrated spatial planning system seems to be favourable for landscape planning. A fragmented division of responsibilities for different environmental assets various laws and administrative units hampers an integrated approach to environmental conservation. A transparent, well-structured and lean planning system depends on reconsideration of the tasks and responsibilities at each political level.

Beyond these observations, the main lesson to be learned from our exploration into the planning-governance contexts of different countries is that there is no 'one size fits all' solution. Differing degrees of economic prosperity, land scarcity and population density, contrasting property regimes and participation rights, as well as the institutional structure of the planning system, all represent diverse starting points for landscape planning or indeed, for any environmental planning. Even if the need for systematic environmental planning is acknowledged, and implementation conditions are favourable, the tasks of data acquisition and particularly of evaluating the state of ES will likely differ across jurisdictions and over time. The methods and approaches laid out in this book will obviously have to be adapted to each particular context.

References

Alexander, E. E. (1993). Interorganizational coordination in theory and practice. *Journal of Planning Literature, 7*(4), 328–343.

Alterman, R. (Ed.). (2001). *National-Level Planning in democratic countries: An international comparison of City and regional policy-making, town planning review book series*. Liverpool: Liverpool University Press.

Alterman, R. (2005). A view from the outside: The role of cross-national learning in land-use law reform in the United States. In D. R. Mandelker (Ed.), *Planning reform in the new century* (pp. 309–320). Chicago: Planners Press.

Alterman, R. (2010). *Takings international: A comparative perspective on land use regulation and compensation rights*. Chicago: American Bar Association.

Alterman, R. (2011). The US regulatory takings debate through international lenses. *Urban Lawyer, 42*, 331.

Bosselman, F., Callies, D., & Banta, J. (1973). *The taxing issue: An analysis of the constitutional limits of land use control*. Washington, DC: Council on Environmental Quality.

Brady, M. E. (2017). The Damagings clauses. University of Virginia Schools of Law, Public Law and Legal Theory Research Paper Series. *Virginia Law Review, 104*, 341.

Cullingworth, B., & Caves, R. W. (2009). *Planning in the USA: Policies, issues, and processes*. New York: Routledge.

de Roo, G., & Silva, E. A. (2010). *A Planner's encounter with complexity*. London: Routledge.

Egoh, B., Reyers, B., Rouget, M., et al. (2008). Mapping ecosystem services for planning and management. *Agriculture, Ecosystems and Environment, 127*(1–2), 135–140.

European Commission. (2016). *No net loss*. http://ec.europa.eu/environment/nature/biodiversity/nnl/index_en.htm. Accessed 7 Dec 2016.

European Environment Agency. (2018). *Natural capital and ecosystem services*. https://www.eea.europa.eu/soer-2015/europe/natural-capital-and-ecosystem-services. Accessed 17 Aug 2018.

Forest Agency. (2014). Shinrin Ringyo Hakusyo (Forest and forestry white paper). http://www.rinya.maff.go.jp/j/kikaku/hakusyo/25hakusyo/index.html. Accessed 7 Dec 2016.

Geißler, G. (2008). *The ballot box – Threat or blessing for planning? – The impact of direct democracy on land use planning in Oregon, USA*. Diploma thesis. Institut für Landschaftsarchitektur und Umweltplanung, Technische Universität Berlin.

Hirt, S. A. (2015). *Zoned in the USA: The origins and implications of American land-use regulation*. Ithaca: Cornell University Press.

Ide, H., & Takeuchi, K. (1985). *Shizen Ricchiteki Tochiriyou Keikaku (Land use planning based on based on geoecological land evaluation)*. Tokyo: University of Tokyo Press.

Japan Country Section. (2008). *Environmental performance of agriculture in OECD countries since 1990*. Paris: OECD.

Johnson, C. (1995). *Japan: Who governs? The rise of the developmental state*. New York: Norton.

Kayden, J. (2001). National land-use planning and regulation in the United States: Understanding its fundamental importance. In R. Alterman (Ed.), *National-level planning in democratic countries* (pp. 43–64). Liverpool: Liverpool University Press.

Lakeview District, Bureau of Land Management (BLM). (2003). *Lakeview resource management plan 2003 as part of national system of public lands*. http://www.blm.gov/or/districts/lakeview/plans/files/signed_scoping_letter.pdf. Accessed 11 June 2018.

Levy, L. W. (2001). *Origins of the bill of rights*. New Haven: Yale University Press.

Liberty, R. (2009). Stopping sprawl in the fifty states. A report by smart growth America. Summary report commissioned by the Wallace Global Fund.

Maes, J., Egoh, B., Willemen, L., et al. (2012). Mapping ecosystem services for policy support and decision making in the European Union. *Ecosystem Services, 1*(1), 31–39.

Mie prefectural Government. (2015). *Mie no Shinrindukuri (Forest development of Mie)*. http://www.pref.mie.lg.jp/common/content/000116668.pdf. Accessed 7 Dec 2016.

Ministry of Land, Infrastructure, Transport and Tourism. (2006). *Landscape Act*. http://www.mlit.go.jp/crd/townscape/keikan/pdf/landscapeact.pdf. Accessed 7 Dec 2016.

Ministry of Land, Infrastructure, Transport and Tourism. (2014). *Special green space Conservation District*. http://www.mlit.go.jp/crd/park/joho/database/toshiryokuchi/ryokuchi_hozen/. Accessed 7 Dec 2016.

Ministry of Land, Infrastructure, Transport and Tourism. (2015). *Urban park data base*. http://www.mlit.go.jp/crd/park/joho/database/t_kouen/pdf/01_h26.pdf. Accessed 7 Dec 2016.

Ministry of Land, Infrastructure, Transport and Tourism. (2018). http://www.mlit.go.jp/common/001251069.pdf. Accessed 20 Mar 2019.

Ministry of the Environment. (2012). *Nature conservation area*. http://www.env.go.jp/nature/hozen/about.html. Accessed 7 Dec 2016.

Ministry of the Environment. (2016). *Table of Natural park areas*. http://www.env.go.jp/park/doc/data/natural/naturalpark_1.pdf. Accessed 7 Dec 2016.

Nagoya City. (2014). *Nagoya green basic plan 2020*. http://www.city.nagoya.jp/shisei/category/53-3-3-2-0-0-0-0-0-0.html. Accessed 7 Dec 2016.

National Land Numerical Information Download Service. (2015). *National information division, National and Regional Policy Bureau, Ministry of Land, Infrastructure, Transport and Tourism*. http://nlftp.mlit.go.jp/ksj-e/index.html. Accessed 7 Dec 2016.

Nelson, E., Mendoza, G., Regetz, J., et al. (2009). Modeling multiple ecosystem services, biodiversity conservation, commodity production, and tradeoffs at landscape scales. *Frontiers in Ecology and the Environment, 7*, 4–11. https://doi.org/10.1890/080023.

OECD. (2017). *Land-use planning systems in the OECD: Country fact sheets*. Paris: OECD Publishing.

Oregon Department of Land Conservation and Development. (2018). *Goal 5 process for aggregate*. https://www.oregon.gov/LCD/pages/goal5agg.aspx. Accessed 16 Aug 2018.

Putter, B. (2010). The special case of Oregon the heated debates regarding measures 37 and 49. In R. Alterman (Ed.), *Takings international* (p. 229). Chicago: American Bar Association Press.

Roberts, T. E. (2010). United States. In R. Alterman (Ed.), *Takings international* (p. 215). Chicago: American Bar Association Press.

Sasada, H. (2008). Japan's new agricultural trade policy and electoral reform: 'Agricultural policy in an offensive posture [seme no nosei]'. *Japanese Journal of Political Science, 9*(2), 121–144.

Seltzer, E., Smith, T., & Cortright, J. et al. (2010). *Making ecodistricts concepts & methods for advancing sustainability in neighborhoods, Portland, OR*. https://pdxscholar.library.pdx.edu/cgi/viewcontent.cgi?article=1035&context=iss_pub. Accessed 29 Aug 2018.

Shibata, K. (2007). The state, planning and the planning and the public interest: The development of city planning in Japan. University of London Press.

Shibata, K. (2008a). Neoliberalism, risk, and spatial governance in the developmental state: Japanese planning in the global economy. *Critical Planning, 15*, 92–118.

Shibata, K. (2008b). *The public interest in planning in Japanese jurisprudence: The limits to participatory democracy*. LSE Research Online. http://eprints.lse.ac.uk/21676/1/The_public_interest_in_planning_in_Japanese_jurisprudence_(LSERO).pdf. Accessed 11 June 2018.

Siems, M. M. (2016). Varieties of legal systems: Towards a new global taxonomy. *Journal of Institutional Economics, 12*(3), 579–602.

Sorensen, A. (2002). *The making of urban Japan: Cities and planning from Edo to the 21st century*. London: Routledge.

Sorensen, A. B. (2005). The developmental state and the extreme narrowness of the public realm: The twentieth century evolution of Japanese planning culture. In B. Sanyal (Ed.), *Comparative planning cultures* (pp. 223–259). New York: Routledge.

Sorensen, A., & Funck, C. (Eds.). (2007). *Living cities in Japan: Citizens' movements, Machizukuri and local environments*. London: Routledge.

Sullivan, E. J. (2007). Through a glass darkly: Measuring. Loss under Oregon's measure. *Urban Lawyer, 39*, 563–618.

Sullivan, E. (2011). The quiet revolution goes West: The Oregon planning program 1961–2011. *John Marshall Law Review, 45*(2012), 357–395.

Takeuchi, K. (1983). Landscape planning methodology based on geoecological land evaluation. *GeoJournal, 7*(2), 167–183.

Urban Development and Improvement Division. (2015). *Tochi Kukaku Seiri Jigyo (Land readjustment project)*. http://www.mlit.go.jp/crd/city/sigaiti/shuhou/kukakuseiri/kukakuseiri01.htm. Accessed 7 Dec 2016.

US Forest Service. (2018). *Ecosystem services*. https://www.fs.fed.us/ecosystemservices/. Accessed 17 Aug 2018.

US Geological Survey. (2018a). *Ecosystem services*. https://www.usgs.gov/ecosystems/status-and-trends-program/science/ecosystem-services?qt-science_center_objects=0#qt-science_center_objects. Accessed 17 Aug 2018.

US Geological Survey. (2018b). *Social Values for Ecosystem Services (SolVES)*. https://www.usgs.gov/centers/gecsc/science/social-values-ecosystem-services-solves?qt-science_center_objects=0#qt-science_center_objects. Accessed 17 Aug 2018.

Waage, S., Stewart, E., & Armstrong, K. (2008). *Measuring corporate impact on ecosystems: A comprehensive review of new tools*. https://www.bsr.org/reports/BSR_EMI_Tools_Application1.pdf. Accessed 13 June 2018.

Synthesis and Prospects for Landscape Planning

31

Christian Albert, Christina von Haaren,
and Andrew A. Lovett

Abstract

This chapter summarises the main themes discussed in previous chapters of the book. In particular, it emphasizes how the assessment and enhancement of ecosystem services can be facilitated though landscape planning. The chapter also outlines possible future roles for landscape planning in Europe and how technological changes might influence how these are carried out. As a conclusion, we offer a short vignette of the activities future landscape planners might be involved in as part of their daily work.

Keywords

Landscape planning · Ecosystem services · Assessment methods · Future planning · Digital technology

31.1 Synthesis

This book illustrates the wide breath of current knowledge regarding concepts and methods in the field of landscape planning to assess and plan for ecosystem services in support of sustainable landscape development. In Part I of the book we set the scene by emphasizing the complementary nature of the landscape planning and ecosystem services (ES) fields, providing definitions of concepts, and introducing the objectives and structure of the book. Although it is clear from the literature that

C. Albert (✉) · C. von Haaren
Institute of Environmental Planning, Leibniz University Hannover, Hannover, Germany
e-mail: albert@umwelt.uni-hannover.de; haaren@umwelt.uni-hannover.de

A. A. Lovett
School of Environmental Sciences, University of East Anglia, Norwich, UK
e-mail: a.lovett@uea.ac.uk

C. von Haaren et al. (eds.), *Landscape Planning with Ecosystem Services*,
Landscape Series 24, https://doi.org/10.1007/978-94-024-1681-7_31

there are different definitions and interpretations of landscape planning and ES around the world, we have particularly tried to provide an orientation for the European context and a focus on how ES can be assessed and enhanced through landscape planning. To this end, a conceptual framework for the consideration of ecosystem services in landscape planning was outlined that draws upon the widely applied Driving forces, Pressures, State, Impacts and Responses (DPSIR) model. We also reviewed data sources and technologies that facilitate the implementation of landscape planning methods in practice.

Part II reviewed drivers and pressures that impact upon ecosystem services in Europe. We focused on EU policies and standards as drivers for ecosystem services provision and provided an overview of the more direct pressures on land and water use that influence ecosystem services. We made the point that landscape planners need to be aware of these drivers and pressures, but that their capacity to actually influence or reverse them is often limited given that many operate at national or international scales. We therefore argued that landscape planning could take three kinds of actions in response. These are to (i) minimize the local impacts of pressures (e.g. through regulation or zoning), (ii) identify those natural assets that are most sensitive to pressures and take actions to increase their resilience and (iii) provide evidence and advice at higher levels of public and private decision making in order to decrease the intensity of adverse drivers and pressures in the first place.

The third and largest part of our book concerns landscape planning methods for assessing and evaluating the state of, and impacts on ecosystem services. This section of the book highlights the great number and diversity of methods and procedures already developed and applied throughout Europe. The chapters also demonstrate that methods exist for many ecosystem services that vary in both their data requirements and the complexity of modelling approaches. This means that landscape planners often have some flexibility to choose an assessment method that is most appropriate for the level of detail required.

Part IV addressed the question of how to set targets for the protection and sustainable use of ecosystem services, as well as identifying measures to attain these objectives in an effective manner. We suggested a strategy for developing targets and actions, and then provided an overview of these for a number of important ecosystem services. We emphasized that such a strategy should address overarching targets, for example those derived from national legislation and EU directives. In addition, the definition of targets and actions should always consider the knowledge and concerns of local stakeholders in order to enhance relevance to the place-specific context and the likelihood of public and private support for implementation. However, the process of developing targets and actions in landscape planning does not merely consist of choosing these from pre-defined lists, but also requires a process of deliberation with experts and stakeholders in order to find a combination of measures that exploits synergies and minimizes conflicts wherever possible. A major task for landscape planners is therefore the development of comprehensive and integrated planning proposals.

These issues of deliberation and integration are further considered in Part V. The chapter on participatory approaches emphasized the importance of actor-specific communication strategies and opportunities to involve decision-makers and stakeholders in the planning process. In addition, design approaches can help with stakeholder engagement and to provide an overview of proposals that helps to achieve an integration of the different elements.

The book concludes with some perspectives on the role of landscape planning in countries outside Europe, in particular the situations in the United States and Japan. These discussions illustrate that despite differences in challenges and governance contexts, the basic principles of landscape planning and methodological approaches applied in Europe could well be transferred and adapted to other settings.

31.2 Prospects: A Vision of Landscape Planning in 2030

The role of landscape planning in the future will depend on changes in both the societal context in which planning takes place, as well as in the tools and methods available to the planner.

In terms of the former, scenario studies such as those reviewed by van Vuuren et al. (2012) can provide insights into potential context situations. The Millennium Ecosystem Assessment (2005) for example, identified two major drivers of change, namely the type of environmental management implemented (proactive vs. reactive) and the type of world development unfolding (globalization vs. regionalization). In a similar vein, the Intergovernmental Panel on Climate Change scenarios (see Nakicenovic et al. 2000) are organized along the dimensions of economic vs. environmental priorities, and global vs. regional development. These dimensions of variability will have very different implications for the role that planning plays in society. More pro-active, environmentally oriented future pathways would strengthen the influence of planning (especially in land and water use decision-making), whereas reactive or economic oriented developments would rather reduce it. Trends towards globalization would emphasize market forces over state-led planning approaches, whereas regional development trends could accentuate the role of planners.

Within planning, the most likely refinements are of tools and methods that are beginning to emerge now. We can anticipate that technological enhancements will give the planner access to real-time data streams from a variety of landscape sensors, as well as the ability to task autonomous devices to visit, view and collect information from specific locations (e.g. Kitchin et al. 2015; Grubesic et al. 2018). This will be combined with more immediate facilities to model ES delivery and assess what-if questions related to different interventions (e.g. Wissen Hayek et al. 2016; Grêt-Regamey et al. 2017). Informed by these data streams and underpinning the modelling capabilities will be much more integrated landscape information models (LIMs) which include digital representations of current and past states (e.g. of land use), as well as hydrological, weather, chemical and biological processes. The online availability of such models also supports citizen involvement, for instance through crowd sourcing of information or forums to discuss different issues or interventions (Galler et al. 2014; Dunkel 2015). All of this will facilitate more feedback in planning systems, with faster adaptation to changing circumstances and potentially more citizen empowerment. The challenges of coordination and integration are still likely to exist, and indeed could become more important if questions of who delivers, and who benefits, from ES become more important in society, but the manner in which such issues are examined and decided upon may well be rather different.

31.3 A Day in the Life of a Future Landscape Planner

The description below is a normative narrative. It reflects our vision of the challenges we think future landscape planners will face and the role we would like to see them playing in society.

Lena sat down in her home workplace and logged in to the municipality server. She frowned as the dashboard of landscape indicators appeared on her screen. Although it was a lovely spring morning the heavy rainstorm the previous day had greatly increase sediment levels in the local river and the flow level on the outskirts of a downstream village was close to triggering an automatic flood warning. There was also a video message from the local councillor asking her to check on the situation further up the catchment and to let him know if any evacuation was likely to be necessary.

She couldn't help thinking how much had changed in the past 10 years. Although the problems arising from an unstable climate system were starting to become more pronounced there was also a greater willingness to think about longer term issues and put measures in place to address them. The information that she and her colleagues could access to help in their role as landscape stewards had also been transformed. Activating a local drone from her dashboard she tasked it to fly over the wood debris dams on the northern slope of the river catchment. A few minutes later she could see that these were working as planned to slow flows, but the intensity of the rainstorm meant that the flow level was still increasing below the confluence of several tributaries. Thankfully, the introduction of 20 year environment plans had provided a framework for investment in wetlands that could be temporarily flooded and there were two of these further downstream. Lena called up a colleague in the local river authority and discussed the situation with him. They agreed that sluices should be opened to divert river water into the wetland and that this should be sufficient to keep the flows within the river banks. She then contacted the local councillor, updated him on the situation and agreed that at present there was no need for an evacuation.

Lena next turned her attention to the online landscape forum. As ever, there were comments about the management of the local lake, particularly the zoning of different parts for recreation activities and conservation priorities. This had long been a difficult issue among local landowners since some received government payments for providing wildlife habitats whilst boating and waterside visitor facilities were an important income source for others. Better ability to quantify the costs and benefits of different activities was helping each sector to appreciate the role of the other, but

Lena had still found it a slow process to build the trust necessary for multifunctional management and it was still something where regular communication with the different parties was important.

Another message concerned plans to develop a 5 hectare field for solar photovoltaic arrays near another village. The local environment plan had accepted the need to increase renewable energy generation in the area, but there was still much debate about where such facilities should be located, what the impacts on visual amenity would be and who would bear the costs or receive the benefits of such a development. Lena had found that her ability to assess the impacts of different sites and provide visualizations of different proposals had helped inform and calm debates but the distribution of benefits was still controversial. There were now proposals to use power from the solar arrays in nearby homes and Lena was helping to facilitate discussions about this community benefit in the hope that it would attract sufficient support to allow the scheme to go ahead. As was so often the case, landscape planning was as much about relationships between people as anything to do with environmental characteristics. This, Lena reflected, was what made her role such a fascinating and satisfying one.

References

Dunkel, A. (2015). Visualizing the perceived environment using crowdsourced photo geodata. *Landscape and Urban Planning, 142*, 173–186.

Galler, C., Krätzig, S., Warren-Kretzschmar, W., et al. (2014). Integrated approaches in digital/interactive landscape planning. In U. Wissen Hayek, P. Fricker, & E. Buhmann (Eds.), *Peer reviewed proceedings of digital landscape architecture 2014* (pp. 70–83). Berlin: Wichmann.

Grêt-Regamey, A., Altwegg, J., Sirén, E. A., et al. (2017). Integrating ecosystem services into spatial planning – A spatial decision support tool. *Landscape and Urban Planning, 165*, 206–219.

Grubesic, T. H., Wallace, D., Chamberlain, A. W., et al. (2018). Using unmanned aerial systems (UAS) for remotely sensing physical disorder in neighborhoods. *Landscape and Urban Planning, 169*, 148–159.

Kitchin, R., Lauriault, T. P., & McArdle, G. (2015). Knowing and governing cities through urban indicators, city benchmarking and real-time dashboards. *Regional Studies, Regional Science, 2*, 6–28. https://doi.org/10.1080/21681376.2014.983149.

Millennium Ecosystem Assessment. (2005). *Ecosystems and human well-being: synthesis.* Washington, DC: Island Press.

Nakicenovic, N., Alcamo, J., Davis, G., et al. (2000). *Special report on emissions scenarios: A special report of Working Group III of the Intergovernmental Panel on Climate Change.* Cambridge: Cambridge University Press.

van Vuuren, D. P., Kok, M. T. J., Girod, B., et al. (2012). Scenarios in global environmental assessments: Key characteristics and lessons for future use. *Global Environmental Change, 22*, 884–895.

Wissen Hayek, U., Teich, M., Klein, T. M., et al. (2016). Bringing ecosystem services indicators into spatial planning practice: Lessons from collaborative development of a web-based visualization platform. *Ecological Indicators, 61*, 90–99.

Correction to: Landscape Planning with Ecosystem Services

Christina von Haaren, Andrew A. Lovett,
and Christian Albert

Correction to:
C. von Haaren et al. (eds.), *Landscape Planning with Ecosystem Services*, Landscape Series 24, https://doi.org/10.1007/978-94-024-1681-7

Fig. 3.3 in Chapter 3 and Fig. 30.1 in Chapter 30 were initially published with errors. The correct presentation is given here.

The updated versions of the chapters can be found at
https://doi.org/10.1007/978-94-024-1681-7_3
https://doi.org/10.1007/978-94-024-1681-7_30

C. von Haaren et al. (eds.), *Landscape Planning with Ecosystem Services*,
Landscape Series 24, https://doi.org/10.1007/978-94-024-1681-7_32

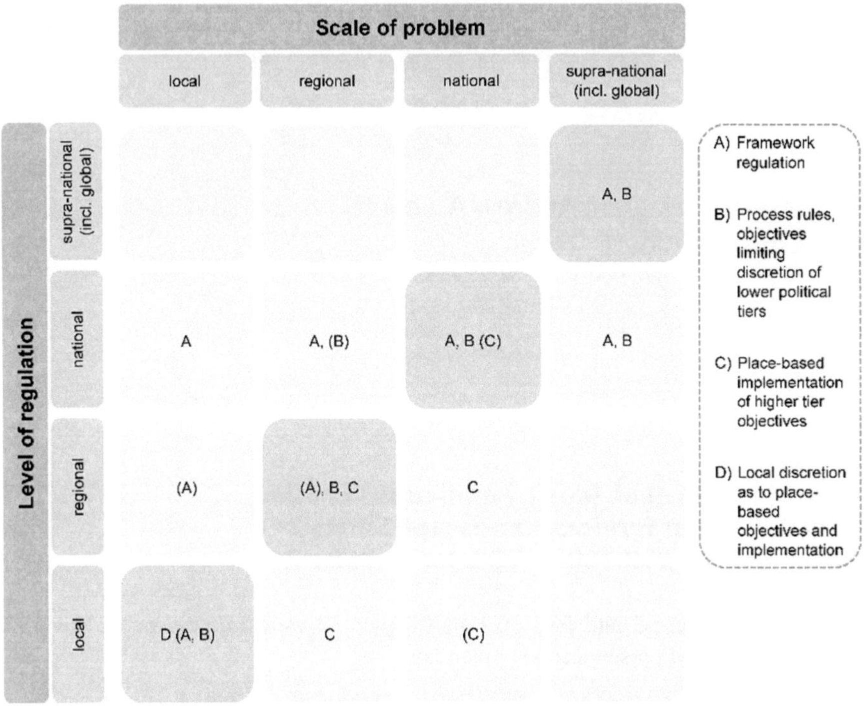

Fig. 3.3 Defining the decision space of landscape planning. Tasks on different planning tiers are determined by the scale of the problem and associated responsibilities. Projects with cross-boundary impacts or trans-boundary ecosystems (such as river catchments) need to be considered at higher planning tiers with authority that covers the whole relevant area (von Haaren 2016: 171, amended)

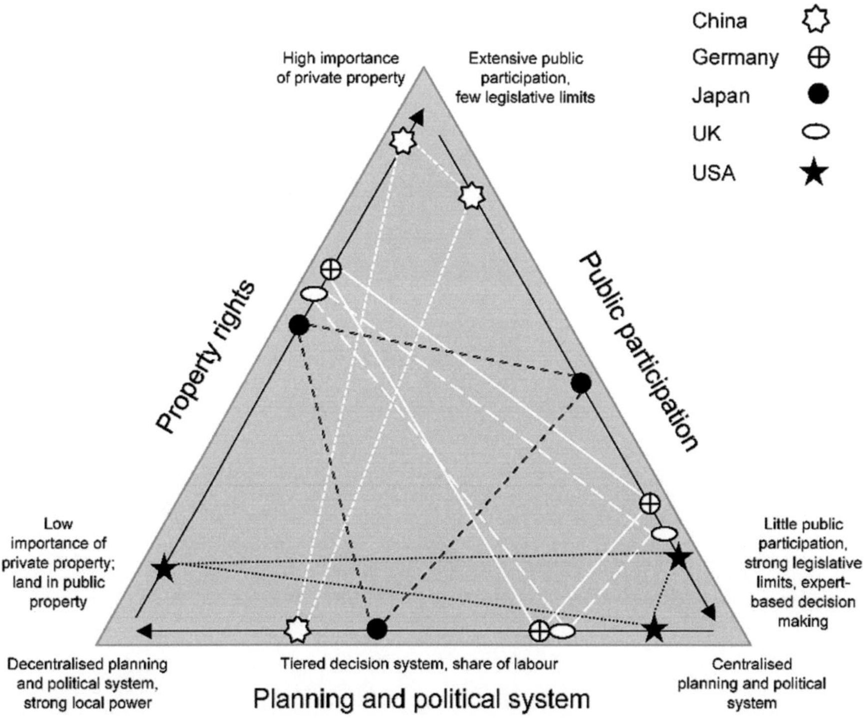

Fig. 30.1 A schematic classification of framing conditions for environmental planning in five example countries. The conditions may be decisive for the importance of economic evaluations, the binding character of objectives, the scope and role of public participation and the scale and delineation of the scope of planning. We are aware of the internal differentiation in the USA, where the preconditions may differ a lot among the states. Nevertheless there are some general frame conditions shared by all US states based on the US legal system and on federal policies

Index

C. von Haaren et al. (eds.), *Landscape Planning with Ecosystem Services*, Landscape Series 24, https://doi.org/10.1007/978-94-024-1681-7

MIX
Papier aus verantwortungsvollen Quellen
Paper from responsible sources
FSC® C105338

FSC
www.fsc.org

If you have any concerns about our products,
you can contact us on
ProductSafety@springernature.com

In case Publisher is established outside the EU,
the EU authorized representative is:
Springer Nature Customer Service Center GmbH
Europaplatz 3, 69115 Heidelberg, Germany

Printed by Libri Plureos GmbH
in Hamburg, Germany